(*continued on back*)

INTRODUCTION TO PROBABILITY AND STATISTICS FOR ENGINEERS AND SCIENTISTS

INTRODUCTION TO PROBABILITY AND STATISTICS FOR ENGINEERS AND SCIENTISTS

Sheldon M. Ross
University of California, Berkeley

JOHN WILEY & SONS

New York • Chichester • Brisbane • Toronto • Singapore

Library of Congress Cataloging-in-Publication Data:

Ross, Sheldon M.
  Introduction to probability and statistics for engineers and scientists.

  (Wiley series in probability and mathematical statistics)
  1. Probabilities.   2. Mathematical statistics.
I. Title. II. Series.

TA340.R67   1987   519.2   87-10406
ISBN 0-471-81752-X

Printed in the United States of America

10 9 8 7 6 5 4 3 2 1

For
Elise

Preface

This book has been written for an introductory course in probability and statistics for students in engineering, computer science, mathematics, statistics, and the physical sciences. As such, it assumes a knowledge of elementary calculus.

Chapters 1 and 2 introduce the fundamental subject matter of probability theory and Chapter 3 considers certain special types of random variables. A variety of examples indicating the wide applicability of these random variables is presented. Also included in our study of these random variables is a consideration of the question of calculating their probabilities. Indeed, computer programs that compute the probability distribution, and in certain cases their inverse of binomial, Poisson, normal, t, F, and chi-square random variables, are presented. These programs, which appear in the Appendix of Programs, are also on a diskette—written for an IBM PC—and are included as part of the text.

Chapter 4, Sampling, begins our study of statistics. We consider such topics as the sample mean, sample variance, sample median, as well as histograms, empirical distribution functions, and stem-and-leaf plots. The distributions of certain of the foregoing statistics are presented when the underlying population is normal. Included in this chapter is a program for computing the sample mean and sample variance for a given data set. (The program for computing the sample variance does not utilize the identity $\sum_1^n (x_i - \bar{x})^2 = \sum x_i^2 - n\bar{x}^2$, which though useful when computing by hand, is potentially hazardous—due to computer round-off error—when computing by machine. Instead, a recursive approach, developed in the chapter, is utilized.)

Chapter 5 considers the problem of statistical parameter estimation. Both point and interval estimates are provided, along with programs for computing interval estimates of desired levels of confidence.

Chapter 6 deals with hypothesis testing. The concept of a statistical hypothesis test, along with its significance level and power, is presented, and a variety of such tests concerning the parameters of both 1 and 2 normal populations are considered. Hypothesis tests concerning Bernoulli and Poisson parameters are also presented. For all of these tests we determine the relevant

p-value by directly computing the appropriate tail probability. Programs that perform the computations are provided. For instance, the Fisher-Irwin test of the equality of 2 Bernoulli parameters is presented, and a program for computing its exact p-value (essentially by computing a hypergeometric distribution function) is provided.

Chapter 7 deals with the important topic of regression. Both simple linear regression—including such subtopics as residual analysis and weighted least squares—and multiple linear regression are considered. Chapter 8 deals with problems in the analysis of variance. Both one-way and two-way (with and without interaction) problems are considered. Programs are provided to determine the regression least squares and the analysis of variance estimates as well as providing (in the analysis of variance case) the p-values of the relevant tests.

In Chapter 9, which deals with goodness of fit and nonparametric testing problems, the influence of the modern computer on statistical analysis really becomes apparent. The classical approach to hypothesis testing in these cases is to determine the value of an appropriate test statistic and then approximate the p-value by determining the limiting distribution (usually either a normal or chi-square) of the test statistic when the null hypothesis is true and the sample sizes are large. However, this usually leaves open the question of "how large is large"? The recent advent of fast and inexpensive computational power, however, has given us a way to finesse the above question by opening up two new approaches to obtaining the relevant p-value. The first of these—especially useful in certain nonparametric testing problems—uses computer solved recursive equations to exactly derive the p-value; the second uses a simulation approach to accurately estimate the p-value. Both of these methods, along with the classical approach, are indicated in Chapter 9. In addition, computer programs for implementing the tests are presented.

Chapter 10 deals with problems related to life testing. In this chapter, the exponential distribution, rather than the normal, plays the key role. Inference problems assuming an underlying Weibull life distribution are also considered. Chapter 11 considers the subject matter of quality control. A variety of control charts, including not only the Shewhart control charts but also the more sophisticated moving average and exponentially weighted moving average control charts, are considered. The final chapter on simulation presents techniques for simulating random variables. Variance reduction techniques are also considered in this chapter.

As mentioned previously, a diskette of 35 programs is included as part of the text. Ideally, every student will have access to a personal computer and thus will be able to utilize these programs. For those students without such access we have provided the appropriate tables to enable them to solve all the problems in the book. A solutions manual is available to instructors.

Preface for the Programs

This book comes with a diskette containing 35 programs. Most of these programs give (except for round-off errors) exact answers. However, those programs that either compute or invert continuous distribution functions give, by necessity, approximations. These approximations, with the exception of the one given by the program for inverting the chi-square distribution (Program 3-8-1-b, which should only be used to solve homework problems in the book), are very accurate.

In order to run the program diskette you need to have the file BASICA.COM. It is a DOS command. To load the diskette put it into disk drive A or B and get your computer into this drive. Now type the four letters "read" and press the enter key. There should now appear a message entitled "How to Start," which will give the appropriate instructions. If the instructions do not enable you to load the diskette (because of some kind of incompatibility problem), you can always first get into BASIC in your usual way, put the diskette into drive (say) A, and access the programs by typing LOAD"A:\ PROGRAMS \ name and then pushing the enter key, where name is the name of the program you want to run. For instance, typing LOAD"A:\ PROGRAMS \ 4-3 and pushing the enter key will load program 4-3, which can then be run by typing RUN and pushing the enter key.

Contents

C H A P T E R   3.

C H A P T E R   4.

C H A P T E R   5.

CHAPTER 6.

Hypothesis Testing 204

CHAPTER 7.

Regression 245

C H A P T E R 8.

Analysis of Variance   **306**

C H A P T E R 9.

Goodness of Fit and Nonparametric Testing   **338**

C H A P T E R 10.

Life Testing   **380**

C H A P T E R 11.

C H A P T E R 12.

C H A P T E R 1

Elements of Probability

1 INTRODUCTION

The concept of the probability of a particular event of an experiment is subject to various meanings or interpretations. For instance, if a geologist is quoted as saying that "there is a 60 percent chance of oil in a certain region," we all probably have some intuitive idea as to what is being said. Indeed, most of us would probably interpret this statement in one of two possible ways. Either by imagining that

(1) The geologist feels that, over the long run, in 60 percent of the regions whose outward environmental conditions are very similar to the conditions that prevail in the region under consideration, there will be oil; or, by imagining that

(2) the geologist believes that it is more likely that the region will contain oil than it is that it will not; and in fact .6 is a measure of the geologist's belief in the hypothesis that the region will contain oil.

   The two foregoing interpretations of the probability of an event are referred to as being the frequency interpretation and the subjective (or personal) interpretation of probability. In the frequency interpretation, the probability of a given outcome of an experiment is considered as being a "property" of that outcome. It is imagined that this property can be operationally determined by continual repetition of the experiment—the probability of the outcome will then be observable as being the proportion of the experiments that result in the outcome. This is the interpretation of probability that is most prevalent among scientists.
   In the subjective interpretation, the probability of an outcome is not thought of as being a property of the outcome but rather is considered a statement about the beliefs of the person who is quoting the probability, concerning the chance that the outcome will occur. Thus, in this interpretation, probability becomes a subjective or personal concept and has no meaning

outside of expressing one's degree of belief. This interpretation of probability is often favored by philosophers and certain economic decision makers.

No matter which interpretation one gives to probability, however, there is a general consensus that the mathematics of probability are the same in either case. For instance, if you think that the probability that it will rain tomorrow is .3 and you feel that the probability that it will be cloudy but without any rain is .2, then you should feel that the probability that it will either be cloudy or rainy is .5 independently of your individual interpretation of the concept of probability. In this chapter we shall present the accepted rules, or axioms, used in probability theory. As a preliminary to this, however, we need to study the concept of the sample space and the events of an experiment.

2 SAMPLE SPACE AND EVENTS

Consider an experiment whose outcome is not predictable with certainty in advance. Although the outcome of the experiment will not be known in advance, however, let us suppose that the set of all possible outcomes is known. This set of all possible outcomes of an experiment is known as the *sample space* of the experiment and is denoted by S. Some examples are the following:

1. If the outcome of an experiment consists in the determination of the sex of a newborn child, then

$$S = \{g, b\}$$

where the outcome g means that the child is a girl and b that it is a boy.
2. If the experiment consists of the running of a race among the 7 horses having post positions $1, 2, 3, 4, 5, 6, 7$, then

$$S = \{\text{all orderings of } (1, 2, 3, 4, 5, 6, 7)\}$$

The outcomes $(2, 3, 1, 6, 5, 4, 7)$ means, for instance, that the number 2 horse is first, then the number 3 horse, then the number 1 horse, and so on.
3. Suppose we are interested in determining the amount of dosage that must be given to a patient until that patient reacts positively. One possible sample space for this experiment is to let S consist of all the positive numbers. That is, let

$$S = (0, \infty)$$

where the outcome would be x if the patient reacts to a dosage of value x but not to any smaller dosage.

Any subset E of the sample space is known as an *event*. That is, an event is a set consisting of possible outcomes of the experiment. If the outcome of the experiment is contained in E, then we say that E has occurred. Some examples of events are the following.

In Example 1 if $E = \{g\}$, then E is the event that the child is a girl. Similarly, if $F = \{b\}$, then F is the event that the child is a boy.

In Example 2 if

$$E = \{\text{all outcomes in } S \text{ starting with a } 3\}$$

then E is the event that the number 3 horse wins the race.

For any two events E and F of a sample space S we define the new event $E \cup F$, called the *union* of the events E and F, to consist of all outcomes that are either in E or in F or in both E and F. That is, the event $E \cup F$ will occur if *either* E or F occurs. For instance, in Example 1 if $E = \{g\}$ and $F = \{b\}$, then $E \cup F = \{g, b\}$. That is, $E \cup F$ would be the whole sample space S. In Example 2 if $E = \{\text{all outcomes starting with 6}\}$ is the event that the number 6 horse wins, and $F = \{\text{all outcomes having 6 in the second position}\}$ is the event that the number 6 horse comes in second, then $E \cup F$ is the event that the number 6 horse comes in either first or second.

Similarly, for any two events E and F, we may also define the new event EF, called the *intersection* of E and F, to consist of all outcomes that are in both E and F. That is, the event EF will occur only if both E and F occur. For instance, in Example 3 if $E = (0, 5)$ is the event that the required dosage is less than 5 and $F = (2, 10)$ is the event that it is between 2 and 10 then $EF = (2, 5)$ is the event that the required dosage is between 2 and 5. In Example 2 if $E = \{\text{all outcomes ending in 5}\}$ is the event that horse number 5 comes in last and $F = \{\text{all outcomes starting with 5}\}$ is the event that horse number 5 comes in first, then the event EF does not contain any outcomes, and hence cannot occur. To give such an event a name we shall refer to it as the null event and denote it by $\varnothing$. Thus $\varnothing$ refers to the event consisting of no outcomes. If $EF = \varnothing$, implying that E and F cannot both occur, then E and F are said to be *mutually exclusive*.

For any event E we define the event E^c, referred to as the *complement* of E, to consist of all outcomes in the sample space S that are not in E. That is, E^c will occur if and only if E does not occur. In Example 1 if $E = \{b\}$ is the event that the child is a boy, then $E^c = \{g\}$ is the event that it is a girl. Also note that since the experiment must result in some outcome, it follows that $S^c = \varnothing$.

For any two events E and F, if all of the outcomes in E are also in F, then we say that E is contained in F and write $E \subset F$ (or equivalently, $F \supset E$). Thus, if $E \subset F$ then the occurrence of E necessarily implies the occurrence of F. If $E \subset F$ and $F \subset E$, then we say that E and F are equal (or identical) and we write $E = F$.

We can also define unions and intersections of more than two events. In particular, the union of the events $E_1, E_2, \ldots, E_n$, denoted either by $E_1 \cup E_2 \cup \cdots \cup E_n$ or by $\bigcup_1^n E_i$, is defined to be the event consisting of all outcomes that are in E_i for at least one $i = 1, 2, \ldots, n$. Similarly the intersection of the events E_i, $i = 1, 2, \ldots, n$, denoted by $E_1 E_2 \cdots E_n$, is defined to be the event consisting of those outcomes that are in all of the events E_i, $i = 1, 2, \ldots, n$. In other words, the union of the E_i occurs when *at least* one of the events E_i occurs; while the intersection occurs when *all* of the events E_i occur.

3   VENN DIAGRAMS AND THE ALGEBRA OF EVENTS

A graphical representation of events that is very useful for illustrating logical relations among them is the Venn diagram. The sample space S is represented as consisting of all the points in a large rectangle, and the events $E, F, G, \ldots$, are represented as consisting of all the points in given circles within the rectangle. Events of interest can then be indicated by shading appropriate regions of the diagram. For instance, in the following three Venn diagrams, the shaded areas represent respectively the events $E \cup F$, EF, and E^c,

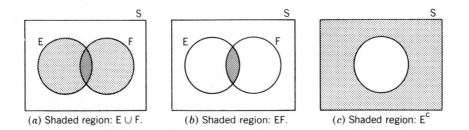

(*a*) Shaded region: E ∪ F.   (*b*) Shaded region: EF.   (*c*) Shaded region: E^c

The Venn diagram in Figure 1.3.1 indicates that $E \subset F$.

The operation of forming unions, intersections, and complements of events obey certain rules not dissimilar to the rules of algebra. We list a few of these

Commutative law	$E \cup F = F \cup E$	$EF = FE$
Associative law	$(E \cup F) \cup G = E \cup (F \cup G)$	$(EF)G = E(FG)$
Distributive law	$(E \cup F)G = EG \cup FG$	$EF \cup G = (E \cup G)(F \cup G)$

These relations are verified by showing that any outcome that is contained in the event on the left side of the equality is also contained in the event on the right side and vice versa. One way of showing this is by means of Venn

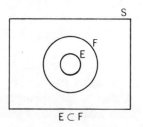

E ⊂ F   **FIGURE 1.3.1** Venn diagrams

diagrams. For instance, the distributive law may be verified by the following sequence of diagrams.

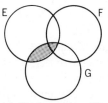

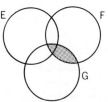

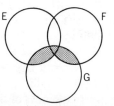

(*a*) Shaded region: EG.　　(*b*) Shaded region: FG.　　(*c*) Shaded region: (E ∪ F)G.
(E ∪ F)G = EG ∪ FG

The following useful relationship between the three basic operations of forming unions, intersections, and complements of events is known as *DeMorgan's laws*.

$$(E \cup F)^c = E^c F^c$$
$$(EF)^c = E^c \cup F^c$$

4 AXIOMS OF PROBABILITY

It appears to be an empirical fact that if an experiment is continually repeated under the exact same conditions, then for any event E, the proportion of time that the outcome is contained in E approaches some constant value as the number of repetitions increases. For instance, if a coin is continually flipped, then the proportion of flips resulting in heads will approach some value as the number of flips increases. It is this constant limiting frequency that we often have in mind when we speak of the probability of an event.

From a purely mathematical viewpoint we shall suppose that for each event E of an experiment having a sample space S there is a number, denoted by $P(E)$, which is in accord with the following three axioms.

AXIOM 1
$$0 \le P(E) \le 1$$

AXIOM 2
$$P(S) = 1$$

AXIOM 3
For any sequence of mutually exclusive events $E_1, E_2, \ldots$ (that is, events for which $E_i E_j = \varnothing$ when $i \ne j$)

$$P\left(\bigcup_{i=1}^{n} E_i \right) = \sum_{i=1}^{n} P(E_i), \qquad n = 1, 2, \ldots, \infty$$

We call $P(E)$ the probability of the event E.

Thus, Axiom 1 states that the probability that the outcome of the experiment is contained in E is some number between 0 and 1. Axiom 2 states that, with probability 1, the outcome will be a member of the sample space S. Axiom 3 states that for any set of mutually exclusive events the probability that at least one of these events occurs is equal to the sum of their respective probabilities.

It should be noted that if we interpret $P(E)$ as the relative frequency of the event E when a large number of repetitions of the experiment are performed, then $P(E)$ would indeed satisfy the above axioms. For instance, the proportion (or frequency) of time that the outcome is in E is clearly between 0 and 1, and the proportion of time that it is in S is 1 (since all outcomes are in S). Also, if E and F have no outcomes in common, then the proportion of time that the outcome is in either E or F is the sum of their respective frequencies. As an illustration of this last statement, suppose the experiment consists of the rolling of a pair of dice and suppose that E is the event that the sum is 2, 3, or 12 and F is the event that the sum is 7 or 11. Then if outcome E occurs 11 percent of the time and outcome F 22 percent of the time (as they would if the dice are perfectly symmetrical), then 33 percent of the time the outcome will be either 2, 3, 12, 7, or 11.

These axioms will now be used to prove two simple propositions concerning probabilities. We first note that E and E^c are always mutually exclusive, and since $E \cup E^c = S$ we have by Axioms 2 and 3 that

$$1 = P(S) = P(E \cup E^c) = P(E) + P(E^c)$$

Or equivalently, we have the following:

Proposition 1.4.1

$$P(E^c) = 1 - P(E)$$

In other words, Proposition 1.4.1 states that the probability that an event does not occur is 1 minus the probability that it does occur. For instance, if the probability of obtaining a head on the toss of a coin is $3/8$, the probability of obtaining a tail must be $5/8$.

Our second proposition gives the relationship between the probability of the union of two events in terms of the individual probabilities and the probability of the intersection.

Proposition 1.4.2

$$P(E \cup F) = P(E) + P(F) - P(EF)$$

Proof This proposition is most easily proven by the use of a Venn diagram as follows:

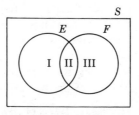

As the regions I, II, III are mutually exclusive, it follows that

$$P(E \cup F) = P(\text{I}) + P(\text{II}) + P(\text{III})$$
$$P(E) = P(\text{I}) + P(\text{III})$$
$$P(F) = P(\text{II}) + P(\text{III})$$

which shows that

$$P(E \cup F) + P(E) + P(F) - P(\text{II})$$

and the proof is complete since II = *EF*.

5 SAMPLE SPACES HAVING EQUALLY LIKELY OUTCOMES

For a large number of experiments it is natural to assume that each point in the sample space is equally likely to occur. That is, for many experiments whose sample space S is a finite set say $S = \{1, 2, \ldots, N\}$ it is often natural to assume that

$$P(\{1\}) = P(\{2\}) = \cdots = P(\{N\}) = p \quad \text{(say)}$$

Now it follows from Axioms 2 and 3 that

$$1 = P(S) = P(\{1\}) + \cdots + P(\{N\}) = Np$$

which shows that

$$P(\{i\}) = p = 1/N$$

From this it follows from Axiom 3 that for any event E

$$P(E) = \frac{\text{Number of points in } E}{N}$$

In words, if we assume that each outcome of an experiment is equally likely to occur, then the probability of any event E equals the proportion of points in the sample space that are contained in E.

Thus, to compute probabilities it is often necessary to be able to effectively count the number of different ways that a given event can occur. To do this we will make use of the following rule.

Basic Principle of Counting

Suppose that two experiments are to be performed. Then if experiment 1 can result in any one of m possible outcomes and if, for each outcome of experiment 1, there are n possible outcomes of experiment 2, then together there are mn possible outcomes of the two experiments.

Proof of the Basic Principle   The Basic Principle can be proven by enumerating all the possible outcomes of the two experiments as follows:

$$(1,1),(1,2),\ldots,(1,n)$$
$$(2,1),(2,2),\ldots,(2,n)$$
$$\vdots$$
$$(m,1),(m,2),\ldots,(m,n)$$

where we say that the outcome is (i, j) if experiment 1 results in its ith possible outcome and experiment 2 then results in the jth of its possible outcomes. Hence, the set of possible outcomes consists of m rows, each row containing n elements, which proves the result.

Example 1.5a Two balls are "randomly drawn" from a bowl containing 6 white and 5 black balls. What is the probability that one of the drawn balls is white and the other black?

Solution If we regard the order in which the balls are selected as being significant, then as the first drawn ball may be any of the 11 and the second any of the remaining 10, it follows that the sample space consists of $11 \cdot 10 = 110$ points. Furthermore there are $6 \cdot 5 = 30$ ways in which the first ball selected is white and the second black, and similarly there are $5 \cdot 6 = 30$ ways in which the first ball is black and the second white. Hence, assuming that "randomly drawn" means that each of the 110 points in the sample space is equally likely to occur, then we see that the desired probability is

$$\frac{30 + 30}{110} = \frac{6}{11} \quad \blacksquare$$

When there are more than two experiments to be performed the basic principle can be generalized as follows:

Generalized Basic Principle of Counting

If r experiments that are to be performed are such that the first one may result in any of n_1 possible outcomes, and if for each of these n_1 possible outcomes there are n_2 possible outcomes of the second experiment, and if for each of the possible outcomes of the first two experiments there are n_3 possible outcomes of the third experiment, and if,..., then there are a total of $n_1 \cdot n_2 \cdots n_r$ possible outcomes of the r experiments.

As an illustration of this let us determine the number of different ways n distinct objects can be arranged in a linear order. For instance how many different ordered arrangements of the letters a, b, c are possible? By direct enumeration we see that there are 6; namely abc, acb, bac, bca, cab, cba. Each one of these ordered arrangements is known as a *permutation*. Thus, there are 6 possible permutations of a set of 3 objects. This result could also have been obtained from the basic principle, since the first object in the permutation can be any of the 3, the second object in the permutation can then be chosen from any of the remaining 2, and the third object in the permutation is then chosen from the remaining one. Thus, there are $3 \cdot 2 \cdot 1 = 6$ possible permutations.

Suppose now that we have n objects. Similar reasoning shows that there are

$$n(n-1)(n-2) \cdots 3 \cdot 2 \cdot 1$$

different permutations of the n objects. It is convenient to introduce the notation $n!$, which is read "n factorial," for the foregoing expression. That is,

$$n! = n(n-1)(n-2) \cdots 3 \cdot 2 \cdot 1$$

Thus, for instance, $1! = 1, 2! = 2 \cdot 1 = 2, 3! = 3 \cdot 2 \cdot 1 = 6, 4! = 4 \cdot 3 \cdot 2 \cdot 1 = 24$, and so on. It is convenient to define $0! = 1$.

Example 1.5b A class in probability theory consists of 6 men and 4 women. An exam is given and the students are ranked according to their performance. Assuming that no two students obtain the same score (a) how many different rankings are possible? (b) If all rankings are considered equally likely what is the probability that women receive the top 4 scores?

Solution

(a) Because each ranking corresponds to a particular ordered arrangement of the 10 people, we see the answer to this part is $10! = 3,628,800$.

(b) As there are $4!$ possible rankings of the women among themselves and $6!$ possible rankings of the men among themselves, it follows from the basic principle that there are $(6!)(4!) = (720)(24) = 17,280$ possible rankings in which the women receive the top 4 scores. Hence the desired probability is

$$\frac{6!4!}{10!} = \frac{4 \cdot 3 \cdot 2 \cdot 1}{10 \cdot 9 \cdot 8 \cdot 7} = \frac{1}{210} \quad \blacksquare$$

Example 1.5c If n people are present in a room, what is the probability that no two of them celebrate their birthday on the same day of the year? How large need n be so that this probability is less than $1/2$?

Solution Because each person can celebrate his birthday on any one of 365 days, there are a total of $(365)^n$ possible outcomes. (We are ignoring the possibility of someone having been born on February 29.) Furthermore there are $(365)(364)(363) \cdots (365 - n + 1)$ possible outcomes that result in no two of the people having the same birthday. This is so because the first person could have any one of 365 birthdays, the next person any of the remaining 364 days, the next any of the remaining 363, and so on. Hence, assuming that each outcome is equally likely, we see that the desired probability is

$$\frac{(365)(364)(363) \cdots (365 - n + 1)}{(365)^n}$$

It is a rather surprising fact that when $n = 23$ the foregoing probability is less than $\frac{1}{2}$. That is, if there are 23 people in a room, then the probability that at least two of them have the same birthday exceeds $\frac{1}{2}$. Many people find this surprising. Perhaps even more surprising, however, is that this probability increases to .970 when there are 50 people in the room. And with 100 persons in the room, the odds are better than three million to one [that is the probability is greater than $(3 \times 10^6)/(3 \times 10^6 + 1)$] that at least two people have the same birthday. ∎

Suppose now that we are interested in determining the number of different groups of r objects that could be formed from a total of n objects. For instance, how many different groups of three could be selected from the five items A, B, C, D, E? To answer this, reason as follows. Since there are 5 ways to select the initial item, 4 ways to then select the next item, and 3 ways to then select the final item, there are thus $5 \cdot 4 \cdot 3$ ways of selecting the group of 3 when the order in which the items are selected is relevant. However, since every group of 3, say the group consisting of items A, B, and C, will be counted 6 times (that is, all of the permutations ABC, ACB, BAC, BCA, CAB, CBA will be counted when the order of selection is relevant), it follows that the total number of different groups that can be formed is $(5 \cdot 4 \cdot 3)/(3 \cdot 2 \cdot 1) = 10$.

In general, as $n(n - 1) \cdots (n - r + 1)$ represents the number of different ways that a group of r items could be selected from n items when the order of selection is considered relevant (since the first one selected can be any one of the n, and the second selected any one of the remaining $n - 1$, etc.), and since each group of r items will be counted $r!$ times in this count, it follows that the number of different groups of r items that could be formed from a set of n items is

$$\frac{n(n - 1) \cdots (n - r + 1)}{r!} = \frac{n!}{(n - r)!r!}$$

Notation and Terminology

We define $\binom{n}{r}$, for $r \leq n$, by

$$\binom{n}{r} = \frac{n!}{(n - r)!r!}$$

and call $\binom{n}{r}$ the number of *combinations* of n objects taken r at a time.

Thus $\binom{n}{r}$ represents the number of different groups of size r that can be selected from a set of size n when the order of selection is not considered relevant. For example, there are

$$\binom{8}{2} = \frac{8 \cdot 7}{2 \cdot 1} = 28$$

different groups of size 2 that can be chosen from a set of 8 people; and

$$\binom{10}{2} = \frac{10 \cdot 9}{2 \cdot 1} = 45$$

different groups of size 2 that can be chosen from a set of 10 people. Also, since $0! = 1$, note that

$$\binom{n}{0} = \binom{n}{n} = 1$$

Example 1.5d A committee of size 5 is to be selected from a group of 6 men and 9 women. If the selection is made randomly, what is the probability that the committee consists of 3 men and 2 women?

Solution Let us assume that "randomly selected" means that each of the $\binom{15}{5}$ possible combinations is equally likely to be selected. Hence, since there are $\binom{6}{3}$ possible choices of 3 men and $\binom{9}{2}$ possible choices of 2 women, it follows that the desired probability is given by

$$\frac{\binom{6}{3}\binom{9}{2}}{\binom{15}{5}} = \frac{240}{1001} \quad \blacksquare$$

Example 1.5e From a set of n items a random sample of size is k to be selected. What is the probability a given item will be among the k selected?

Solution The number of different selections that contain the given item is $\binom{1}{1}\binom{n-1}{k-1}$. Hence the probability that a particular item is among the k selected is

$$\binom{n-1}{k-1} \bigg/ \binom{n}{k} = \frac{(n-1)!}{(n-k)!(k-1)!} \bigg/ \frac{n!}{(n-k)!k!} = \frac{k}{n} \quad \blacksquare$$

Example 1.5f A basketball team consists of 6 black and 6 white players. The players are to be paired in groups of two for the purpose of determining roommates. If the pairings are done at random, what is the probability that none of the black players has a white roommate?

Solution Let us start by imagining that the 6 pairs are numbered—that is, there is a first pair, a second pair, and so on. Since there are $\binom{12}{2}$ different choices of a first pair; and for each choice of a first pair there are $\binom{10}{2}$ different choices of a second pair; and for each choice of the first 2 pairs there are $\binom{8}{2}$ choices for a third pair; and so on, if follows from the generalized basic principle of counting that there are

$$\binom{12}{2}\binom{10}{2}\binom{8}{2}\binom{6}{2}\binom{4}{2}\binom{2}{2} = \frac{12!}{(2!)^6}$$

ways of dividing the players into a *first* pair, a *second* pair, and so on. Hence there are $(12)!/2^6 6!$ ways of dividing the players into 6 (unordered) pairs of 2 each. Furthermore, since there are, by the same reasoning, $6!/2^3 3!$ ways of pairing the white players among themselves and $6!/2^3 3!$ ways of pairing the black players among themselves, it follows that there are $(6!/2^3 3!)^2$ pairings that do not result in any black–white roommate pairs. Hence, if the pairings are done at random (so that all outcomes are equally likely), then the desired probability is

$$\left(\frac{6!}{2^3 3!}\right)^2 \bigg/ \frac{(12)!}{2^6 6!} = \frac{5}{231} = .0216$$

Hence, there are roughly only two chances in a hundred that a random pairing will not result in any of the white and black players rooming together. ∎

6   CONDITIONAL PROBABILITY

In this section we introduce one of the most important concepts in all of probability theory—that of conditional probability. Its importance is twofold. In the first place we are often interested in calculating probabilities when some partial information concerning the result of the experiment is available, or in recalculating them in light of additional information. In such situations the desired probabilities are conditional ones. Second, as a kind of a bonus, it often turns out that the easiest way to compute the probability of an event is to first "condition" on the occurrence or nonoccurrence of a secondary event.

As an illustration of a conditional probability, suppose that one rolls a pair of dice. The sample space S of this experiment can be taken to be the following set of 36 outcomes

$$S = \{(i, j), \quad i = 1, 2, 3, 4, 5, 6, \quad j = 1, 2, 3, 4, 5, 6\}$$

where we say that the outcome is (i, j) if the first die lands on side i and the second on side j. Suppose now that each of the 36 possible outcomes is equally likely to occur and thus has probability $1/36$. (In such a situation we say that the dice are fair). Suppose further that we observe that the first die lands on side 3. Then, given this information, what is the probability that the sum of the two dice equals eight? To calculate this probability we reason as follows: given that the initial die is a three, there can be at most six possible outcomes of our experiment, namely, $(3, 1)$, $(3, 2)$, $(3, 3)$, $(3, 4)$, $(3, 5)$, and $(3, 6)$. In addition, as each of these outcomes originally had the same probability of occurring, they should still have equal probabilities. That is, given that the first die is a three, then the (conditional) probability of each of the outcomes $(3,1), (3,2), (3,3), (3,4), (3,5), (3,6)$ is $1/6$, whereas the (conditional) probability of the other 30 points in the sample space is 0. Hence, the desired probability will be $1/6$.

If we let E and F denote respectively the event that the sum of the dice is eight and the event that the first die is a three, then the probability just obtained is called the conditional probability of E given that F has occurred, and is denoted by

$$P(E|F).$$

A general formula for $P(E|F)$ that is valid for all events E and F is derived in the same manner as just described. Namely, if the event F occurs, then in order for E to occur it is necessary that the actual occurrence be a point in both E and F; that is, it must be in EF. Now, since we know that F has occurred, it follows that F becomes our new (reduced) sample space and hence the probability that the event EF occurs will equal the probability of EF relative to the probability of F. That is

$$P(E|F) = \frac{P(EF)}{P(F)} \qquad (1.6.1)$$

Note that Equation 1.6.1 is well defined only when $P(F) > 0$ and hence $P(E|F)$ is defined only when $P(F) > 0$. (See Figure 1.6.1.)

Example 1.6a A bin contains 5 defective (that immediately fail when put in use), 10 partially defective (that fail after a couple of hours of use), and 25 acceptable transistors. A transistor is chosen at random from the bin and put into use. If it does not immediately fail, what is the probability it is acceptable?

Solution Since the transistor did not immediately fail, we know that it is not one of the 5 defectives and so the desired probability is:

$$P\{\text{acceptable}|\text{not defective}\}$$

$$= \frac{P\{\text{acceptable, not defective}\}}{P\{\text{not defective}\}}$$

$$= \frac{P\{\text{acceptable}\}}{P\{\text{not defective}\}}$$

where the last equality follows since the transistor will be both acceptable and

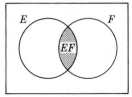

FIGURE 1.6.1 $P(E|F) = \dfrac{P(EF)}{P(F)}$

not defective if it is acceptable. Hence, assuming that each of the 40 transistors is equally likely to be chosen, we obtain that

$$P\{\text{acceptable}|\text{not defective}\} = \frac{25/40}{35/40} = 5/7$$

It should be noted that we could also have derived this probability by working directly with the reduced sample space. That is, since we know that the chosen transistor is not defective, the problem reduces to computing the probability that a transistor, chosen at random from a bin containing 25 acceptable and 10 partially defective transistors, is acceptable. This is clearly equal to 25/35. ∎

Example 1.6b The organization that Jones works for is running a father–son dinner for those employees having at least one son. Each of these employees is invited to attend along with his youngest son. If Jones is known to have two children, what is the conditional probability that they are both boys given that he is invited to the dinner? Assume that the sample space S is given by $S = \{(b, b), (b, g), (g, b), (g, g)\}$ and all outcomes are equally likely [(b, g) means, for instance, that the younger child is a boy and the older child is a girl].

Solution The knowledge that Jones has been invited to the dinner is equivalent to knowing that he has at least one son. Hence, letting B denote the event that both children are boys, and A the event that at least one of them is a boy, we have that the desired probability $P(B|A)$ is given by

$$P(B|A) = \frac{P(BA)}{P(A)}$$

$$= \frac{P(\{(b, b)\})}{P(\{(b, b), (b, g), (g, b)\})}$$

$$= \frac{\frac{1}{4}}{\frac{3}{4}} = \frac{1}{3}$$

Many readers incorrectly reason that the conditional probability of two boys given at least one is $1/2$, as opposed to the correct $1/3$, since they reason that the Jones child not attending the dinner is equally likely to be a boy or a girl. Their mistake, however, is in assuming that these two possibilities are equally likely. For initially there were 4 equally likely outcomes. Now the information that at least one child is a boy is equivalent to knowing that the outcome is not (g, g). Hence we are left with the 3 equally likely outcomes $(b, b), (b, g), (g, b)$, thus showing that the Jones child not attending the dinner is twice as likely to be a girl as a boy. ∎

By multiplying both sides of Equation 1.6.1 by $P(F)$ we obtain that

$$P(EF) = P(F)P(E|F) \tag{1.6.2}$$

In words, Equation 1.6.2 states that the probability both E and F occur is equal to the probability that F occurs multiplied by the conditional probability of E given that F occurred. Equation 1.6.2 is often quite useful in computing the probability of the intersection of events. This is illustrated by the following example.

Example 1.6c Mr. Jones figures that there is a 30 percent chance that his company will set up a branch office in Phoenix. If it does, he is 60 percent certain that he will be made manager of this new operation. What is the probability that Jones will be a Phoenix branch office manager?

Solution If we let B denote the event that the company sets up a branch office in Phoenix and M the event that Jones is made the Phoenix manager, then the desired probability is $P(BM)$, which is obtained as follows:

$$P(BM) = P(B)P(M|B)$$
$$= (.3)(.6)$$
$$= .18$$

Hence, there is an 18 percent chance that Jones will be the Phoenix manager. ∎

7 BAYES' FORMULA

Let E and F be events. We may express E as

$$E = EF \cup EF^c$$

for, in order for a point to be in E, it must either be in both E and F or be in E but not in F. (See Figure 1.7.1.) As EF and EF^c are clearly mutually exclusive, we have by Axiom 3 that

$$P(E) = P(EF) + P(EF^c) \tag{1.7.1}$$
$$= P(E|F)P(F) + P(E|F^c)P(F^c)$$
$$= P(E|F)P(F) + P(E|F^c)[1 - P(F)].$$

Equation 1.7.1 states that the probability of the event E is a weighted average of the conditional probability of E given that F has occurred and the conditional probability of E given that F has not occurred: each conditional probability being given as much weight as the event it is conditioned on has of occurring. It is an extremely useful formula, for its use often enables us to determine the probability of an event by first "conditioning" on whether or

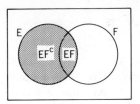

FIGURE 1.7.1 $E = EF \cup EF^c$. EF = shaded area; EF^c = striped area

not some second event has occurred. That is, there are many instances where it is difficult to compute the probability of an event directly, but it is straightforward to compute it once we know whether or not some second event has occurred.

Example 1.7a An insurance company believes that people can be divided into two classes—those that are accident prone and those that are not. Their statistics show that an accident-prone person will have an accident at some time within a fixed 1-year period with probability .4, whereas this probability decreases to .2 for a non-accident-prone person. If we assume that 30 percent of the population is accident prone, what is the probability that a new policy holder will have an accident within a year of purchasing a policy?

Solution We shall obtain the desired probability by first conditioning on whether or not the policy holder is accident prone. Let A_1 denote the event that the policy holder will have an accident within a year of purchase; and let A denote the event that the policy holder is accident prone. Hence the desired probability, $P(A_1)$, is given by

$$P(A_1) = P(A_1|A)P(A) + P(A_1|A^c)P(A^c)$$
$$= (.4)(.3) + (.2)(.7) = .26 \quad \blacksquare$$

In the next series of examples we will indicate how to reevaluate an initial probability assessment in the light of additional (or new) information. That is, we will show how to incorporate new information with an initial probability assessment to obtain an updated probability.

Example 1.7b Reconsider Example 1.7a and suppose that a new policy holder has an accident within a year of purchasing his policy. What is the probability that he is accident prone?

Solution Initially, at the moment when the policy holder purchased his policy, we assumed there was a 30 percent chance that he was accident prone. That is, $P(A) = .3$. However, based on the fact that he has had an accident within a year, we now reevaluate his probability of being accident prone as follows.

$$P(A|A_1) = \frac{P(AA_1)}{P(A_1)}$$
$$= \frac{P(A)P(A_1|A)}{P(A_1)}$$
$$= \frac{(.3)(.4)}{.26} = \frac{6}{13} = .4615 \quad \blacksquare$$

Example 1.7c In answering a question on a multiple choice test, a student either knows the answer or he guesses. Let p be the probability that he knows the answer and $1 - p$ the probability that he guesses. Assume that a student who guesses at the answer will be correct with probability $1/m$, where m is the number of multiple choice alternatives. What is the conditional probability that a student knew the answer to a question given that he answered it correctly?

Solution Let C and K denote respectively the events that the student answers the question correctly, and the event that he actually knows the answer. To compute

$$P(K|C) = \frac{P(KC)}{P(C)}$$

we first note that

$$P(KC) = P(K)P(C|K)$$
$$= p \cdot 1$$
$$= p$$

To compute the probability that the student answers correctly, we condition on whether or not he knows the answer. That is,

$$P(C) = P(C|K)P(K) + P(C|K^c)P(K^c)$$
$$= p + (1/m)(1 - p)$$

Hence, the desired probability is given by

$$P(K|C) = \frac{p}{p + (1/m)(1 - p)} = \frac{mp}{1 + (m - 1)p}$$

Thus, for example, if $m = 5$, $p = 1/2$, then the probability that a student knew the answer to a question he correctly answered is $5/6$. ∎

Example 1.7d A laboratory blood test is 99 percent effective in detecting a certain disease when it is, in fact, present. However, the test also yields a "false positive" result for 1 percent of the healthy persons tested. (That is, if a healthy person is tested, then, with probability .01, the test result will imply he has the disease.) If .5 percent of the population actually has the disease, what is the probability a person has the disease given that his test result is positive?

Solution Let D be the event that the tested person has the disease and E the event that his test result is positive. The desired probability $P(D|E)$ is obtained by

$$P(D|E) = \frac{P(DE)}{P(E)}$$

$$= \frac{P(E|D)P(D)}{P(E|D)P(D) + P(E|D^c)P(D^c)}$$

$$= \frac{(.99)(.005)}{(.99)(.005) + (.01)(.995)}$$

$$= .3322.$$

Thus, only 33 percent of those persons whose test results are positive actually have the disease. Since many students are often surprised at this result (because they expected this figure to be much higher since the blood test seems to be a good one), it is probably worthwhile to present a second argument which, though less rigorous than the foregoing, is probably more revealing. We now do so.

Since .5 percent of the population actually has the disease, it follows that, on the average, 1 person out of every 200 tested will have it. The test will correctly confirm that this person has the disease with probability .99. Thus, on the average, out of every 200 persons tested, the test will correctly confirm that .99 persons have the disease. On the other hand, out of the (on the average) 199 healthy people, the test will incorrectly state that (199)(.01) of these people have the disease. Hence, for every .99 diseased persons that the test correctly states are ill, there are (on the average) 1.99 healthy persons that the test incorrectly states are ill. Hence, the proportion of time that the test result is correct when it states that a person is ill is

$$\frac{.99}{.99 + 1.99} = .3322 \quad \blacksquare$$

Equation 1.7.1 is also useful when one has to reassess one's (personal) probabilities in the light of additional information. For instance, consider the following examples.

Example 1.7e At a certain stage of a criminal investigation the inspector in charge is 60 percent convinced of the guilt of a certain suspect. Suppose now that a *new* piece of evidence that shows that the criminal has a certain characteristic (such as left-handedness, baldness, brown hair, etc.) is uncovered. If 20 percent of the population possesses this characteristic, how certain of the guilt of the suspect should the inspector now be if it turns out that the suspect is among this group?

Solution Letting G denote the event that the suspect is guilty and C the event that he possesses the characteristic of the criminal, we have

$$P(G|C) = \frac{P(GC)}{P(C)}$$

Now

$$\begin{aligned} P(GC) &= P(G)P(C|G) \\ &= (.6)(1) \\ &= .6 \end{aligned}$$

To compute the probability that the suspect has the characteristic, we condition on whether or not he is guilty. That is,

$$\begin{aligned} P(C) &= P(C|G)P(G) + P(C|G^c)P(G^c) \\ &= (1)(.6) + (.2)(.4) \\ &= .68 \end{aligned}$$

where we have supposed that the probability of the suspect having the characteristic if he is, in fact, innocent is equal to .2, the proportion of the population possessing the characteristic. Hence

$$P(G|C) = \frac{60}{68} = .882$$

and so the inspector should now be 88 percent certain of the guilt of the suspect. ∎

Example 1.7e (continued) Let us now suppose that the new evidence is subject to different possible interpretations, and in fact only shows that it is 90 percent likely that the criminal possesses this certain characteristic. In this case how likely would it be that the suspect is guilty (assuming, as before, that he has this characteristic)?

Solution In this case the situation is as before with the exception that the probability of the suspect having the characteristic given that he is guilty is now .9 (rather than 1). Hence

$$P(G|C) = \frac{P(GC)}{P(C)}$$

$$= \frac{P(G)P(C|G)}{P(C|G)P(G) + P(C|G^c)P(G^c)}$$

$$= \frac{(.6)(.9)}{(.9)(.6) + (.2)(.4)}$$

$$= \frac{54}{62} = .871$$

which is slightly less than in the previous case (why?). ∎

Equation 1.7.1 may be generalized in the following manner. Suppose that $F_1, F_2, \ldots, F_n$ are mutually exclusive events such that

$$\bigcup_{i=1}^{n} F_i = S$$

In other words, exactly one of the events $F_1, F_2, \ldots, F_n$ must occur. By writing

$$E = \bigcup_{i=1}^{n} EF_i$$

and using the fact that the events EF_i, $i = 1, \ldots, n$ are mutually exclusive, we obtain that

$$P(E) = \sum_{i=1}^{n} P(EF_i) \tag{1.7.2}$$

$$= \sum_{i=1}^{n} P(E|F_i)P(F_i)$$

Thus, Equation 1.7.2 shows how, for given events $F_1, F_2, \ldots, F_n$ of which one and only one must occur, we can compute $P(E)$ by first "conditioning" on which one of the F_i occurs. That is, it states that $P(E)$ is equal to a weighted average of $P(E|F_i)$, each term being weighted by the probability of the event on which it is conditioned.

Suppose now that E has occurred and we are interested in determining which one of the F_j also occurred. By Equation 1.7.2, we have that

$$P(F_j|E) = \frac{P(EF_j)}{P(E)} \tag{1.7.3}$$

$$= \frac{P(E|F_j)P(F_j)}{\sum_{i=1}^{n}P(E|F_i)P(F_i)}$$

Equation 1.7.3 is known as Bayes' formula, after the English philosopher Thomas Bayes. If we think of the events F_j as being possible "hypotheses" about some subject matter, then Bayes' formula may be interpreted as showing us how opinions about these hypotheses held before the experiment [that is, the $P(F_j)$] should be modified by the evidence of the experiment.

Example 1.7f A plane is missing and it is presumed that it was equally likely to have gone down in any of 3 possible regions. Let $1 - \alpha_i$ denote the probability the plane will be found upon a search of the ith region when the plane is, in fact, in that region, $i = 1, 2, 3$. (The constants α_i are called *overlook probabilities* because they represent the probability of overlooking the plane; they are generally attributable to the geographical and environmental conditions of the regions.) What is the conditional probability that the plane is in the ith region, given that a search of region 1 is unsuccessful, $i = 1, 2, 3$?

Solution Let R_i, $i = 1, 2, 3$, be the event that the plane is in region i; and let E be the event that a search of region 1 is unsuccessful. From Bayes' formula we obtain

$$P(R_1|E) = \frac{P(ER_1)}{P(E)}$$

$$= \frac{P(E|R_1)P(R_1)}{\sum_{i=1}^{3}P(E|R_i)P(R_i)}$$

$$= \frac{(\alpha_1)(1/3)}{(\alpha_1)(1/3) + (1)1/3 + (1)(1/3)}$$

$$= \frac{\alpha_1}{\alpha_1 + 2}$$

For $j = 2, 3$

$$P(R_j|E) = \frac{P(E|R_j)P(R_j)}{P(E)}$$

$$= \frac{(1)(1/3)}{(\alpha_1)1/3 + 1/3 + 1/3}$$

$$= \frac{1}{\alpha_1 + 2}, \qquad j = 2, 3$$

Thus, for instance, if $\alpha_1 = .4$ then the conditional probability that the plane is in region 1 given that a search of that region did not uncover it is $1/6$. ∎

8 INDEPENDENT EVENTS

The previous examples in this chapter show that $P(E|F)$, the conditional probability of E given F, is not generally equal to $P(E)$, the unconditional probability of E. In other words, knowing that F has occurred generally changes the chances of E's occurrence. In the special cases where $P(E|F)$ does in fact equal $P(E)$, we say that E is independent of F. That is, E is independent of F if knowledge that F has occurred does not change the probability that E occurs.

Since $P(E|F) = P(EF)/P(F)$, we see that E is independent of F if

$$P(EF) = P(E)P(F) \qquad (1.8.1)$$

Since this equation is symmetric in E and F, it shows that whenever E is independent of F so is F of E. We thus have the following.

> **Definition.** Two events E and F are said to be *independent* if Equation 1.8.1 holds. Two events E and F that are not independent are said to be *dependent*.

Example 1.8a A card is selected at random from an ordinary deck of 52 playing cards. If A is the event that the selected card is an ace and H is the event that it is a heart, then A and H are independent, since $P(AH) = 1/52$, while $P(A) = 4/52$ and $P(H) = 13/52$.

Example 1.8b If we let E denote the event that the next president is a Republican and F the event that there will be a major earthquake within the next year, then most people would probably be willing to assume that E and F are independent. However, there would probably be some controversy over whether it is reasonable to assume that E is independent of G, where G is the event that there will be a recession within the next two years. ∎

We now show that if E is independent of F then E is also independent of F^c.

Proposition 1.8.1

If E and F are independent then so are E and F^c.

Proof Assume that E and F are independent. Since $E = EF \cup EF^c$, and EF and EF^c are obviously mutually exclusive we have that

$$P(E) = P(EF) + P(EF^c)$$
$$= P(E)P(F) + P(EF^c) \qquad \text{by the independence of } E \text{ and } F$$

or equivalently

$$P(EF^c) = P(E)(1 - P(F))$$
$$= P(E)P(F^c)$$

and the result is proven.

Thus if E is independent of F, then the probability of E's occurrence is unchanged by information as to whether or not F has occurred.

Suppose now that E is independent of F and is also independent of G. Is E then necessarily independent of FG? The answer, somewhat surprisingly, is no. For consider the following example.

Example 1.8c Two fair dice are thrown. Let E_7 denote the event that the sum of the dice is 7. Let F denote the event that the first die equals 4 and let T be the event that the second die equals 3. Now it can be shown (see Problem 28) that E_7 is independent of F and that E_7 is also independent of T; but clearly E_7 is not independent of FT (since $P(E_7|FT) = 1$). ∎

It would appear to follow from the foregoing example that an appropriate definition of the independence of three events E, F, and G would have to go further then merely assuming that all of the $\binom{3}{2}$ pairs of events are independent. We are thus led to the following definition.

Definition. The three events E, F, and G are said to be independent if

$$P(EFG) = P(E)P(F)P(G)$$
$$P(EF) = P(E)P(F)$$
$$P(EG) = P(E)P(G)$$
$$P(FG) = P(F)P(G)$$

It should be noted that if the events E, F, G are independent, then E will be independent of any event formed from F and G. For instance, E is independent of $F \cup G$ since

$$P(E(F \cup G)) = P(EF \cup EG)$$
$$= P(EF) + P(EG) - P(EFG)$$
$$= P(E)P(F) + P(E)P(G) - P(E)P(FG)$$
$$= P(E)[P(F) + P(G) - P(FG)]$$
$$= P(E)P(F \cup G)$$

Of course we may also extend the definition of independence to more than three events. The events $E_1, E_2, \ldots, E_n$ are said to be independent if for every subset $E_{1'}, E_{2'}, \ldots, E_{r'}$, $r \leq n$, of these events

$$P(E_{1'}E_{2'} \cdots E_{r'}) = P(E_{1'})P(E_{2'}) \cdots P(E_{r'})$$

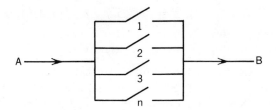

FIGURE 1.8.1 Parallel system: functions if current flows from A to B

It is sometimes the case that the probability experiment under considera-tion consists of performing a sequence of subexperiments. For instance, if the experiment consists of continually tossing a coin, then we may think of each toss as being a subexperiment. In many cases it is reasonable to assume that the outcomes of any group of the subexperiments have no effect on the probabilities of the outcomes of the other subexperiments. If such is the case, then we say that the subexperiments are independent.

Example 1.8d A system composed of n separate components is said to be a parallel system if it functions when at least one of the components functions. (See Figure 1.8.1.) For such a system, if component i, independent of other components, functions with probability p_i, $i = 1, \ldots, n$, what is the probabil-ity the system functions?

Solution Let A_i denote the event that component i functions. Then

$$P\{\text{system functions}\} = 1 - P\{\text{system does not function}\}$$

$$= 1 - P\{\text{all components do not function}\}$$

$$= 1 - P\left(\bigcap_i A_i^c\right)$$

$$= 1 - \prod_{i=1}^{n}(1 - p_i) \qquad \text{by independence} \quad \blacksquare$$

PROBLEMS

1. A box contains 3 marbles, one red, one green, and one blue. Consider an experiment that consists of taking one marble from the box, then replacing it in the box and drawing a second marble from the box. Describe the sample space. Repeat when the second marble is drawn without first replacing the first marble.

2. An experiment consists of tossing a coin 3 times. What is the sample space of this experiment? Which event corresponds to the experiment resulting in more heads than tails?

3. Let $S = \{1, 2, 3, 4, 5, 6, 7\}$, $E = \{1, 3, 5, 7\}$, $F = \{7, 4, 6\}$, $G = \{1, 4\}$. Find

   (a) EF (c) EG^c (e) $E^c(F \cup G)$

   (b) $E \cup FG$ (d) $EF^c \cup G$ (f) $EG \cup FG$

4. Two dice are thrown. Let E be the event that the sum of the dice is odd, let F be the event that the first die lands on 1, and let G be the event that the sum is 5. Describe the events EF, $E \cup F$, FG, EF^c, EFG.

5. Let E, F, G be three events. Find expressions for the events that of E, F, G

   (a) only E occurs

   (b) both E and G but not F occurs

   (c) at least one of the events occurs

   (d) at least two of the events occur

   (e) all three occur

   (f) none of the events occur

   (g) at most one of them occurs

   (h) at most two of them occur

   (i) exactly two of them occur

   (j) at most three of them occur

6. Find simple expressions for the events

   (a) $E \cup E^c$

   (b) EE^c

   (c) $(E \cup F)(E \cup F^c)$

   (d) $(E \cup F)(E^c \cup F)(E \cup F^c)$

   (e) $(E \cup F)(F \cup G)$

7. Use Venn diagrams (or any other method) to show that

   (a) $EF \subset E$, $E \subset E \cup F$

   (b) if $E \subset F$ then $F^c \subset E^c$

   (c) the commutative laws are valid

   (d) the associative laws are valid

   (e) $F = FE \cup FE^c$

   (f) $E \cup F = E \cup E^c F$

   (g) DeMorgans laws are valid

8. For the following Venn diagram

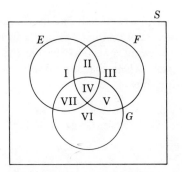

describe in terms of E, F, and G the events denoted in the diagram by the Roman numerals I through VII.

9. Show that if $E \subset F$ then $P(E) \le P(F)$.
Hint: Write F as the union of two mutually exclusive events, one of them being E.

10. Prove Boole's inequality, namely that

$$P\left(\bigcup_{i=1}^{n} E_i\right) \le \sum_{i=1}^{n} P(E_i)$$

11. If $P(E) = .9$ and $P(F) = .9$ show that $P(EF) \ge .8$. In general, prove Bonferroni's inequality, namely that

$$P(EF) \ge P(E) + P(F) - 1$$

12. Prove that

(a) $$P(EF^c) = P(E) - P(EF)$$
(b) $$P(E^c F^c) = 1 - P(E) - P(F) + P(EF)$$

13. Show that the probability that exactly one of the events E or F occurs is equal to $P(E) + P(F) - 2P(EF)$.

14. Calculate $\binom{9}{3}$, $\binom{9}{6}$, $\binom{7}{2}$, $\binom{7}{5}$, $\binom{10}{7}$

15. Show that

$$\binom{n}{r} = \binom{n}{n-r}$$

Now present a combinatorial argument for the foregoing by explaining why a choice of r items from a set of size n is equivalent to a choice of $n - r$ items from that set.

16. Show that

$$\binom{n}{r} = \binom{n-1}{r-1} + \binom{n-1}{r}$$

For a combinatorial argument, consider a set of n items and fix attention on one of these items. How many different sets of size r contain this item? and how many do not?

17. A group of 5 boys and 10 girls is lined up in random order—that is, each of the 15! permutations is assumed to be equally likely.

(a) What is the probability that the person in the 4th position is a boy?
(b) What about the person in the 12th position?
(c) What is the probability that a particular boy is in the 3rd position?

18. There are 5 hotels in a certain town. If 3 people check into hotels in a day, what is the probability they each check into a different hotel? What assumptions are you making?

19. A town contains 4 television repairmen. If 4 sets break down, what is the probability that exactly 2 of the repairmen are called. What assumptions are you making?

20. A woman has n keys, of which one will open her door. If she tries the keys at random, discarding those that do not work, what is the probability that she will open the door on her kth try? What if she does not discard previously tried keys?

21. A closet contains 8 pairs of shoes. If 4 shoes are randomly selected, what is the probability that there will be (a) no complete pair, and (b) exactly 1 complete pair?

22. The king comes from a family of 2 children. What is the probability that the other child is his sister?

23. A couple has 2 children. What is the probability that both are girls if the eldest is a girl?

24. Each of 2 balls is painted black or gold and then placed in an urn. Suppose that each ball is colored black, with probability $\frac{1}{2}$, and that these events are independent.

    (a) Suppose that you obtain information that the gold paint has been used (and thus at least one of the balls is painted gold). Compute the conditional probability that both balls are painted gold.

    (b) Suppose, now, that the urn tips over and 1 ball falls out. It is painted gold. What is the probability that both balls are gold in this case? Explain.

25. Each of 2 cabinets identical in appearance has 2 drawers. Cabinet A contains a silver coin in each drawer, and cabinet B contains a silver coin in one of its drawers and a gold coin in the other. A cabinet is randomly selected, one of its drawers is opened, and a silver coin is found. What is the probability that there is a silver coin in the other drawer?

26. Suppose that there was a cancer diagnostic test that was 95 percent accurate both on those that do and those that do not have the disease. If .4 percent of the population have cancer, compute the probability that a tested person has cancer, given that his or her test result indicates so.

27. Suppose that an insurance company classifies people into one of three classes—good risks, average risks, and bad risks. Their records indicate that the probabilities that good, average, and bad risk persons will be involved in an accident over a 1-year span are, respectively, .05, .15, and .30. If 20 percent of the population are "good risks," 50 percent are "average risks," and 30 percent are "bad risks," what proportion of people have accidents in a fixed year? If policy holder A had no accidents in 1987, what is the probability that he or she is a good (average) risk?

28. A pair of fair dice are rolled. Let E denote the event that the sum of the dice is equal to 7.

(a) Show that E is independent of the event that the first die lands on 4.

(b) Show that E is independent of the event that the second die lands on 3.

29. The probability of the closing of the ith relay in the circuits shown is given by p_i, $i = 1, 2, 3, 4, 5$. If all relays function independently, what is the probability that a current flows between A and B for the respective circuits?

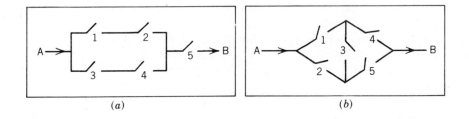

(a)   (b)

30. An engineering system consisting of n components is said to be a k-out-of-n system ($k \leq n$) if the system functions if and only if at least k of the n components function. Suppose that all components function independently of each other.

(a) If the ith component functions with probability P_i, $i = 1, 2, 3, 4$, compute the probability that a 2-out-of-4 system functions.

(b) Repeat (a) for a 3-out-of-5 system.

31. A certain organism possesses a pair of each of 5 different genes (which we will designate by the first 5 letters of the English alphabet). Each gene appears in 2 forms (which we designate by lowercase and capital letters). The capital letter will be assumed to be the dominant gene in the sense that if an organism possesses the gene pair xX, then it will outwardly have the appearance of the X gene. For instance, if X stands for brown eyes and x for blue eyes, then an individual having either gene pair XX or xX will have brown eyes, whereas one having gene pair xx will be blue eyed. The characteristic appearance of an organism is called its phenotype, whereas its genetic constitution is called its genotype. (Thus 2 organisms with respective genotypes aA, bB, cc, dD, ee and AA, BB, cc, DD, ee would have different genotypes but the same phenotype.) In a mating between 2 organisms each one contributes, at random, one of its gene pairs of each type. The 5 contributions of an organism (one of each of the 5 types) are assumed to be independent and are also independent of the contributions of its mate. In a mating between organisms having genotypes aA, bB, cC, dD, eE, and aa, bB, cc, Dd, ee, what is the probability that the progeny will (1) phenotypically, (2) genotypically resemble

(a) The first parent
(b) The second parent
(c) Either parent
(d) Neither parent?

32. Three prisoners are informed by their jailer that one of them has been chosen at random to be executed, and the other two are to be freed. Prisoner A asks the jailer to tell him privately which of his fellow prisoners will be set free, claiming that there would be no harm in divulging this information because he already knows that at least one of the two will go free. The jailer refuses to answer this question, pointing out that if A knew which of his fellow prisoners were to be set free, then his own probability of being executed would rise from $\frac{1}{3}$ to $\frac{1}{2}$ because he would then be one of two prisoners. What do you think of the jailer's reasoning?

CHAPTER 2

Random Variables and Expectation

1 RANDOM VARIABLES

When a random experiment is performed, we are often not interested in all of the details of the experimental result but only in the value of some numerical quantity determined by the result. For instance, in tossing dice we are often interested in the sum of the two dice and are not really concerned about the values of the individual dice. That is, we may be interested in knowing that the sum is seven and not be concerned over whether the actual outcome was $(1, 6)$ or $(2, 5)$ or $(3, 4)$ or $(4, 3)$ or $(5, 2)$ or $(6, 1)$. Also a civil engineer may not be directly concerned with the daily risings and declines of the water level of a reservoir (which we can take as the experimental result) but may only care about the level at the end of a rainy season. These quantities of interest that are determined by the result of the experiment are known as *random variables*.

Since the value of a random variable is determined by the outcome of the experiment, we may assign probabilities to its possible values.

Example 2.1a Letting X denote the random variable that is defined as the sum of two fair dice, then

$$P\{X = 2\} = P\{(1,1)\} = \tfrac{1}{36} \tag{2.1.1}$$

$$P\{X = 3\} = P\{(1,2),(2,1)\} = \tfrac{2}{36}$$

$$P\{X = 4\} = P\{(1,3),(2,2),(3,1)\} = \tfrac{3}{36}$$

$$P\{X = 5\} = P\{(1,4),(2,3),(3,2),(4,1)\} = \tfrac{4}{36}$$

$$P\{X = 6\} = P\{(1,5),(2,4),(3,3),(4,2),(5,1)\} = \tfrac{5}{36}$$

$$P\{X = 7\} = P\{(1,6),(2,5),(3,4),(4,3),(5,2),(6,1)\} = \tfrac{6}{36}$$

$$P\{X = 8\} = P\{(2,6),(3,5),(4,4),(5,3),(6,2)\} = \tfrac{5}{36}$$

$$P\{X = 9\} = P\{(3,6),(4,5),(5,4),(6,3)\} = \tfrac{4}{36}$$

$$P\{X = 10\} = P\{(4,6),(5,5),(6,4)\} = \tfrac{3}{36}$$

$$P\{X = 11\} = P\{(5,6),(6,5)\} = \tfrac{2}{36}$$

$$P\{X = 12\} = P\{(6,6)\} = \tfrac{1}{36}$$

In other words, the random variable X can take on any integral value between 2 and 12 and the probability that it takes on each value is given by Equation 2.1.1. Since X must take on some value we must have

$$1 = P(S) = P\left(\bigcup_{i=2}^{12} \{X = i\}\right) = \sum_{i=2}^{12} P\{X = i\}$$

which is easily verified from Equation 2.1.1.

Another random variable of possible interest in this experiment is the value of the first die. Letting Y denote this random variable, then Y is equally likely to take on any of the values 1 through 6. That is,

$$P\{Y = i\} = 1/6, \qquad i = 1,2,3,4,5,6 \quad \blacksquare$$

Example 2.1b Suppose that an individual purchases 2 electronic components each of which may be either defective or acceptable. In addition, suppose that the 4 possible results—$(d, d), (d, a), (a, d), (a, a)$—have respective probabilities .09, .21, .21, .49 [where (d, d) means that both components are defective; (d, a) that the first component is defective and the second acceptable, and so on]. If we let X denote the number of acceptable components obtained in the purchase, then X is a random variable taking on one of the values $0, 1, 2$ with respective probabilities

$$P\{X = 0\} = .09$$

$$P\{X = 1\} = .42$$

$$P\{X = 2\} = .49$$

If we were mainly concerned with whether there was at least 1 acceptable component, we could define the random variable I by

$$I = \begin{cases} 1 & \text{if } X = 1 \text{ or } 2 \\ 0 & \text{if } X = 0 \end{cases}$$

If A denotes the event that at least 1 acceptable component is obtained, then the random variable I is called the indicator random variable for the event A, since I will equal 1 or 0 depending upon whether A occurs. The probabilities attached to the possible values of I are

$$P\{I = 1\} = .91$$

$$P\{I = 0\} = .09 \quad \blacksquare$$

In the two foregoing examples, the random variables of interest took on either a finite or a countable number of possible values. Such random variables are called *discrete*. However, there also exist random variables that take on a continuum of possible values. These are known as *continuous* random variables. One example is the random variable denoting the lifetime of a car, when the car's lifetime is assumed to take on any value in some interval (a, b).

The *cumulative distribution function*, or more simply the *distribution function*, F of the random variable X is defined for any real number x by

$$F(x) = P\{X \leq x\}$$

That is, $F(x)$ is the probability that the random variable X takes on a value that is less than or equal to x.

Notation: We will use the notation $X \sim F$ to signify that F is the distribution function of X.

All probability questions about X can be answered in terms of its distribution function F. For example, suppose we wanted to compute $P\{a < X \leq b\}$. This can be accomplished by first noting that the event $\{X \leq b\}$ can be expressed as the union of the 2 mutually exclusive events $\{X \leq a\}$ and $\{a < X \leq b\}$. Therefore, applying Axiom 3, we obtain that

$$P\{X \leq b\} = P\{X \leq a\} + P\{a < X \leq b\}$$

or

$$P\{a < X \leq b\} = F(b) - F(a)$$

Example 2.1c Suppose the random variable X has distribution function

$$F(x) = \begin{cases} 0 & x \leq 0 \\ 1 - \exp\{-x^2\} & x > 0 \end{cases}$$

What is the probability that X exceeds 1?

Solution The desired probability is computed as follows:

$$\begin{aligned} P\{X > 1\} &= 1 - P\{X \leq 1\} \\ &= 1 - F(1) \\ &= e^{-1} \\ &= .368 \quad \blacksquare \end{aligned}$$

2 TYPES OF RANDOM VARIABLES

As was previously mentioned, a random variable that can take on at most a countable number of possible values is said to be *discrete*. For a discrete random variable X, we define the *probability mass function $p(a)$* of X by

$$p(a) = P\{X = a\}$$

The probability mass function $p(a)$ is positive for at most a countable number of values of a. That is, if X must assume one of the values $x_1, x_2, \ldots$, then

$$p(x_i) > 0, \qquad i = 1, 2, \ldots$$
$$p(x) = 0, \qquad \text{all other values of } x$$

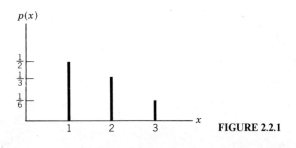

FIGURE 2.2.1

Since X must take on one of the values x_i, we have

$$\sum_{i=1}^{\infty} p(x_i) = 1$$

Example 2.2a Consider a random variable X that is either equal to 1, 2, or 3. If we know that

$$p(1) = \tfrac{1}{2} \quad \text{and} \quad p(2) = \tfrac{1}{3}$$

then it follows (since $p(1) + p(2) + p(3) = 1$) that

$$p(3) = \tfrac{1}{6}.$$

A graph of $p(x)$ is presented in Figure 2.2.1. ∎

The cumulative distribution function F can be expressed in terms of $p(x)$ by

$$F(a) = \sum_{\text{all } x \leq a} p(x)$$

If X is a discrete random variable whose set of possible values are $x_1, x_2, x_3, \ldots$, where $x_1 < x_2 < x_3 < \ldots$, then its distribution function F is a step function. That is, the value of F is constant in the intervals $[x_{i-1}, x_i)$ and then takes a step (or jump) of size $p(x_i)$ at x_i.

For instance, suppose X has a probability mass function given (as in Example 2.2a) by

$$p(1) = \tfrac{1}{2}, \qquad p(2) = \tfrac{1}{3}, \qquad p(3) = \tfrac{1}{6}$$

then, the cumulative distribution function F of X is given by

$$F(a) = \begin{cases} 0, & a < 1 \\ \tfrac{1}{2}, & 1 \leq a < 2 \\ \tfrac{5}{6}, & 2 \leq a < 3 \\ 1, & 3 \leq a \end{cases}$$

This is graphically presented in Figure 2.2.2.

Whereas the set of possible values of a discrete random variable is countable, we often must consider random variables whose set of possible values is uncountable. Let X be such a random variable. We say that X is a *continuous*

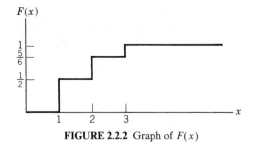

FIGURE 2.2.2 Graph of $F(x)$

random variable is there exists a nonnegative function $f(x)$, defined for all real $x \in (-\infty, \infty)$, having the property that for any set B of real numbers

$$P\{X \in B\} = \int_B f(x)\, dx \qquad (2.1)$$

The function $f(x)$ is called the *probability density function* of the random variable X.

In words, Equation 2.2 states that the probability that X will be in B may be obtained by integrating the probability density function over the set B. Since X must assume some value, $f(x)$ must satisfy

$$1 = P\{X \in (-\infty, \infty)\} = \int_{-\infty}^{\infty} f(x)\, dx$$

All probability statements about X can be answered in terms of $f(x)$. For instance, letting $B = [a, b]$, we obtain from Equation 2.2.1 that

$$P\{a \le X \le b\} = \int_a^b f(x)\, dx \qquad (2.2.2)$$

If we let $a = b$ in the above, then

$$P\{X = a\} = \int_a^a f(x)\, dx = 0$$

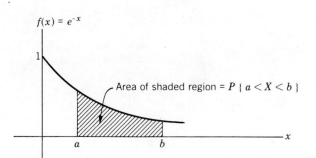

FIGURE 2.2.3 The probability density function

$$f(x) = \begin{cases} e^{-x} & x \ge 0 \\ 0 & x < 0 \end{cases}$$

In words, this equation states that the probability that a continuous random variable will assume any *particular* value is zero. (See Figure 2.2.3.)

The relationship between the cumulative distribution $F(\cdot)$ and the probability density $f(\cdot)$ is expressed by

$$F(a) = P\{ X \in (-\infty, a] \} = \int_{-\infty}^{a} f(x)\, dx$$

Differentiating both sides yields

$$\frac{d}{da} F(a) = f(a)$$

That is, the density is the derivative of the cumulative distribution function. A somewhat more intuitive interpretation of the density function may be obtained from Equation 2.2.2 as follows:

$$P\left\{ a - \frac{\epsilon}{2} \leq X \leq a + \frac{\epsilon}{2} \right\} = \int_{a-\epsilon/2}^{a+\epsilon/2} f(x)\, dx \approx \epsilon f(a)$$

when ϵ is small. In other words, the probability that X will be contained in an interval of length ϵ around the point a is approximately $\epsilon f(a)$. From this, we see that $f(a)$ is a measure of how likely it is that the random variable will be near a.

Example 2.2b Suppose that X is a continuous random variable whose probability density function is given by

$$f(x) = \begin{cases} C(4x - 2x^2) & 0 < x < 2 \\ 0 & \text{otherwise} \end{cases}$$

1. What is the value of C?

2. Find $P\{ X > 1 \}$.

Solution Since f is a probability density function, we must have that $\int_{-\infty}^{\infty} f(x)\, dx = 1$, implying that

$$C \int_{0}^{2} (4x - 2x^2)\, dx = 1$$

or

$$C \left[2x^2 - \frac{2x^3}{3} \right] \Bigg|_{x=0}^{x=2} = 1$$

or

$$C = \tfrac{3}{8}$$

Hence

$$P\{ X > 1 \} = \int_{1}^{\infty} f(x)\, dx = \tfrac{3}{8} \int_{1}^{2} (4x - 2x^2)\, dx = \tfrac{1}{2} \quad \blacksquare$$

3   JOINTLY DISTRIBUTED RANDOM VARIABLES

For a given experiment we are often interested not only in probability distribution functions of individual random variables but also in the relationships between two or more random variables. For instance, in an experiment into the possible causes of cancer, we might be interested in the relationship between the average number of cigarettes smoked daily and the age at which an individual contracts cancer. Similarly, an engineer might be interested in the relationship between the shear strength and the diameter of a spot weld in a fabricated sheet steel specimen.

In order to specify the relationship between two random variables, we define the joint cumulative probability distribution function of X and Y by

$$F(x, y) = P\{X \le x, Y \le y\}$$

A knowledge of the joint probability distribution function enables one, at least in theory, to compute the probability of any statement concerning the values of X and Y. For instance, the distribution function of X—call it F_X—can be obtained from the joint distribution function F of X and Y as follows:

$$F_X(x) = P\{X \le x\}$$
$$= P\{X \le x, Y \le \infty\}$$
$$= F(x, \infty)$$

Similarly, the cumulative distribution function of Y is given by

$$F_Y(y) = F(\infty, y)$$

In the case where X and Y are both discrete random variables whose possible values are, respectively, $x_1, x_2, \ldots,$ and $y_1, y_2, \ldots,$ we define the *joint probability mass function* of X and Y, $p(x_i, y_j)$, by

$$p(x_i, y_j) = P\{X = x_i, Y = y_j\}$$

The individual probability mass functions of X and Y are easily obtained from the joint probability mass function by the following reasoning. Since Y must take on some value y_j, it follows that the event $\{X = x_i\}$ can be written as the union, over all j, of the mutually exclusive events $\{X = x_i, Y = y_j\}$. That is,

$$\{X = x_i\} = \bigcup_j \{X = x_i, Y = y_j\}$$

and so, using Axiom 3 of the probability function, we see that

$$P\{X = x_i\} = P\left(\bigcup_j \{X = x_i, Y = y_j\}\right) \tag{2.3.1}$$
$$= \sum_j P\{X = x_i, Y = y_j\}$$
$$= \sum_j p(x_i, y_j)$$

Similarly we can obtain $P\{Y = y_j\}$ by summing $p(x_i, y_j)$ over all possible values of x_i, that is,

$$P\{Y = y_j\} = \sum_i P\{X = x_i, Y = y_j\} \tag{2.3.2}$$

$$= \sum_i p(x_i, y_j)$$

Hence, specifying the joint probability mass function always determines the individual mass functions. However, it should be noted that the reverse is not true. Namely, knowledge of $P\{X = x_i\}$ and $P\{Y = y_j\}$ does not determine the value of $P\{X = x_i, Y = y_j\}$.

Example 2.3a Suppose that 3 batteries are randomly chosen from a group of 3 new, 4 used but still working, and 5 defective batteries. If we let X and Y denote, respectively, the number of new and used but still working batteries that are chosen, then the joint probability mass function of X and Y, $p(i, j) = P\{X = i, Y = j\}$, is given by

$$p(0,0) = \binom{5}{3} \bigg/ \binom{12}{3} = 10/220$$

$$p(0,1) = \binom{4}{1}\binom{5}{2} \bigg/ \binom{12}{3} = 40/220$$

$$p(0,2) = \binom{4}{2}\binom{5}{1} \bigg/ \binom{12}{3} = 30/220$$

$$p(0,3) = \binom{4}{3} \bigg/ \binom{12}{3} = 4/220$$

$$p(1,0) = \binom{3}{1}\binom{5}{2} \bigg/ \binom{12}{3} = 30/220$$

$$p(1,1) = \binom{3}{1}\binom{4}{1}\binom{5}{1} \bigg/ \binom{12}{3} = 60/220$$

$$p(1,2) = \binom{3}{1}\binom{4}{2} \bigg/ \binom{12}{3} = 18/220$$

$$p(2,0) = \binom{3}{2}\binom{5}{1} \bigg/ \binom{12}{3} = 15/220$$

$$p(2,1) = \binom{3}{2}\binom{4}{1} \bigg/ \binom{12}{3} = 12/220$$

$$p(3,0) = \binom{3}{3} \bigg/ \binom{12}{3} = 1/220$$

These probabilities can most easily be expressed in tabular form as follows.

TABLE 2.3.1

$P\{X = i, Y = j\}$

i \ j	0	1	2	3	Row Sums = $P\{X = i\}$
0	$\frac{10}{220}$	$\frac{40}{220}$	$\frac{30}{220}$	$\frac{4}{220}$	$\frac{84}{220}$
1	$\frac{30}{220}$	$\frac{60}{220}$	$\frac{18}{220}$	0	$\frac{108}{220}$
2	$\frac{15}{220}$	$\frac{12}{220}$	0	0	$\frac{27}{220}$
3	$\frac{1}{220}$	0	0	0	$\frac{1}{220}$
Column Sums = $P\{Y = j\}$	$\frac{56}{220}$	$\frac{112}{220}$	$\frac{48}{220}$	$\frac{4}{220}$	

The reader should note that the probability mass function of X is obtained by computing the row sums, in accordance with the Equation 2.3.1, whereas the probability mass function of Y is obtained by computing the column sums, in accordance with Equation 2.3.2. As the individual probability mass functions of X and Y thus appear in the margin of such a table, they are often referred to as being the marginal probability mass functions of X and Y respectively. It should be noted that to check the correctness of such a table we could sum the marginal row (or the marginal column) and verify that its sum is 1. (Why must the sum of the entries in the marginal row (or column) equal 1?) ■

Example 2.3b Suppose that 15 percent of the families in a certain community have no children, 20 percent have 1, 35 percent have 2, and 30 percent have 3 children; and suppose, further, that in each family, each child is equally likely (and independently) to be a boy or a girl. If a family is chosen at random from this community then B, the number of boys, and G, the number of girls, in this family will have the following joint probability mass function.

TABLE 2.3.2

$P\{B = i, G = j\}$

i \ j	0	1	2	3	Row Sum = $P\{B = i\}$
0	.15	.10	.0875	.0375	.3750
1	.10	.175	.1125	0	.3875
2	.0875	.1125	0	0	.2000
3	.0375	0	0	0	.0375
Column Sum = $P\{G = j\}$	.3750	.3875	.2000	.0375	

These probabilities are obtained as follows:

$$P\{B = 0, G = 0\} = P\{\text{no children}\}$$
$$= .15$$
$$P\{B = 0, G = 1\} = P\{1 \text{ girl and total of 1 child}\}$$
$$= P\{1 \text{ child}\} P\{1 \text{ girl}|1 \text{ child}\}$$
$$= (.20)(\tfrac{1}{2}) = .1$$
$$P\{B = 0, G = 2\} = P\{2 \text{ girls and total of 2 children}\}$$
$$= P\{2 \text{ children}\} P\{2 \text{ girls}|2 \text{ children}\}$$
$$= (.35)(\tfrac{1}{2})^2 = .0875$$
$$P\{B = 0, G = 3\} = P\{3 \text{ girls and total of 3 children}\}$$
$$= P\{3 \text{ children}\} P\{3 \text{ girls}|3 \text{ children}\}$$
$$= (.30)(\tfrac{1}{2})^3 = .0375$$

We leave it to the reader to verify the remainder of Table 2.3.2, which tells us, among other things, that the family chosen will have at least 1 girl with probability .625. ∎

We say that X and Y are *jointly continuous* if there exists a function $f(x, y)$ defined for all real x and y, having the property that for every set C of pairs of real numbers (that is, C is a set in the two-dimensional plane)

$$P\{(X, Y) \in C\} = \iint\limits_{(x, y) \in C} f(x, y) \, dx \, dy \qquad (2.3.3)$$

The function $f(x, y)$ is called the *joint probability density function* of X and Y. If A and B are any sets of real numbers, then by defining $C = \{(x, y): x \in A, y \in B\}$, we see from Equation 2.3.3 that

$$P\{X \in A, Y \in B\} = \int_B \int_A f(x, y) \, dx \, dy \qquad (2.3.4)$$

Because

$$F(a, b) = P\{X \in (-\infty, a], Y \in (-\infty, b]\}$$
$$= \int_{-\infty}^b \int_{-\infty}^a f(x, y) \, dx \, dy$$

it follows, upon differentiation, that

$$f(a, b) = \frac{\partial^2}{\partial a \, \partial b} F(a, b)$$

wherever the partial derivatives are defined. Another interpretation of the joint density function is obtained from Equation 2.3.4 as follows:

$$P\{a < X < a + da, b < Y < b + db\} = \int_b^{d+db} \int_a^{a+da} f(x, y) \, dx \, dy$$
$$\approx f(a, b) \, da \, db$$

when da and db are small and $f(x, y)$ is continuous at a, b. Hence $f(a, b)$ is a measure of how likely it is that the random vector (X, Y) will be near (a, b).

If X and Y are jointly continuous, they are individually continuous, and their probability density functions can be obtained as follows:

$$P\{X \in A\} = P\{X \in A, Y \in (-\infty, \infty)\} \qquad (2.3.5)$$

$$= \int_A \int_{-\infty}^{\infty} f(x, y) \, dy \, dx$$

$$= \int_A f_X(x) \, dx$$

where

$$f_X(x) = \int_{-\infty}^{\infty} f(x, y) \, dy$$

is thus the probability density function of X. Similarly, the probability density function of Y is given by

$$f_Y(y) = \int_{-\infty}^{\infty} f(x, y) \, dx \qquad (2.3.6)$$

Example 2.3c The joint density function of X and Y is given by

$$f(x, y) = \begin{cases} 2e^{-x}e^{-2y} & 0 < x < \infty, 0 < y < \infty \\ 0 & \text{otherwise} \end{cases}$$

Compute (1) $P\{X > 1, Y < 1\}$; (2) $P\{X < Y\}$; and (3) $P\{X < a\}$.

Solution

$$P\{X > 1, Y < 1\} = \int_0^1 \int_1^{\infty} 2e^{-x}e^{-2y} \, dx \, dy$$

$$= \int_0^1 2e^{-2y}\left(-e^{-x}\big|_1^{\infty}\right) dy$$

$$= e^{-1} \int_0^1 2e^{-2y} \, dy$$

$$= e^{-1}(1 - e^{-2})$$

$$P\{X < Y\} = \iint\limits_{(x, y):\, x<y} 2e^{-x}e^{-2y} \, dx \, dy$$

$$= \int_0^{\infty} \int_0^{y} 2e^{-x}e^{-2y} \, dx \, dy$$

$$= \int_0^{\infty} 2e^{-2y}(1 - e^{-y}) \, dy$$

$$= \int_0^{\infty} 2e^{-2y} \, dy - \int_0^{\infty} 2e^{-3y} \, dy$$

$$= 1 - \tfrac{2}{3}$$

$$= \tfrac{1}{3}$$

$$P\{X < a\} = \int_0^a \int_0^{\infty} 2e^{-2y}e^{-x} \, dy \, dx$$

$$= \int_0^a e^{-x} \, dx$$

$$= 1 - e^{-a} \quad \blacksquare$$

The random variables X and Y are said to be independent if for any two sets of real numbers A and B

$$P\{X \in A, Y \in B\} = P\{X \in A\}P\{Y \in B\} \qquad (2.3.7)$$

In other words, X and Y are independent if, for all A and B, the events $E_A = \{X \in A\}$ and $F_B = \{Y \in B\}$ are independent.

It can be shown by using the three axioms of probability that Equation 2.3.7 will follow if and only if for all a, b

$$P\{X \le a, Y \le b\} = P\{X \le a\}P\{Y \le b\}$$

Hence, in terms of the joint distribution function F of X and Y, we have that X and Y are independent if

$$F(a, b) = F_X(a)F_Y(b) \qquad \text{for all } a, b$$

When X and Y are discrete random variables, the condition of independence Equation 2.3.7 is equivalent to

$$p(x, y) = p_X(x)p_Y(y) \qquad \text{for all } x, y \qquad (2.3.8)$$

where p_X and p_Y are the probability mass functions of X and Y. The equivalence follows because, if Equation 2.3.7 is satisfied, then we obtain Equation 2.3.8 by letting A and B be, respectively, the one point sets $A = \{x\}$, $B = \{y\}$. Furthermore, if Equation 2.3.8 is valid, then for any sets A, B

$$\begin{aligned}
P\{X \in A, Y \in B\} &= \sum_{y \in B} \sum_{x \in A} p(x, y) \\
&= \sum_{y \in B} \sum_{x \in A} p_X(x)p_Y(y) \\
&= \sum_{y \in B} p_Y(y) \sum_{x \in A} p_X(x) \\
&= P\{Y \in B\}P\{X \in A\}
\end{aligned}$$

and thus Equation 2.3.7 is established.

In the jointly continuous case the condition of independence is equivalent to

$$f(x, y) = f_X(x)f_Y(y) \qquad \text{for all } x, y$$

Loosely speaking, X and Y are independent if knowing the value of one does not change the distribution of the other. Random variables that are not independent are said to be dependent.

Example 2.3d Suppose that X and Y are independent random variables having the common density function

$$f(x) = \begin{cases} e^{-x} & x > 0 \\ 0 & \text{otherwise} \end{cases}$$

Find the density function of the random variable X/Y.

Solution We start by determining the distribution function of X/Y. For $a > 0$

$$F_{X/Y}(a) = P\{X/Y \le a\}$$

$$= \iint\limits_{x/y \le a} f(x, y) \, dx \, dy$$

$$= \iint\limits_{x/y \le a} e^{-x} e^{-y} \, dx \, dy$$

$$= \int_0^\infty \int_0^{ay} e^{-x} e^{-y} \, dx \, dy$$

$$= \int_0^\infty (1 - e^{-ay}) e^{-y} \, dy$$

$$= \left[-e^{-y} + \frac{e^{-(a+1)y}}{a+1} \right]\Big|_0^\infty$$

$$= 1 - \frac{1}{a+1}$$

Differentiation yields that the density function of X/Y is given by

$$f_{X/Y}(a) = 1/(a+1)^2, \quad 0 < a < \infty. \quad \blacksquare$$

We can also define joint probability distributions for n random variables in exactly the same manner as we did for $n = 2$. For instance, the joint cumulative probability distribution function $F(a_1, a_2, \ldots, a_n)$ of the n random variables $X_1, X_2, \ldots, X_n$ is defined by

$$F(a_1, a_2, \ldots, a_n) = P\{X_1 \le a_1, X_2 \le a_2, \ldots, X_n \le a_n\}$$

If these random variables are discrete, we define their joint probability mass function $p(x_1, x_2, \ldots, x_n)$ by

$$p(x_1, x_2, \ldots, x_n) = P\{X_1 = x_1, X_2 = x_2, \ldots, X_n = x_n\}$$

Further, the n random variables are said to be jointly continuous if there exists a function $f(x_1, x_2, \ldots, x_n)$, called the joint probability density function, such that for any set C in n-space

$$P\{(X_1, X_2, \ldots, X_n) \in C\} = \iint\limits_{(x_1, \ldots, x_n) \in C} \cdots \int f(x_1, \ldots, x_n) \, dx_1 \, dx_2 \, \cdots \, dx_n$$

In particular, for any n sets of real numbers $A_1, A_2, \ldots, A_n$

$$P\{X_1 \in A_1, X_2 \in A_2, \ldots, X_n \in A_n\}$$

$$= \int_{A_n} \int_{A_{n-1}} \cdots \int_{A_1} f(x_1, \ldots, x_n) \, dx_1 \, dx_2 \, \cdots \, dx_n$$

The concept of independence may, of course, also be defined for more than two random variables. In general, the n random variables $X_1, X_2, \ldots, X_n$ are said to be independent if, for all sets of real numbers $A_1, A_2, \ldots, A_n$,

$$P\{ X_1 \in A_1, X_2 \in A_2, \ldots, X_n \in A_n\} = \prod_{i=1}^{n} P\{ X_i \in A_i\}$$

As before, it can be shown that this condition is equivalent to

$$P\{ X_1 \le a_1, X_2 \le a_2, \ldots, X_n \le a_n\}$$
$$= \prod_{i=1}^{n} P\{ X_1 \le a_i\} \qquad \text{for all } a_1, a_2, \ldots, a_n$$

Finally, we say that an infinite collection of random variables is independent if every finite subcollection of them is independent.

Example 2.3e Suppose that the successive daily changes of the price of a given stock are assumed to be independent and identically distributed random variables with probability mass function given by

$$P\{\text{daily change is } i \} = \begin{cases} -3 & \text{with probability .05} \\ -2 & \text{with probability .10} \\ -1 & \text{with probability .20} \\ 0 & \text{with probability .30} \\ 1 & \text{with probability .20} \\ 2 & \text{with probability .10} \\ 3 & \text{with probability .05} \end{cases}$$

Then the probability that the stock's price will increase successively by 1, 2, and 0 points in the next three days is

$$P\{ X_1 = 1, X_2 = 2, X_3 = 0\} = (.20)(.10)(.30) = .006$$

where we have let X_i denote the change on the ith day. ∎

3.1 Conditional Distributions

The relationship between two random variables can often be clarified by consideration of the conditional distribution of one given the value of the other.

Recall that for any two events E and F, the conditional probability of E given F is defined, provided that $P(F) > 0$, by

$$P(E|F) = \frac{P(EF)}{P(F)}$$

Hence, if X and Y are discrete random variables, it is natural to define the conditional probability mass function of X given that $Y = y$, by

$$p_{X|Y}(x|y) = P\{ X = x | Y = y\}$$
$$= \frac{P\{ X = x, Y = y\}}{P\{Y = y\}}$$
$$= \frac{p(x, y)}{p_Y(y)}$$

for all values of y such that $p_Y(y) > 0$.

Example 2.3f If we know, in Example 2.3b, that the family chosen has 1 girl, compute the conditional probability mass function of the number of boys in the family.

Solution We first note, from Table 2.3.2 that

$$P\{G = 1\} = .3875$$

Hence,

$$P\{B = 0|G = 1\} = \frac{P\{B = 0, G = 1\}}{P\{G = 1\}} = \frac{.10}{.3875} = 8/31$$

$$P\{B = 1|G = 1\} = \frac{P\{B = 1, G = 1\}}{P\{G = 1\}} = \frac{.175}{.3875} = 14/31$$

$$P\{B = 2|G = 1\} = \frac{P\{B = 2, G = 1\}}{P\{G = 1\}} = \frac{.1125}{.3875} = 9/31$$

$$P\{B = 3|G = 1\} = \frac{P\{B = 3, G = 1\}}{P\{G = 1\}} = 0$$

Thus, for instance, given 1 girl, there are 23 chances out of 31 that there will also be at least 1 boy. ∎

Example 2.3g Suppose that $p(x, y)$, the joint probability mass function of X and Y is given by

$$p(0,0) = .4, \qquad p(0,1) = .2, \qquad p(1,0) = .1, \qquad p(1,1) = .3.$$

Calculate the conditional probability mass function of X given that $Y = 1$.

Solution We first note that

$$P\{Y = 1\} = \sum_x p(x, 1) = p(0, 1) + p(1, 1) = .5$$

Hence,

$$P\{X = 0|Y = 1\} = \frac{p(0, 1)}{P\{Y = 1\}} = 2/5$$

$$P\{X = 1|Y = 1\} = \frac{p(1, 1)}{P\{Y = 1\}} = 3/5 \quad ∎$$

If X and Y have a joint probability density function $f(x, y)$, then the conditional probability density function of X, given that $Y = y$, is defined for all values of y such that $f_Y(y) > 0$, by

$$f_{X|Y}(x|y) = \frac{f(x, y)}{f_Y(y)}$$

To motivate this definition, multiply the left-hand side by dx and the right-hand side by $(dx\, dy)/dy$ to obtain

$$f_{X|Y}(x|y)\, dx = \frac{f(x, y)\, dx\, dy}{f_Y(y)\, dy}$$

$$\approx \frac{P\{x \le X \le x + dx, \, y \le Y \le y + dy\}}{P\{y \le Y \le y + dy\}}$$

$$= P\{x \le X \le x + dy | y \le Y \le y + dy\}$$

In other words, for small values of dx and dy, $f_{X|Y}(x|y)\,dx$ represents the conditional probability that X is between x and $x + dx$, given that Y is between y and $y + dy$.

The use of conditional densities allows us to define conditional probabilities of events associated with one random variable when we are given the value of a second random variable. That is, if X and Y are jointly continuous, then, for any set A,

$$P\{X \in A \mid Y = y\} = \int_A f_{X|Y}(x|y)\,dx$$

Example 2.3h The joint density of X and Y is given by

$$f(x, y) = \begin{cases} \frac{12}{5}x(2 - x - y) & 0 < x < 1, 0 < y < 1 \\ 0 & \text{otherwise} \end{cases}$$

Compute the conditional density of X, given that $Y = y$, where $0 < y < 1$.

Solution For $0 < x < 1, 0 < y < 1$, we have

$$f_{X|Y}(x|y) = \frac{f(x, y)}{f_Y(y)}$$

$$= \frac{f(x, y)}{\int_{-\infty}^{\infty} f(x, y)\,dx}$$

$$= \frac{x(2 - x - y)}{\int_0^1 x(2 - x - y)\,dx}$$

$$= \frac{x(2 - x \cdots y)}{\frac{2}{3} - y/2}$$

$$= \frac{6x(2 - x - y)}{4 - 3y} \quad \blacksquare$$

4 EXPECTATION

One of the most important concepts in probability theory is that of the expectation of a random variable. If X is a discrete random variable taking on the possible values $x_1, x_2, \ldots$, then the *expectation* or *expected value* of X, denoted by $E[X]$, is defined by

$$E[X] = \sum_i x_i P\{X = x_i\}$$

In words, the expected value of X is a weighted average of the possible values that X can take on, each value being weighted by the probability that X assumes it. For instance, if the probability mass function of X is given by

$$p(0) = \tfrac{1}{2} = p(1)$$

then

$$E[X] = 0\left(\tfrac{1}{2}\right) + 1\left(\tfrac{1}{2}\right) = \tfrac{1}{2}$$

is just the ordinary average of the two possible values 0 and 1 that X can assume. On the other hand, if

$$p(0) = \tfrac{1}{3}, \qquad p(1) = \tfrac{2}{3}$$

then

$$E[X] = 0\left(\tfrac{1}{3}\right) + 1\left(\tfrac{2}{3}\right) = \tfrac{2}{3}$$

is a weighted average of the two possible values 0 and 1 where the value 1 is given twice as much weight as the value 0 since $p(1) = 2p(0)$.

Another motivation of the definition of expectation is provided by the frequency interpretation of probabilities. This interpretation assumes that if an infinite sequence of independent replications of an experiment are performed, then for any event E, the proportion of time that E occurs will be $P(E)$. Now, consider a random variable X that must take on one of the values $x_1, x_2, \ldots, x_n$ with respective probabilities $p(x_1), p(x_2), \ldots, p(x_n)$; and think of X as representing our winnings in a single game of chance. That is, with probability $p(x_i)$ we shall win x_i units $i = 1, 2, \ldots, n$. Now by the frequency interpretation, it follows that if we continually play this game, then the proportion of time that we win x_i will be $p(x_i)$. Since this is true for all i, $i = 1, 2, \ldots, n$, it follows that our average winnings per game will be

$$\sum_{i=1}^{n} x_i p(x_i) = E[X]$$

(To see this argument more clearly, suppose that we play N games where N is very large. Then in approximately $Np(x_i)$ of these games, we shall win x_i, and thus our total winnings in the N games will be

$$\sum_{i=1}^{n} x_i N p(x_i)$$

implying that our average winnings per game are

$$\sum_{i=1}^{n} \frac{x_i N p(x_i)}{N} = \sum_{i=1}^{n} x_i p(x_i) = E[X].)$$

Example 2.4a Find $E[X]$ where X is the outcome when we roll a fair die.

Solution Since $p(1) = p(2) = p(3) = p(4) = p(5) = p(6) = \tfrac{1}{6}$, we obtain that

$$E[X] = 1\left(\tfrac{1}{6}\right) + 2\left(\tfrac{1}{6}\right) + 3\left(\tfrac{1}{6}\right) + 4\left(\tfrac{1}{6}\right) + 5\left(\tfrac{1}{6}\right) + 6\left(\tfrac{1}{6}\right) = \tfrac{7}{2}$$

The reader should note that, for this example, the expected value of X is not a value that X could possibly assume. (That is, rolling a die cannot possibly lead to an outcome of $7/2$.) Thus, even though we call $E[X]$ the *expectation* of X, it should not be interpreted as the value that we *expect* X to have but rather as the average value of X in a large number of repetitions of the experiment. That is, if we continually roll a fair die, then after a large number of rolls the average of all the outcomes will be approximately $7/2$. (The interested reader should try this as an experiment.) ∎

Example 2.4b If I is an indicator random variable for the event A, that is, if

$$I = \begin{cases} 1 & \text{if } A \text{ occurs} \\ 0 & \text{if } A \text{ does not occur} \end{cases}$$

then,

$$E[I] = 1P(A) + 0P(A^c) = P(A)$$

Hence, the expectation of the indicator random variable for the event A is just the probability that A occurs. ∎

Example 2.4c Entropy For a given random variable X, how much information is conveyed in the message that $X = x$? Let us begin our attempts at quantifying this statement by agreeing that the amount of information in the message that $X = x$ should depend on how likely it was that X would equal x. In addition, it seems reasonable that the more unlikely it was that X would equal x, the more informative would be the message. For instance, if X represents the sum of two fair dice, then there seems to be more information in the message that X equals 12 than there would be in the message that X equals 7, since the former event has probability $1/36$ and the latter $1/6$.

Let us denote by $I(p)$ the amount of information contained in the message that an event, whose probability is p, has occurred. Clearly $I(p)$ should be a nonnegative, decreasing function of p. To determine its form, let X and Y be independent random variables, and suppose that $P\{X = x\} = p$ and $P\{Y = y\} = q$. How much information is contained in the message that X equals x and Y equals y? To answer this, note first that the amount of information in the statement that X equals x is $I(p)$. Also, since knowledge of the fact that X is equal to x does not affect the probability that Y will equal y (since X and Y are independent), it seems reasonable that the additional amount of information contained in the statement that $Y = y$ should equal $I(q)$. Thus, it seems that the amount of information in the message that X equals x and Y equals y is $I(p) + I(q)$. On the other hand, however, we have that

$$P\{X = x, Y = y\} = P\{X = x\}P\{Y = y\} = pq$$

which implies that the amount of information in the message that X equals x and Y equals y is $I(pq)$. Therefore, it seems that the function I should satisfy the identity

$$I(pq) = I(p) + I(q)$$

However, if we define the function G by

$$G(p) = I(2^{-p})$$

then we see from the above that

$$\begin{aligned} G(p + q) &= I(2^{-(p+q)}) \\ &= I(2^{-p}2^{-q}) \\ &= I(2^{-p}) + I(2^{-q}) \\ &= G(p) + G(q) \end{aligned}$$

However, it can be shown that the only (monotone) functions G that satisfy the foregoing functional relationship are those of the form

$$G(p) = cp$$

for some constant c. Therefore, we must have that

$$I(2^{-p}) = cp$$

or, letting $q = 2^{-p}$

$$I(q) = -c \log_2(q)$$

for some positive constant c. It is traditional to let $c = 1$ and to say that the information is measured in units of *bits* (short for binary digits).

Consider now a random variable X, which must take on one of the values $x_1, \ldots, x_n$ with respective probabilities $p_1, \ldots, p_n$. As $-\log(p_i)$ represents the information conveyed by the message that X is equal to x_i, it follows that the expected amount of information that will be conveyed when the value of X is transmitted is given by

$$H(X) = -\sum_{i=1}^{n} p_i \log_2(p_i)$$

The quantity $H(X)$ is known in information theory as the *entropy* of the random variable X. ∎

We can also define the expectation of a continuous random variable. Suppose that X is a continuous random variable with probability density function f. Since, for dx small

$$f(x) \, dx \approx P\{x < X < x + dx\}$$

it follows that a weighted average of all possible values of X, with the weight given to x equal to the probability that X is near x, is just the integral over all x of $xf(x) \, dx$. Hence, it is natural to define the expected value of X by

$$E[X] = \int_{-\infty}^{\infty} xf(x) \, dx$$

Example 2.4d Suppose that you are expecting a message at some time past 5 P.M. From experience you know that X, the number of hours after 5 P.M. until the message arrives, is a random variable with the following probability density function:

$$f(x) = \frac{1}{1.5} \quad \text{if } 0 < x < 1.5$$
$$0 \quad \text{otherwise}$$

The expected amount of time past 5 P.M. until the message arrives is given by

$$E[X] = \int_{0}^{1.5} \frac{x}{1.5} \, dx = .75$$

Hence, on average, you would have to wait three-fourths of an hour. ∎

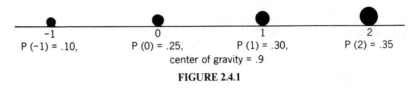

$$P(-1) = .10, \qquad P(0) = .25, \qquad P(1) = .30, \qquad P(2) = .35$$

center of gravity = .9

FIGURE 2.4.1

REMARKS

(a) The concept of expectation is analogous to the physical concept of the center of gravity of a distribution of mass. Consider a discrete random variable X having probability mass function $P(x_i)$, $i \geq 1$. If we now imagine a weightless rod in which weights with mass $P(x_i)$, $i \geq 1$ are located at the points x_i, $i \geq 1$ (see Figure 2.4.1), then the point at which the rod would be in balance is known as the center of gravity. For those readers acquainted with elementary statics, it is now a simple matter to show that this point is at $E[X]$.[†]

(b) $E[X]$ has the same units of measurement as does X.

5 PROPERTIES OF THE EXPECTED VALUE

Suppose now that we are given a random variable X and its probability distribution (that is, its probability mass function in the discrete case or its probability density function in the continuous case). Suppose also that we are interested in calculating, not the expected value of X, but the expected value of some function of X, say $g(X)$. How do we go about doing this? One way is as follows. Since $g(X)$ is itself a random variable, it must have a probability distribution, which should be computable from a knowledge of the distribution of X. Once we have obtained the distribution of $g(X)$, we can then compute $E[g(X)]$ by the definition of the expectation.

Example 2.5a Suppose X has the following probability mass function

$$p(0) = .2, \qquad p(1) = .5, \qquad p(2) = .3$$

Calculate $E[X^2]$.

Solution Letting $Y = X^2$, we have that Y is a random variable that can take on one of the values $0^2, 1^2, 2^2$ with respective probabilities

$$p_Y(0) = P\{Y = 0^2\} = .2$$

$$p_Y(1) = P\{Y = 1^2\} = .5$$

$$p_Y(4) = P\{Y = 2^2\} = .3$$

[†] To prove this, we must show that the sum of the torques tending to turn the point around $E[X]$ is equal to 0. That is, we must show that $0 = \Sigma_i(x_i - E[X])P(x_i)$, which is immediate.

Hence,

$$E[X^2] = E[Y] = 0(.2) + 1(.5) + 4(.3) = 1.7 \quad \blacksquare$$

Example 2.5b The time, in hours, it takes to locate and repair an electrical breakdown in a certain factory is a random variable—call it X—whose density function is given by

$$f_X(x) = \begin{cases} 1 & \text{if } 0 < x < 1 \\ 0 & \text{otherwise} \end{cases}$$

If the cost involved in a breakdown of duration x is x^3, what is the expected cost of such a breakdown?

Solution Letting $Y = X^3$ denote the cost, we first calculate its distribution function as follows. For $0 \le a \le 1$,

$$\begin{aligned} F_Y(a) &= P\{Y \le a\} \\ &= P\{X^3 \le a\} \\ &= P\{X \le a^{1/3}\} \\ &= \int_0^{a^{1/3}} dx \\ &= a^{1/3} \end{aligned}$$

By differentiating $F_Y(a)$, we obtain the density of Y,

$$f_Y(a) = \tfrac{1}{3}a^{-2/3}, \qquad 0 \le a < 1$$

Hence,

$$\begin{aligned} E[X^3] = E[Y] &= \int_{-\infty}^{\infty} af_Y(a)\, da \\ &= \int_0^1 a\tfrac{1}{3}a^{-2/3}\, da \\ &= \tfrac{1}{3}\int_0^1 a^{1/3}\, da \\ &= \tfrac{1}{3}\tfrac{3}{4}a^{4/3}\big|_0^1 \\ &= \tfrac{1}{4} \quad \blacksquare \end{aligned}$$

While the foregoing procedure will, in theory, always enable us to compute the expectation of any function of X from a knowledge of the distribution of X, there is an easier way of doing this. Suppose, for instance, that we wanted to compute the expected value of $g(X)$. Since $g(X)$ takes on the value $g(x)$ when $X = x$, it seems intuitive that $E[g(X)]$ should be a weighted average of the possible values $g(x)$ with, for a given x, the weight given to $g(x)$ being equal to the probability (or probability density in the continuous case) that X will equal x. Indeed the foregoing can be shown to be true and we thus have the following proposition.

Proposition 2.5.1 Expectation of a Function of a Random Variable

(a) If X is a discrete random variable with probability mass function $p(x)$, then for any real-valued function g,

$$E[g(X)] = \sum_x g(x)p(x)$$

(b) If X is a continuous random variable with probability density function $f(x)$, then for any real-valued function g,

$$E[g(X)] = \int_{-\infty}^{\infty} g(x)f(x)\, dx$$

Example 2.5c Applying the Proposition 2.5.1 to Example 2.5a yields

$$E[X^2] = 0^2(0.2) + (1^2)(0.5) + (2^2)(0.3) = 1.7$$

which, of course, checks with the result derived in Example 2.5a. ∎

Example 2.5d Applying the proposition to Example 2.5b yields

$$E[X^3] = \int_0^1 x^3\, dx \qquad (\text{since } f(x) = 1, \;\; 0 < x < 1)$$

$$= \tfrac{1}{4} \quad ∎$$

An immediate corollary of Proposition 2.5.1 is the following.

Corollary 2.5.2 If a and b are constants, then

$$E[aX + b] = aE[X] + b$$

Proof In the discrete case,

$$E[aX + b] = \sum_x (ax + b)p(x)$$

$$= a\sum_x xp(x) + b\sum_x p(x)$$

$$= aE[X] + b$$

In the continuous case,

$$E[aX + b] = \int_{-\infty}^{\infty} (ax + b)f(x)\, dx$$

$$= a\int_{-\infty}^{\infty} xf(x)\, dx + b\int_{-\infty}^{\infty} f(x)\, dx$$

$$= aE[X] + b \quad ∎$$

If we take $a = 0$ in Corollary 2.5.2, we see that

$$E[b] = b$$

That is, the expected value of a constant is just its value (is this intuitive?).

Also, if we take $b = 0$, then we obtain

$$E[aX] = aE[X]$$

or, in words, the expected value of a constant multiplied by a random variable is just the constant times the expected value of the random variable. The expected value of a random variable X, $E[X]$, is also referred to as the *mean* or the *first moment* of X. The quantity $E[X^n]$, $n \geq 1$, is called the nth moment of X. By Proposition 2.5.1, we note that

$$E[X^n] = \begin{cases} \sum_x x^n p(x) & \text{if } X \text{ is discrete} \\ \int_{-\infty}^{\infty} x^n f(x) \, dx & \text{if } X \text{ is continuous} \end{cases}$$

The two-dimensional version of Proposition 2.5.1 states that if X and Y are random variables and g is a function of two variables, then

$$E[g(X, Y)] = \sum_y \sum_x g(x, y) p(x, y) \qquad \text{in the discrete case}$$

$$= \int_{-\infty}^{\infty} \int_{-\infty}^{\infty} g(x, y) f(x, y) \, dx \, dy \qquad \text{in the continuous case}$$

For example, if $g(X, Y) = X + Y$, then, in the continuous case,

$$E[X + Y] = \int_{-\infty}^{\infty} \int_{-\infty}^{\infty} (x + y) f(x, y) \, dx \, dy$$

$$= \int_{-\infty}^{\infty} \int_{-\infty}^{\infty} x f(x, y) \, dx \, dy + \int_{-\infty}^{\infty} \int_{-\infty}^{\infty} y f(x, y) \, dx \, dy$$

$$= \int_{-\infty}^{\infty} x \left(\int_{-\infty}^{\infty} f(x, y) \, dy \right) dx + \int_{-\infty}^{\infty} y \left(\int_{-\infty}^{\infty} f(x, y) \, dx \right) dy$$

$$= \int_{-\infty}^{\infty} x f_X(x) \, dx + \int_{-\infty}^{\infty} y f_Y(y) \, dy$$

$$= E[X] + E[Y]$$

A similar result can be shown in the discrete case and indeed for any random variables X and Y

$$E[X + Y] = E[X] + E[Y] \qquad (2.5.1)$$

By repeatedly applying Equation 2.5.1 we can show that the expected value of the sum of any number of random variables equals the sum of their individual expectations. For instance,

$$E[X + Y + Z] = E[(X + Y) + Z]$$

$$= E[X + Y] + E[Z] \qquad \text{by Equation 2.5.1}$$

$$= E[X] + E[Y] + E[Z] \qquad \text{again by Equation 2.5.1}$$

And in general, for any n,

$$E[X_1 + X_2 \cdots + X_n] = E[X_1] + E[X_2] + \cdots + E[X_n] \qquad (2.5.2)$$

Equation 2.5.2 is an extremely useful formula whose utility will now be illustrated by a series of examples.

Example 2.5e A construction firm has recently sent in bids for 3 jobs worth (in profits) respectively 10, 20, and 40 (thousand) dollars. If their probabilities of winning the jobs are respectively .2, .8, .3, what is their expected total profit?

Solution Letting X_i, $i = 1, 2, 3$ denote the firm's profit from job i, then

$$\text{total profit} = X_1 + X_2 + X_3$$

and so

$$E[\text{total profit}] = E[X_1] + E[X_2] + E[X_3]$$

Now

$$E[X_1] = 10(.2) + 0(.8) = 2$$
$$E[X_2] = 20(.8) + 0(.2) = 16$$
$$E[X_3] = 40(.3) + 0(.7) = 12$$

and thus the firm's expected total profit is 30 thousand dollars. ∎

Example 2.5f A secretary has typed N letters along with their respective envelopes. If the envelopes were to get rearranged leading to the letters being distributed in them in a completely random manner (that is, each letter is equally likely to end up in any of the envelopes), what is the expected number of letters that are placed in their correct envelope?

Solution Letting X denote the number of letters that are placed in the correct envelope, we can most easily compute $E[X]$ by noting that

$$X = X_1 + X_2 + \cdots + X_N$$

where

$$X_i = \begin{cases} 1 & \text{if the } i\text{th letter is placed in its proper envelope} \\ 0 & \text{otherwise} \end{cases}$$

Now, since the ith letter is equally likely to be put in any of the N envelopes it follows that

$$P\{X_i = 1\} = P\{i\text{th letter is in its proper envelope}\} = 1/N$$

and so

$$E[X_i] = 1P\{X_i = 1\} + 0P\{X_i = 0\} = 1/N$$

Hence, from Equation 2.5.2 we obtain that

$$E[X] = E[X_1] + \cdots + E[X_N] = \left(\frac{1}{N}\right)N = 1$$

Hence, no matter how many letters there are, on the average, exactly one of the letters will be in its own envelope. ∎

Example 2.5g Suppose there are 20 different types of coupons and suppose that each time one obtains a coupon it is equally likely to be any one of the types. Compute the expected number of different types that are contained in a set of 10 coupons.

Solution Let X denote the number of different types in the set of 10 coupons. We compute $E[X]$ by using the representation

$$X = X_1 + \cdots + X_{20}$$

where

$$X_i = \begin{cases} 1 & \text{if at least one type } i \text{ coupon is contained in the set of 10} \\ 0 & \text{otherwise} \end{cases}$$

Now,

$$\begin{aligned} E[X_i] &= P\{X_i = 1\} \\ &= P\{\text{at least one type } i \text{ coupon is in the set of 10}\} \\ &= 1 - P\{\text{no type } i \text{ coupons are contained in the set of 10}\} \\ &= 1 - \left(\tfrac{19}{20}\right)^{10} \end{aligned}$$

when the last equality follows since each of the 10 coupons will (independently) not be a type i with probability 19/20. Hence,

$$E[X] = E[X_1] + \cdots + E[X_{20}] = 20\left[1 - \left(\tfrac{19}{20}\right)^{10}\right] = 8.025 \quad \blacksquare$$

6 VARIANCE

Given a random variable X along with its probability distribution function, it would be extremely useful if we were able to summarize the essential properties of the mass function by certain suitably defined measures. One such measure would be $E[X]$, the expected value of X. However, while $E[X]$ yields the weighted average of the possible values of X, it does not tell us anything about the variation, or spread, of these values. For instance, while the following random variables W, Y, and Z having probability mass functions determined by

$$W = 0 \text{ with probability } 1$$

$$Y = \begin{cases} -1 & \text{with probability } \tfrac{1}{2} \\ 1 & \text{with probability } \tfrac{1}{2} \end{cases}$$

$$Z = \begin{cases} -100 & \text{with probability } \tfrac{1}{2} \\ 100 & \text{with probability } \tfrac{1}{2} \end{cases}$$

all have the same expectation—namely 0—there is much greater spread in the possible values of Y than in those of W (which is a constant) and in the possible values of Z than in those of Y.

As we expect X to take on values around its mean $E[X]$, it would appear that a reasonable way of measuring the possible variation of X would be to look at how far apart X would be from its mean on the average. One possible way to measure this would be to consider the quantity $E[|X - \mu|]$, where $\mu = E[X]$ and $|X - \mu|$ represents the absolute value of $X - \mu$. However, it turns out to be mathematically inconvenient to deal with this quantity and so a more tractable quantity is usually considered—namely the expectation of the square of the difference between X and its mean. We thus have the following definition.

Definition. If X is a random variable with mean μ then the *variance* of X, denoted by $\text{Var}(X)$, is defined by

$$\text{Var}(X) = E\left[(X - \mu)^2\right]$$

An alternative formula for $\text{Var}(X)$ can be derived as follows:

$$\begin{aligned}
\text{Var}(X) &= E\left[(X - \mu)^2\right] \\
&= E\left[X^2 - 2\mu X + \mu^2\right] \\
&= E[X^2] - E[2\mu X] + E\left[\mu^2\right] \\
&= E[X^2] - 2\mu E[X] + \mu^2 \\
&= E[X^2] - \mu^2
\end{aligned}$$

That is

$$\text{Var}(X) = E[X^2] - (E[X])^2 \qquad (2.6.1)$$

or, in words, the variance of X is equal to the expected value of the square of X minus the square of the expected value of X. This is, in practice, often the easiest way to compute $\text{Var}(X)$.

Example 2.6a Compute $\text{Var}(X)$ when X represents the outcome when we roll a fair die.

Solution Since $P\{X = i\} = 1/6$, $i = 1, 2, 3, 4, 5, 6$, we obtain

$$\begin{aligned}
E[X^2] &= \sum_{i=1}^{6} i^2 P\{X = i\} \\
&= 1^2\left(\tfrac{1}{6}\right) + 2^2\left(\tfrac{1}{6}\right) + 3^2\left(\tfrac{1}{6}\right) + 4^2\left(\tfrac{1}{6}\right) + 5^2\left(\tfrac{1}{6}\right) + 6^2\left(\tfrac{1}{6}\right) \\
&= \tfrac{91}{6}
\end{aligned}$$

Hence, since it was shown in Example 2.4a that $E[X] = 7/2$, we obtain from Equation 2.6.1 that

$$\begin{aligned}
\text{Var}(X) &= E[X^2] - (E[X])^2 \\
&= \tfrac{91}{6} - \left(\tfrac{7}{2}\right)^2 = \tfrac{35}{12} \quad \blacksquare
\end{aligned}$$

Example 2.6b Variance of an Indicator Random Variable. If, for some event A

$$I = \begin{cases} 1 & \text{if event } A \text{ occurs} \\ 0 & \text{if event } A \text{ does not occur} \end{cases}$$

then

$$\begin{aligned}
\text{Var}(I) &= E[I^2] - (E[I])^2 \\
&= E[I] - (E[I])^2 \quad \text{since } I^2 = I \text{ (as } 1^2 = 1 \text{ and } 0^2 = 0) \\
&= E[I](1 - E[I]) \\
&= P(A)[1 - P(A)] \quad \text{since } E[I] = P(A) \text{ from Example 2.4b} \quad \blacksquare
\end{aligned}$$

A useful identity concerning variances is that for any constants a and b

$$\text{Var}(aX + b) = a^2 \text{Var}(X) \tag{2.6.2}$$

To prove Equation 2.6.2, let $\mu = E[X]$ and recall that $E[aX + b] = a\mu + b$. Thus, by the definition of variance, we have

$$\begin{aligned}
\text{Var}(aX + b) &= E\left[(aX + b - E[aX + b])^2\right] \\
&= E\left[(aX + b - a\mu - b)^2\right] \\
&= E\left[(aX - a\mu)^2\right] \\
&= E\left[a^2(X - \mu)^2\right] \\
&= a^2 E\left[(X - \mu)^2\right] \\
&= a^2 \text{Var}(X)
\end{aligned}$$

Specifying particular values for a and b in Equation 2.6.2 leads to some interesting corollaries. For instance, by setting $a = 0$ in Equation 2.6.2 we obtain that

$$\text{Var}(b) = 0$$

That is, the variance of a constant is 0 (is this intuitive?). Similarly by setting $a = 1$ we obtain

$$\text{Var}(X + b) = \text{Var}(X)$$

That is, the variance of a constant plus a random variable is equal to the variance of the random variable (is this intuitive? Think about it). Finally, setting $b = 0$ yields

$$\text{Var}(aX) = a^2 \text{Var}(X)$$

The quantity $\sqrt{\text{Var}(X)}$ is called the *standard deviation* of X. The standard deviation has the same units as does the mean.

REMARK
Analogous to the mean's being the center of gravity of a distribution of mass, the variance represents, in the terminology of mechanics, the moment of inertia.

7   COVARIANCE AND VARIANCE OF SUMS OF RANDOM VARIABLES

It was shown in Section 5 that the expectation of a sum of random variables is equal to the sum of their expectations. The corresponding result for variances is, however, not generally valid. Consider

$$\begin{aligned} \text{Var}(X + X) &= \text{Var}(2X) \\ &= 2^2\,\text{Var}(X) \\ &= 4\,\text{Var}(X) \\ &\neq \text{Var}(X) + \text{Var}(X) \end{aligned}$$

There is, however, an important case in which the variance of a sum of random variables is equal to the sum of the variances; and this is when the random variables are independent. Before proving this, however, let us define the concept of the covariance of two random variables.

Definition. The *covariance* of two random variables X and Y, written $\text{Cov}(X, Y)$ is defined by

$$\text{Cov}(X, Y) = E\big[(X - \mu_x)(Y - \mu_y)\big]$$

where μ_x and μ_y are the means of X and Y respectively.

A useful expression for $\text{Cov}(X, Y)$ can be obtained by expanding the right side of the definition. This yields

$$\begin{aligned} \text{Cov}(X, Y) &= E\big[XY - \mu_x Y - \mu_y X + \mu_x \mu_y\big] \\ &= E[XY] - \mu_x E[Y] - \mu_y E[X] + \mu_x \mu_y \\ &= E[XY] - \mu_x \mu_y - \mu_y \mu_x + \mu_x \mu_y \\ &= E[XY] - E[X]E[Y] \end{aligned} \tag{2.7.1}$$

From its definition we see that covariance satisfies the following properties:

$$\text{Cov}(X, Y) = \text{Cov}(Y, X) \tag{2.7.2}$$

and

$$\text{Cov}(X, X) = \text{Var}(X) \tag{2.7.3}$$

Another property of covariance, which immediately follows from its definition, is that, for any constant a,

$$\text{Cov}(aX, Y) = a\,\text{Cov}(X, Y) \tag{2.7.4}$$

The proof of Equation 2.7.4 is left as an exercise.

Covariance, like expectation, possesses an additive property.

Lemma 2.7.1

$$\text{Cov}(X + Z, Y) = \text{Cov}(X, Y) + \text{Cov}(Z, Y)$$

Proof

$$\begin{aligned}
\text{Cov}\,(X + Z, Y) \\
&= E[(X + Z)Y] - E[X + Z]E[Y] \quad \text{from Equation 2.7.1} \\
&= E[XY] + E[ZY] - (E[X] + E[Z])E[Y] \\
&= E[XY] - E[X]E[Y] + E[ZY] - E[Z]E[Y] \\
&= \text{Cov}\,(X, Y) + \text{Cov}\,(Z, Y) \quad \blacksquare
\end{aligned}$$

Lemma 2.7.1 can be easily generalized (see Problem 42) to show that

$$\text{Cov}\left(\sum_{i=1}^{n} X_i, Y\right) = \sum_{i=1}^{n} \text{Cov}\,(X_i, Y) \qquad (2.7.5)$$

which gives rise to the following.

Proposition 2.7.2

$$\text{Cov}\left(\sum_{i=1}^{n} X_i, \sum_{j=1}^{m} Y_j\right) = \sum_{i=1}^{n} \sum_{j=1}^{m} \text{Cov}\,(X_i, Y_j)$$

Proof

$$\begin{aligned}
\text{Cov}&\left(\sum_{i=1}^{n} X_i, \sum_{j=1}^{m} Y_j\right) \\
&= \sum_{i=1}^{n} \text{Cov}\left(X_i, \sum_{j=1}^{m} Y_j\right) \quad \text{from Equation 2.7.5} \\
&= \sum_{i=1}^{n} \text{Cov}\left(\sum_{j=1}^{m} Y_j, X_i\right) \quad \text{by the symmetry property Equation 2.7.2} \\
&= \sum_{i=1}^{n} \sum_{j=1}^{m} \text{Cov}\,(Y_j, X_i) \quad \text{again from Equation 2.7.5}
\end{aligned}$$

and the result now follows by again applying the property Equation 2.7.2. $\blacksquare$

Using Equation 2.7.3 gives rise to the following formula for the variance of a sum of random variables.

Corollary 2.7.3

$$\text{Var}\left(\sum_{i=1}^{n} X_i\right) = \sum_{i=1}^{n} \text{Var}\,(X_i) + \sum_{i=1}^{n} \sum_{\substack{j=1 \\ j \neq i}}^{m} \text{Cov}\,(X_i, X_j)$$

Proof The proof follows directly from Proposition 2.7.2 upon setting $m = n$, and $Y_j = X_j$ for $j = 1, \ldots, n$. $\blacksquare$

In the case of $n = 2$ Corollary 2.7.3 yields that

$$\text{Var}(X + Y) = \text{Var}(X) + \text{Var}(Y) + \text{Cov}(X, Y) + \text{Cov}(Y, X)$$

or using Equation 2.7.2

$$\text{Var}(X + Y) = \text{Var}(X) + \text{Var}(Y) + 2\,\text{Cov}(X, Y) \qquad (2.7.6)$$

Theorem 2.7.4

If X and Y are independent random variables then

$$\text{Cov}(X, Y) = 0$$

and so for independent $X_1, \ldots, X_n$

$$\text{Var}\left(\sum_{i=1}^{n} X_i\right) = \sum_{i=1}^{n} \text{Var}(X_i)$$

Proof We need to prove that $E[XY] = E[X]E[Y]$. Now, in the discrete case

$$E[XY] = \sum_j \sum_i x_i y_j P\{X = x_i, Y = y_j\}$$

$$= \sum_j \sum_i x_i y_j P\{X = x_i\} P\{Y = y_j\} \qquad \text{by independence}$$

$$= \sum_y y_j P\{Y = y_j\} \sum_i x_i P\{X = x_i\}$$

$$= E[Y]E[X]$$

Because a similar argument holds in all other cases, the result is proven.

Example 2.7a Compute the variance of the sum obtained when 10 independent rolls of a fair die are made.

Solution Letting X_i denote the outcome of the ith roll, we have that

$$\text{Var}\left(\sum_{1}^{10} X_i\right) = \sum_{1}^{10} \text{Var}(X_i)$$

$$= 10\tfrac{35}{12} \qquad \text{from Example 2.6a}$$

$$= \tfrac{175}{6} \quad \blacksquare$$

Example 2.7b Compute the variance of the number of heads resulting from 10 independent tosses of a fair coin.

Solution Letting

$$I_j = \begin{cases} 1 & \text{if the } j\text{th toss lands heads} \\ 0 & \text{if the } j\text{th toss lands tails} \end{cases}$$

then the total number of heads is equal to

$$\sum_{j=1}^{10} I_j$$

Hence, from Theorem 2.7.4,

$$\text{Var}\left(\sum_{j=1}^{10} I_j\right) = \sum_{j=1}^{10} \text{Var}\left(I_j\right)$$

Now, since I_j is an indicator random variable for an event having probability $\frac{1}{2}$, it follows from Example 2.6b that

$$\text{Var}\left(I_j\right) = \frac{1}{2}\left(1 - \frac{1}{2}\right) = \frac{1}{4}$$

and thus

$$\text{Var}\left(\sum_{j=1}^{10} I_j\right) = \frac{10}{4} \quad \blacksquare$$

The covariance of two random variables is important as an indicator of the relationship between them. For instance, consider the situation where X and Y are indicator variables for whether or not the events A and B occur. That is, for events A and B, define

$$X = \begin{cases} 1 & \text{if } A \text{ occurs} \\ 0 & \text{otherwise} \end{cases}, \qquad Y = \begin{cases} 1 & \text{if } B \text{ occurs} \\ 0 & \text{otherwise} \end{cases}$$

and note that

$$XY = \begin{cases} 1 & \text{if } X = 1, Y = 1 \\ 0 & \text{otherwise} \end{cases}$$

Thus,

$$\begin{aligned} \text{Cov}(X, Y) &= E[XY] - E[X]E[Y] \\ &= P\{X = 1, Y = 1\} - P\{X = 1\}P\{Y = 1\} \end{aligned}$$

From this we see that

$$\text{Cov}(X, Y) > 0 \Leftrightarrow P\{X = 1, Y = 1\} > P\{X = 1\}P\{Y = 1\}$$

$$\Leftrightarrow \frac{P\{X = 1, Y = 1\}}{P\{X = 1\}} > P\{Y = 1\}$$

$$\Leftrightarrow P\{Y = 1 \mid X = 1\} > P\{Y = 1\}$$

That is, the covariance of X and Y is positive if the outcome $X = 1$ makes it more likely that $Y = 1$ (which, as is easily seen by symmetry, also implies the reverse).

In general it can be shown that a positive value of $\text{Cov}(X, Y)$ is an indication that Y tends to increase as X does, whereas a negative value indicates that Y tends to decrease as X increases. The strength of the relationship between X and Y is indicated by the correlation between X and Y, a dimensionless quantity obtained by dividing the covariance by the product of the standard deviations of X and Y. That is,

$$\text{Corr}(X, Y) = \frac{\text{Cov}(X, Y)}{\sqrt{\text{Var}(X)\text{Var}(Y)}}$$

It can be shown (see Problem 43) that this quantity always has a value between -1 and $+1$.

8 MOMENT GENERATING FUNCTIONS

The moment generating function $\phi(t)$ of the random variable X is defined for all values t by

$$\phi(t) = E[e^{tX}]$$

$$= \begin{cases} \sum_x e^{tx} p(x) & \text{if } X \text{ is discrete} \\ \int_{-\infty}^{\infty} e^{tx} f(x)\, dx & \text{if } X \text{ is continuous} \end{cases}$$

We call $\phi(t)$ the moment generating function because all of the moments of X can be obtained by successively differentiating $\phi(t)$. For example,

$$\phi'(t) = \frac{d}{dt} E[e^{tX}]$$

$$= E\left[\frac{d}{dt}(e^{tX})\right]$$

$$= E[Xe^{tX}]$$

Hence,

$$\phi'(0) = E[X]$$

Similarly,

$$\phi''(t) = \frac{d}{dt}\phi'(t)$$

$$= \frac{d}{dt} E[Xe^{tX}]$$

$$= E\left[\frac{d}{dt}(Xe^{tX})\right]$$

$$= E[X^2 e^{tX}]$$

and so

$$\phi''(0) = E[X^2]$$

In general, the nth derivative of $\phi(t)$ evaluated at $t = 0$ equals $E[X^n]$, that is,

$$\phi^n(0) = E[X^n] \qquad n \geq 1$$

An important property of moment generating functions is that the *moment generating function of the sum of independent random variables is just the product of the individual moment generating functions*. To see this, suppose that X and Y are independent and have moment generating functions $\phi_X(t)$ and $\phi_Y(t)$, respectively. Then $\phi_{X+Y}(t)$, the moment generating function of $X + Y$, is given by

$$\phi_{X+Y}(t) = E[e^{t(X+Y)}]$$
$$= E[e^{tX}e^{tY}]$$
$$= E[e^{tX}]E[e^{tY}]$$
$$= \phi_X(t)\phi_Y(t)$$

where the next to the last equality follows from Theorem 2.7.4 since X and Y, and thus e^{tX} and e^{tY}, are independent.

Another important result is that the *moment generating function uniquely determines the distribution.* That is, there exists a one-to-one correspondence between the moment generating function and the distribution function of a random variable.

9   CHEBYSHEV'S INEQUALITY AND THE WEAK LAW OF LARGE NUMBERS

We start this section by proving a result known as Markov's inequality.

Proposition 2.9.1 Markov's Inequality

If X is a random variable that takes only nonnegative values, then for any value $a > 0$

$$P\{X \ge a\} \le \frac{E[X]}{a}$$

Proof   We give a proof for the case where X is continuous with density f.

$$E[X] = \int_0^\infty xf(x)\, dx$$
$$= \int_0^a xf(x)\, dx + \int_a^\infty xf(x)\, dx$$
$$\ge \int_a^\infty xf(x)\, dx$$
$$\ge \int_a^\infty af(x)\, dx$$
$$= a\int_a^\infty f(x)\, dx$$
$$= aP\{X \ge a\}$$

and the result is proved.   ∎

As a corollary, we obtain Proposition 2.9.2

Proposition 2.9.2 Chebyshev's Inequality

If X is a random variable with mean μ and variance σ^2, then for any value $k > 0$

$$P\{|X - \mu| \geq k\} \leq \frac{\sigma^2}{k^2}$$

Proof Since $(X - \mu)^2$ is a nonnegative random variable, we can apply Markov's inequality (with $a = k^2$) to obtain

$$P\{(X - \mu)^2 \geq k^2\} \leq \frac{E\left[(X - \mu)^2\right]}{k^2} \qquad (2.9.1)$$

But since $(X - \mu) \geq k^2$ if and only if $|X - \mu| \geq k$, Equation 2.9.1 is equivalent to

$$P\{|X - \mu| \geq k\} \leq \frac{E\left[(X - \mu)^2\right]}{k^2} = \frac{\sigma^2}{k^2}$$

and the proof is complete.

The importance of Markov's and Chebyshev's inequalities is that they enable us to derive bounds on probabilities when only the mean, or both the mean and the variance, of the probability distribution are known. Of course, if the actual distribution were known, then the desired probabilities could be exactly computed and we would not need to resort to bounds.

Example 2.9a Suppose that it is known that the number of items produced in a factory during a week is a random variable with mean 50.

1. What can be said about the probability that this week's production will exceed 75?

2. If the variance of a week's production is known to equal 25, then what can be said about the probability that this week's production will be between 40 and 60?

Solution Let X be the number of items that will be produced in a week:

1. By Markov's inequality

$$P\{X > 75\} \leq \frac{E[X]}{75} = \frac{50}{75} = \frac{2}{3}$$

2. By Chebyshev's inequality

$$P\{|X - 50| \geq 10\} \leq \frac{\sigma^2}{10^2} = \frac{1}{4}$$

Hence

$$P\{|X - 50| < 10\} \geq 1 - \tfrac{1}{4} = \tfrac{3}{4}$$

and so the probability that this week's production will be between 40 and 60 is at least .75. ∎

By replacing k by $k\sigma$ in Equation 2.9.1 we can write Chebyshev's inequality as

$$P\{|X - \mu| > k\sigma\} \le 1/k^2$$

Thus it states that the probability a random variable differs from its mean by more than k standard deviations is bounded by $1/k^2$.

We will end this section by using Chebyshev's inequality to prove the weak law of large numbers, which states that the probability that the average of the first n terms in a sequence of independent and identically distributed random variables differs by its mean by more than ϵ goes to 0 as n goes to infinity.

Theorem 2.9.3 The Weak Law of Large Numbers

Let $X_1, X_2, \ldots$ be a sequence of independent and identically distributed random variables, each having mean $E[X_i] = \mu$. Then, for any $\epsilon > 0$,

$$P\left\{\left|\frac{X_1 + \cdots + X_n}{n} - \mu\right| > \epsilon\right\} \to 0 \qquad \text{as } n \to \infty$$

Proof We shall prove the result only under the additional assumption that the random variables have a finite variance σ^2. Now, as

$$E\left[\frac{X_1 + \cdots + X_n}{n}\right] = \mu \quad \text{and} \quad \text{Var}\left(\frac{X_1 + \cdots + X_n}{n}\right) = \frac{\sigma^2}{n}$$

it follows from Chebyshev's inequality that

$$P\left\{\left|\frac{X_1 + \cdots + X_n}{n} - \mu\right| > \epsilon\right\} \le \frac{\sigma^2}{n\epsilon^2}$$

and the result is proved.

For an application of the above, suppose that a sequence of independent trials are performed. Let E be a fixed event and denote by $P(E)$ the probability that E occurs on a given trial. Letting

$$X_i = \begin{cases} 1 & \text{if } E \text{ occurs on trial } i \\ 0 & \text{if } E \text{ does not occur on trial } i \end{cases}$$

it follows that $X_1 + X_2 + \cdots + X_n$ represents the number of times that E occurs in the first n trials. As $E[X_i] = P(E)$ it thus follows from the weak law of large numbers that for any positive number ϵ, no matter how small, the probability that the proportion of the first n trials in which E occurs differs from $P(E)$ by more than ϵ goes to 0 as n increases.

PROBLEMS

1. Five men and 5 women are ranked according to their scores on an examination. Assume that no two scores are alike and all 10! possible rankings are equally likely. Let X denote the highest ranking achieved by

a woman (for instance, $X = 2$ if the top ranked person was male and the next ranked person was female). Find $P\{X = i\}$, $i = 1, 2, 3, \ldots, 8, 9, 10$.

2. Let X represent the difference between the number of heads and the number of tails obtained when a coin is tossed n times. What are the possible values of X?

3. In Problem 2, if the coin is assumed fair, for $n = 3$, what are the probabilities associated with the values that X can take on?

4. The distribution function of the random variable X is given

$$F(x) = \begin{cases} 0 & x < 0 \\ \dfrac{x}{2} & 0 \le x < 1 \\ \dfrac{2}{3} & 1 \le x < 2 \\ \dfrac{11}{12} & 2 \le x < 3 \\ 1 & 3 \le x \end{cases}$$

(a) Plot this distribution function
(b) What is $P\{X > \frac{1}{2}\}$?
(c) What is $P\{2 < X \le 4\}$?
(d) What is $P\{X < 3\}$?
(e) What is $P\{X = 1\}$?

5. Suppose you are given the distribution function F of a random variable X. Explain how you could determine $P\{X = 1\}$.
   Hint: You will need to use the concept of a limit.

6. The amount of time, in hours, that a computer functions before breaking down is a continuous random variable with probability density function given by

$$f(x) = \begin{cases} \lambda e^{-x/100} & x \ge 0 \\ 0 & x < 0 \end{cases}$$

What is the probability that a computer will function between 50 and 150 hours before breaking down? What is the probability that it will function less than 100 hours?

7. The lifetime in hours of a certain kind of radio tube is a random variable having a probability density function given by

$$f(x) = \begin{cases} 0 & x \le 100 \\ \dfrac{100}{x^2} & x > 100 \end{cases}$$

What is the probability that exactly 2 of 5 such tubes in a radio set will have to be replaced within the first 150 hours of operation? Assume that the events E_i, $i = 1, 2, 3, 4, 5$, that the ith such tube will have to be replaced within this time, are independent.

8. If the density function of X equals

$$f(x) = \begin{cases} ce^{-2x}, & 0 < x < \infty \\ 0, & x < 0 \end{cases}$$

find c. What is $P\{X > 2\}$?

9. A bin of 5 transistors is known to contain 3 that are defective. The transistors are to be tested, one at a time, until the defective ones are identified. Denote by N_1 the number of tests made until the first defective is spotted and by N_2 the number of additional tests until the second defective is spotted; find the joint probability mass function of N_1 and N_2.

10. The joint probability density function of X and Y is given by

$$f(x, y) = \frac{6}{7}\left(x^2 + \frac{xy}{2}\right) \qquad 0 < x < 1, 0 < y < 2$$

(a) Verify that this is indeed a joint density function.
(b) Compute the density function of X.
(c) Find $P\{X > Y\}$.

11. Let $X_1, X_2, \ldots, X_n$ be independent random variables, each having a uniform distribution over $(0, 1)$. Let $M = \text{maximum } (X_1, X_2, \ldots, X_n)$. Show that the distribution function of M, $F_M(\cdot)$, is given by

$$F_M(x) = x^n \qquad 0 \le x \le 1$$

What is the probability density function of M?

12. The joint density of X and Y is given by

$$f(x, y) = \begin{cases} xe^{-(x+y)} & x > 0, y > 0 \\ 0 & \text{otherwise} \end{cases}$$

(a) Compute the density of X
(b) Compute the density of Y
(c) Are X and Y independent?

13. The joint density of X and Y is

$$f(x, y) = \begin{cases} 2 & 0 < x < y, 0 < y < 1 \\ 0 & \text{otherwise} \end{cases}$$

(a) Compute the density of X
(b) Compute the density of Y
(c) Are X and Y independent?

14. If the joint density function of X and Y factors into a part depending only on x and one depending only on y, show that X and Y are independent. That is, if

$$f(x, y) = k(x)l(y), \qquad -\infty < x < \infty, \quad -\infty < y < \infty$$

show that X and Y are independent.

15. Is Problem 14 consistent with the results of Problems 12 and 13?

16. Suppose that X and Y are independent continuous random variables. Show that

(a)
$$P\{X + Y \le a\} = \int_{-\infty}^{\infty} F_X(a - y)f_Y(y)\, dy$$

(b)
$$P\{X \le Y\} = \int_{-\infty}^{\infty} F_X(y)f_Y(y)\, dy$$

where f_Y is the density function of Y, and F_X is the distribution function of X.

17. When a current I (measured in amperes) flows through a resistance R (measured in ohms), the power generated (measured in watts) is given by $W = I^2 R$. Suppose that I and R are independent random variables with densities

$$f_I(x) = 6x(1 - x) \qquad 0 \le x \le 1$$
$$f_R(x) = 2x \qquad\qquad 0 \le x \le 1$$

Determine the density function of W.

18. In Example 2.3b, determine the conditional probability mass function of the size of a randomly chosen family containing 2 girls.

19. Compute the conditional density function of X given $Y = y$ in (a) Problem 10 and (b) Problem 13.

20. Show that X and Y are independent if and only if
(a) $P_{X|Y}^{(x|y)} = p_X(x)$ in the discrete case
(b) $f_{X|Y}^{(x|y)} = f_X(x)$ in the continuous case

21. Compute the expected value of the random variable in Problem 1.

22. Compute the expected value of the random variable in Problem 3.

23. Each night different meterologists give us the "probability" that it will rain the next day. To judge how well these people predict, we will score each of them as follows: If a meterologist says that it will rain with probability p, then he or she will receive a score of

$$1 - (1 - p)^2 \qquad \text{if it does rain}$$

$$1 - p^2 \qquad \text{if it does not rain}$$

We will then keep track of scores over a certain time span and conclude that the meterologist with the highest average score is the best predictor of weather. Suppose now that a given meterologist is aware of this and so wants to maximize his or her expected score. If this individual truly believes that it will rain tomorrow with probability p^*, what value of p should he or she assert so as to maximize the expected score?

24. The density function of X is given by

$$f(x) = \begin{cases} a + bx^2 & 0 \le x \le 1 \\ 0 & \text{otherwise} \end{cases}$$

If $E[X] = \frac{3}{5}$, find a, b.

25. The lifetime in hours of electronic tubes is a random variable having a probability density function given by

$$f(x) = \alpha^2 x e^{-\alpha x} \qquad x \ge 0$$

Compute the expected lifetime of such a tube.

26. Let $X_1, X_2, \ldots, X_n$ be independent random variables having the common density function

$$f(x) = \begin{cases} 1 & 0 < x < 1 \\ 0 & \text{otherwise} \end{cases}$$

Find (a) $E[\text{Max}(X_1, \ldots, X_n)]$ and (b) $E[\text{Min}(X_1, \ldots, X_n)]$.

27. Suppose that X has density function

$$f(x) = \begin{cases} 1 & 0 < x < 1 \\ 0 & \text{otherwise} \end{cases}$$

Compute $E[X^n]$ first by computing the density of X^n and then using the definition of expectation; and second by using Proposition 2.5.1.

28. The time it takes to repair a personal computer is a random variable whose density, in hours, is given by

$$f(x) = \begin{cases} \frac{1}{2} & 0 < x < 2 \\ 0 & \text{otherwise} \end{cases}$$

The cost of the repair depends on the time it takes and is equal to $40 + 30\sqrt{x}$ when the time is x. Compute the expected cost to repair a personal computer.

29. If $E[X] = 2$ and $E[X^2] = 8$ calculate (a) $E[(2 + 4X)^2]$ and (b) $E[X^2 + (X + 1)^2]$.

30. Ten balls are randomly chosen from an urn containing 17 white and 23 black balls. Let X denote the number of white balls chosen. Compute $E[X]$

(a) by defining appropriate indicator variables X_i, $i = 1, \ldots, 10$ so that

$$X = \sum_{i=1}^{10} X_i$$

(b) by defining appropriate indicator variables Y_i, $i = 1, \ldots, 17$ so that

$$X = \sum_{i=1}^{17} Y_i.$$

31. A community consists of 100 married couples. If during a given year 50 of the members of the community die, what is the expected number of marriages that remain intact? Assume that the 50 people that die is equally to be any of the $\binom{200}{50}$ groups of size 50.

Hint: For $i = 1, \ldots, 100$ let

$$X_i = \begin{cases} 1 & \text{if neither member of couple } i \text{ dies} \\ 0 & \text{otherwise} \end{cases}$$

32. Compute the expectation and variance of the number of successes in n independent trials, each of which results in a success with probability p. Is independence necessary?

33. Suppose that X is equally likely to take on any of the values $1, 2, 3, 4$. Compute (a) $E[X]$ and (b) $\text{Var}(X)$.

34. Let $p_i = P\{X = i\}$ and suppose that $p_1 + p_2 + p_3 = 1$. If $E[X] = 2$, what values of p_1, p_2, p_3 (a) maximize and (b) minimize $\text{Var}(X)$?

35. Compute the mean and variance of the number of heads that appear in 3 flips of a fair coin.

36. Argue that for any random variable X

$$E[X^2] \geq (E[X])^2$$

When does one have equality?

37. A random variable X, which represents the weight (in ounces) of an article, has density function given by $f_X(z)$,

$$f_X(z) = \begin{cases} (z - 8) & \text{for } 8 \leq z \leq 9 \\ (10 - z) & \text{for } 9 < z \leq 10 \\ 0 & \text{otherwise} \end{cases}$$

(a) Calculate the mean and variance of the random variable X. (b) The manufacturer sells the article for a fixed price of $2.00. He guarantees to refund the purchase money to any customer who finds the weight of his article to be less than 8.25 oz. His cost of production is related to the weight of the article by the relation $x/15 + 0.35$. Find the expected profit per article.

38. Suppose that the Rockwell hardness X and abrasion loss Y of a specimen (coded data) have a joint density given by

$$f_{XY}(u, v) = \begin{cases} (u + v) & \text{for } 0 \leq u, v \leq 1 \\ 0 & \text{otherwise.} \end{cases}$$

(a) Find the marginal densities of X and Y. (b) Find $E(X)$ and $\text{Var}(X)$.

39. A product is classified according to the number of defects it contains and the factory that produces it. Let X_1 and X_2 be the random variables that represent the number of defects per unit (taking on possible values of 0, 1, 2, or 3) and the factory number (taking on possible values 1 or 2), respectively. The entries in the table represent the joint probability mass function.

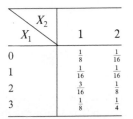

X_1 \ X_2	1	2
0	$\frac{1}{8}$	$\frac{1}{16}$
1	$\frac{1}{16}$	$\frac{1}{16}$
2	$\frac{3}{16}$	$\frac{1}{8}$
3	$\frac{1}{8}$	$\frac{1}{4}$

(a) Find the marginal probability distributions of X_1 and X_2. (b) Find $E(X_1)$, $E(X_2)$, $\mathrm{Var}(X_1)$, and $\mathrm{Var}(X_2)$.

40. A machine makes a product that is screened (inspected 100 percent) before being shipped. The measuring instrument is such that it is difficult to read between 1 and $1\frac{1}{3}$ (coded data). After the screening process takes place, the measured dimension has density

$$f_X(z) = \begin{cases} kz^2 & \text{for } 0 \le z \le 1 \\ 1 & \text{for } 1 < z \le 1\frac{1}{3} \\ 0 & \text{otherwise} \end{cases}$$

(a) Find the value of k. (b) What fraction of the items will fall outside the twilight zone (fall between 0 and 1)? (c) Find the mean and variance of this random variable.

41. Verify Equation 2.7.4

42. Prove Equation 2.7.5 by using mathematical induction.

43. Let X have variance σ_x^2 and let Y have variance σ_y^2. Starting with

$$0 \le \mathrm{Var}\left(X/\sigma_x + Y/\sigma_y\right)$$

show that

$$-1 \le \mathrm{Corr}(X, Y)$$

Now using that

$$0 \le \mathrm{Var}\left(X/\sigma_x - Y/\sigma_y\right)$$

conclude that

$$-1 \le \mathrm{Corr}(X, Y) \le 1$$

Using the result that $\mathrm{Var}(Z) = 0$ implies that Z is a constant, argue that if $\mathrm{Corr}(X, Y) = 1$ or -1 then X and Y are related by

$$Y = a + bx$$

where the sign of b is positive when the correlation is 1 and negative when it is -1.

44. In Example 2.5f compute $\text{Cov}(X_i, X_j)$ and use this result to show that $\text{Var}(X) = 1$.

45. If X_1 and X_2 have the same probability distribution function, show that
$$\text{Cov}(X_1 - X_2, X_1 + X_2) = 0.$$
Note that independence is not being assumed.

46. Suppose that X has density function
$$f(x) = e^{-x} \qquad x > 0$$
Compute the moment generating function of X and use your result to determine its mean and variance. Check your answer for the mean by a direct calculation.

47. If the density function of X is
$$f(x) = 1 \qquad 0 < x < 1$$
determine $E[e^{tX}]$. Differentiate to obtain $E[X^n]$ and then check your answer.

48. Suppose that X is a random variable with mean and variance both equal to 20. What can be said about $P\{0 \le X \le 40\}$?

49. From past experience a professor knows that the test score of a student taking his final examination is a random variable with mean 75.

(a) Give an upper bound to the probability that a student's test score will exceed 85.
   Suppose in addition the professor knows that the variance of a student's test score is equal to 25.
(b) What can be said about the probability that a student will score between 65 and 85?
(c) How many students would have to take the examination so as to ensure, with probability at least .9, that the class average would be within 5 of 75?

C H A P T E R 3

Special Random Variables

0 INTRODUCTION

There are certain types of random variables that occur over and over again in applications. In this chapter we will study a variety of them.

1 THE BERNOULLI AND BINOMIAL RANDOM VARIABLES

Suppose that a trial, or an experiment, whose outcome can be classified as either a "success" or as a "failure" is performed. If we let $X = 1$ when the outcome is a success and $X = 0$ when it is a failure, then the probability mass function of X is given by

$$P\{X = 0\} = 1 - p \qquad (3.1.1)$$
$$P\{X = 1\} = p$$

where p, $0 \leq p \leq 1$, is the probability that the trial is a "success."

A random variable X is said to be a Bernoulli random variable (after the Swiss mathematician James Bernoulli) if its probability mass function is given by Equations 3.1.1 for some $p \in (0, 1)$. Its expected value is

$$E[X] = 1 \cdot P\{X = 1\} + 0 \cdot P\{X = 0\} = p$$

That is, the expectation of a Bernoulli random variable is the probability that the random variable equals 1.

Suppose now that n independent trials, each of which results in a "success" with probability p and in a "failure" with probability $1 - p$, are to be performed. If X represents the number of successes that occur in the n trials, then X is said to be a *binomial* random variable with parameters (n, p).

The probability mass function of a binomial random variable having parameters (n, p) is given by

$$P\{X = i\} = \binom{n}{i} p^i (1 - p)^{n-i}, \qquad i = 0, 1, \ldots, n \qquad (3.1.2)$$

where $\binom{n}{i} = n! / [(n - i)! i!]$ equals the number of different groups of i objects

71

that can be chosen from a set of n objects. The validity of Equation 1.2 may be verified by first noting that the probability of any particular sequence of the n outcomes containing i successes and $n - i$ failures is, by the assumed independence of trials, $p^i(1 - p)^{n-i}$. Equation 3.1.2 then follows since there are $\binom{n}{i}$ different sequences of the n outcomes leading to i successes and $n - i$ failures—which can perhaps most easily be seen by noting that there are $\binom{n}{i}$ different selections of the i trials that result in successes. For instance, if $n = 5$, $i = 2$ then there are $\binom{5}{2}$ choices of the two trials that are to result in successes—namely, any of the outcomes

$$
\begin{array}{lll}
(s, s, f, f, f) & (f, s, s, f, f) & (f, f, s, f, s) \\
(s, f, s, f, f) & (f, s, f, s, f) & \\
(s, f, f, s, f) & (f, s, f, f, s) & (f, f, f, s, s) \\
(s, f, f, f, s) & (f, f, s, s, f) &
\end{array}
$$

where the outcome (f, s, f, s, f) means, for instance, that the two successes appeared on trials 2 and 4. Since each of the $\binom{5}{2}$ outcomes has probability $p^2(1 - p)^3$, we see that the probability of a total of 2 successes in 5 independent trials is $\binom{5}{2}p^2(1 - p)^3$. Note that, by the binomial theorem, the probabilities sum to one, that is,

$$
\sum_{i=0}^{\infty} p(i) = \sum_{i=0}^{n} \binom{n}{i} p^i (1 - p)^{n-i} = [p + (1 - p)]^n = 1
$$

The probability mass function of three binomial random variables with respective parameters $(10, .5)$, $(10, .3)$, and $(10, .6)$ are presented in Figure 3.1.1. The first of these is symmetric about the value .5, whereas the second is somewhat weighted, or *skewed*, to lower values and the third to higher values.

Example 3.1a It is known that diskettes produced by a certain company will be defective with probability .01 independently of each other. The company sells the diskettes in packages of 10 and offers a money-back guarantee that at most 1 of the 10 diskettes is defective. What proportion of packages are returned? If someone buys 3 packages what is the probability that he will return exactly 1 of them?

Solution If X is the number of defective diskettes in a package, then X is a binomial random variable with parameters $(10, .01)$. Hence the probability that a package will have to be replaced is

$$
P\{X > 1\} = 1 - P\{X = 0\} - P\{X = 1\}
$$

$$
= 1 - \binom{10}{0}(.01)^0(.99)^{10} - \binom{10}{1}(.01)^1(.99)^9 \approx .005
$$

As each package will, independently, have to be replaced with probability .005, it follows from the law of large numbers that in the long run .5 percent of the packages will have to be replaced.

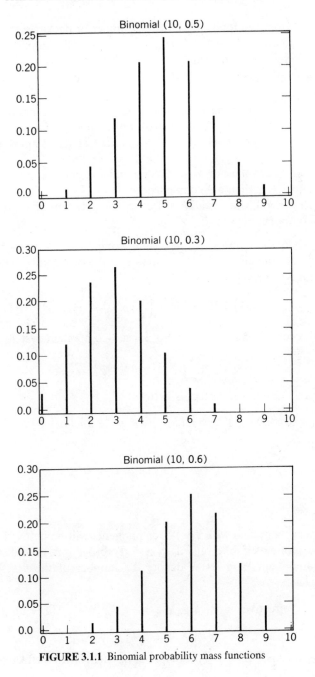

FIGURE 3.1.1 Binomial probability mass functions

It follows from the foregoing that the number of packages that the person will have to return is a binomial random variable with parameters $n = 3$ and $p = .005$. Therefore, the probability that exactly one of the three packages will have to be returned is $\binom{3}{1}(.005)(.995)^2 = .015$. ∎

Example 3.1b A communication system consists of n components each of which will, independently, function with probability p. The total system will be able to operate effectively if at least one-half of its components function.

    (a) For what values of p is a 5-component system more likely to operate effectively than a 3-component system?
    (b) In general, when is a $2k + 1$ component system better than a $2k - 1$ component system?

Solution (a) As the number of functioning components is a binomial random variable with parameters (n, p), it follows that the probability that a 5-component system will be effective is

$$\binom{5}{3}p^3(1 - p)^2 + \binom{5}{4}p^4(1 - p) + p^5$$

whereas the corresponding probability for a 3-component system is

$$\binom{3}{2}p^2(1 - p) + p^3$$

Hence, the 5-component system is better if

$$10p^3(1 - p)^2 + 5p^4(1 - p) + p^5 \geq 3p^2(1 - p) + p^3$$

which reduces to

$$3(p - 1)^2(2p - 1) \geq 0$$

or

$$p \geq \tfrac{1}{2}$$

(b) In general, a system with $2k + 1$ components will be better than one with $2k - 1$ components if (and only if) $p \geq \tfrac{1}{2}$. To prove this, consider a system of $2k + 1$ components and let X denote the number of the first $2k - 1$ that function. Then

$$P_{2k+1}(\text{effective}) = P\{ X \geq k + 1\} + P\{ X = k \}\left(1 - (1 - p)^2\right)$$
$$+ P\{ X = k - 1\} p^2$$

which follows since the $2k + 1$ component system will be effective if either

    (a) $X \geq k + 1$
    (b) $X = k$ and at least one of the remaining 2 components function
    or
    (c) $X = k - 1$ and both of the next 2 function

As

$$P_{2k-1} \text{(effective)} = P\{ X \geq k \}$$
$$= P\{ X = k \} + P\{ X \geq k + 1 \}$$

we obtain that

$$P_{2k+1} \text{(effective)} - P_{2k-1} \text{(effective)}$$

$$= P\{ X = k - 1 \} p^2 - (1 - p)^2 P\{ X = k \}$$

$$= \binom{2k-1}{k-1} p^{k-1}(1-p)^k p^2 - (1-p)^2 \binom{2k-1}{k} p^k (1-p)^{k-1}$$

$$= \binom{2k-1}{k} p^k (1-p)^k [p - (1-p)] \qquad \text{since } \binom{2k-1}{k-1} = \binom{2k-1}{k}$$

$$\geq 0 \Leftrightarrow p \geq \tfrac{1}{2} \quad \blacksquare$$

Example 3.1c Suppose that 10 percent of the chips produced by a computer hardware manufacturer are defective. If we order 100 such chips, will X, the number of defective ones we receive, be a binomial random variable?

Solution The random variable X will be a binomial random variable with parameters $(100, .1)$ if each chip has probability .9 of being functional and if the functioning of successive chips are independent. Whether this is a reasonable assumption when we know that 10 percent of the chips produced are defective depends on additional factors. For instance, suppose that all the chips produced on a given day are always either functional or defective (with 90 percent of the days resulting in functional chips). In this case, if we know that all of our 100 chips were manufactured on the same day, then X will not be a binomial random variable. This is so since the independence of successive chips is not valid. In fact, in this case, we would have

$$P\{ X = 100 \} = .9$$
$$P\{ X = 0 \} = .1 \quad \blacksquare$$

Since a binomial random variable X, with parameters n and p, represents the number of successes in n independent trials, each having success probability p, we can represent X as follows:

$$X = \sum_{i=1}^{n} X_i \qquad (3.1.3)$$

where

$$X_i = \begin{cases} 1 & \text{if the } i\text{th trial is a success} \\ 0 & \text{otherwise} \end{cases}$$

Because the X_i, $i = 1, \ldots, n$ are independent Bernoulli random variables, we have that

$$E[X_i] = P\{ X_i = 1 \} = p$$
$$\text{Var}(X_i) = E[X_i^2] - p^2$$
$$= p(1 - p)$$

where the last equality follows since $X_i^2 = X_i$ and so $E[X_i^2] = E[X_i] = p$. Using the representation Equation 3.1.3, it is now an easy matter to compute the mean and variance of X:

$$E[X] = \sum_{i=1}^{n} E[X_i]$$

$$= np$$

$$\text{Var}(X) = \sum_{i=1}^{n} \text{Var}(X_i) \qquad \text{since the } X_i \text{ are independent}$$

$$= np(1 - p)$$

If X_1 and X_2 are independent binomial random variables having respective parameters (n_i, p), $i = 1, 2$, then their sum is binomial with parameters $(n_1 + n_2, p)$. This can most easily be seen by noting that as X_i, $i = 1, 2$, represents the number of successes in n_i independent trials each of which is a success with probability p, then $X_1 + X_2$ represents the number of successes in $n_1 + n_2$ independent trials each of which is a success with probability p. Therefore, $X_1 + X_2$ is binomial with parameters $(n_1 + n_2, p)$.

1.1   Computing the Binomial Distribution Function

Suppose that X is binomial with parameters (n, p). The key to computing its distribution function

$$P\{X \le i\} = \sum_{k=0}^{i} \binom{n}{k} p^k (1 - p)^{n-k}, \qquad i = 0, 1, \ldots, n$$

is to utilize the following relationship between $P\{X = k + 1\}$ and $P\{X = k\}$:

$$P\{X = k + 1\} = \frac{p}{1 - p} \frac{n - k}{k + 1} P\{X = k\} \tag{3.1.4}$$

The proof of this equation is left as an exercise.

Example 3.1d Let X be a binomial random variable with parameters $n = 6$, $p = .4$. Then, starting with $P\{X = 0\} = (.6)^6$ and recursively employing Equation 3.1.4, we obtain

$$P\{X = 0\} = (.6)^6 = .0467$$

$$P\{X = 1\} = \tfrac{4}{6}\tfrac{6}{1} P\{X = 0\} = .1866$$

$$P\{X = 2\} = \tfrac{4}{6}\tfrac{5}{2} P\{X = 1\} = .3110$$

$$P\{X = 3\} = \tfrac{4}{6}\tfrac{4}{3} P\{X = 2\} = .2765$$

$$P\{X = 4\} = \tfrac{4}{6}\tfrac{3}{4} P\{X = 3\} = .1382$$

$$P\{X = 5\} = \tfrac{4}{6}\tfrac{2}{5} P\{X = 4\} = .0369$$

$$P\{X = 6\} = \tfrac{4}{6}\tfrac{1}{6} P\{X = 5\} = .0041. \quad \blacksquare$$

Basic Program 3-1, which is presented in the Appendix of Programs, computes $P\{X \le i\}$. This program attempts to first compute $P\{X = 0\} = (1 - p)^n$ and then utilize Equation 3.1.4 to successively compute $P\{X = 1\}, \ldots, P\{X = i\}$. However, this will be successful only for moderate values of n, since in the case of large n, due to computer round-off error, $P\{X = 0\} = (1 - p)^n$ will be computed to equal 0. If this error is made, then all subsequent terms $P\{X = k\}$, $k = 1, \ldots, i$ will also be taken to equal 0 and so the program would incorrectly conclude that $P\{X \le i\} = 0$. To guard against this possibility, in cases where $(1 - p)^n$ is computed to equal 0, the basic program is instructed to begin calculations not with $P\{X = 0\}$ but with $P\{X = J\}$ where

$$J = \begin{cases} i & \text{if } i \le np \\ [np] & \text{if } i > np \end{cases}$$

where $[np]$—called $\text{Int}(np)$ in the program—is the largest integer less than or equal to np. Of all the probabilities $P\{X = k\}$, $k = 0, 1, \ldots, i$ that need be computed, $P\{X = J\}$ will be either the largest or (at worse) second largest (see Problem 9). The program then recursively computes $P\{X = J - 1\}$, $P\{X = J - 2\}, \ldots, P\{X = 0\}$. In addition if $J < i$, it also computes $P\{X = J + 1\}, \ldots, P\{X = i\}$.

The computation of

$$P\{X = J\} = \binom{n}{J} p^J (1 - p)^{n-J}$$

$$= \frac{n(n - 1) \cdots (n - J + 1)}{J(J - 1) \cdots 1} p^J (1 - p)^{n-J}$$

is accomplished by first taking logarithms to compute

$$\log P\{X = J\} = \sum_{k=1}^{J} \log(n + 1 - k)$$

$$- \sum_{k=1}^{J} \log(k) + J \log p + (n - J) \log(1 - p)$$

and then taking

$$P\{X = J\} = \exp\{\log P\{X = J\}\}$$

Example 3.1e

(a) Compute $P\{X \le 50\}$ when $X \sim \text{Bin}(100, .4)$
(b) Compute $P\{X \le 325\}$ when $X \sim \text{Bin}(500, .7)$

Solution We run Program 3-1.

```
RUN
THIS PROGRAM COMPUTES THE PROBABILITY THAT A BINOMIAL(n,p) RANDOM VARIABLE IS LE
SS THAN OR EQUAL TO i
ENTER n
? 100
ENTER p
? .4
ENTER i
? 50
THE PROBABILITY IS .9832359
Ok
```

```
RUN
THIS PROGRAM COMPUTES THE PROBABILITY THAT A BINOMIAL(n,p) RANDOM VARIABLE IS LE
SS THAN OR EQUAL TO i
ENTER n
? 500
ENTER p
? .7
ENTER i
? 325
THE PROBABILITY IS 9.055146E-03        ■
Ok
```

2 THE POISSON RANDOM VARIABLE

A random variable X, taking on one of the values $0, 1, 2, \ldots$, is said to be a Poisson random variable with parameter λ if for some $\lambda > 0$, its probability mass function is given by

$$P\{ X = i\} = e^{-\lambda} \frac{\lambda^i}{i!} \qquad i = 0, 1, \ldots . \tag{3.2.1}$$

The symbol e stands for a constant approximately equal to 2.7183. It is a famous constant in mathematics, named after the Swiss mathematician L. Euler and it is also the base of the so-called natural logarithm.

Equation 3.2.1 defines a probability mass function, since

$$\sum_{i=0}^{\infty} p(i) = e^{-\lambda} \sum_{i=0}^{\infty} \lambda^i/i! = e^{-\lambda}e^{\lambda} = 1$$

a graph of this mass function when $\lambda = 4$ is given in Figure 3.2.1.

The Poisson probability distribution was introduced by S. D. Poisson in a book he wrote dealing with the application of probability theory to lawsuits, criminal trials, and the like. This book, published in 1837, was entitled *Recherches sur la probabilité des jugements en matière criminelle et en matière civile.*

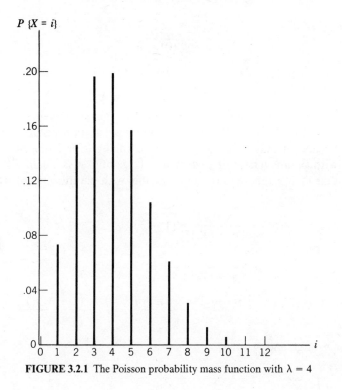

FIGURE 3.2.1 The Poisson probability mass function with $\lambda = 4$

As a prelude to obtaining the mean and variance of a Poisson random variable, let us first determine its moment generating function. This is accomplished as follows.

$$\psi(t) = E[e^{tX}]$$

$$= \sum_{i=0}^{\infty} e^{ti} e^{-\lambda} \frac{\lambda^i}{i!}$$

$$= e^{-\lambda} \sum_{i=0}^{\infty} \frac{(\lambda e^t)^i}{i!}$$

$$= e^{-\lambda} e^{\lambda e^t}$$

$$= \exp\left\{\lambda(e^t - 1)\right\}$$

Differentiation yields

$$\psi'(t) = \lambda e^t \exp\left\{\lambda(e^t - 1)\right\}$$

$$\psi''(t) = (\lambda e^t)^2 \exp\left\{\lambda(e^t - 1)\right\} + \lambda e^t \exp\left\{\lambda(e^t - 1)\right\}$$

Therefore,

$$E[X] = \psi'(0) = \lambda$$

$$\text{Var}(X) = \psi''(0) - (E[X])^2$$

$$= \lambda^2 + \lambda - \lambda^2 = \lambda$$

Hence, the mean and variance are both equal to the parameter λ.

The Poisson random variable has a wide range of applications in a variety of areas because it may be used as an approximation for a binomial random variable with parameters (n, p) when n is large and p is small. To see this, suppose that X is a binomial random variable with parameters (n, p) and let $\lambda = np$. Then

$$P\{X = i\} = \frac{n!}{(n-i)!i!} p^i (1-p)^{n-i}$$

$$= \frac{n!}{(n-i)!i!} \left(\frac{\lambda}{n}\right)^i \left(1 - \frac{\lambda}{n}\right)^{n-i}$$

$$= \frac{n(n-1)\cdots(n-i+1)}{n^i} \frac{\lambda^i}{i!} \frac{(1-\lambda/n)^n}{(1-\lambda/n)^i}$$

Now, for n large and p small

$$\left(1 - \frac{\lambda}{n}\right)^n \approx e^{-\lambda} \qquad \frac{n(n-1)\cdots(n-i+1)}{n^i} \approx 1 \qquad \left(1 - \frac{\lambda}{n}\right)^i \approx 1$$

Hence, for n large and p small

$$P\{X = i\} \approx e^{-\lambda} \frac{\lambda^i}{i!}$$

In other words, if n independent trials, each of which results in a "success" with probability p, are performed, then when n is large and p small, the number of successes occurring is approximately a Poisson random variable with mean $\lambda = np$.

Some examples of random variables that usually obey, to a good approximation, the Poisson probability law (that is, they usually obey Equation 3.2.1 for some value of λ) are:

1. The number of misprints on a page (or a group of pages) of a book.
2. The number of people in a community living to 100 years of age.
3. The number of wrong telephone numbers that are dialed in a day.
4. The number of transistors that fail on their first day of use.
5. The number of customers entering a post office on a given day.
6. The number of α-particles discharged in a fixed period of time from some radioactive particle.

Each of the foregoing, and numerous other random variables, are approximately Poisson for the same reason—namely, because of the Poisson approximation to the binomial. For instance, we can suppose that there is a small probability p that each letter typed on a page will be misprinted, and so the number of misprints on a given page will be approximately Poisson with mean $\lambda = np$ where n is the (presumably) large number of letters on that page. Similarly, we can suppose that each person in a given community, independently, has a small probability p of reaching the age 100, and so the number of people that do will approximately have a Poisson distribution with mean np where n is the large number of people in the community. We leave it for the reader to reason out why the remaining random variables in (3) through (6) should approximately have a Poisson distribution.

Example 3.2a Suppose that the average number of accidents occurring weekly on a particular stretch of a highway equals 3. Calculate the probability that there is at least one accident this week.

Solution Let X denote the number of accidents occurring on the stretch of highway in question during this week. As it is reasonable to suppose that there are a large number of cars passing along that stretch, each having a small probability of being involved in an accident, the number of such accidents should be approximately Poisson distributed. Hence,

$$P\{X \geq 1\} = 1 - P\{X = 0\}$$

$$= 1 - e^{-3}\frac{3^0}{0!}$$

$$= 1 - e^{-3}$$

$$\approx .9502 \quad \blacksquare$$

Example 3.2b Suppose the probability that an item produced by a certain machine will be defective is .1. Find the probability that a sample of 10 items will contain at most one defective item. Assume that the quality of successive items is independent.

Solution The desired probability is $\binom{10}{0}(.1)^0(.9)^{10} + \binom{10}{1}(.1)^1(.9)^9 = .7361$, whereas the Poisson approximation yields the value

$$e^{-1}\frac{1^0}{0!} + e^{-1}\frac{1^1}{1!} = 2e^{-1} \approx .7358 \quad \blacksquare$$

Example 3.2c Consider an experiment that consists of counting the number of alpha particles given off in a one-second interval by one gram of radioactive material. If we know from past experience that, on the average, 3.2 such α-particles are given off, what is a good approximation to the probability that no more than 2 α-particles will appear?

Solution If we think of the gram of radioactive material as consisting of a large number n of atoms each of which has probability $3.2/n$ of disintegrating and sending off an α-particle during the second considered, then we see that, to a very close approximation, the number of α-particles given off will be a Poisson random variable with parameter $\lambda = 3.2$. Hence the desired probability is

$$P\{X \leq 2\} = e^{-3.2} + 3.2e^{-3.2} + \frac{(3.2)^2}{2}e^{-3.2}$$

$$= .382 \quad \blacksquare$$

Example 3.2d If the average number of claims handled daily by an insurance company is 5, what proportion of days have less than 3 claims? What is the probability that there will be 4 claims in exactly 3 of the next 5 days? Assume that the number of claims on different days is independent.

Solution As the company probably insures a large number of clients, each having a small probability of making a claim on any given day, it is reasonable to suppose that the number of claims handled daily, call it X, is a Poisson random variable. Since $E(X) = 5$, the probability that there will be fewer than 3 claims on any given day is

$$P\{X < 3\} = P\{X = 0\} + P\{X = 1\} + P\{X = 2\}$$

$$= e^{-5} + e^{-5}\frac{5^1}{1!} + e^{-5}\frac{5^2}{2!}$$

$$= \frac{37}{2}e^{-5}$$

$$\approx .1247$$

Since any given day will have fewer than 3 claims with probability .125, it follows, from the law of large numbers, that over the long run $12\frac{1}{2}$ percent of days will have fewer than 3 claims.

It follows from the assumed independence of the number of claims over successive days that the number of days in a 5-day span that have exactly 4 claims is a binomial random variable with parameters 5 and $P\{X = 4\}$. Because

$$P\{X = 4\} = e^{-5}\frac{5^4}{4!} \approx .1755$$

it follows that the probability that 3 of the next 5 days will have 4 claims is equal to

$$\binom{5}{3}(.1755)^3(.8245)^2 \approx .0367 \quad \blacksquare$$

The Poisson approximation result can be shown to be valid under even more general conditions than those so far mentioned. For instance, suppose that n independent trials are to be performed, with the ith trial resulting in a success with probability p_i, $i = 1, \ldots, n$. Then it can be shown that if n is large and each p_i is small, then the number of successful trials is approximately Poisson distributed with mean equal to $\sum_{i=1}^{n} p_i$. In fact, this result will sometimes remain true even when the trials are not independent, provided that their dependence is "weak." For instance, consider the following example.

Example 3.2e At a party n people put their hats in the center of a room, where they are mixed together. Each person then randomly chooses a hat. If X denotes the number of people that select their own hat, then for large n, it can be shown that X has approximately a Poisson distribution with mean 1. To see why this might be true, let

$$X_i = \begin{cases} 1 & \text{if the } i\text{th person selects his own hat} \\ 0 & \text{otherwise} \end{cases}$$

Then we can express X as

$$X = X_1 + \cdots + X_n$$

and so X can be regarded as representing the number of "successes" in n "trials" where trial i is said to be a success if the ith person chooses his own hat. Now, since the ith person is equally likely to end up with any of the n hats, one of which is his own, it follows that

$$P\{ X_i = 1 \} = \frac{1}{n} \tag{3.2.2}$$

Suppose now that $i \neq j$ and consider the conditional probability that the ith person chooses his own hat given that the jth person does—that is, consider $P\{ X_i = 1 | X_j = 1 \}$. Now given that the jth person indeed selects his own hat, it follows that the ith individual is equally likely to end up with any of the remaining $n - 1$, one of which is his own. Hence, it follows that

$$P\{ X_i = 1 | X_j = 1 \} = \frac{1}{n - 1} \tag{3.2.3}$$

Thus, we see from Equations 3.2.2 and 3.2.3 that whereas the trials are not independent, their dependence is rather weak (since, if the above conditional probability were equal to $1/n$ rather than $1/(n - 1)$, then trials i and j would be independent); and thus it is not at all surprising that X has approximately a Poisson distribution. The fact that $E[X] = 1$ follows since

$$E[X] = E[X_1 + \cdots + X_n]$$

$$= E[X_1] + \cdots + E[X_n]$$

$$= n\left(\frac{1}{n}\right) = 1$$

The last equality following since, from Equation 3.2.2,

$$E[X_i] = P\{ X_i = 1 \} = \frac{1}{n} \quad \blacksquare$$

The Poisson distribution possesses the reproductive property that the sum of independent Poisson random variables is also a Poisson random variable. To see this, suppose that X_1 and X_2 are independent Poisson random variables having respective means λ_1 and λ_2. Then the moment generating function of $X_1 + X_2$ is s follows:

$$E[e^{t(X_1+X_2)}] = E[e^{tX_1}e^{tX_2}]$$

$$= E[e^{tX_1}]E[e^{tX_2}] \qquad \text{by independence}$$

$$= \exp\{\lambda_1(e^t - 1)\} \exp\{\lambda_2(e^t - 1)\}$$

$$= \exp\{(\lambda_1 + \lambda_2)(e^t - 1)\}$$

As $\exp\{(\lambda_1 + \lambda_2)(e^t - 1)\}$ is the moment generating function of a Poisson random variable having mean $\lambda_1 + \lambda_2$, we may conclude, from the fact that the moment generating function uniquely specifies the distribution, that $X_1 + X_2$ is Poisson with mean $\lambda_1 + \lambda_2$.

Example 3.2f It has been established that the number of defective stereos produced daily at a certain plant is Poisson distributed with mean 4. Over a 2-day span, what is the probability that the number of defective stereos does not exceed 3?

Solution Assuming that X_1, the number of defectives produced during the first day, is independent of X_2, the number produced during the second day, then $X_1 + X_2$ is Poisson with mean 8. Hence,

$$P\{X_1 + X_2 \le 3\} = \sum_{i=0}^{3} e^{-8} \frac{8^i}{i!} = .04238 \quad \blacksquare$$

2.1 Computing the Poisson Distribution Function

If X is Poisson with mean λ, then

$$\frac{P\{X = i + 1\}}{P\{X = i\}} = \frac{e^{-\lambda}\lambda^{i+1}/(i+1)!}{e^{-\lambda}\lambda^i/i!} = \frac{\lambda}{i+1} \tag{3.2.4}$$

Starting with $P\{X = 0\} = e^{-\lambda}$, we can use Equation 3.2.4 to successively compute

$$P\{X = 1\} = \lambda P\{X = 0\}$$

$$P\{X = 2\} = \frac{\lambda}{2} P\{X = 1\}$$

$$\vdots$$

$$P\{X = i + 1\} = \frac{\lambda}{i+1} P\{X = i\}$$

Basic Program 3-2 computes, by using Equation 3.2.4, the probability that a Poisson random variable with a given mean does not exceed i for any i. It starts the computation with $P\{X = 0\} = e^{-\lambda}$. However, if as will be the case when λ is large, the computer returns the value 0 for $e^{-\lambda}$, the program calls for beginning with $P\{X = J\}$ where

$$J = \begin{cases} i & \text{if } i \leq \lambda \\ \text{Int}\,(\lambda) & \text{if } i > \lambda \end{cases}$$

where $\text{Int}\,(\lambda)$ is the largest integer less than or equal to λ. The reason for this choice is that of all the values $P\{X = k\}$, $k = 0, 1, \ldots, i$ that we must compute, the largest one is $P\{X = J\}$ (see Problem 16). The program computes $P\{X = J\}$ by first computing

$$\log P\{X = J\} = -\lambda + J \log(\lambda) - \sum_{k=1}^{J} \log k$$

and then takes $P\{X = J\} = \exp\{\log P\{X = J\}\}$.

Once $P\{X = J\}$ has been computed, the program uses Equations 3.2.4 to recursively compute $P\{X = J - 1\}$, $P\{X = J - 2\}, \ldots, P\{X = 0\}$. If $J < i$ it again uses Equation 3.2.4 to recursively compute $P\{X = J + 1\}, \ldots,$ $P\{X = i\}$. Summing all of these quantities yields $P\{X \leq i\}$.

Example 3.2g

(a) Compute $P\{X \leq 90\}$ when $X \sim \text{Poisson}\,(100)$.
(b) Compute $P\{X \leq 1087\}$ when $X \sim \text{Poisson}\,(1000)$.

Solution We run Program 3-2:

```
RUN
THIS PROGRAM COMPUTES THE PROBABILITY THAT A POISSON RANDOM VARIABLE
IS LESS THAN OR EQUAL TO i
ENTER THE MEAN OF THE RANDOM VARIABLE
? 100
ENTER THE DESIRED VALUE OF i
? 90
THE PROBABILITY THAT A POISSON RANDOM VARIABLE WITH MEAN   100
IS LESS THAN OR EQUAL TO 90 IS .1713914
Ok
```

```
RUN
THIS PROGRAM COMPUTES THE PROBABILITY THAT A POISSON RANDOM VARIABLE
IS LESS THAN OR EQUAL TO i
ENTER THE MEAN OF THE RANDOM VARIABLE
? 1000
ENTER THE DESIRED VALUE OF i
? 1087
THE PROBABILITY THAT A POISSON RANDOM VARIABLE WITH MEAN   1000
IS LESS THAN OR EQUAL TO 1087 IS .9952581
Ok
```

3 THE HYPERGEOMETRIC RANDOM VARIABLE

A bin contains $N + M$ batteries of which N are of acceptable quality and the other M are defective. A sample of size n is to be randomly chosen (without replacements) in the sense that the set of sampled batteries is equally likely to be any of the $\binom{N+M}{n}$ subsets of size n. If we let X denote the number of acceptable batteries in the sample, then

$$P\{X = i\} = \frac{\binom{N}{i}\binom{M}{n-i}}{\binom{N+M}{n}}, \qquad i = 0, 1, \ldots, \min(N, n) \qquad (3.3.1)$$

Any random variable X whose probability mass function is given by Equation 3.3.1 is said to be a *hypergeometric* random variable with parameters N, M, n.

To compute the mean and variance of a hypergeometric random variable whose probability mass function is given by Equation 3.3.1 imagine that the batteries are drawn sequentially and let

$$X_i = \begin{cases} 1 & \text{if the } i\text{th selection is acceptable} \\ 0 & \text{otherwise} \end{cases}$$

Now, since the ith selection is equally likely to be any of the $N + M$ batteries, of which N are acceptable, it follows that

$$P\{X_i = 1\} = \frac{N}{N + M} \qquad (3.3.2)$$

Also, for $i \neq j$,

$$P\{X_i = 1, \quad X_j = 1\} = P\{X_i = 1\}P\{X_j = 1 | X_i = 1\}$$

$$= \frac{N}{N + M}\frac{N - 1}{N + M - 1} \qquad (3.3.3)$$

which follows since, given that the ith selection is acceptable, the jth selection is equally likely to be any of the other $N + M - 1$ batteries of which $N - 1$ are acceptable.

To compute the mean and variance of X, the number of acceptable batteries in the sample of size n, use the representation

$$X = \sum_{i=1}^{n} X_i$$

This gives

$$E[X] = \sum_{i=1}^{n} E[X_i] = \sum_{i=1}^{n} P\{X_i = 1\} = \frac{nN}{N + M} \qquad (3.3.4)$$

Also, Corollary 2.7.3 of Chapter 2 for the variance of a sum of random variables gives

$$\text{Var}(X) = \sum_{i=1}^{n} \text{Var}(X_i) + 2 \sum\sum_{1 \le i < j \le n} \text{Cov}(X_i, X_j) \qquad (3.3.5)$$

Now, X_i is a Bernoulli random variable and so

$$\text{Var}(X_i) = P\{X_i = 1\}(1 - P\{X_i = 1\}) = \frac{N}{N+M}\frac{M}{N+M} \qquad (3.3.6)$$

Also, for $i < j$,

$$\text{Cov}(X_i, X_j) = E[X_i X_j] - E[X_i]E[X_j]$$

Now, as both X_i and X_j are Bernoulli (that is $0 - 1$) random variables, it follows that $X_i X_j$ is a Bernoulli random variable, and so

$$\begin{aligned} E[X_i X_j] &= P\{X_i X_j = 1\} \\ &= P\{X_i = 1, \quad X_j = 1\} \\ &= \frac{N(N-1)}{(N+M)(N+M-1)} \qquad \text{from Equation 3.3.3} \quad (3.3.7) \end{aligned}$$

So from Equation 3.3.2 and the foregoing we see that for $i \neq j$

$$\begin{aligned} \text{Cov}(X_i, X_j) &= \frac{N(N-1)}{(N+M)(N+M-1)} - \left(\frac{N}{N+M}\right)^2 \\ &= \frac{-NM}{(N+M)^2(N+M-1)} \end{aligned}$$

Hence, since there are $\binom{n}{2}$ terms in the second sum on the right side of Equation 3.3.5, we obtain from Equation 3.3.6

$$\begin{aligned} \text{Var}(X) &= \frac{nNM}{(N+M)^2} - \frac{n(n-1)NM}{(N+M)^2(N+M-1)} \\ &= \frac{nNM}{(N+M)^2}\left(1 - \frac{n-1}{N+M-1}\right) \qquad (3.3.8) \end{aligned}$$

If we let $p = N/(N+M)$ denote the proportion of acceptable batteries in the bin, we can rewrite Equations 3.4 and 3.8 as

$$E(X) = np$$

$$\text{Var}(X) = np(1-p)\left[1 - \frac{n-1}{N+M-1}\right]$$

It should be noted that, for fixed p, as $N + M$ increases to ∞, $\text{Var}(X)$ converges to $np(1 - p)$, which is the variance of a binomial random variable with parameters (n, p). (Why was this to be expected?)

Example 3.3a An unknown number, say N, of animals inhabit a certain region. To obtain some information about the population size, ecologists often perform the following experiment: They first catch a number, say r, of these animals, mark them in some manner, and release them. After allowing the marked animals time to disperse throughout the region, a new catch of size, say, n is made. Let X denote the number of marked animals in this second capture. If we assume that the population of animals in the region remained fixed between the time of the two catches and that each time an animal was caught it was equally likely to be any of the remaining uncaught animals, it follows that X is a hypergeometric random variable such that

$$P\{X=i\} = \frac{\binom{r}{i}\binom{N-r}{n-i}}{\binom{N}{n}} \equiv P_i(N)$$

Suppose now that X is observed to equal i. That is, the fraction i/n of the animals in the second catch were marked. By taking this as an approximation of r/N, the proportion of animals in the region that are marked, we obtain the estimate rn/i of the number of animals in the region. For instance if $i = 50$ animals are initially caught, marked, and then released; and a subsequent catch of $n = 100$ animals revealed $X = 25$ of them that were marked then we would estimate the number of animals in the region to be about 200 ∎

There is a relationship between binomial random variables and the hypergeometric distribution that will be useful to us in developing a statistical test concerning two binomial populations.

Example 3.3b Let X and Y be independent binomial random variables having respective parameters (n, p) and (m, p). The conditional probability mass function of X given that $X + Y = k$ is obtained as follows:

$$P\{X=i|X+Y=k\} = \frac{P\{X=i, \quad X+Y=k\}}{P\{X+Y=k\}}$$

$$= \frac{P\{X=i, \quad Y=k-i\}}{P\{X+Y=k\}}$$

$$= \frac{P\{X=i\}P\{Y=k-i\}}{P\{X+Y=k\}}$$

$$= \frac{\binom{n}{i}p^i(1-p)^{n-i}\binom{m}{k-i}p^{k-i}(1-p)^{m-(k-i)}}{\binom{n+m}{k}p^k(1-p)^{n+m-k}}$$

$$= \frac{\binom{n}{i}\binom{m}{k-i}}{\binom{n+m}{k}}$$

where the next to last equality used the fact that $X + Y$ is binomial with parameters $(n + m, p)$. Hence, we see that the conditional distribution of X given the value of $X + Y$ is hypergeometric. (The fact that this distribution does not depend on p is, as will be seen, significant for statistical applications.) ∎

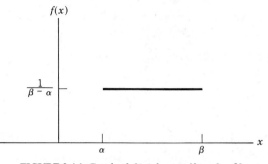

FIGURE 3.4.1 Graph of $f(x)$ for a uniform $[\alpha, \beta]$

4 THE UNIFORM RANDOM VARIABLE

A random variable X is said to be uniformly distributed over the interval $[\alpha, \beta]$ if its probability density function is given by

$$f(x) = \begin{cases} \dfrac{1}{\beta - \alpha} & \text{if } \alpha \le x \le \beta \\ 0 & \text{otherwise} \end{cases}$$

A graph of this function is given in Figure 3.4.1. Note that the foregoing meets the requirements of being a probability density function since

$$\frac{1}{\beta - \alpha} \int_\alpha^\beta dx = 1$$

The uniform distribution arises in practice when we suppose a certain random variable is equally likely to be near any value in the interval $[\alpha, \beta]$.

The probability that X lies in any subinterval of $[\alpha, \beta]$ is equal to the length of that subinterval divided by the length of the interval $[\alpha, \beta]$. This follows since when $[a, b]$ is a subinterval of $[\alpha, \beta]$

$$P\{a < X < b\}$$

$$= \frac{1}{\beta - \alpha} \int_a^b dx$$

$$= \frac{b - a}{\beta - \alpha}$$

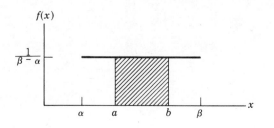

Example 3.4a If X is uniformly distributed over the interval $[0, 10]$, compute the probability that (a) $2 < X < 9$, (b) $1 < X < 4$, (c) $X < 5$, (d) $X > 6$.

Solution The respective answers are (a) $7/10$, (b) $3/10$, (c) $5/10$, (d) $4/10$. ∎

Example 3.4b Buses arrive at a specified stop at 15-minute intervals starting at 7 A.M. That is, they arrive at 7, 7:15, 7:30, 7:45, and so on. If a passenger arrives at the stop at a time that is uniformly distributed between 7 and 7:30, find the probability that he waits

(a) less than 5 minutes for a bus.

(b) at least 12 minutes for a bus.

Solution Let X denote the time in minutes past 7 A.M. that the passenger arrives at the stop. Since X is a uniform random variable over the interval $(0, 30)$, it follows that the passenger will have to wait less than 5 minutes if he arrives between 7:10 and 7:15 or between 7:25 and 7:30. Hence, the desired probability for (a) is

$$P\{10 < X < 15\} + P\{25 < X < 30\} = \tfrac{5}{30} + \tfrac{5}{30} = \tfrac{1}{3}$$

Similarly, he would have to wait at least 12 minutes if he arrives between 7 and 7:03 or between 7:15 and 7:18, and so the probability for (b) is

$$P\{0 < X < 3\} + P\{15 < X < 18\} = \tfrac{3}{30} + \tfrac{3}{30} = \tfrac{1}{5} \quad ∎$$

The mean of a uniform $[\alpha, \beta]$ random variable is

$$E[X] = \int_{\alpha}^{\beta} \frac{x}{\beta - \alpha} \, dx$$

$$= \frac{\beta^2 - \alpha^2}{2(\beta - \alpha)}$$

$$= \frac{(\beta - \alpha)(\beta - \alpha)}{2(\beta - \alpha)}$$

or

$$E[X] = \frac{\alpha + \beta}{2}$$

Or, in other words, the expected value of a uniform $[\alpha, \beta]$ random variable is equal to the midpoint of the interval $[\alpha, \beta]$, which is clearly what one would expect (why?).

The variance is computed as follows.

$$E[X^2] = \frac{1}{\beta - \alpha} \int_{\alpha}^{\beta} x^2 \, dx$$

$$= \frac{\beta^3 - \alpha^3}{3(\beta - \alpha)}$$

$$= \frac{\beta^2 + \alpha\beta + \alpha^2}{3}$$

TABLE 3.4.1

A Random Number Table

0.68587	0.25848	0.85227	0.78724	0.05302	0.70712	0.76552	0.70326	0.80402	0.49479
0.73253	0.41629	0.37913	0.00236	0.60196	0.59048	0.59946	0.75657	0.61849	0.90181
0.84448	0.42477	0.94829	0.86678	0.14030	0.04072	0.45580	0.36833	0.10783	0.33199
0.49564	0.98590	0.92880	0.69970	0.83898	0.21077	0.71374	0.85967	0.20857	0.51433
0.68304	0.46922	0.14218	0.63014	0.50116	0.33569	0.97793	0.84637	0.27681	0.04354
0.76992	0.70179	0.75568	0.21792	0.50646	0.07744	0.38064	0.06107	0.41481	0.93919
0.37604	0.27772	0.75615	0.51157	0.73821	0.29928	0.62603	0.06259	0.21552	0.72977
0.43898	0.06592	0.44474	0.07517	0.44831	0.01337	0.04538	0.15198	0.50345	0.65288
0.86039	0.28645	0.44931	0.59203	0.98254	0.56697	0.55897	0.25109	0.47585	0.59524
0.28877	0.84966	0.97319	0.66633	0.71350	0.28403	0.28265	0.61379	0.13886	0.78325
0.44973	0.12332	0.16649	0.88908	0.31019	0.33358	0.68401	0.10177	0.92873	0.13065
0.42529	0.37593	0.90208	0.50331	0.37531	0.72208	0.42884	0.07435	0.58647	0.84972
0.82004	0.74696	0.10136	0.35971	0.72014	0.08345	0.49366	0.68501	0.14135	0.15718
0.67090	0.08493	0.47151	0.06464	0.14425	0.28381	0.40455	0.87302	0.07135	0.04507
0.62825	0.83809	0.37425	0.17693	0.69327	0.04144	0.00924	0.68246	0.48573	0.24647
0.10720	0.89919	0.90448	0.80838	0.70997	0.98438	0.51651	0.71379	0.10830	0.69984
0.69854	0.89270	0.54348	0.22658	0.94233	0.08889	0.52655	0.83351	0.73627	0.39018
0.71460	0.25022	0.06988	0.64146	0.69407	0.39125	0.10090	0.08415	0.07094	0.14244
0.69040	0.33461	0.79399	0.22664	0.68810	0.56303	0.65947	0.88951	0.40180	0.87943
0.13452	0.36642	0.98785	0.62929	0.88509	0.64690	0.38981	0.99092	0.91137	0.02411
0.94232	0.91117	0.98610	0.71605	0.89560	0.92921	0.51481	0.20016	0.56769	0.60462
0.99269	0.98876	0.47254	0.93637	0.83954	0.60990	0.10353	0.13206	0.33480	0.29440
0.75323	0.86974	0.91355	0.12780	0.01906	0.96412	0.61320	0.47629	0.33890	0.22099
0.75003	0.98538	0.63622	0.94890	0.96744	0.73870	0.72527	0.17745	0.01151	0.47200

and so

$$\mathrm{Var}(X) = \frac{\beta^2 + \alpha\beta + \alpha^2}{3} - \left(\frac{\alpha + \beta}{2}\right)^2$$

$$= \frac{\alpha^2 + \beta^2 - 2\alpha\beta}{12}$$

$$= \frac{(\beta - \alpha)^2}{12}$$

The value of a uniform $(0, 1)$ random variable is called a *random number*. Most computer systems have a built-in subroutine for generating (to a high level of approximation) sequences of independent random numbers—for instance, Table 3.4.1 presents a set of independent random numbers generated by an IBM Personal Computer. Random numbers are quite useful in probability and statistics because their use enables one to empirically estimate various probabilities and expectations. The following example illustrates how random numbers can be used to solve a problem in combinatorics.

Example 3.4c Estimating the Number of Distinct Entries in a Large List
Consider a list of n entries where n is very large and suppose we are interested in estimating d, the number of distinct elements in the list. If we let $m(i)$

denote the number of times that the element in position i appears on the list, then we can express d by

$$d = \sum_{i=1}^{n} 1/m(i)$$

For instance, if $n = 9$ and the entries are $3, 2, 1, 2, 4, 1, 5, 1, 3$, then $m(1) = m(9) = 2$, $m(2) = m(4) = 2$, $m(3) = m(6) = m(8) = 3$, $m(5) = 1$, $m(7) = 1$, and $d = \sum_{i=1}^{9} 1/m(i) = 2(\frac{1}{2}) + 2(\frac{1}{2}) + 3(\frac{1}{3}) + 1 + 1 = 5$.

To estimate d, suppose that we generate a random variable X that is equally likely to be either $1, 2, \ldots, n$—that is, X is such that

$$P\{X = i\} = \frac{1}{n} \qquad i = 1, \ldots, n$$

Then, with $m(X)$ equaling the number of times that the entry in position X appears on the list, we have that

$$E\left[\frac{1}{m(X)}\right] = \sum_{i=1}^{n} \frac{1}{m(i)} P\{X = i\}$$

$$= \frac{1}{n} \sum_{i=1}^{n} \frac{1}{m(i)} = \frac{d}{n}$$

Therefore, if we let

$$Y = n/m(X)$$

then

$$E[Y] = d$$

From the foregoing it follows that if we generate k independent random variables $X_1, X_2, \ldots, X_k$, each being equally likely to take on any of the values $1, 2, \ldots, n$, and set $Y_i = n/m(X_i)$, then $Y_1, \ldots, Y_k$ are independent and identically distributed random variables with mean equal to d. From the law of large numbers, it now follows that, for k large, $\sum_{i=1}^{k} Y_i/k$ will approximately equal d. That is,

$$d \approx \frac{n}{k} \sum_{i=1}^{k} 1/m(X_i)$$

and so we can approximate d by determining $m(X_1), \ldots, m(X_k)$ rather than having to determining $m(i)$ for each i.

It remains only to show how one can generate a random variable X that is equally likely to be either $1, 2, \ldots, n$. To do this, we start by letting U denote a random number [that is, U is uniformly distributed over $(0, 1)$]. Then nU is uniform on $(0, n)$ and so

$$P\{i - 1 < nU < i\} = \frac{1}{n} \qquad i = 1, \ldots, n$$

Hence, if we set

$$X = i \qquad \text{if } i - 1 < nU < i \qquad\qquad (3.4.1)$$

then X has the appropriate distribution. We can more succinctly express X in terms of U by using the greatest integer notation [ ], where $[x]$ is defined to be the greatest integer less than or equal to x. Using this notation we have that Equation 3.4.1 may be written as

$$X = [nU] + 1$$

So, summing up, we can estimate d by generating k random numbers $U_1, \ldots, U_k$; setting $X_i = [nU_i] + 1$, and then approximating d by

$$d \approx \frac{n}{k} \sum_{i=1}^{k} 1/m(X_i) \quad \blacksquare$$

For another illustration of the use of random numbers, suppose that a medical center is planning to test a new drug designed to reduce its user's blood cholesterol level. To test its effectiveness, the medical center has recruited 1000 volunteers to be subjects in the test. To take into account the possibility that the subjects blood cholesterol levels may be affected by factors external to the test (such as changing weather conditions), it has been decided to split the volunteers into 2 groups of size 500—a *treatment* group that will be given the drug and a *control* that will be given a placebo. Both the volunteers and the administrators of the drug will not be told who is in each group (such a test is called *double-blind*). It remains to determine which of the volunteers should be chosen to constitute the treatment group. Clearly, one would want the treatment group and the control group to be as similar as possible in all respects with the exception that members in the first group are to receive the drug while those in the other group receive a placebo; for then it would be possible to conclude that any difference in response between the groups is indeed due to the drug. There is general agreement that the best way to accomplish this is to choose the 500 volunteers to be in the treatment group in a completely random fashion. That is, the choice should be made so that each of the $\binom{1000}{500}$ subsets of 500 volunteers is equally likely to constitute the control group. How can this be accomplished?

Example 3.4d Choosing a Random Subset From a set of n elements—numbered $1, 2, \ldots, n$—suppose we want to generate a random subset of size k that is to be chosen in such a manner so that each of the $\binom{n}{k}$ subsets is equally likely to be the subset chosen. How can we do this?

To answer this question, let us work backwards and suppose that we have indeed randomly generated such a subset of size k. Now for each $j = 1, \ldots, n$ set

$$I_j = \begin{cases} 1 & \text{if element } j \text{ is in the subset} \\ 0 & \text{otherwise} \end{cases}$$

and let us compute the conditional distribution of I_j given $I_1, \ldots, I_{j-1}$. To start, note that the probability that element 1 is in the subset of size k is clearly k/n (which can be seen either by noting that there is probability $1/n$

that element 1 would have been the jth element chosen, $j = 1, \ldots, k$; or by noting that the proportion of outcomes of the random selection that result in element 1 being chosen is $\binom{1}{1}\binom{n-1}{k-1} / \binom{n}{k} = k/n$). Therefore, we have that

$$P\{I_1 = 1\} = k/n \qquad (3.4.2)$$

To compute the conditional probability that element 2 is in the subset given I_1, note that if $I_1 = 1$, then aside from element 1 the remaining $k - 1$ members of the subset would have been chosen "at random" from the remaining $n - 1$ elements (in the sense that each of the subsets of size $k - 1$ of the numbers $2, \ldots, n$ is equally likely to be the other elements of the subset). Hence, we have that

$$P\{I_2 = 1 | I_1 = 1\} = \frac{k - 1}{n - 1} \qquad (3.4.3)$$

Similarly, if element 1 is not in the subgroup, then the k members of the subgroup would have been chosen "at random" from the other $n - 1$ elements, and thus

$$P\{I_2 = 1 | I_1 = 0\} = \frac{k}{n - 1} \qquad (3.4.4)$$

From Equations 3.4.3 and 3.4.4, we see that

$$P\{I_2 = 1 | I_1\} = \frac{k - I_1}{n - 1}$$

In general, we have that

$$P\{I_j = 1 | I_1, \ldots, I_{j-1}\} = \frac{k - \sum_{i=1}^{j-1} I_i}{n - j + 1}, \qquad j = 2, \ldots, n \qquad (3.4.5)$$

The preceding formula follows since $\sum_{i=1}^{j-1} I_i$ represents the number of the first $j - 1$ elements that are included in the subset, and so given $I_1, \ldots, I_{j-1}$ there remain $k - \sum_{i=1}^{j-1} I_i$ elements to be selected from the remaining $n - (j - 1)$.

Since $P\{U < a\} = a$, $0 \le a \le 1$, when U is a uniform $(0,1)$ random variable, Equations 3.4.2 and 3.4.5 lead to the following method for generating a random subset of size k from a set of n elements: namely, generate a sequence of (at most n) random numbers $U_1, U_2, \ldots$ and set

$$I_1 = \begin{cases} 1 & \text{if } U_1 < \dfrac{k}{n} \\ 0 & \text{otherwise} \end{cases}$$

$$I_2 = \begin{cases} 1 & \text{if } U_2 < \dfrac{k - I_1}{n - 1} \\ 0 & \text{otherwise} \end{cases}$$

$$\vdots$$

$$I_j = \begin{cases} 1 & \text{if } U_j < \dfrac{k - I_1 - \cdots - I_{j-1}}{n - j + 1} \\ 0 & \text{otherwise} \end{cases}$$

This process stops when $I_1 + \cdots + I_j = k$ and the random subset consists of the k elements whose I-value equals 1. That is, $S = \{i : I_i = 1\}$ is the subset.

For instance, if $k = 2$, $n = 5$, then the following tree diagram illustrates the foregoing technique. The random subset S is given by the final position on the tree.

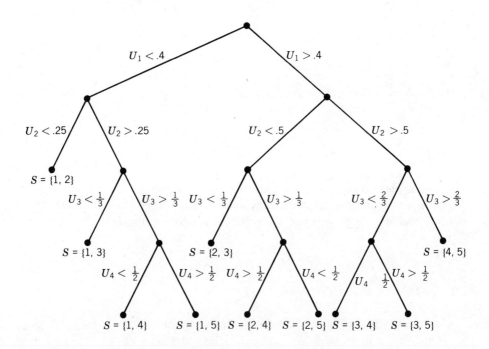

Note that the probability of ending up in any given final position is equal to $1/10$, which can be seen by multiplying the probabilities of moving through the tree to the desired endpoint. For instance, the probability of ending at the point labeled $S = \{2, 4\}$ is $P\{U_1 > .4\}P\{U_2 < .5\}P\{U_3 > \frac{1}{3}\}P\{U_4 > \frac{1}{2}\} = (.6)(.5)(\frac{2}{3})(\frac{1}{2}) = .1$.

As indicated in the tree diagram (see the rightmost branches that result in $S = \{4, 5\}$), we can stop generating random numbers when the number of remaining places in the subset to be chosen is equal to the remaining number of elements. That is, the general procedure would stop whenever either $\sum_{i=1}^{j} I_i = k$ or $\sum_{i=1}^{j} I_i = k - (n - j)$. In the latter case, $S = \{i \leq j : I_i = 1, j + 1, \ldots, n\}$. ∎

REMARK

The foregoing method for generating a random subset has a very low memory requirement. A faster algorithm that requires somewhat more memory is presented in Section 1 of Chapter 12. (This latter algorithm uses the first k elements of a random permutation of $1, 2, \ldots, n$).

Program 3-4, which uses the preceding algorithm to generate a random subset, is presented in the Appendix of Programs.

Example 3.4e Suppose we wanted to generate a random subset of size 5 from the intergers $1, \ldots, 25$. To do so we run Program 3-4:

```
RUN
THIS PROGRAM GENERATES A RANDOM SUBSET OF SIZE K FROM THE SET 1,2,...N
ENTER THE VALUE OF N
? 25
ENTER THE VALUE OF K
? 5
Random number seed (-32768 to 32767)? 4762
THE RANDOM SUBSET CONSISTS OF THE FOLLOWING 5 VALUES
                7
               10
               13
               16
               19
     Ok
```

The reader should note that when the Program calls for a random number seed input, the user should input any number within the indicated limits. This input enables the program to use different random numbers each time it is run.

5   NORMAL RANDOM VARIABLES

A random variable is said to be normally distributed with mean μ and variance σ^2, and we write $X \sim \mathcal{N}(\mu, \sigma^2)$, if its density is

$$ f(x) = \frac{1}{\sqrt{2\pi}\,\sigma}\, e^{-(x-\mu)^2/2\sigma^2} \qquad -\infty < x < \infty^\dagger $$

The normal density $f(x)$ is a bell-shaped curve that is symmetric about μ and that attains its maximum value of $1/\sigma\sqrt{2\pi} \approx 0.399/\sigma$ at $x = \mu$ (see Figure 3.5.1).

The normal distribution was introduced by the French mathematician Abraham de Moivre in 1733 and was used by him to approximate probabilities associated with binomial random variables when the binomial parameter n is large. This result was later extended by Laplace and others and is now encompassed in a probability theorem known as the central limit theorem, which gives a theoretical base to the often noted empirical observation that, in practice, many random phenomena obey, at least approximately, a normal probability distribution. Some examples of this behavior are the height of a person, the velocity in any direction of a molecule in gas, and the error made in measuring a physical quantity.

---

† To verify that this is indeed a density function, see Problem 29.

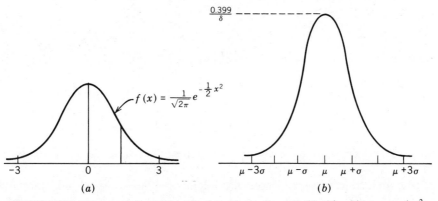

FIGURE 3.5.1 The normal density function (a) with $\mu = 0$, $\sigma = 1$ (b) with arbitrary μ and σ^2

The moment generating function of a normal random variable with parameters μ and σ^2 is derived as follows:

$$\phi(t) = E[e^{tX}] \tag{3.5.1}$$

$$= \frac{1}{\sqrt{2\pi}\,\sigma} \int_{-\infty}^{\infty} e^{tx} e^{-(x-\mu)^2/2\sigma^2} \, dx$$

$$= \frac{1}{\sqrt{2\pi}} e^{\mu t} \int_{-\infty}^{\infty} e^{t\sigma y} e^{-y^2/2} \, dy \qquad \text{by letting } y = \frac{x - \mu}{\sigma}$$

$$= \frac{e^{\mu t}}{\sqrt{2\pi}} \int_{-\infty}^{\infty} \exp\left\{ -\left[\frac{y^2 - 2t\sigma y}{2} \right] \right\} \, dy$$

$$= \frac{e^{\mu t}}{\sqrt{2\pi}} \int_{-\infty}^{\infty} \exp\left\{ -\frac{(y - t\sigma)^2}{2} + \frac{t^2\sigma^2}{2} \right\} \, dy$$

$$= \exp\left\{ \mu t + \frac{\sigma^2 t^2}{2} \right\} \frac{1}{\sqrt{2\pi}} \int_{-\infty}^{\infty} e^{-(y - t\sigma)^2/2} \, dy$$

$$= \exp\left\{ \mu t + \frac{\sigma^2 t^2}{2} \right\}$$

where the last equality follows since

$$\frac{1}{\sqrt{2\pi}} e^{-(y - t\sigma)^2/2}$$

is the density of a normal random variable (having parameters $t\sigma$ and 1) and its integral must thus equal 1.

Upon differentiating Equation 3.5.1, we obtain

$$\phi'(t) = (\mu + t\sigma^2) \exp\left\{ \mu t + \sigma^2 \frac{t^2}{2} \right\}$$

$$\phi''(t) = \sigma^2 \exp\left\{ \mu t + \sigma^2 \frac{t^2}{2} \right\} + \exp\left\{ \mu t + \sigma^2 \frac{t^2}{2} \right\} (\mu + t\sigma^2)^2$$

Hence,

$$E[X] = \phi'(0) = \mu$$

$$E[X^2] = \phi''(0) = \sigma^2 + \mu^2$$

and so

$$E[X] = \mu$$

$$\text{Var}(X) = E[X^2] - (E[X])^2 = \sigma^2$$

Thus μ and σ^2 represent respectively the mean and variance of the distribution.

An important fact about normal random variables is that if X is normally distributed with parameters μ and σ^2, then $Y = \alpha X + \beta$ is normally distributed with parameters $\alpha\mu + \beta$ and $\alpha^2\sigma^2$. This can easily be seen by moment generating functions as follows:

$$E\left[e^{t(\alpha X + \beta)}\right] = e^{t\beta}E\left[e^{\alpha t X}\right]$$

$$= e^{t\beta}\exp\left\{\mu\alpha t + \sigma^2(\alpha t)^2/2\right\} \qquad \text{from Equation 3.5.1}$$

$$= \exp\left\{t\beta + \mu\alpha t + \sigma^2\frac{\alpha^2 t^2}{2}\right\}$$

$$= \exp\left\{(\beta + \alpha\mu)t + \alpha^2\sigma^2\frac{t^2}{2}\right\}$$

As the last expression is the moment generating function of a normal random variable with parameters $\beta + \alpha\mu$ and $\alpha^2\sigma^2$, the result follows.

It follows from the foregoing that if $X \sim \mathcal{N}(\mu, \sigma^2)$, then

$$Z = \frac{X - \mu}{\sigma}$$

is a normal random variable with mean 0 and variance 1. Such a random variable Z is said to have a *standard*, or *unit*, normal distribution. Let $\Phi(\cdot)$ denote its distribution function. That is,

$$\Phi(x) = \frac{1}{\sqrt{2\pi}}\int_{-\infty}^{x} e^{-y^2/2}\,dy \qquad -\infty < x < \infty$$

This result that $Z = (X - \mu)/\sigma$ has a standard normal distribution when X is normal with parameters μ and σ^2 is quite important for it enables us to write all probability statements about X in terms of probabilities for Z. For instance, to obtain $P\{X < b\}$, we note that X will be less than b if and only if $(X - \mu)/\sigma$ is less than $(b - \mu)/\sigma$, and so

$$P\{X < b\} = P\left\{\frac{X - \mu}{\sigma} < \frac{b - \mu}{\sigma}\right\}$$

$$= \Phi\left(\frac{b - \mu}{\sigma}\right)$$

Similarly, for any $a < b$,

$$P\{a < X < b\} = P\left\{\frac{a - \mu}{\sigma} < \frac{X - \mu}{\sigma} < \frac{b - \mu}{\sigma}\right\}$$

$$= P\left\{\frac{a - \mu}{\sigma} < Z < \frac{b - \mu}{\sigma}\right\}$$

$$= P\left\{Z < \frac{b - \mu}{\sigma}\right\} - P\left\{Z < \frac{a - \mu}{\sigma}\right\}$$

$$= \Phi\left(\frac{b - \mu}{\sigma}\right) - \Phi\left(\frac{a - \mu}{\sigma}\right)$$

It remains to compute $\Phi(x)$. This has been accomplished by an approximation and the results are presented in Table A3-5-1 of the Appendix, which tabulates $\Phi(x)$ (to a 4-digit level of accuracy) for a wide range of nonnegative values of x. In addition, a basic program for the computation is outlined at the end of this section and presented in the Appendix of Programs.

While Table A3-5-1 tabulates $\Phi(x)$ only for nonnegative values of x, we can also obtain $\Phi(-x)$ from the table by making use of the symmetry (about 0) of the unit normal probability density function. That is, for $x > 0$, if Z represents a standard normal random variable, then

$$\Phi(-x) = P\{Z < -x\}$$

$$= P\{Z > x\} \qquad \text{by symmetry}$$

$$= 1 - \Phi(x)$$

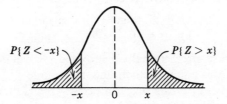

Thus, for instance,

$$P\{Z < -1\} = \Phi(-1) = 1 - \Phi(1) = 1 - .8413 = .1587$$

Example 3.5a If X is a normal random variable with mean $\mu = 3$ and variance $\sigma^2 = 16$, find

1. $P\{X < 11\}$
2. $P\{X > -1\}$
3. $P\{2 < X < 7\}$

Solution

$$P\{X < 11\} = P\left\{\frac{X - 3}{4} < \frac{11 - 3}{4}\right\}$$

$$= \Phi(2)$$

$$= .9772$$

$$P\{X > -1\} = P\left\{\frac{X - 3}{4} > \frac{-1 - 3}{4}\right\}$$

$$= P\{Z > -1\}$$

$$= P\{Z < 1\}$$

$$= .8413$$

$$P\{2 < X < 7\} = P\left\{\frac{2 - 3}{4} < \frac{X - 3}{4} < \frac{7 - 3}{4}\right\}$$

$$= \Phi(1) - \Phi(-1/4)$$

$$= \Phi(1) - (1 - \Phi(1/4))$$

$$= .8413 + .5987 - 1 = .4400 \quad \blacksquare$$

Example 3.5b Suppose that a binary message—either "0" or "1"—must be transmitted by wire from location A to location B. However, the data sent over the wire is subject to a channel noise disturbance and so to reduce the possibility of error, the value 2 is sent over the wire when the message is "1" and the value -2 is sent when the message is "0." If x, $x = \pm 2$, is the value sent at location A then R, the value received at location B, is given by $R = x + N$, where N is the channel noise disturbance. When the message is received at location B, the receiver decodes it according to the following rule:

if $R \geq .5$, then "1" is concluded

if $R < .5$, then "0" is concluded

As the channel noise is often normally distributed, we will determine the error probabilities when N is a unit normal random variable.

There are two types of errors that can occur: One is that the message "1" can be incorrectly concluded to be "0" and the other that "0" is concluded to be "1." The first type of error will occur if the message is "1" and $2 + N < .5$, whereas the second will occur if the message is "0" and $-2 + N \geq .5$.

Hence,

$$P\{\text{error}|\text{message is "1"}\} = P\{N < -1.5\}$$

$$= 1 - \Phi(1.5) = .0668$$

and

$$P\{\text{error}|\text{message is "0"}\} = P\{N > 2.5\}$$

$$= 1 - \Phi(2.5) = .0062 \quad \blacksquare$$

Another important result is that the sum of independent normal random variables remains normal. To see this, suppose that X_i, $i = 1, \ldots, n$, are independent and X_i is normal with mean μ_i and variance σ_i^2. Then the moment generating function of $\sum_{i=1}^{n} X_i$ is as follows:

$$E\left[e^{t\sum_{i=1}^{n} X_i}\right] = E\left[e^{tX_1} e^{tX_2} \ldots e^{tX_n}\right]$$

$$= \prod_{i=1}^{n} E\left[e^{tX_i}\right] \qquad \text{by independence}$$

$$= \prod_{i=1}^{n} \left(e^{\mu_i t + \sigma_i^2 t^2/2}\right)$$

$$= e^{\mu t + \sigma^2 t^2/2}$$

where

$$\mu \equiv \sum_{i=1}^{n} \mu_i, \qquad \sigma^2 \equiv \sum_{i=1}^{n} \sigma_i^2$$

Therefore, $\sum_{i=1}^{n} X_i$ have the same moment generating function as a normal random variable having mean μ and variance σ^2. Hence, from the one-to-one correspondence between moment generating functions and distributions, we can conclude that $\sum_{i=1}^{n} X_i$ is normal.

For $\alpha \in (0, 1)$ let z_α be such that

$$P\{Z > z_\alpha\} = 1 - \Phi(z_\alpha) = \alpha$$

That is, a unit normal will be greater than z_α with probability α (see Figure 3.5.2). The value of z_α can, for any α, be obtained from Table A3.5.1. For instance, since

$$1 - \Phi(1.64) = .05$$

$$1 - \Phi(1.96) = .025$$

$$1 - \Phi(2.33) = .01$$

it follows that $z_{.05} = 1.64$, $z_{.025} = 1.96$, $z_{.01} = 2.33$.

Program 3-5-1-B, which approximates z_α to a 4-digit level of accuracy, is presented in the Appendix of Programs.

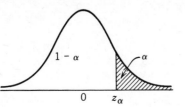

FIGURE 3.5.2 $P\{Z > z_\alpha\} = \alpha$

5.1 Computing the Unit Normal Distribution Function and Its Inverse

Program 3-5-1-A approximates the integral from 0 to x, $x > 0$, of the function $e^{-y^2/2}$ by integrating, term by term, the first 40 terms of the Taylor series expansion

$$e^{-y^2/2} = \sum_{i=0}^{\infty} \left(-y^2/2\right)^i / i!$$

It then computes the unit normal distribution function $\Phi(x)$ by noting that

$$\Phi(x) = \begin{cases} \frac{1}{2} + \dfrac{1}{\sqrt{2\pi}} \displaystyle\int_0^x e^{-y^2/2}\, dy & \text{if } x > 0 \\[2ex] \frac{1}{2} - \dfrac{1}{\sqrt{2\pi}} \displaystyle\int_0^{-x} e^{-y^2/2}\, dy & \text{if } x < 0 \end{cases}$$

Program 3-5-1-B computes the approximate value of z_α for any desired value of α.

Example 3.5c

  (a) Compute $\Phi(2.12)$.
  (b) Compute $\Phi(-1.64)$.

Solution Run Program 3-5-1-A.

```
RUN
THIS PROGRAM COMPUTES THE PROBABILITY THAT A UNIT NORMAL RANDOM VARIABLE IS LESS
   THAN X
ENTER THE DESIRED VALUE OF X
? 2.12
THE PROBABILITY IS .9829972
Ok
```

```
RUN
THIS PROGRAM COMPUTES THE PROBABILITY THAT A UNIT NORMAL RANDOM VARIABLE IS LESS
   THAN X
ENTER THE DESIRED VALUE OF X
? -1.64
THE PROBABILITY IS 5.050236E-02
Ok
```

REMARK

A quick approximation to the normal distribution function, that is valid to 2 decimal places is the following[†]:

$$\frac{1}{\sqrt{2\pi}} \int_0^x e^{-y^2/2}\, dy \approx \begin{cases} x(4.4 - x)/10 & 0 \le x \le 2.2 \\ .49 & 2.2 < x < 2.6 \\ .50 & 2.6 \le x \end{cases}$$

[†]Shah, A., "A Simpler Approximation for Areas under the Standard Normal Curve," *American Statistician*, Vol. 39, No. 1, Feb. 1985.

Example 3.5d Compute $z_{.05}$ and $z_{.09}$.

Solution Run Program 3-5-1-B twice:

```
RUN
FOR A GIVEN INPUT a, 0<a<.5, THIS PROGRAM COMPUTES THE VALUE z  SUCH  THAT THE
PROBABILITY THAT A UNIT NORMAL EXCEEDS z IS EQUAL TO a
ENTER THE DESIRED VALUE OF a
? .05
THE VALUE IS 1.645212
Ok
RUN
FOR A GIVEN INPUT a, 0<a<.5, THIS PROGRAM COMPUTES THE VALUE z  SUCH  THAT THE
PROBABILITY THAT A UNIT NORMAL EXCEEDS z IS EQUAL TO a
ENTER THE DESIRED VALUE OF a
? .09
THE VALUE IS 1.340969
Ok
```

5.2   The Central Limit Theorem

In this section we will consider one of the most remarkable results in probability—namely, the *central limit theorem*. Loosely speaking, this theorem asserts that the sum of a large number of independent random variables has a distribution that is approximately normal. Hence, it not only provides a simple method for computing approximate probabilities for sums of independent random variables, but it also helps explain the remarkable fact that the empirical frequencies of so many natural populations exhibit a bell-shaped (that is, a normal) curve.

In its simplest form the central limit theorem is as follows:

Theorem 3.5.1. The Central Limit Theorem
Let $X_1, X_2, \ldots, X_n$ be a sequence of independent and identically distributed random variables each having mean μ and variance σ^2. Then for n large, the distribution of

$$\frac{X_1 + \cdots + X_n - n\mu}{\sigma\sqrt{n}}$$

is approximately that of a unit normal random variable. That is, for n large,

$$P\left\{\frac{X_1 + \cdots + X_n - n\mu}{\sigma\sqrt{n}} < x\right\} \approx P\{Z < x\}$$

where Z is a unit normal random variable. ■

Example 3.5e An astronomer is interested in measuring, in light years, the distance from his observatory to a distant star. Although the astronomer has a measuring technique, he knows that, due to changing atmospheric conditions and normal error, each time a measurement is made it will not yield the exact distance but merely an estimate of it. As a result the astronomer plans to make

a series of measurements and then use the average value of these measurements as his estimated value of the actual distance. If the astronomer believes
that the values of each of the measurements are independent and identically
distributed random variables having a common mean d (the actual distance)
and a common variance of 4 light years, how many measurements need he
make to be reasonably sure that his estimated distance is accurate to within
± 0.5 light years?

Solution Suppose the astronomer decides to make n observations. If
$X_1, X_2, \ldots, X_n$ are the n measurements, then, from the central limit theorem,
it follows that

$$Z_n = \frac{\sum_{i=1}^{n} X_i - nd}{2\sqrt{n}}$$

has approximately a unit normal distribution. Now the astronomer wants to
choose the number n of measurements so as to be reasonably certain that

$$-0.5 < \frac{\sum_{i=1}^{n} X_i}{n} - d < 0.5$$

However,

$$P\left\{ -0.5 < \frac{\sum_{i=1}^{n} X_i}{n} - d < 0.5 \right\}$$

$$= P\left\{ -0.5 < \frac{2}{\sqrt{n}} Z_n < 0.5 \right\}$$

$$= P\left\{ -0.5 \frac{\sqrt{n}}{2} < Z_n < 0.5 \frac{\sqrt{n}}{2} \right\}$$

$$\approx P\left\{ Z < \frac{\sqrt{n}}{4} \right\} - P\left\{ Z < -\frac{\sqrt{n}}{4} \right\}$$

$$= P\left\{ Z < \frac{\sqrt{n}}{4} \right\} - P\left\{ Z > \frac{\sqrt{n}}{4} \right\}$$

$$= 2P\left\{ Z < \frac{\sqrt{n}}{4} \right\} - 1$$

Therefore, if the astronomer wanted, for instance, to be 95 percent certain that
his estimated value is accurate to within 0.5 light years, he should make n^*
measurements where n^* is such that

$$2P\left\{ Z < \sqrt{n^*}/4 \right\} - 1 = .95$$

or

$$P\left\{ Z < \sqrt{n^*}/4 \right\} = .975$$

and thus from Table A3.5.1, he should choose n^* so that

$$\frac{\sqrt{n^*}}{4} = 1.96$$

or

$$n^* = (7.84)^2 = 64.47$$

Since n^* is not integral valued, he should make 65 observations. ∎

Example 3.5f Civil engineers believe that W, the amount of weight (in units of 1000 pounds) that a certain span of a bridge can withstand without structural damage resulting, is normally distributed with mean 400 and standard deviation 40. Suppose that the weight (again, in units of 1000 pounds) of a car is a random variable with mean 3 and standard deviation .3. How many cars would have to be on the bridge span for the probability of structural damage to exceed .1?

Solution Let P_n denote the probability of structural damage when there are n cars on the bridge. That is,

$$P_n = P\{ X_1 + \cdots + X_n \geq W \}$$
$$= P\{ X_1 + \cdots + X_n - W \geq 0\}$$

where X_i is the weight of the ith car, $i = 1, \ldots, n$. Now it follows from the central limit theorem that $\sum_{i=1}^n X_i$ is approximately normal with mean $3n$ and variance $.09n$. Hence, since W is independent of the X_i, $i = 1, \ldots, n$, and is also normal, it follows that $\sum_{i=1}^n X_i - W$ is approximately normal, with mean and variance given by

$$E\left[\sum_1^n X_i - W \right] = 3n - 400$$

$$\text{Var}\left(\sum_1^n X_i - W \right) = \text{Var}\left(\sum_1^n X_i \right) + \text{Var}(W) = .09n + 1600$$

Therefore, if we let

$$Z = \frac{\sum_{i=1}^n X_i - W - (3n - 400)}{\sqrt{.09n + 1600}}$$

then

$$P_n = P\left\{ Z \geq \frac{-(3n - 400)}{\sqrt{.09n + 1600}} \right\}$$

where Z is approximately a unit normal random variable. Now $P\{ Z \geq 1.28\} \approx .1$, and so if the number of cars n is such that

$$\frac{400 - 3n}{\sqrt{.09n + 1600}} \leq 1.28$$

or

$$n \geq 117$$

then there is at least 1 chance in 10 that structural damage will occur. ∎

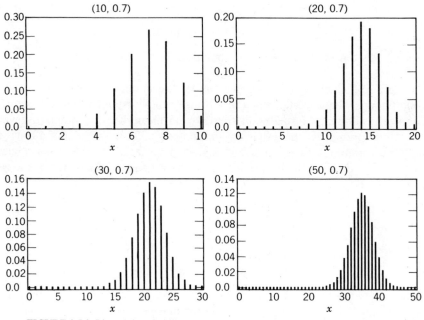

FIGURE 3.5.3 Binomial probability mass functions converging to the normal density

One of the most important applications of the central limit theorem is in regard to binomial random variables. Since such a random variable X having parameters (n, p) represents the number of successes in n independent trials when each trial is a success with probability p, we can express it as

$$X = X_1 + \cdots + X_n$$

where

$$X_i = \begin{cases} 1 & \text{if the } i\text{th trial is a success} \\ 0 & \text{otherwise} \end{cases}$$

Because

$$E[X_i] = p, \qquad \text{Var}(X_i) = p(1 - p)$$

it follows, from the central limit theorem that for n large

$$\frac{X - np}{\sqrt{np(1 - p)}}$$

will be approximately a unit normal random variable [see Figure 3.5.3, which graphically illustrates how the probability mass function of a binomial (n, p) random variable becomes more and more "normal" as n becomes larger and larger].

Example 3.5g The ideal size of a first-year class at a particular college is 150 students. The college, knowing from past experience that, on the average, only

30 percent of those accepted for admission will actually attend, uses a policy of approving the applications of 450 students. Compute the probability that more than 150 first-year students attend this college.

Solution Let X denote the number of students that attend; then X is a binomial random variable with parameters $n = 450$ and $p = .3$. Since the binomial is a discrete and the normal a continuous distribution, it is best to compute $P\{X = i\}$ as $P\{i - .5 < X < i + .5\}$ when applying the normal approximation (this is called the continuity correction). This yields the approximation

$$P\{X > 150.5\} = P\left\{ \frac{X - (450)(.3)}{\sqrt{450(.3)(.7)}} \geq \frac{150.5 - (450)(.3)}{\sqrt{450(.3)(.7)}} \right\}$$

$$\approx P\{Z > 1.65\} = .0495$$

Hence, only 5 percent of the time do more than 150 of the first 450 accepted actually attend. ∎

It should be noted that we now have two possible approximations to binomial probabilities: the Poisson approximation, which yields a good approximation when n is large and p small, and the normal approximation, which can be shown to be quite good when $np(1 - p)$ is large. [The normal approximation will, in general, be quite good for values of n satisfying $np(1 - p) \geq 10$.]

Example 3.5h The central limit theorem provides a method for simulating normal random variables. If $U_1, \ldots, U_{12}$ represent independent uniform $(0, 1)$ random variables, then because

$$E[U_i] = 1/2, \quad \text{Var}(U_i) = 1/12$$

it follows that

$$Z \equiv \sum_{i=1}^{12} U_i - 6$$

will have mean 0 and variance 1. From the central limit theorem, Z should also be approximately normally distributed—and in fact it can be shown that summing 12 uniform $(0, 1)$ random variables gives rise to a random variable that for all practical purposes can be regarded as being normally distributed. Hence, Z can be regarded as the value of a unit normal random variable.

A normal random variable X with mean μ and variance σ^2 can be generated by first subtracting 6 from the sum of 12 random numbers (to obtain a unit normal random variable Z) and then setting $X = \sigma Z + \mu$. That is, if $U_1, \ldots, U_{12}$ represent the random numbers, then

$$X = \sigma \sum_{i=1}^{12} U_i + \mu$$

approximately has a normal distribution with mean μ and variance σ^2. ∎

6 EXPONENTIAL RANDOM VARIABLES

A continuous random variable whose probability density function is given, for some $\lambda > 0$, by

$$f(x) = \begin{cases} \lambda e^{-\lambda x} & \text{if } x \geq 0 \\ 0 & \text{if } x < 0 \end{cases}$$

is said to be an *exponential* random variable (or, more simply, is said to be exponentially distributed) with parameter λ. The cumulative distribution function $F(x)$ of an exponential variable is given by

$$F(x) = P\{X \leq x\}$$

$$= \int_0^x \lambda e^{-\lambda y} \, dy$$

$$= 1 - e^{-\lambda x}, \qquad x \geq 0$$

The exponential distribution often arises, in practice, as being the distribution of the amount of time until some specific event occurs. For instance, the amount of time (starting from now) until an earthquake occurs, or until a new war breaks out, or until a telephone call you receive turns out to be a wrong number are all random variables that tend in practice to have exponential distributions (see Section 6.1 for an explanation).

The moment generating function of the exponential is given by

$$\Psi(t) = E[e^{tX}]$$

$$= \int_0^\infty e^{tx} \lambda e^{-\lambda x} \, dx$$

$$= \lambda \int_0^\infty e^{-(\lambda - t)x} \, dx$$

$$= \frac{\lambda}{\lambda - t} \qquad t < \lambda$$

Differentiation yields

$$\Psi'(t) = \frac{\lambda}{(\lambda - t)^2}$$

$$\Psi''(t) = \frac{2\lambda}{(\lambda - t)^3}$$

and so

$$E[X] = \Psi'(0) = \frac{1}{\lambda}$$

$$\text{Var}(X) = E[X^2] - E^2[X]$$

$$= \Psi''(0) - \frac{1}{\lambda^2} = \frac{1}{\lambda^2}$$

Hence λ is the reciprocal of the mean.

The key property of an exponential random variable is that it is memory-less, where we say that a nonnegative random variable X is *memoryless* if

$$P\{X > s + t | X > t\} = P\{X > s\} \qquad \text{for all } s, t \geq 0 \qquad (3.6.1)$$

To understand why Equation 3.6.1 is called the *memoryless property*, imagine that X represents the length of time that a certain item functions before failing. Now let us consider the probability that an item that is still function-ing at time t will continue to function for at least an additional time s. Since this will be the case if the total functional lifetime of the item exceeds $t + s$ given that the item is still functioning at t, we see that

$$P\{\text{additional functional life of } t \text{ unit old item exceeds } s\}$$

$$= P\{X > t + s | X > t\}$$

Thus, we see that Equation 3.6.1 states that the distribution of additional functional life of an item of age t is the same as that of a new item—in, other words, when Equation 3.6.1 is satisfied, there is no need to remember the age of a functional item since as long as it is still functional it is "as good as new."

The condition in Equation 3.6.1 is equivalent to

$$\frac{P\{X > s + t, \quad X > t\}}{P\{X > t\}} = P\{X > s\}$$

or

$$P\{X > s + t\} = P\{X > s\} P\{X > t\} \qquad (3.6.2)$$

When X is an exponential random variable, then

$$P\{X > x\} = e^{-\lambda x}, \qquad x > 0$$

and so Equation 3.6.2 is satisfied (since $e^{-\lambda(s+t)} = e^{-\lambda s} e^{-\lambda t}$). Hence, *exponen-tially distributed random variables are memoryless* (and in fact it can be shown then they are the only random variables that are memoryless).

Example 3.6a Suppose that a number of miles that a car can run before its battery wears out is exponentially distributed with an average value of 10,000 miles. If a person desires to take a 5000-mile trip, what is the probability that he will be able to complete his trip without having to replace the car battery? What can be said when the distribution is not exponential?

Solution It follows, by the memoryless property of the exponential distribu-tion, that the remaining lifetime (in thousands of miles) of the battery is exponential with parameter $\lambda = 1/10$. Hence the desired probability is

$$P\{\text{remaining lifetime} > 5\} = 1 - F(5)$$

$$= e^{-5\lambda}$$

$$= e^{-1/2} \approx .604$$

However, if the lifetime distribution F is not exponential, then the relevant probability is

$$P\{\text{lifetime} > t + 5 | \text{lifetime} > t\} = \frac{1 - F(t + 5)}{1 - F(t)}$$

where t is the number of miles that the battery had been in use prior to the start of the trip. Therefore, if the distribution is not exponential, additional information is needed (namely, t) before the desired probability can be calculated. ■

For another illustration of the memoryless property, consider the following example.

Example 3.6b A crew of workers have 3 interchangeable machines of which 2 must be working for the crew to do their job. When in use each machine will function for an exponentially distributed time having parameter λ before breaking down. The workers decide to initially use machines A and B and keep machine C in reserve to replace whichever of A and B first breaks down. They will then be able to continue working until one of the remaining machines break down. When the crew is forced to stop working because only one of the machines has not yet broken down, what is the probability that the still operable machine is machine C?

This can be easily answered, without any need for computations, by invoking the memoryless property of the exponential distribution. The argument is as follows: Consider the moment at which machine C is first put in use. At that time either A or B would have just broken down and the other one —call it machine 0—will still be functioning. Now even though 0 would have already been functioning for some time, by the lack of memory property of the exponential distribution, it follows that its remaining lifetime has the same distribution as that of a machine that is just being put into use. Thus, the remaining lifetimes of machine 0 and machine C have the same distribution and so, by symmetry, the probability that 0 will fail before C is $1/2$. ■

The following proposition presents another property of the exponential distribution

Proposition 3.6.1

If $X_1, X_2, \ldots, X_n$ are independent exponential random variables having respective parameters $\lambda_1, \lambda_2, \ldots, \lambda_n$, then $\min(X_1, X_2, \ldots, X_n)$ is exponential with parameter $\sum_{i=1}^{n} \lambda_i$.

Proof Since the smallest value of a set of numbers is greater than x if and only if all values are greater than x, we have

$$P\{\min(X_1, X_2, \ldots, X_n) > x\} = P\{X_1 > x, X_2 > x, \ldots, X_n > x\}$$

$$= \prod_{i=1}^{n} P\{X_i > x\} \qquad \text{by independence}$$

$$= \prod_{i=1}^{n} e^{-\lambda_i x}$$

$$= e^{-\sum_{i=1}^{n} \lambda_i x}$$

Example 3.6c A series system is one that needs all of its components to function in order for the system itself to be functional. For an n-component series system in which the component lifetimes are independent exponential random variables with respective parameters $\lambda_1, \lambda_2, \ldots, \lambda_n$, what is the probability that the system survives for a time t?

Solution Since the system life is equal to the minimal component life, it follows from Proposition 3.6.1 that

$$P\{\text{system life exceeds } t\} = e^{-\Sigma_i \lambda_i t} \quad \blacksquare$$

Another useful property of exponential random variables is that cX is exponential with parameter λ/c when X is exponential with parameter λ, and $c > 0$. This follows since

$$P\{cX \le x\} = P\{X \le x/c\}$$

$$= 1 - e^{-\lambda x/c}.$$

The parameter λ is called the *rate* of the distribution.

6.1 The Poisson Process

Suppose that "events" are occurring at random time points and let $N(t)$ denote the number of events that occur in the time interval $[0, t]$. These events are said to constitute a *Poisson process having rate* λ, $\lambda > 0$ if

(a) $N(0) = 0$
(b) The number of events that occur in disjoint time intervals are independent.
(c) The distribution of the number of events that occur in a given interval depends only on the length of the interval and not on its location.
(d) $\lim_{h \to 0} \dfrac{P\{N(h) = 1\}}{h} = \lambda$
(e) $\lim_{h \to 0} \dfrac{P\{N(h) \ge 2\}}{h} = 0$

Thus, Condition (a) states that the process begins at time 0. Condition (b), the *independent increment* assumption, states for instance that the number of events by time t [that is $N(t)$] is independent of the number of events that occur between t and $t + s$ [that is $N(t + s) - N(t)$]. Condition (c), the *stationary increment* assumption, states that the probability distribution of $N(t + s) - N(t)$ is the same for all values of t. Conditions (d) and (e) state that in a small interval of length h, the probability of one event occurring is approximately λh, whereas the probability of 2 or more is approximately 0.

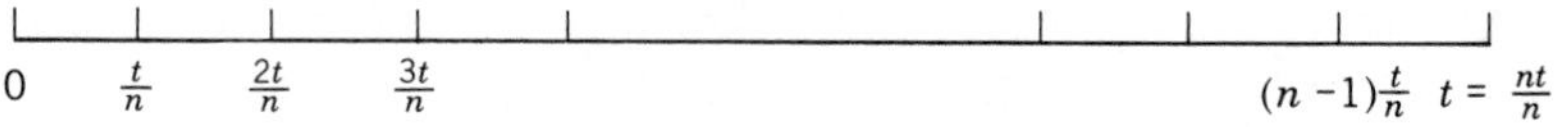

FIGURE 3.6.1

We will now show that these assumptions imply that the number of events occurring in any interval of length t is a Poisson random variable with parameter λt. To be precise, let us call the interval $[0, t]$ and denote by $N(t)$ the number of events occurring in that interval. To obtain an expression for $P\{N(t) = k\}$, we start by breaking the interval $[0, t]$ into n nonoverlapping subintervals each of length t/n (Figure 3.6.1). Now there will be k events in $[0, t]$ if either

   (a) $N(t)$ equals k and there is at most one event in each subinterval
   (b) $N(t)$ equals k and at least one of the subintervals contains 2 or more events

Since these two possibilities are clearly mutually exclusive, and since (a) is equivalent to the statement that k of the n subintervals contain exactly 1 event and the other $n - k$ contain 0 events, we have that

$$P\{N(t) = k\} = P\{k \text{ of the } n \text{ subintervals contain exactly 1} \qquad (3.6.3)$$

$$\text{event and the other } n - k \text{ contain 0 events}\}$$

$$+ P\{N(t) = k \text{ and at least 1 subinterval contains}$$

$$2 \text{ or more events}\}$$

Now it can be shown, using Condition (e), that

$$P\{N(t) = k \text{ and at least 1 subinterval contains 2 or} \qquad (3.6.4)$$

$$\text{more events}\} \to 0 \text{ as } n \to \infty$$

Also, it follows from Conditions (d) and (e) that

$$P\{\text{exactly 1 event in a subinterval}\} \approx \frac{\lambda t}{n}$$

$$P\{0 \text{ events in a subinterval}\} \approx 1 - \frac{\lambda t}{n}$$

Hence, since the number of events in the different subintervals are independent [from Condition (b)], it follows that

$$P\{k \text{ of the subintervals contain exactly 1 event} \qquad (3.6.5)$$

$$\text{and the other } n - k \text{ contain 0 events}\}$$

$$\approx \binom{n}{k}\left(\frac{\lambda t}{n}\right)^k\left(1 - \frac{\lambda t}{n}\right)^{n-k}$$

with the approximation becoming exact as the number of subintervals, n, goes to ∞. However, the probability in Equation 3.6.5 is just the probability that a binomial random variable with parameters n and $p = \lambda t/n$ equals k. Hence, as n becomes larger and larger, this approaches the probability that a Poisson random variable with mean $n\lambda t/n = \lambda t$ equals k. Hence, from Equations 3.6.3, 3.6.4, and 3.6.5 we see upon letting n approach ∞ that

$$P\{N(t) = k\} = e^{-\lambda t}\frac{(\lambda t)^k}{k!}$$

We have shown:

Proposition 3.6.2

For a Poisson process having rate λ

$$P\{N(t) = k\} = e^{-\lambda t}\frac{(\lambda t)^k}{k!}, \qquad k = 0, 1, \ldots$$

That is, the number of events in any interval of length t has a Poisson distribution with mean λt.

For a Poisson process let us denote by X_1, the time of the first event. Further, for $n > 1$, let X_n denote the elapsed time between $(n - 1)$st and the nth event. The sequence $\{X_n, n = 1, 2, \ldots\}$ is called the *sequence of inter-arrival times*. For instance, if $X_1 = 5$ and $X_2 = 10$, then the first event of the Poisson process would have occurred at time 5 and the second at time 15.

We shall now determine the distribution of the X_n. To do so, we first note that the event $\{X_1 > t\}$ takes place if and only if no events of the Poisson process occur in the interval $[0, t]$ and thus,

$$P\{X_1 > t\} = P\{N(t) = 0\} = e^{-\lambda t}$$

Hence, X_1 has an exponential distribution with mean $1/\lambda$. To obtain the distribution of X_2, note that

$$P\{X_2 > t | X_1 = s\} = P\{0 \text{ events in } (s, \quad s + t] | X_1 = s\}$$
$$= P\{0 \text{ events in } (s, \quad s + t]\}$$
$$= e^{-\lambda t}$$

where the last two equations followed from independent and stationary increments. Therefore, from the foregoing we conclude that X_2 is also an exponential random variable with mean $1/\lambda$, and furthermore, that X_2 is independent of X_1. Repeating the same argument yields:

Proposition 3.6.3

$X_1, X_2, \ldots$ are independent exponential random variables each with mean $1/\lambda$.

7   THE GAMMA DISTRIBUTION

A random variable is said to have a gamma distribution with parameters
(α, λ), $\lambda > 0$, $\alpha > 0$, if its density function is given by

$$f(x) = \begin{cases} \dfrac{\lambda e^{-\lambda x}(\lambda x)^{\alpha-1}}{\Gamma(\alpha)} & x \geq 0 \\ 0 & x < 0 \end{cases}$$

where

$$\Gamma(\alpha) = \int_0^\infty \lambda e^{-\lambda x}(\lambda x)^{\alpha-1} \, dx$$

$$= \int_0^\infty e^{-y} y^{\alpha-1} \, dy \qquad \text{(by letting } y = \lambda x)$$

The integration by parts formula $\int u \, dv = uv - \int v \, du$ yields, with $u = y^{\alpha-1}$,
$dv = e^{-y} \, dy$, $v = -e^{-y}$, that

$$\int_0^\infty e^{-y} y^{\alpha-1} \, dy = \left. -e^{-y} y^{\alpha-1} \right|_{y=0}^{y=\infty} + \int_0^\infty e^{-y}(\alpha-1) y^{\alpha-2} \, dy$$

$$= (\alpha - 1) \int_0^\infty e^{-y} y^{\alpha-2} \, dy$$

or

$$\Gamma(\alpha) = (\alpha - 1)\Gamma(\alpha - 1) \qquad (3.7.1)$$

When α is an integer—say, $\alpha = n$—we can iterate the foregoing to obtain that

$\Gamma(n) = (n - 1)\Gamma(n - 1)$

$\qquad = (n - 1)(n - 2)\Gamma(n - 2) \qquad \text{by letting } \alpha = n - 1 \text{ in Equation 3.7.1}$

$\qquad = (n - 1)(n - 2)(n - 3)\Gamma(n - 3) \quad \text{by letting } \alpha = n - 2 \text{ in Equation 3.7.1}$

$$\vdots$$

$\qquad = (n - 1)!\Gamma(1)$

As

$$\Gamma(1) = \int_0^\infty e^{-y} \, dy = 1$$

we see that

$$\Gamma(n) = (n - 1)!$$

The function $\Gamma(\alpha)$ is called the *gamma* function.

It should be noted that when $\alpha = 1$, the gamma distribution reduces to the
exponential with mean $1/\lambda$.

The moment generating function of a gamma random variable X with parameters (α, λ) is obtained as follows:

$$\Psi(t) = E[e^{tX}] \tag{3.7.2}$$

$$= \frac{\lambda^\alpha}{\Gamma(\alpha)} \int_0^\infty e^{tx} e^{-\lambda x} x^{\alpha-1} \, dx$$

$$= \frac{\lambda^\alpha}{\Gamma(\alpha)} \int_0^\infty e^{-(\lambda-t)x} x^{\alpha-1} \, dx$$

$$= \left(\frac{\lambda}{\lambda-t}\right)^\alpha \frac{1}{\Gamma(\alpha)} \int_0^\infty e^{-y} y^{\alpha-1} \, dy \quad [\text{by } y = (\lambda - t)x]$$

$$= \left(\frac{\lambda}{\lambda-t}\right)^\alpha$$

Differentiation of Equation 3.7.2 yields

$$\Psi'(t) = \frac{\alpha\lambda^\alpha}{(\lambda-t)^{\alpha+1}}$$

$$\Psi''(t) = \frac{\alpha(\alpha+1)\lambda^\alpha}{(\lambda-t)^{\alpha+2}}$$

Hence,

$$E[X] = \Psi'(0) = \frac{\alpha}{\lambda} \tag{3.7.3}$$

$$\operatorname{Var}(X) = E[X^2] - (E[X])^2 \tag{3.7.4}$$

$$= \Psi''(0) - \left(\frac{\alpha}{\lambda}\right)^2$$

$$= \frac{\alpha(\alpha+1)}{\lambda^2} - \frac{\alpha^2}{\lambda^2} = \frac{\alpha}{\lambda^2}$$

An important property of the gamma is that if X_1 and X_2 are independent gamma random variables having respective parameters (α_1, λ) and (α_2, λ), then $X_1 + X_2$ is gamma with parameters $(\alpha_1 + \alpha_2, \lambda)$. This result easily follows since

$$\Psi_{X_1+X_2}(t) = E[e^{t(X_1+X_2)}] \tag{3.7.5}$$

$$= \Psi_{X_1}(t)\Psi_{X_2}(t)$$

$$= \left(\frac{\lambda}{\lambda-t}\right)^{\alpha_1}\left(\frac{\lambda}{\lambda-t}\right)^{\alpha_2} \quad \text{from Equation 3.7.2}$$

$$= \left(\frac{\lambda}{\lambda-t}\right)^{\alpha_1+\alpha_2}$$

which is seen to be the moment generating function of a gamma $(\alpha_1 + \alpha_2, \lambda)$ random variable. Since a moment generating function uniquely characterizes a distribution, the result entails.

The foregoing result easily generalizes to yield the following proposition.

Proposition 3.7.1

If X_i, $i = 1, \ldots, n$ are independent gamma random variables with respective parameters (α_i, λ), then $\sum_{i=1}^{n} X_i$ is gamma with parameters $\sum_{i=1}^{n} \alpha_i, \lambda$.

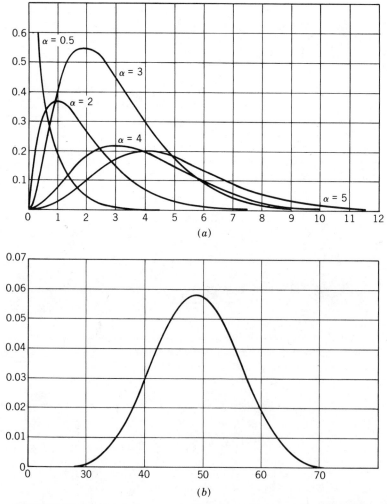

FIGURE 3.7.1 Graphs of the gamma $(\alpha, 1)$ density for (a) $\alpha = 0.5, 2, 3, 4, 5$ and (b) $\alpha = 50$

Since the gamma distribution with parameters $(1, \lambda)$ reduces to the exponential with rate λ, we have thus shown the following useful result.

Corollary 3.7.2 If $X_1, \ldots, X_n$ are independent exponential random variables, each having rate λ, then $\sum_{i=1}^{n} X_i$ is a gamma random variable with parameters (n, λ).

Example 3.7a The lifetime of a battery is exponentially distributed with rate λ. If a stereo cassette requires one battery to operate, then the total playing time one can obtain from a total of n batteries is a gamma random variable with parameters (n, λ). ∎

Figure 3.7.1 presents a graph of the gamma $(\alpha, 1)$ density for a variety of values of α. It should be noted that as α becomes large, the density starts to resemble the normal density. This is theoretically explained by the central limit theorem since a gamma (α, λ) random variable X can be thought of as being the sum of $[\alpha]$ independent exponentials each with rate λ and a gamma $(\alpha - [\alpha], \lambda)$ random variable (where $[\alpha]$ is the largest integer less than or equal to α). That is, we can use the preceding proposition to express X as

$$X = \sum_{i=1}^{[\alpha]} X_i + Y$$

where the X_i, $i = 1, \ldots, [\alpha]$ are independent exponential random variables each having mean 1, and are independent of Y, which is gamma with parameters $(\alpha - [\alpha], 1)$. The central limit theorem now implies that, when α is large, the distribution of $X/[\alpha]$ is approximately normal.

8 DISTRIBUTIONS ARISING FROM THE NORMAL

8.1 The Chi-Square Distribution

Definition. If $Z_1, Z_2, \ldots, Z_n$ are independent unit normal random variables, then X, defined by

$$X = Z_1^2 + Z_2^2 + \cdots + Z_n^2 \tag{3.8.1}$$

is said to have a *chi-square distribution with n degrees of freedom*. We will use the notation

$$X \sim \chi_n^2$$

to signify that X has a chi-square distribution with n degrees of freedom.

Let us compute the moment generating function of a chi-square random variable with n degrees of freedom. To begin, we have when $n = 1$ that

$$E[e^{tX}] = E[e^{tZ^2}] \qquad \text{where } Z \sim \mathcal{N}(0,1) \qquad (3.8.2)$$

$$= \int_{-\infty}^{\infty} e^{tx^2} f_Z(x)\, dx$$

$$= \frac{1}{\sqrt{2\pi}} \int_{-\infty}^{\infty} e^{tx^2} e^{-x^2/2}\, dx$$

$$= \frac{1}{\sqrt{2\pi}} \int_{-\infty}^{\infty} e^{-x^2(1-2t)/2}\, dx$$

$$= \frac{1}{\sqrt{2\pi}} \int_{-\infty}^{\infty} e^{-x^2/2\bar{\sigma}^2}\, dx \qquad \text{where } \bar{\sigma}^2 = (1-2t)^{-1}$$

$$= (1-2t)^{-1/2} \frac{1}{\sqrt{2\pi}\,\bar{\sigma}} \int_{-\infty}^{\infty} e^{-x^2/2\bar{\sigma}^2}\, dx$$

$$= (1-2t)^{-1/2}$$

where the last equality follows since the integral of the normal $(0, \bar{\sigma}^2)$ density equals 1. Hence, in the general case of n degrees of freedom

$$E[e^{tX}] = E\left[e^{t\Sigma_{i=1}^n Z_i^2}\right]$$

$$= E\left[\prod_{i=1}^{n} e^{tZ_i^2}\right]$$

$$= \prod_{i=1}^{n} E[e^{tZ_i^2}] \qquad \text{by independence of the } Z_i$$

$$= (1-2t)^{-n/2} \qquad \text{from Equation 3.8.2}$$

However, we recognize $[1/(1-2t)]^{n/2}$ as being the moment generating function of a gamma random variable with parameters $(n/2, 1/2)$. Hence, by the uniqueness of moment generating functions, it follows that these two distributions—chi-square with n degrees of freedom and gamma with parameters $n/2$ and $1/2$—are identical and thus we can conclude that the density of X is given by

$$f(x) = \frac{\frac{1}{2} e^{-x/2} \left(\frac{x}{2}\right)^{(n/2)-1}}{\Gamma\left(\frac{n}{2}\right)} \qquad x > 0$$

The chi-square density functions having 1, 3, and 10 degrees of freedom respectively are plotted in Figure 3.8.1.

The chi-square distribution has the additive property that if X_1 and X_2 are independent chi-square random variables with n_1 and n_2 degrees of freedom respectively, then $X_1 + X_2$ is chi-square with $n_1 + n_2$ degrees of freedom. This can be formally shown either by the use of moment generating

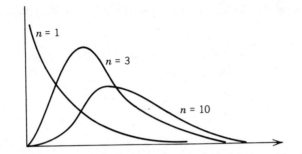

FIGURE 3.8.1 The chi-square density function with n degrees of freedom

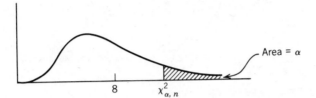

FIGURE 3.8.2 The chi-square density function with 8 degrees of freedom

functions or, most easily, by noting that $X_1 + X_2$ is the sum of squares of $n_1 + n_2$ independent unit normals and thus has a chi-square distribution with $n_1 + n_2$ degrees of freedom.

Since the chi-square distribution with n degrees of freedom is identical to the gamma distribution with parameters $\alpha = n/2$ and $\lambda = 1/2$, it follows from Equations 3.7.3 and 3.7.4 that the mean and variance of a random variable X having this distribution is

$$E[X] = n, \quad \text{Var}(X) = 2n$$

If X is a chi-square random variable with n degrees of freedom then for any $\alpha \in (0, 1)$, the quantity $\chi^2_{\alpha, n}$ is defined to be such that

$$P\{ X \geq \chi^2_{\alpha, n} \} = \alpha$$

This is illustrated in Figure 3.8.2.

In Table A3-8-1 of the Appendix, we list $\chi^2_{\alpha, n}$ for a variety of values of α and n (including all those needed to solve problems and examples in this text). The reason we have not provided more extensive tables is that Basic programs for approximating the chi-square distribution function (Program 3-8-1-A) and the values $\chi^2_{\alpha, n}$ (Program 3-8-1-B) are provided.

Example 3.8a Compute $P\{\chi^2_{25} \leq 30\}$ when χ^2_{25} is a chi-square random variable with 25 degrees of freedom.

Solution We run Program 3-8-1-A:

```
RUN
THIS PROGRAM COMPUTES THE PROBABILITY THAT A CHI-SQUARED RANDOM VARIABLE WITH N
DEGREES OF FREEDOM IS LESS THAN X
ENTER THE DEGREE OF FREEDOM PARAMETER
? 25
ENTER THE DESIRED VALUE OF X
? 30
THE PROBABILITY IS .7757181
Ok
```

Example 3.8b Find $\chi^2_{.05,15}$.

Solution Run Program 3-8-1-B:

```
RUN
FOR A GIVEN INPUT a, 0<a<.5, THIS PROGRAM COMPUTES THE VALUE           chisq(a,n)
  SUCH THAT THE PROBABILITY THAT A CHI SQUARE RANDOM VARIABLE WITH n     DEGREES OF
    FREEDOM EXCEEDS chisq(a,n) IS EQUAL TO a
ENTER THE DEGREE OF FREEDOM PARAMETER n
? 15
ENTER THE DESIRED VALUE OF a
? .05
THE VALUE IS 24.99751
Ok
```

8.2 The *t*-Distribution

If Z and X are independent random variables with Z having a unit normal distribution and X a chi-square distribution with n degrees of freedom, then T_n defined by

$$T_n = \frac{Z}{\sqrt{X/n}}$$

is said to have a *t-distribution with n degrees of freedom*. It can be shown that its density function is

$$f_{T_n}(x) = \frac{1}{\sqrt{n\pi}} \frac{\Gamma\left(\dfrac{n+1}{2}\right)}{\Gamma\left(\dfrac{n}{2}\right)} \left(1 + \frac{x^2}{n}\right)^{-(n+1)/2} \qquad -\infty < x < \infty$$

where Γ is the gamma function defined in Section 7. A graph of this density is given in Figure 3.8.3 for $n = 1, 5,$ and 10.

Like the unit normal density, the *t*-density is symmetric about zero. In addition, as n becomes larger, the density becomes more and more like a unit normal density. The reason for this is that, since a chi-square random variable

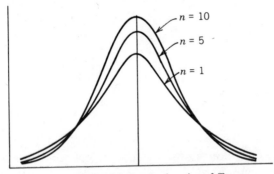

FIGURE 3.8.3 Density function of T_n

X with n degrees of freedom can be expressed as the sum of the squares of n unit normals, it follows that we can express X/n as

$$\frac{X}{n} = \frac{Z_1^2 + \cdots + Z_n^2}{n}$$

where Z_i, $i = 1, \ldots, n$ are independent unit normals. It now follows, by the weak law of large numbers, that for large n, X/n will, with probability close to 1, be approximately equal to $E[Z_i^2] = 1$. Hence, for large n, $Z/\sqrt{X/n}$ will approximately have the same distribution as Z, a unit normal random variable.

Figure 3.8.4 shows a graph of the t-density function with 5 degrees of freedom compared with the standard normal density. Notice that the t-density has thicker "tails," indicating greater variability, than does the normal density.

The mean and variance of T_n can be shown to equal

$$E[T_n] = 0, \qquad n > 1$$

$$\text{Var}(T_n) = \frac{n}{n-2}, \qquad n > 2$$

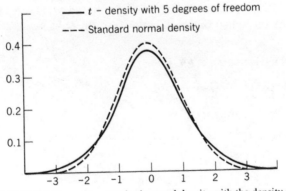

FIGURE 3.8.4 Comparing standard normal density with the density of T_5

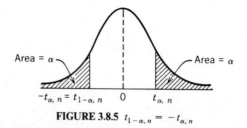

FIGURE 3.8.5 $t_{1-\alpha,\,n} = -t_{\alpha,\,n}$

Thus the variance of T_n decreases to 1—the variance of a unit normal random variable—as n increases to ∞.

For $\alpha, 0 < \alpha < 1$, let $t_{\alpha,\,n}$ be such that

$$P\{T_n \ge t_{\alpha,\,n}\} = \alpha$$

It follows from the symmetry about zero of the t-density that $-T_n$ has the same distribution as T_n and so

$$\alpha = P\{-T_n \ge t_{\alpha,\,n}\}$$
$$= P\{T_n \le -t_{\alpha,\,n}\}$$
$$= 1 - P\{T_n \ge -t_{\alpha,\,n}\}$$

Therefore,

$$P\{T_n \ge -t_{\alpha,\,n}\} = 1 - \alpha$$

leading to the conclusion that

$$-t_{\alpha,\,n} = t_{1-\alpha,\,n}$$

which is illustrated in Figure 3.8.5.

The values of $t_{\alpha,\,n}$ for a variety of values of n and α have been tabulated in Table A3.8.2 of the Appendix. In addition Basic programs 3-8-2-A and 3-8-2-B compute the t-distribution function and the values $t_{\alpha,\,n}$ respectively.

Example 8.3c Determine $P\{T_{12} \le 1.4\}$.

Solution Run Program 3-8-2-A:

```
RUN
THIS PROGRAM COMPUTES THE PROBABILITY THAT A t-RANDOM VARIABLE WITH N DEGREES OF
 FREEDOM   IS LESS THAN X
ENTER THE DEGREES OF FREEDOM
? 12
ENTER THE VALUE OF X
? 1.4
THE PROBABILITY IS .9066057
```

Example 8.3d Find $t_{.025,15}$.

Solution Run Program 3-8-2-B:

```
RUN
FOR A GIVEN INPUT a, 0<a<.5, THIS PROGRAM COMPUTES THE VALUE t(a,n)    SUCH THAT
  THE PROBABILITY THAT A t-RANDOM VARIABLE WITH n DEGREES OF FREEDOM   EXCEEDS
  t(a,n) IS EQUAL TO a
ENTER THE DEGREES OF FREEDOM PARAMETER n
? 9
ENTER THE DESIRED VALUE OF a
? .025
THE VALUE IS 2.262517
Ok
```

8.3   The *F* Distribution

If χ_n^2 and χ_m^2 are independent chi-square random variables with n and m degrees of freedom respectively, then the random variable $F_{n,m}$ defined by

$$F_{n,m} = \frac{\chi_n^2/n}{\chi_m^2/m}$$

is said to have an *F distribution with n and m degrees of freedom*. The probability density function of $F_{n,m}$ is given by

$$f(x) = \frac{\Gamma\!\left(\dfrac{n+m}{2}\right) n^{n/2} m^{m/2}}{\Gamma\!\left(\dfrac{n}{2}\right)\Gamma\!\left(\dfrac{m}{2}\right)} \frac{x^{(n-2)/2}}{(m+nx)^{(n+m)/2}}, \qquad x > 0$$

A sketch of this density function is given in Figure 3.8.6.

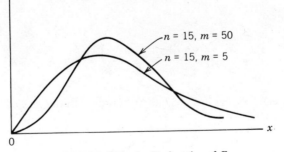

FIGURE 3.8.6 The density function of $F_{n,m}$

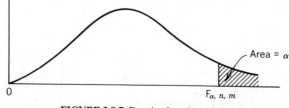

FIGURE 3.8.7 Density function of $F_{n,\,m}$

The expectation and variance of $F_{n,\,m}$ are given by

$$E[F_{n,\,m}] = \frac{m}{m-2} \qquad \text{for } m > 2$$

$$\text{Var}[F_{n,\,m}] = \frac{m^2(2m+2n-4)}{n(m-2)^2(m-4)} \qquad \text{for } m > 4$$

For any $\alpha \in (0,1)$, let $F_{\alpha,\,n,\,m}$ be such that

$$P\{F_{n,\,m} > F_{\alpha,\,n,\,m}\} = \alpha$$

This is illustrated in Figure 3.8.7.

The quantities $F_{\alpha,\,n,\,m}$ are tabulated in Table A3.8.3 of the Appendix for different values of n, m, and $\alpha \leq \frac{1}{2}$. If $F_{\alpha,\,n,\,m}$ is desired when $\alpha > \frac{1}{2}$, it can be obtained by using the following equalities:

$$\alpha = P\left\{ \frac{\chi_n^2/n}{\chi_m^2/m} > F_{\alpha,\,n,\,m} \right\}$$

$$= P\left\{ \frac{\chi_m^2/m}{\chi_n^2/n} < \frac{1}{F_{\alpha,\,n,\,m}} \right\}$$

$$= 1 - P\left\{ \frac{\chi_m^2/m}{\chi_n^2/n} \geq \frac{1}{F_{\alpha,\,n,\,m}} \right\}$$

or, equivalently

$$P\left\{ \frac{\chi_m^2/m}{\chi_n^2/n} \geq \frac{1}{F_{\alpha,\,n,\,m}} \right\} = 1 - \alpha \qquad (3.8.2)$$

But as $(\chi_m^2/m)/(\chi_n^2/n)$ has an F distribution with degrees of freedom m and n, it follows that

$$1 - \alpha = P\left\{ \frac{\chi_m^2/m}{\chi_n^2/n} \geq F_{1-\alpha,\,m,\,n} \right\}$$

implying, from Equation 3.8.2, that

$$\frac{1}{F_{\alpha, n, m}} = F_{1-\alpha, m, n}$$

Thus, for instance $F_{.9, 5, 7} = 1/F_{.1, 7, 5} = 1/3.37 = .2967$
where the value of $F_{.1, 7, 5}$ was obtained from Table A3.8.3 of the Appendix of
Tables.

Program 3-8-3-A computes the distribution function of $F_{n, m}$.

Example 3.8e Determine $P\{F_{6, 14} \leq 1.5\}$.

Solution Run Program 3-8-3-A:

```
RUN
THIS PROGRAM COMPUTES THE PROBABILITY THAT AN F RANDOM VARIABLE WITH  DEGREES OF
 FREEDOM N AND M IS LESS THAN X
ENTER THE FIRST DEGREE OF FREEDOM PARAMETER
? 6
ENTER THE SECOND DEGREE OF FREEDOM PARAMETER
? 14
ENTER THE DESIRED VALUE OF X
? 1.5
THE PROBABILITY IS .7518277
Ok
```

PROBLEMS

1. A satellite system consists of 4 components and can function adequately if at least 2 of the 4 components is in working condition. If each component is, independently, in working condition with probability .6, what is the probability the system functions adequately?

2. A communications channel transmits the digits 0 and 1. However, due to static, the digit transmitted is incorrectly received with probability .2. Suppose that we want to transmit an important message consisting of one binary digit. To reduce the chance of error, we transmit 00000 instead of 0 and 11111 instead of 1. If the receiver of the message uses "majority" decoding, what is the probability that the message will be incorrectly decoded? What independence assumptions are you making? (By majority decoding we mean that the message is decoded as "0" if there are at least 3 zeroes in the message received and as "1" otherwise.)

3. If each voter is for Proposition A with probability .7, what is the probability that exactly 7 of the first 10 voters are for this proposition?

4. Suppose that a particular trait (such as eye color or left handedness) of a person is classified on the basis of one pair of genes and suppose that *d* represents a dominant gene and *r* a recessive gene. Thus, a person with *dd* genes is pure dominance, one with *rr* is pure recessive, and one with *rd* is hybrid. The pure dominance and the hybrid are alike in appearance. Children receive 1 gene from each parent. If, with respect to a particular trait, 2 hybrid parents have a total of 4 children, what is the probability

that 3 of the 4 children have the outward appearance of the dominant gene?

5. At least one-half of an airplane's engines are required to function in order for it to operate. If each engine independently functions with probability p, for what values of p is a 4-engine plane more likely to operate than a 2-engine plane?

6. If X is binomial with parameters $n = 50$ and $p = .3$, compute
   (a) $P\{X < 17\}$
   (b) $P\{13 \le X \le 17\}$

7. Use the basic program to compute
   (a) $P\{X \le 75\}$
   (b) $P\{90 \le X \le 100\}$
   (c) $P\{X > 105\}$
   when X is a binomial random variable with parameters $n = 300$, $p = .3$.

8. If X and Y are binomial random variables with respective parameters (n, p) and $(n, 1 - p)$ verify and explain the following identities:
   (a) $P\{X \le i\} = P\{Y \ge n - i\}$
   (b) $P\{X = k\} = P\{Y = n - k\}$

9. If X is a binomial random variable with parameters n and p, where $0 < p < 1$, show that
   (a) $P\{X = k + 1\} = \dfrac{p}{1 - p}\dfrac{n - k}{k + 1}P\{X = k\}$, $k = 0, 1, \ldots, n - 1$.
   (b) As k goes from 0 to n, $P\{X = k\}$ first increases and then decreases, reaching its largest value when k is the largest integer less than or equal to $(n + 1)p$.

10. Derive the moment generating function of a binomial random variable and then use your result to verify the formulas for the mean and variance given in the text.

11. Compare the Poisson approximation with the correct binomial probability for the following cases:
    (a) $P\{X = 2\}$ when $n = 10$, $p = .1$
    (b) $P\{X = 0\}$ when $n = 10$, $p = .1$
    (c) $P\{X = 4\}$ when $n = 9$, $p = .2$

12. If you buy a lottery ticket in 50 lotteries, in each of which your chance of winning a prize is $1/100$, what is the (approximate) probability that you will win a prize (a) at least once, (b) exactly once, and (c) at least twice?

13. The number of times that an individual contracts a cold in a given year is a Poisson random variable with parameter $\lambda = 3$. Suppose a new wonder drug (based on large quantities of vitamin C) has just been marketed that reduces the Poisson parameter to $\lambda = 2$ for 75 percent of the population. For the other 25 percent of the population the drug has no appreciable

effect on colds. If an individual tries the drug for a year and has 0 colds in that time, how likely is it that the drug is beneficial for him or her?

14. The suicide rate in a certain state is 1 suicide per 100,000 inhabitants per month.
    (a) Find the probability that in a city of 400,000 inhabitants within this state, there will be 8 or more suicides in a given month.
    (b) What is the probability that there will be at least 2 months during the year that will have 8 or more suicides?

15. The probability of error in the transmission of a binary digit over a communication channel is $1/10^3$. Write an expression for the exact probability of more than 3 errors when transmitting a block of 10^3 bits. What is its approximate value? Assume independence.

16. If X is a Poisson random variable with mean λ, show that $P\{X = i\}$ first increases and then decreases as i increases, reaching its maximum value when k is the largest integer less than or equal to λ.

17. If X is Poisson with mean 600, compute
    (a) $P\{X > 575\}$
    (b) $P\{590 \le x \le 610\}$

18. Let X_1, X_2, X_3 be binomial random variables having respective parameters $(20, .2)$, $(40, .1)$, and $(80, .05)$. Compute $P\{X_i \le 5\}$ for $i = 1, 2, 3$ and compare the results with $P\{Y \le 5\}$ where Y is a Poisson random variable having mean 4. Use the programs.

19. A contractor purchases a shipment of 100 transistors. It is his policy to test 10 of these transistors and to keep the shipment only if at least 9 of the 10 are in working condition. If the shipment contains 20 defective transistors, what is the probability it will be kept?

20. Let X denote a hypergeometric random variable with parameters n, m, and k. That is,

$$P\{X = i\} = \frac{\binom{n}{i}\binom{m}{k-i}}{\binom{n+m}{k}}, \qquad i = 0, 1, \ldots, \min(k, n)$$

    (a) Derive a formula for $P\{X = i\}$ in terms of $P\{X = i - 1\}$.
    (b) Use (a) to compute $P\{X = i\}$ for $i = 0, 1, 2, 3, 4, 5$ when $n = m = 10$, $k = 5$, by starting with $P\{X = 0\}$.
    (c) Based on the recursion in (a), write a program to compute the hypergeometric distribution function.
    (d) Use your program from (c) to compute $P\{X \le 10\}$ when $n + m = 30$, $k = 15$.

21. Independent trials, each of which is a success with probability p are successively performed. Let X denote the first trial resulting in a success. That is, X will equal k if the first $k - 1$ trials are all failures and the kth

a success. X is called a *geometric* random variable. Compute

(a) $P\{X = k\}$, $k = 1, 2, \ldots$

(b) $E[X]$

Let Y denote the number of trials needed to obtain r successes. Y is called a *negative binomial random variable*. Compute

(c) $P\{Y = k\}$, $k = r, r + 1, \ldots$

Hint: In order for Y to equal k, how many successes must result in the first $k - 1$ trials and what must be the outcome of trial k?

(d) Show that

$$E[Y] = r/p$$

Hint: Write $Y = Y_1 + \cdots + Y_r$ where Y_i is the number of trials needed to go from a total of $i - 1$ to a total of i successes.

22. If U is uniformly distributed on $(0, 1)$, show that $a + (b - a)U$ is uniform on (a, b).

23. You arrive at a bus stop at 10 o'clock, knowing that the bus will arrive at some time uniformly distributed between 10 and 10:30. What is the probability that you will have to wait longer than 10 minutes? If at 10:15 the bus has not yet arrived, what is the probability that you will have to wait at least an additional 10 minutes?

24. If X is a normal random variable with parameters $\mu = 10$, $\sigma^2 = 36$, compute

(a) $P\{X > 5\}$ (d) $P\{X < 20\}$

(b) $P\{4 < X < 16\}$ (e) $P\{X > 16\}$

(c) $P\{X < 8\}$

25. The annual rainfall (in inches) in a certain region is normally distributed with $\mu = 40$, $\sigma = 4$. What is the probability that in 2 of the next 4 years the rainfall will exceed 50 inches? Assume that the rainfalls in different years are independent.

26. The width of a slot of a duralumin forging is (in inches) normally distributed with $\mu = .9000$ and $\sigma = .0030$. The specification limits were given as $.9000 \pm .0050$. What percentage of forgings will be defective? What is the maximum allowable value of σ that will permit no more than 1 in 100 defectives when the widths are normally distributed with $\mu = .9000$ and σ?

27. A certain type of light bulb has an output that is normally distributed with mean 2000 end foot candles and standard deviation 85 end foot candles. Determine a lower specification limit L so that only 5 percent of the light bulbs produced will be defective. (That is, determine L so that $P\{X \geq L\} = .95$ where X is the output of a bulb.)

28. A manufacturer produces bolts that are specified to be between 1.19 and 1.21 inches in diameter. If his production process results in a bolt's diameter being normally distributed with mean 1.20 inches and standard deviation .005, what percentage of bolts will not meet specifications?

29. Let $I = \int_{-\infty}^{\infty} e^{-x^2/2} \, dx$.

(a) Show that for any μ and σ

$$\frac{1}{\sqrt{2\pi}\,\sigma} \int_{-\infty}^{\infty} e^{-(x-\mu)^2/2\sigma^2} \, dx = 1$$

is equivalent to $I = \sqrt{2\pi}$.

(b) Show that $I = \sqrt{2\pi}$ by writing

$$I^2 = \int_{-\infty}^{\infty} e^{-x^2/2} \, dx \int_{-\infty}^{\infty} e^{-y^2/2} \, dy = \int_{-\infty}^{\infty}\int_{-\infty}^{\infty} e^{-(x^2+y^2)/2} \, dx \, dy$$

and then evaluating the double integral by means of a change of variables to polar coordinates. (That is, let $x = r\cos\theta$, $y = r\sin\theta$, $dx \, dy = r \, dr \, d\theta$.)

30. A random variable X is said to have a lognormal distribution if $\log X$ is normally distributed. If X is lognormal with $E[\log X] = \mu$ and $\mathrm{Var}(\log X) = \sigma^2$ determine the distribution function of X. That is, what is $P\{X \le x\}$?

31. The lifetimes of interactive computer chips produced by a certain semi-conductor manufacturer are normally distributed having mean 1.4×10^6 hours with a standard deviation of 3×10^5 hours. If a mainframe manufacturer requires that at least 90 percent of the chips from a large batch will have lifetimes of at least 3.8×10^6 hours, should he contract with the semiconductor firm?

32. In Problem 31, what is the probability that a batch of 100 chips will contain at least 20 whose lifetimes are less than 4.2×10^6 hours?

33. If X is binomial with parameters $n = 150$, $p = .6$, compute the exact value of $P\{X \le 80\}$ and compare with its normal approximation both (a) making use of and (b) not making use of the continuity correction.

34. Each computer chip made in a certain plant will, independently, be defective with probability $\frac{1}{4}$. If a sample of 1000 chips are tested, what is the approximate probability that less than 200 will be defective?

35. Argue, based on the central limit theorem, that a Poisson random variable having mean λ will approximately have a normal distribution with mean and variance both equal to λ when λ is large. If X is Poisson with mean 100, compute the exact probability that X is less than or equal to 116 and compare it with its normal approximation both when a continuity correction is utilized and when it is not. The convergence of the Poisson to the normal is indicated in Figure 3.9.1.

36. Use the computer programs given in the text to compute $P\{X \le 10\}$ when X is a binomial random variable with parameters $n = 100$, $p = .1$. Now compare this with its (a) Poisson and (b) normal approximation. In using the normal approximation, write the desired probability as $P\{X < 10.5\}$ so as to utilize the continuity correction.

SPECIAL RANDOM VARIABLES

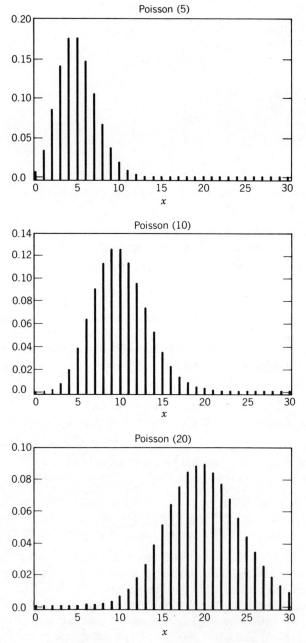

FIGURE 3.9.1 Poisson probability mass functions

37. The time (in hours) required to repair a machine is an exponentially distributed random variable with parameter $\lambda = \frac{1}{2}$.

   (a) What is the probability that a repair time exceeds 2 hours?
   (b) What is the conditional probability that a repair takes at least 10 hours, given that its duration exceeds 9 hours?

38. The number of years a radio functions is exponentially distributed with parameter $\lambda = \frac{1}{8}$. If Jones buys a used radio, what is the probability that it will be working after an additional 10 years?

39. Jones figures that the total number of thousands of miles that an auto can be driven before it would need to be junked is an exponential random variable with parameter $1/20$. Smith has a used car that he claims has been driven only 10,000 miles. If Jones purchases the car, what is the probability that she would get at least 20,000 additional miles out of it? Repeat under the assumption that the lifetime mileage of the car is not exponentially distributed but rather is (in thousands of miles) uniformly distributed over $(0, 40)$.

40. Let $X_1, X_2, \ldots, X_n$ denote the first n interarrival times of a Poisson process and set $S_n = \sum_{i=1}^{n} X_i$.

   (a) What is the interpretation of S_n?
   (b) Argue that the 2 events $\{S_n \leq t\}$ and $\{N(t) \geq n\}$ are identical.
   (c) Use (b) to show that

$$P\{S_n \leq t\} = 1 - \sum_{j=0}^{n-1} e^{-\lambda t}(\lambda t)^j / j!$$

   (d) By differentiating the distribution function of S_n given in (c), conclude that S_n is a gamma random variable with parameters n and λ. (This result also follows from Corollary 3.7.2.)

41. Earthquakes occur in a given region in accordance with a Poisson process with rate 5 per year.

   (a) What is the probability there will be at least 2 earthquakes in the first half of 2010?
   (b) Assuming that the event in (a) occurs, what is the probability that there will be no earthquakes during the first 9 months of 2011?
   (c) Assuming that the event in (a) occurs, what is the probability that there will be at least 4 earthquakes over the first 9 months of the year 2010?

42. When shooting at a target in a 2-dimensional plane, suppose that the horizontal miss distance is normally distributed with mean 0 and variance 4 and is independent of the vertical miss distance, which is also normally distributed with mean 0 and variance 4. Let D denote the distance between the point at which the shot leads and the target. Compute $P\{D > 3.3\}$.

43. If X is a chi-square random variable with 6 degrees of freedom, find

(a) $P\{X \le 6\}$

(b) $P\{3 \le X \le 9\}$

44. If X and Y are independent chi-square random variables with 3 and 6 degrees of freedom respectively, determine the probability that $X + Y$ will exceed 10.

45. Show $\Gamma(1/2) = \sqrt{\pi}$. *Hint:* Evaluate $\int_0^\infty e^{-x} x^{-1/2} \, dx$ by letting $x = y^2/2$, $dx = y \, dy$.

46. If T has a t-distribution with 8 degrees of freedom, find (a) $P\{T \ge 1\}$, (b) $P\{T \le 2\}$, (c) $P\{-1 < T < 1\}$.

47. If T_n has a t-distribution with n degrees of freedom, show that T_n^2 has an F distribution with 1 and n degrees of freedom.

48. In Section 8.2 a probabilistic argument was given to conclude that the t-density approaches the unit normal density as the degree of freedom parameter n increases. Give an analytic verification of this by using the fact that

$$\lim_{n \to \infty} \left(1 + \frac{y}{n}\right)^n = e^y \qquad \text{for any } y$$

Hint: Start by writing the t-density as

$$f_{T_n}(x) = C_n \left(1 + \frac{x^2}{n}\right)^{-(n+1)/2}$$

where C_n does not depend on x. Hence,

$$f_{T_n}(x) = \frac{C_n \left(1 + \dfrac{x^2}{n}\right)^{-1/2}}{\left(1 + \dfrac{x^2/2}{n/2}\right)^{n/2}}$$

Now let $n \to \infty$ to conclude that

$$f_{T_n}(x) \to c e^{-x^2/2}$$

where c does not depend on x. As the right-hand side integrates to 1, conclude that $c = 1/\sqrt{2\pi}$.

49. If $F_{n,m}$ has an F distribution with n and m degrees of freedom determine

(a) $P\{F_{2,4} > 3\}$

(b) $P\{F_{3,6} > 6\}$

(c) $P\{F_{4,5} < 8\}$

50. Argue that

$$\lim_{m \to \infty} F_{\alpha, n, m} = \frac{\chi^2_{\alpha, n}}{n}$$

Sampling

1 INTRODUCTION

The science of statistics deals with drawing conclusions from observed data. For instance, a typical situation in an engineering study arises when one is confronted with a large collection, or *population*, of items that have measurable values associated with them. By suitably *sampling* from this collection, and then analyzing the sampled items, one hopes to be able to draw some conclusions about the collection as a whole.

In order to use sample data to make inferences about an entire population it is necessary to make some assumptions about the relationship between the two. One such assumption, which is often quite reasonable, is that there is an underlying (population) probability distribution such that the measurable values of the items in the population can be thought of as being independent random variables having this distribution.

> **Definition.** If $X_1, \ldots, X_n$ are independent random variables having a common distribution F, then we say that they constitute a *sample* (sometimes called a *random sample*) from the distribution F.

In most applications, the population distribution F will not be completely specified and one will attempt to use the data to make inferences about F. Sometimes it will be supposed that F is specified up to some unknown parameters (for instance, one might suppose that F was a normal distribution function having an unknown mean and variance, or that it is a Poisson distribution function whose mean is not given), and at other times it might be assumed that almost nothing is known about F (except maybe for assuming that it is a continuous, or a discrete, distribution). Problems in which the form of the underlying distribution is specified up to a set of unknown parameters are called *parametric* inference problems, whereas those in which nothing is assumed about the form of F are called *nonparametric* inference problems.

Example 4.1a Suppose that a new process has just been installed to produce computer chips, and suppose that the successive chips produced by this new process will have useful lifetimes that are independent with a common unknown distribution F. Sometimes physical reasons suggest that F should be an exponential distribution.[†] If this were the case, then we would be confronted with a parametric statistical problem in which we would want to use the observed data (i.e., the lifetimes of the successive chips produced) to estimate the unknown mean of the (exponential) life distribution. On the other hand, there may be no physical justification for supposing that F has any particular form; in this case, the problem of making inferences about F would be a nonparametric inference problem. ∎

In Section 2, we consider some measures of central tendency of a population distribution—in particular, the mean, median, and mode. We also introduce the sample mean and sample median—functions of the sampled data that are natural "estimators" of the (population) mean and median respectively. In Section 3, we discuss the sample variance and sample range—functions of the data that can be used as indicators of the degree of "spread" in the population. In Section 4, we show how, by use of empirical distribution functions, histograms, and stem and leaf plots, one can use the sample to learn about the population distribution in its entirety. In Section 5, we consider the joint distribution of the sample mean and variance when one is sampling from a normal population.

In the final section of this chapter, we consider the problem of learning about a population by sampling some of its members, under a different assumption about the relationship between the population and its sampled members. Specifically, rather than continuing to suppose that all population values are independently chosen from a common population distribution, we now suppose that the population is of finite size with each element of the population having a fixed unknown value. The problem of interest is to use the data values from a "randomly" chosen subset to make inferences about the set of all values.

2 MEASURES OF CENTRAL TENDENCY

Let $X_1, \ldots, X_n$ denote a random sample whose distribution function F is unknown. Rather than attempting to make inferences about F in its entirety, we often try to make inferences about its properties that are described by certain suitably defined measures. For instance, 3 common measures of the *location*, or *central tendency*, of a distribution function are its mean, median, and mode.

The mean (sometimes called the population mean) μ of the distribution function F is the expected value of a random variable X whose distribution is

---

[†] The reason for this is usually the fact that the exponential distribution has a constant failure rate (Chapter 11).

F. That is,

$$\mu = E[X]$$

$$= \begin{cases} \sum x_i P\{X = x_i\} & \text{if } F \text{ is a discrete distribution} \\ \int x f(x)\,dx & \text{if } F \text{ is a continuous distribution with density } f \end{cases}$$

The mean is a particularly useful measure of central tendency since it is a weighted average of the possible values of a random variable, whose distribution function is F, with the weight given to any value being equal to the probability (or probability density in the continuous case) that the random variable assumes that value.

An important property of the mean arises when one must predict the value of a random variable. That is, suppose that the value of a random variable X is to be predicted. If we predict that X will equal c, then the square of the "error" involved will be $(X - c)^2$. We will now show that the average squared error is minimized when we predict that X will equal its mean μ. To see this, note that for any constant c

$$\begin{aligned} E\left[(X - c)^2\right] &= E\left[(X - \mu + \mu - c)^2\right] \\ &= E\left[(X - \mu)^2 + 2(\mu - c)(X - \mu) + (\mu - c)^2\right] \\ &= E\left[(X - \mu)^2\right] + 2(\mu - c)E[X - \mu] + (\mu - c)^2 \\ &= E\left[(X - \mu)^2\right] + (\mu - c)^2 \qquad \text{since } E[X - \mu] = E[X] - \mu = 0 \\ &\geq E\left[(X - \mu)^2\right] \end{aligned}$$

Hence, the best predictor of a random variable, in terms of minimizing its mean square error, is just its mean.

Let $X_1, \ldots, X_n$ denote a random sample from the distribution F.

Definition. Any quantity that depends only on the sample and not on any unknown parameters of the population distribution is called a *statistic*.

Thus a statistic is completely determined when the sample is obtained.

Definition. The statistic $\overline{X}$ defined by

$$\overline{X} = \frac{X_1 + \cdots + X_n}{n}$$

is called the *sample mean*.

It should be noted that the sample mean, being the arithmetic mean of the sample values, is a random variable. In fact, its mean and variance are obtained as follows:

$$E[\bar{X}] = E\left[\frac{X_1 + \cdots + X_n}{n}\right]$$

$$= \frac{1}{n}(E[X_1] + \cdots + E[X_n])$$

$$= \mu$$

and

$$\text{Var}(\bar{X}) = \text{Var}\left(\frac{X_1 + \cdots + X_n}{n}\right)$$

$$= \frac{1}{n^2}[\text{Var}(X_1) + \cdots + \text{Var}(X_n)] \qquad \text{by independence}$$

$$= \frac{n\sigma^2}{n^2}$$

$$= \frac{\sigma^2}{n}$$

where μ and σ^2 are the population mean and variance respectively. Hence, the expected value of the sample mean is the population mean μ, whereas its variance is $1/n$ times the population variance. As we shall see, in cases where μ is unknown, the sample mean $\bar{X}$ is a natural estimator of μ.

Another measure of central tendency of a distribution is its median.

> **Definition.** For an arbitrary distribution F, we say that m is a median of this distribution if
>
> $$P\{X \geq m\} \geq \tfrac{1}{2} \qquad \text{and} \qquad P\{X \leq m\} \geq \tfrac{1}{2}$$
>
> when $X \sim F$.

Thus, in the case of a continuous distribution function, the median m is the unique value such that $F(m) = \tfrac{1}{2}$. That is, it is the midpoint of the distribution in the sense that a random variable having this distribution is just as likely to be greater than m as it is to be less than m.

The median, like the mean, is important when one wants to predict the value of a random variable X. For suppose that we want to choose c, as a predictor of X, so as to minimize the expected value (not of the square but) of the absolute error involved. That is, suppose we want to choose c so as to minimize $E[|X - c|]$. We now show that $c = m$ does the trick.

We need to prove that when $X \sim F$, $E[|X - c|]$ is minimized when c is equal to m, the median of F. To prove this, let us suppose that F is a continuous distribution whose density is f [that is, $F'(t) = f(t)$]. Then,

$$E[|X - c|] = \int_{-\infty}^{\infty} |x - c| f(x) \, dx$$

$$= \int_{-\infty}^{c} |x - c| f(x) \, dx + \int_{c}^{\infty} |x - c| f(x) \, dx$$

$$= \int_{-\infty}^{c} (c - x) f(x) \, dx + \int_{c}^{\infty} (x - c) f(x) \, dx$$

$$= cF(c) - \int_{-\infty}^{c} xf(x) \, dx + \int_{c}^{\infty} xf(x) \, dx - c[1 - F(c)]$$

Differentiation yields

$$\frac{d}{dc} E[|X - c|] = F(c) + cf(c) - cf(c) - cf(c) - 1 + F(c) + cf(c)$$

$$= 2F(c) - 1$$

Hence, the minimum value is obtained when c is such that

$$2F(c) - 1 = 0$$

or

$$F(c) = \tfrac{1}{2}$$

That is, the minimum value is obtained when c is chosen to equal the median of F. (To see that it is a minimum, just note that $(d^2/dc^2)E[|X - c|] = 2f(c) \geq 0$).

Example 4.2a The distribution function

$$F(x) = 1 - e^{-x^2}, \qquad 0 < x < \infty$$

is an instance of the so-called Weibull distribution,[†] which has been widely employed in engineering practice because of its versatility. It was originally proposed for the interpretation of fatigue data, but now its use has extended to many other engineering problems. In particular, it is widely used, in the field of life phenomena, as the distribution of the lifetime of some object, particularly when the "weakest link" model is appropriate for the object. That is, consider an object consisting of many parts and suppose that the object experiences death (failure) when any of its parts fail. Under these conditions, it has been shown (both theoretically and empirically) that a Weibull distribution provides a close approximation to the distribution of the lifetime of the item.

[†] The general form of the Weibull distribution function is

$$F(x) = 1 - e^{-\alpha x^\beta} \qquad 0 < x < \infty$$

where $\alpha > 0, \beta > 0$.

We compute the median m of the foregoing distribution as follows:

$$1/2 = F(m) = 1 - e^{-m^2}$$

or

$$e^{-m^2} = \tfrac{1}{2}$$

or, taking logarithms (base e)

$$-m^2 = \log\left(\tfrac{1}{2}\right)$$

or

$$m^2 = \log 2$$

or, finally

$$m = \sqrt{\log 2} \approx .83255.$$

The graph of F is presented in Figure 4.2.1. ∎

A median is sometimes referred to as the 50-percentile, where, in general, we say that ζ_p is a 100 p-percentile, $0 < p < 1$, of the continuous distribution function F if

$$F(\zeta_p) = p$$

The percentiles $\zeta_{1/4}$, $\zeta_{1/2}$, $\zeta_{3/4}$ are called the quartiles of the distributions. Thus the middle quartile is just the median of F. If we were to plot the density function $F' = f$, then 25 percent of the area under f would be to the left of $\zeta_{1/4}$, 25 percent between $\zeta_{1/4}$ and $\zeta_{1/2}$, 25 percent between $\zeta_{1/2}$ and $\zeta_{3/4}$, and 25 percent to the right of $\zeta_{3/4}$.

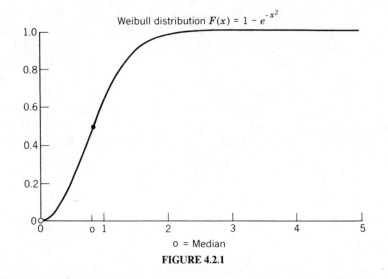

o = Median

FIGURE 4.2.1

Example 4.2b Consider the exponential distribution with density function

$$f(x) = \begin{cases} 2e^{-2x} & x > 0 \\ 0 & x < 0 \end{cases}$$

Since the distribution function is

$$F(y) = \int_0^y f(x)\,dx = 1 - e^{-2y} \qquad y > 0$$

we can compute its percentiles as follows:

$$p = F(\zeta_p) = 1 - e^{-2\zeta_p}$$

or

$$e^{-2\zeta_p} = 1 - p$$

or

$$-2\zeta_p = \log(1 - p)$$

or

$$\zeta_p = \frac{-\log(1 - p)}{2}$$

Thus, the quartiles are

$$\zeta_{1/4} = \frac{-\log(3/4)}{2} = \frac{\log(4/3)}{2} \approx .1438$$

$$\zeta_{1/2} = \frac{-\log(1/2)}{2} = \frac{\log 2}{2} \approx .3466$$

$$\zeta_{3/4} = \frac{-\log(1/4)}{2} = \frac{\log 4}{2} \approx .6931 \quad \blacksquare$$

The quartiles are indicated in the graph of the density function presented in Figure 4.2.2.

To define the sample median, we first need to introduce the concept of an order statistic. To do so, let $X_1, X_2, \ldots, X_n$ be a random sample from the distribution F, and suppose that we order the X's so that

$$X_{(1)} = \text{smallest value of } X_1, \ldots, X_n$$

$$X_{(2)} = \text{2nd smallest value of } X_1, \ldots, X_n$$

$$\vdots$$

$$X_{(i)} = i\text{th smallest value of } X_1, \ldots, X_n$$

$$\vdots$$

$$X_{(n)} = \text{largest value of } X_1, \ldots, X_n$$

Definition. The ordered values $X_{(1)} \le X_{(2)} \le \cdots \le X_{(n)}$ are called the *order statistics* of the sample. If $n = 2k - 1$ is an odd integer, then $X_{(k)}$ is called the *sample median*, whereas if $n = 2k$, then $\frac{1}{2}[X_{(k)} + X_{(k+1)}]$ is called the *sample median*.

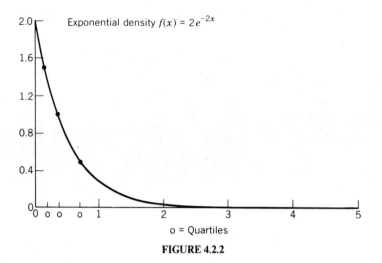

FIGURE 4.2.2

In other words, from a sample of size 7, the sample median is equal to the 4th smallest value, whereas from a sample of size 8 it is an average of the 4th and 5th smallest values.

The sample median is a "natural" estimator of the population median.

A third measure of central tendency of a distribution F is its mode.

Definition. Let X be a random variable having distribution function F. We say that a is a *mode* of F if

$$P(a) = \max_{x} P(x)$$

where

$$P(x) = \begin{cases} P\{X = x\} & \text{if } F \text{ is discrete} \\ f(x) & \text{if } F \text{ is continuous with density function } f \end{cases}$$

In other words, the mode represents the *most likely value* of the random variable. For instance, the distribution whose density is plotted in Figure 4.2.3 has a unique mode at $x = 1$.

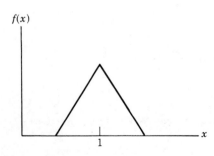

FIGURE 4.2.3 Density with mode at $x = 1$

$f(x)$

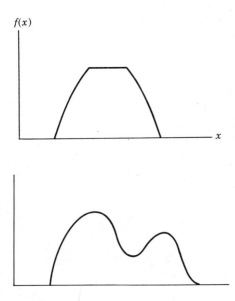

FIGURE 4.2.4 A unimodal distribution (with an infinite number of modal values)

FIGURE 4.2.5 A bimodal distribution (with a unique mode)

It is also common to use the term *relative mode* of a discrete (continuous) distribution to describe a value for which the probability mass function (probability density function) is *locally* the highest. The descriptive terms "unimodal" and "bimodal" when used in connection with a distribution means that the distribution has a single region of relative modes or, in the bimodal case, two such regions (see Figures 4.2.4 and 4.2.5).

Definition. If $X_1, \ldots, X_n$ is a random sample, then the value that occurs most frequently is called the sample mode.

Example 4.2c If the sample is $5, 8, 3, 2, 3$, then the sample mode is equal to 3. If the sample were $5, 8, 8, 3, 2, 3$ then there would be two sample modes—3 and 8. ∎

3 SAMPLE VARIANCE AND SAMPLE RANGE

Let $X_1, \ldots, X_n$ be a random sample from the distribution F.

Definition. The statistic S^2, defined by

$$S^2 = \sum_{i=1}^{n} \frac{\left(X_i - \overline{X} \right)^2}{n - 1}$$

where $\overline{X} = \sum_{i=1}^{n} X_i / n$ is called the *sample variance*. The statistic $S = \sqrt{S^2}$ is called the *sample standard deviation*.

A useful computational formula for S^2 is obtained as follows:

$$(n-1)S^2 = \sum_{i=1}^{n} (X_i - \bar{X})^2 \qquad (4.3.1)$$

$$= \sum_{i=1}^{n} (X_i^2 - 2X_i\bar{X} + \bar{X}^2)$$

$$= \sum_{i=1}^{n} X_i^2 - 2\bar{X}\sum_{i=1}^{n} X_i + \sum_{i=1}^{n} \bar{X}^2$$

$$= \sum_{i=1}^{n} X_i^2 - 2\bar{X}(n\bar{X}) + n\bar{X}^2$$

$$= \sum_{i=1}^{n} X_i^2 - n\bar{X}^2$$

If the variance of the distribution F (sometimes called the population variance) is σ^2, then we can use Equation 4.3.1 to compute $E[S^2]$.

$$E[(n-1)S^2] = E\left[\sum_{i=1}^{n} X_i^2\right] - nE[\bar{X}^2] \qquad \text{from Equation 4.3.1} \quad (4.3.2)$$

$$= nE[X_1^2] - nE[\bar{X}^2]$$

Now, as shown in Section 2, with $\mu = E[X_i]$,

$$E[\bar{X}] = \mu$$

$$\text{Var}(\bar{X}) = \sigma^2/n$$

and so, using the result that for any random variable U, $\text{Var}(U) = E[U^2] - (E[U])^2$, or equivalently, $E[U^2] = \text{Var}(U) + (E[U])^2$, we see that

$$E[\bar{X}^2] = \frac{\sigma^2}{n} + \mu^2$$

Hence, as $E[X_1^2] = \mu^2 + \sigma^2$, we obtain from Equation 4.3.2 that

$$E[(n-1)S^2] = n(\mu^2 + \sigma^2) - n\left(\frac{\sigma^2}{n} + \mu^2\right)$$

$$= (n-1)\sigma^2$$

or

$$E[S^2] = \sigma^2$$

That is, the expected value of the sample variance S^2 is equal to the population variance σ^2.

Definition. If $X_{(i)}$, $i = 1, \ldots, n$, is the ith smallest value of $X_1, \ldots, X_n$ then R defined by

$$R = X_{(n)} - X_{(1)}$$

is called the *sample range*.

Thus R measures the size of the interval that contains the random sample. Both R and S^2 give information about the spread of the distribution F.

3.1 COMPUTING THE SAMPLE VARIANCE

When computing "by hand," Equation 4.3.1 provides an efficient method for computing the sample variance. However, if there is a large data set and the computation is being performed on a computer, then because of computer "round-off error" the use of Equation 4.3.1 can be perilous, especially when the individual data values are large. That is, a computer will compute only to a given level of accuracy—usually 8 or 16 digits—and then ignore any remainder. However, if the individual terms are large, then when subtracting one positive number from another, as in Equation 4.3.1, this round-off error can be critical. Indeed, individuals have sometimes been unpleasantly surprised when utilizing Equation 4.3.1 by having their computer return a negative value for S^2. We thus need a different (from using Equation 4.3.1) computer algorithm for computing S^2.

The straightforward approach of having a computer first determine $\overline{X}$ and then sum the terms $(X_i - \overline{X})^2$, $i = 1, \ldots, n$ is fairly accurate (since it is always adding nonnegative terms) but not particularly efficient since it requires that the computer go through the set of data values twice [first to compute $\overline{X}$ and then to sum the terms $(X_i - \overline{X})^2$]. A better algorithm is based on a recursive formula, which we now develop.

To begin, let

$$\overline{X}_j = \frac{\sum_{i=1}^{j} X_i}{j} \qquad j \geq 1$$

$$S_j^2 = \frac{\sum_{i=1}^{j}(X_i - \overline{X}_j)^2}{j - 1} \qquad j \geq 2$$

That is, $\overline{X}_j$ and S_j^2 are the sample mean and sample variance respectively of the first j data values $X_1, \ldots, X_j$. The following proposition now shows how these quantities can be recursively computed.

Proposition 4.3.1

With $S_1^2 \equiv 0$, for $j = 1, \ldots, n$

$$\overline{X}_{j+1} = \overline{X}_j + \frac{X_{j+1} - \overline{X}_j}{j + 1} \tag{4.3.3}$$

$$S_{j+1}^2 = \left(1 - \frac{1}{j}\right)S_j^2 + (j + 1)\left(\overline{X}_{j+1} - \overline{X}_j\right)^2 \tag{4.3.4}$$

Proof The verification of Equation 4.3.3, which is equivalent to

$$X_{j+1} - \overline{X}_j = (j + 1)\left(\overline{X}_{j+1} - \overline{X}_j\right) \tag{4.3.5}$$

is quite straightforward and will be left as an exercise for the reader.

To prove the identity 4.3.4, note that

$$jS_{j+1}^2 = \sum_{i=1}^{j+1}\left(X_i - \bar{X}_{j+1}\right)^2 \tag{4.3.6}$$

$$= \sum_{i=1}^{j+1}\left[\left(X_i - \bar{X}_j\right) + \left(\bar{X}_j - \bar{X}_{j+1}\right)\right]^2$$

$$= \sum_{i=1}^{j+1}\left(X_i - \bar{X}_j\right)^2 + \sum_{i=1}^{j+1}\left(\bar{X}_j - \bar{X}_{j+1}\right)^2$$

$$+ 2\sum_{i=1}^{j+1}\left(X_i - \bar{X}_j\right)\left(\bar{X}_j - \bar{X}_{j+1}\right)$$

Now,

$$\sum_{i=1}^{j+1}\left(X_i - \bar{X}_j\right)^2 = \sum_{i=1}^{j}\left(X_i - \bar{X}_j\right)^2 + \left(X_{j+1} - \bar{X}_j\right)^2$$

$$= (j-1)S_j^2 + \left(X_{j+1} - \bar{X}_j\right)^2$$

$$= (j-1)S_j^2 + (j+1)^2\left(\bar{X}_{j+1} - \bar{X}_j\right)^2 \quad \text{from Equation 4.3.5}$$

$$\sum_{i=1}^{j+1}\left(\bar{X}_j - \bar{X}_{j+1}\right)^2 = (j+1)\left(\bar{X}_j - \bar{X}_{j+1}\right)^2$$

$$\sum_{i=1}^{j+1}\left(X_i - \bar{X}_j\right)\left(\bar{X}_j - \bar{X}_{j+1}\right) = \left(\bar{X}_j - \bar{X}_{j+1}\right)\sum_{i=1}^{j+1}\left(X_i - \bar{X}_j\right)$$

$$= \left(\bar{X}_j - \bar{X}_{j+1}\right)\left[\sum_{i=1}^{j}\left(X_i - \bar{X}_j\right) + X_{j+1} - \bar{X}_j\right]$$

$$= \left(\bar{X}_j - \bar{X}_{j+1}\right)\left[\sum_{i=1}^{j}X_i - j\bar{X}_j + X_{j+1} - \bar{X}_j\right]$$

$$= \left(\bar{X}_j - \bar{X}_{j+1}\right)\left(X_{j+1} - \bar{X}_j\right)$$

$$\text{since } \sum_{i=1}^{j}X_i - j\bar{X}_j = 0$$

$$= \left(\bar{X}_j - \bar{X}_{j+1}\right)(j+1)\left(\bar{X}_{j+1} - \bar{X}_j\right)$$

$$\text{from Equation 4.3.5}$$

$$= -(j+1)\left(\bar{X}_j - \bar{X}_{j+1}\right)^2$$

Upon substituting the foregoing equations into Equation 4.3.6, we see that

$$jS_{j+1}^2 = (j-1)S_j^2 + \left[(j+1)^2 + j + 1 - 2(j+1)\right]\left(\bar{X}_j - \bar{X}_{j+1}\right)^2$$

or, equivalently

$$jS_{j+1}^2 = (j-1)S_j^2 + (j+1)j(\overline{X}_j - \overline{X}_{j+1})^2$$

which, when divided through by j, yields the desired Equation 4.3.4. ∎

The recursive Equations 4.3.3 and 4.3.4 can be used to compute $\overline{X} = \overline{X}_n$ and $S^2 = S_n^2$ by starting with $\overline{X}_1 = X_1$.

Example 4.3a If $n = 4$ and the data values are $X_1 = 5$, $X_2 = 14$, $X_3 = 9$, $X_4 = 6$, then

$$\overline{X}_1 = X_1 = 5$$

$$\overline{X}_2 = \overline{X}_1 + \frac{X_2 - \overline{X}_1}{2} = 5 + \frac{9}{2} = \frac{19}{2}$$

$$S_2^2 = \left(1 - \frac{1}{1}\right)S_1^2 + 2(\overline{X}_2 - \overline{X}_1)^2 = \frac{81}{2}$$

$$\overline{X}_3 = \overline{X}_2 + \frac{X_3 - \overline{X}_2}{3} = \frac{19}{2} + \frac{9 - 19/2}{3} = \frac{28}{3}$$

$$S_3^2 = \left(1 - \frac{1}{2}\right)S_2^2 + 3(\overline{X}_3 - \overline{X}_2)^2 = \frac{61}{3}$$

$$\overline{X} = \overline{X}_4 = \overline{X}_3 + \frac{X_4 - \overline{X}_3}{4} = \frac{28}{3} + \frac{6 - \frac{28}{3}}{4} = \frac{17}{2}$$

$$S^2 = S_4^2 = \left(1 - \frac{1}{3}\right)S_3^2 + 4(\overline{X}_4 - \overline{X}_3)^2 = \frac{49}{3} \quad ∎$$

Program 4-3 uses the recursion Equations 4.3.3 and 4.3.4 to compute the sample mean, sample variance, and sample standard deviation of a data set.

Example 3.4b Compute the sample mean, variance, and standard deviation from the following set of data: 143, 147, 154, 158, 175, 139, 130, 157, 163, 166, 174, 169.

Solution Run Program 4-3:

```
RUN
THIS PROGRAM COMPUTES THE SAMPLE MEAN, SAMPLE VARIANCE, AND SAMPLE STANDARD DEVI
ATION OF A DATA SET
ENTER THE SAMPLE SIZE
? 12
ENTER THE DATA VALUES ONE AT A TIME
? 143
? 147
? 154
? 158
? 175
? 139
? 130
? 157
? 163
? 166
? 174
? 169
SAMPLE MEAN IS 156.25
SAMPLE VARIANCE IS 202.3868
SAMPLE STANDARD DEVIATION IS 14.22627
```

4 EMPIRICAL DISTRIBUTION FUNCTIONS, HISTOGRAMS, AND STEM-AND-LEAF PLOTS

Let $X_1, \ldots, X_n$ denote a random sample from the unknown distribution function F; and suppose we are interested in using the data to estimate F in its entirety. Since

$$F(x) = P\{X_i \leq x\}$$

a natural estimate for $F(x)$ is the proportion of the X_i that are less than or equal to x.

Definition. The function $F_n(x)$, $-\infty < x < \infty$, defined by

$$F_n(x) = \frac{\text{number of } i: X_i \leq x}{n}$$

is called the *empirical distribution function*.

Example 4.4a If a sample of size 10 yields the values $9, 5, 2, 1, 8, 2, 4, 8, 9, 8$, then the graph of the empirical distribution function is as follows:

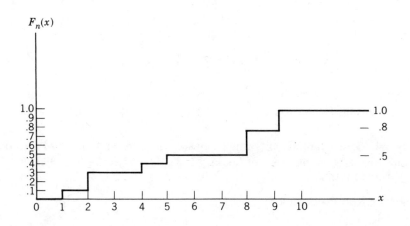

Example 4.4b The following table represents the lifetime of 200 incandescent lamps.

The corresponding empirical distribution function is presented in Figure 4.4.1.

It should be noted that, for each x, $F_n(x)$ is a random variable—in fact

$$nF_n(x) = \text{number } i: X_i \leq x$$

is a binomial random variable with parameters n and $F(x)$. This latter

TABLE 4.4.1

Life in Hours of 200 Incandescent Lamps

				Item Lifetimes					
1,067	919	1,196	785	1,126	936	918	1,156	920	948
855	1,092	1,162	1,170	929	950	905	972	1,035	1,045
1,157	1,195	1,195	1,340	1,122	938	970	1,237	956	1,102
1,022	978	832	1,009	1,157	1,151	1,009	765	958	902
923	1,333	811	1,217	1,085	896	958	1,311	1,037	702
521	933	928	1,153	946	858	1,071	1,069	830	1,063
930	807	954	1,063	1,002	909	1,077	1,021	1,062	1,157
999	932	1,035	944	1,049	940	1,122	1,115	833	1,320
901	1,324	818	1,250	1,203	1,078	890	1,303	1,011	1,102
996	780	900	1,106	704	621	854	1,178	1,138	951
1,187	1,067	1,118	1,037	958	760	1,101	949	992	966
824	653	980	935	878	934	910	1,058	730	980
844	814	1,103	1,000	788	1,143	935	1,069	1,170	1,067
1,037	1,151	863	990	1,035	1,112	931	970	932	904
1,026	1,147	883	867	990	1,258	1,192	922	1,150	1,091
1,039	1,083	1,040	1,289	699	1,083	880	1,029	658	912
1,023	984	856	924	801	1,122	1,292	1,116	880	1,173
1,134	932	938	1,078	1,180	1,106	1,184	954	824	529
998	996	1,133	765	775	1,105	1,081	1,171	705	1,425
610	916	1,001	895	709	860	1,110	1,149	972	1,002

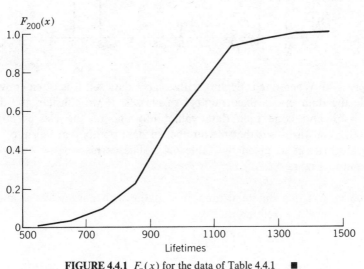

FIGURE 4.4.1 $F_n(x)$ for the data of Table 4.4.1 ∎

statement follows since each one of the X_i will, independently, be less than or equal to x with probability $F(x)$. Hence,

$$E[nF_n(x)] = nF(x)$$

and so

$$E[F_n(x)] = F(x)$$

When F is known to be a discrete distribution—say giving mass to the nonnegative integers—then it is often illuminating to look at $p_n(j)$, the empirical probability mass function, defined by

$$p_n(j) = \frac{\text{number } i : X_i = j}{n}$$

That is, $p_n(j)$ is the proportion of the n values that equal j. A graph of $p_n(j)$, $j = 0, 1, \ldots$ is called a *histogram*.

Example 4.4c For the data of Example 4.4a, the histogram is as follows:

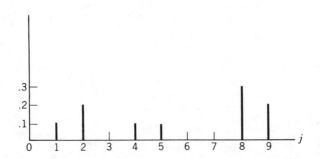

Example 4.4d When one is dealing with a large data set, it is often convenient to code the data into a fixed number of classes. If we consider the data of Table 4.4.1 and code each data value into one of the classes 501–600, 601–700, ..., it turns out that the number of data points that fall into each of these categories is as given by Table 4.4.2. The corresponding histogram is presented in Figure 4.4.2.

Since $np_n(j)$ is a binomial random variable with parameters n and $p = P\{X = j\}$ (why?), it follows that

$$E[p_n(j)] = P\{X = j\}$$

Another way of graphically representing data—similar to the histogram but with the added benefit of retaining all information about within-group data—is the so-called stem-and-leaf plot. We illustrate it by the following example.

TABLE 4.4.2

Class Interval	Frequency (Number of Data Values in the Interval)
501–600	2
601–700	5
701–800	12
801–900	25
901–1000	58
1001–1100	41
1101–1200	43
1201–1300	7
1301–1400	6
1401–1500	1

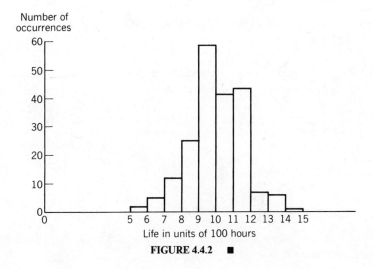

FIGURE 4.4.2   ■

Example 4.4e Suppose we have the following 40 data values:

92	91	72	56	81
74	67	75	62	77
68	66	74	68	72
81	69	88	80	62
86	70	58	78	59
75	71	57	65	94
66	90	83	81	83
54	87	69	76	81

These values can be graphically represented in the following stem-and-leaf plot.

Stem	Leaves
9	0, 1, 2, 4
8	0, 1, 1, 1, 1, 3, 3, 6, 7, 8
7	0, 1, 2, 2, 4, 4, 5, 5, 6, 7, 8
6	2, 2, 5, 6, 6, 7, 8, 8, 9, 9
5	4, 6, 7, 8, 9

Thus, for instance, the foregoing stem-and-leaf plot indicates 4 values in the nineties—namely, $90, 91, 92, 94$. ∎

When turned on its side, a stem-and-leaf plot looks like a histogram with the additional plus that it also presents the within-group data values.

5   SAMPLING DISTRIBUTIONS FROM A NORMAL POPULATION

Let $X_1, X_2, \ldots, X_n$ be a sample from a normal population having mean μ and variance σ^2. That is, they are independent and $X_i \sim \mathcal{N}(\mu, \sigma^2)$, $i = 1, \ldots, n$. Also let

$$\bar{X} = \sum_{i=1}^{n} X_i / n$$

and

$$S^2 = \frac{\sum_{i=1}^{n}(X_i - \bar{X})^2}{n - 1}$$

denote the sample mean and sample variance respectively. We would like to compute their distributions.

5.1   Distribution of Sample Mean

Since the sum of independent normal random variables is normally distributed (see Section 5 of Chapter 3), it follows that $\bar{X}$ is normal with mean

$$E[\bar{X}] = \sum_{i=1}^{n} \frac{E[X_i]}{n} = \mu$$

and variance

$$\text{Var}(\bar{X}) = \frac{1}{n^2} \sum_{i=1}^{n} \text{Var}(X_i) = \sigma^2 / n$$

That is, $\bar{X}$, the average of the sample, is normal with a mean equal to the population mean but with a variance reduced by a factor of $1/n$. It follows from this that $(\bar{X} - \mu)/(\sigma/\sqrt{n})$ is a unit normal random variable—that is,

$$\sqrt{n}\,\frac{(\bar{X} - \mu)}{\sigma} \sim \mathcal{N}(0,1)$$

5.2   Joint Distribution of $\bar{X}$ and S^2

In this section, we will not only obtain the distribution of the sample variance S^2 but we will also discover a fundamental fact about normal samples—namely, that $\bar{X}$ and S^2 are independent with $(n - 1)S^2/\sigma^2$ having a chi-square distribution with $n - 1$ degrees of freedom.

   To start, let us note the following algebraic identity: If for numbers $x_1, \ldots, x_n$ we set $\bar{x} = \sum_{i=1}^{n} x_i/n$, then for any constant μ we have that

$$\sum_{i=1}^{n} (x_i - \mu)^2 = \sum_{i=1}^{n} (x_i - \bar{x} + \bar{x} - \mu)^2 \tag{4.5.1}$$

$$= \sum_{i=1}^{n} (x_i - \bar{x})^2 + 2 \sum_{i=1}^{n} (x_i - \bar{x})(\bar{x} - \mu) + \sum_{i=1}^{n} (\bar{x} - \mu)^2$$

$$= \sum_{i=1}^{n} (x_i - \bar{x})^2 + 2(\bar{x} - \mu) \sum_{i=1}^{n} (x_i - \bar{x}) + n(\bar{x} - \mu)^2$$

$$= \sum_{i=1}^{n} (x_i - \bar{x})^2 + n(\bar{x} - \mu)^2$$

where the final step used the relationship that

$$\sum_{i=1}^{n} (x_i - \bar{x}) = \sum_{i=1}^{n} x_i - n\bar{x} = 0$$

   Now let $X_1, \ldots, X_n$ denote a sample from a normal population having mean μ and variance σ^2. From Equation 4.5.1, we obtain that

$$\frac{\sum_{i=1}^{n}(X_i - \mu)^2}{\sigma^2} = \frac{\sum_{i=1}^{n}(X_i - \bar{X})^2}{\sigma^2} + \frac{n(\bar{X} - \mu)^2}{\sigma^2}$$

or, equivalently

$$\sum_{i=1}^{n} \left(\frac{X_i - \mu}{\sigma}\right)^2 = \frac{\sum_{i=1}^{n}(X_i - \bar{X})^2}{\sigma^2} + \left[\frac{\sqrt{n}\,(\bar{X} - \mu)}{\sigma}\right]^2 \tag{4.5.2}$$

As $(X_i - \mu)/\sigma$, $i = 1, \ldots, n$ are independent unit normals, it follows that the left side of Equation 4.5.2 is a chi-square random variable with n degrees of freedom. Also, as shown in Section 5.1, $\sqrt{n}\,(\bar{X} - \mu)/\sigma$ is a unit normal random variable and so its square is a chi-square random variable with 1 degree of freedom. Thus Equation 4.5.2 equates a chi-square random variable having n degrees of freedom to the sum of 2 random variables, one of which is chi-square with 1 degree of freedom. Now it has been established that the sum of 2 independent chi-square random variables is also chi-square with a degree of freedom equal to the sum of the two degrees of freedom. Hence, it would seem that there is a reasonable possibility that the two terms on the right side of Equation 4.5.2 are independent with $\sum_{i=1}^{n}(X_i - \bar{X})^2/\sigma^2$ having a chi-square distribution with $n - 1$ degrees of freedom. Since this result can indeed be established, we thus have the following fundamental result.

Theorem 4.5.1

If $X_1, \ldots, X_n$ is a sample from a normal population having mean μ and variance σ^2, then $\bar{X}$ and S^2 are independent random variables, with $\bar{X}$ being normal with mean μ and variance σ^2/n and $(n - 1)S^2/\sigma^2$ being chi-square with $n - 1$ degrees of freedom.

Theorem 4.5.1 not only provides the distributions of $\bar{X}$ and S^2 for a normal population but also establishes the important fact that they are independent. In fact, it turns out that this independence of $\bar{X}$ and S^2 is a unique property of the normal distribution. Its importance will become evident in the following chapters.

Example 4.5a The time it takes a central processing unit to process a certain type of job is normally distributed with mean 20 seconds and standard deviation 3 seconds. If a sample of 15 such jobs is observed, what is the probability that the sample variance will exceed 12?

Solution Since the sample is of size $n = 15$ and $\sigma^2 = 9$, write

$$P\{S^2 > 12\} = P\left\{\frac{14S^2}{9} > \frac{14}{9} \cdot 12\right\}$$

$$= P\{\chi_{14}^2 > 18.67\}$$

$$= 1 - .8221 \qquad \text{from Program 3-8-1-A}$$

$$= .1779 \quad \blacksquare$$

The following corollary of Theorem 4.5.1 will be quite useful in the following chapters. Its proof is left as an exercise.

Corollary 4.5.2 Let $X_i, \ldots, X_n$ be a sample from a normal population with mean μ. If $\bar{X}$ denotes the sample mean and S the sample standard deviation, then

$$\sqrt{n}\,\frac{(\bar{X} - \mu)}{S} \sim t_{n-1}$$

That is, $\sqrt{n}(\bar{X} - \mu)/S$ has a t distribution with $n - 1$ degrees of freedom.

6 SAMPLING FROM A FINITE SET

Consider a set of N items each having some measurable characteristic and let v_i denote the measurable characteristic of the ith item, $i = 1, \ldots, N$. A common situation is one in which the set of values $\{v_1, v_2, \ldots, v_N\}$ is unknown and one wants to make inferences about this set by looking at the values of a randomly chosen sample.

Definition. A choice of a subset of size n from a larger set of size N is called a *random sample* if each of the $\binom{N}{n}$ subsets of size n is equally likely to be chosen.[†]

If we imagine that the choice is done sequentially, then a random sample would result if at each choice each of the remaining items not yet selected is equally likely to be chosen.

A quantity of some importance is $\bar{v} = \sum_{i=1}^{N} v_i/N$, the average of the N values. A natural way to estimate $\bar{v}$ is to take a random sample of n items: determine T, the sum of the values of these n items, and then use T/n as an estimator of $\bar{v}$. For instance, an important application of the foregoing is to a forthcoming election in which each individual in the population is either for or against a certain candidate or proposition. If we take v_i to equal 1 if individual i is in favor and 0 if he is against, then $\bar{v} = \sum_{i=1}^{N} v_i/N$ represents the proportion of the population that is in favor. To estimate $\bar{v}$, a random sample of n individuals is chosen and these individuals are polled. The proportion of those polled that are in favor—that is, T/n—is often used as an estimate of $\bar{v}$.

It is of interest to compute the mean and variance of T/n. To do so, define for each $i = 1, \ldots, n$ an indicator variable I_i to indicate whether or not item i is included in the sample. That is,

$$I_i = \begin{cases} 1 & \text{if item } i \text{ is in the random sample} \\ 0 & \text{otherwise} \end{cases}$$

Now T can be expressed by

$$T = \sum_{i=1}^{N} v_i I_i$$

and so

$$E[T] = \sum_{i=1}^{N} v_i E[I_i] \tag{4.6.1}$$

$$\text{Var}(T) = \sum_{i=1}^{N} \text{Var}(v_i I_i) + 2 \sum \sum_{i<j} \text{Cov}(v_i I_i, v_j I_j) \tag{4.6.2}$$

$$= \sum_{i=1}^{N} v_i^2 \text{Var}(I_i) + 2 \sum \sum_{i<j} v_i v_j \text{Cov}(I_i, I_j)$$

Since item i will be included in the sample with probability n/N, it follows

---

[†] Program 3-4 can be employed to generate a random subset.

that I_i is a Bernoulli random variable with

$$E[I_i] = \frac{n}{N}$$

Hence, from Equation 4.6.1,

$$E[T] = \frac{n}{N} \sum_{i=1}^{N} v_i$$

and so

$$E[T/n] = \sum_{i=1}^{N} \frac{v_i}{N} = \bar{v} \qquad (4.6.3)$$

Also, as the variance of a Bernoulli random variable having mean p is just $p(1 - p)$, it follows that

$$\text{Var}\,(I_i) = \frac{n}{N}\left(1 - \frac{n}{N}\right) \qquad (4.6.4)$$

Also, since

$$I_i I_j = \begin{cases} 1 & \text{if both items } i \text{ and } j \text{ are in the sample} \\ 0 & \text{otherwise} \end{cases}$$

we have that

$$E\left[I_i I_j\right] = P\{i \text{ and } j \text{ are in the sample}\}$$
$$= P\{i \text{ is the sample}\}\,P\{\,j \text{ is in } | i \text{ is in}\}$$
$$= \frac{n}{N}\frac{n-1}{N-1}$$

implying that

$$\text{Cov}\,(I_i, I_j) = E\left[I_i I_j\right] - E[I_i]E[I_j] \qquad (4.6.5)$$
$$= \frac{n(n-1)}{N(N-1)} - \left(\frac{n}{N}\right)^2$$
$$= \frac{-n(N-n)}{N^2(N-1)}$$

Hence, from Equations 4.6.2, 4.6.4, and 4.6.5 we see that

$$\text{Var}\,(T) = \frac{n}{N}\left(\frac{N-n}{N}\right)\sum_{i=1}^{N} v_i^2 - \frac{2n(N-n)}{N^2(N-1)}\sum\sum_{i<j} v_i v_j \qquad (4.6.6)$$

This expression can be simplified somewhat by using the identity $(v_1 + \cdots + v_N)^2 = \sum_{i=1}^{N} v_i^2 + 2\sum\sum_{i<j} v_i v_j$ to give, after some simplification,

$$\text{Var}\,(T/n) = \frac{1}{n^2}\,\text{Var}\,(T) = \frac{N-n}{n(N-1)}\left(\sum_{i=1}^{N} \frac{v_i^2}{N} - \bar{v}^2\right) \qquad (4.6.7)$$

Special case

(a) Consider now the special case in which Np of the v's are equal to 1 and the remainder equal to 0. Then in this case T (which is a hypergeometric random variable) will have mean and variance given by

$$E[T] = n\bar{v} = np \qquad \text{since } \bar{v} = \frac{Np}{N} = p$$

$$\text{Var}(T) = n^2 \text{Var}(T/n) = \frac{n(N-n)}{(N-1)}\left(\frac{Np}{N} - p^2\right)$$

$$= \frac{n(N-n)}{N-1}p(1-p)$$

where we have used that $\Sigma_1^N v_i^2 = \Sigma_1^N v_i = Np$. The quantity T/n, equal to the proportion of those sampled that have values equal to 1, is such that

$$E[T/n] = p$$

$$\text{Var}(T/n) = \frac{(N-n)}{n(N-1)}p(1-p)$$

(b) Another important special case is when $v_i = i$, $i = 1,\ldots, N$ (see Section 7 of Chapter 9). Using the identities

$$\sum_{i=1}^{N} i = N(N+1)/2 \qquad \sum_{i=1}^{N} i^2 = N(N+1)(2N+1)/6$$

it follows from Equations 4.6.3 and 4.6.7, after some algebraic manipulations, that

$$E[T] = n(N+1)/2$$

$$\text{Var}(T) = n(N-n)(N+1)/12$$

PROBLEMS

1. A data set of size 500 was determined. Each value was one of the numbers 1, 2, 3, 4, or 5 and the data were summarized as follows:

Value	Frequency (Number of Occurrences)
1	84
2	92
3	116
4	110
5	98

Determine the sample mean. In general, if the data are summarized as

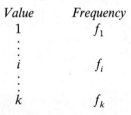

Value	Frequency
1	f_1
$\vdots$	
i	f_i
$\vdots$	
k	f_k

what is the sample mean?

2. Repeat Problem 1 first for the sample median and then for the sample mode.

3. Medians and modes, unlike means, are not additive. To verify this, let X and Y be independent exponential random variables, each having parameter $\lambda = 1$. Show that the median of $X + Y$ is not equal to the sum of the medians and then do the same for the mode.

4. Compute the quartiles and mode of the distribution

$$F(x) = 1 - e^{-x^2}, \qquad 0 < x < \infty$$

5. Compute the median and mode of the distribution whose density function is given by

$$f(x) = \begin{cases} xe^{-x}, & x > 0 \\ 0, & x < 0 \end{cases}$$

6. An economist is studying the incomes of U.S. citizens. What do you think is a more important figure—the average or the median income?

7. Let $\text{Med}(X)$ denote the median of the distribution of X. Show that

$$\text{Med}(aX + b) = a\,\text{Med}(X) + b$$

where a and b are constants.

8. Repeat Problem 7 when $\text{Mode}(X)$, which denotes the mode of the distribution of X, replaces $\text{Med}(X)$. Assume that $\text{Mode}(X)$ is unique.

9. Compute the mean, median, and mode of the following distributions:
   (a) Exponential with parameter λ
   (b) Uniform on (a, b)

10. Studies have shown that X, the distance in miles traveled by a passenger on a certain bus route of length 2 miles, has a probability density function given by

$$f(x) = \begin{cases} \frac{3}{8}(2 - x)^2, & 0 < x < 2 \\ 0, & \text{otherwise} \end{cases}$$

   (a) Graph the density function.
   (b) Determine the mode of the distribution.
   (c) Determine the median.
   (d) Determine the mean.

11. Suppose that X is a nonnegative random variable whose mean and median are equal to 4. (More precisely the mean and median of the distribution function of X are equal to 4.)

   (a) What is the mean and median of the distribution of $Y = 2X + 5$?
   (b) Do you have enough information to determine the mean and the median of the distribution of $Y = X^2$?

12. Show that the mean, median, and mode of a normal distribution are all equal.

13. Use the identity $\text{Var}(X) = E[X^2] - (E[X])^2$ to prove that

$$\sum_{i=1}^{n} (x_i - \bar{x})^2 = \sum_{i=1}^{n} x_i^2 - n\bar{x}^2$$

14. Compute by hand the sample variance of the following data set first by using Equation 4.3.1 and second, as a check, by using the recursion of Section 3.1.

$$8, 12, 14, 10, 11, 9, 7, 13$$

15. Verify Equation 4.3.5.

16. The following data represents the lifetimes (in hours) of a sample of 40 transistors:

$$112, 121, 126, 108, 141, 104, 136, 134$$
$$121, 118, 143, 116, 108, 122, 127, 140$$
$$113, 117, 126, 130, 134, 120, 131, 133$$
$$118, 125, 151, 147, 137, 140, 132, 119$$
$$110, 124, 132, 152, 135, 130, 136, 128$$

   (a) Determine the sample mean, median, and mode.
   (b) Plot the empirical distribution function.
   (c) Group the data into the 8 classes 108–113, 114–119, 120–125, 126–131, 132–137, 138–143, 144–149, 150–155, and plot the resulting histogram.

17. An experiment measuring the percent shrinkage on drying of 50 clay specimens produced the following data:

18.2	21.2	23.1	18.5	15.6
20.8	19.4	15.4	21.2	13.4
16.4	18.7	18.2	19.6	14.3
16.6	24.0	17.6	17.8	20.2
17.4	23.6	17.5	20.3	16.6
19.3	18.5	19.3	21.2	13.9
20.5	19.0	17.6	22.3	18.4
21.2	20.4	21.4	20.3	20.1
19.6	20.6	14.8	19.7	20.5
18.0	20.8	15.8	23.1	17.0

(a) Draw a stem-and-leaf plot of these data.

(b) Compute the sample mean, median, and mode.

(c) Compute the sample variance.

(d) Draw the empirical distribution function.

(e) Group the data into class intervals of size 1 percent starting with the interval 13.0–13.9, and draw the resulting histogram.

(f) For the grouped data acting as if each of the data points in an interval was actually located at the midpoint of that interval, compute the sample mean and sample variance and compare this with the result obtained above. Why do they differ?

18. Use Program 4-3 to compute the sample mean and sample variance for the data given in Problem 16.

19. In the foregoing problem, if each data value were increased by 10, what would the effect be on (a) the sample mean, (b) the sample median, (c) the sample mode, (d) the sample variance, (e) the sample standard deviation.

20. Soil acidity in a certain region is measured weekly with successive measurements constituting a sample from a normal population with mean 50 and standard deviation 8. For measurements taken over a 12-week period, what is the probability that

(a) The sample mean exceeds 55?

(b) The sample standard deviation exceeds 10?

(c) Repeat (b) when it is given that the sample mean is equal to 48.

21. The lifetime of a certain electrical part is, according to its manufacturer, supposed to be normally distributed with mean 100 (hours) and standard deviation 20 (hours). A sample of 10 such parts is to be tested.

(a) What is the probability that the sample mean will be less than or equal to 85?

(b) Suppose you are somewhat skeptical about the manufacturer's claim. What conclusion would you draw if it turns out that the sample mean is equal to 85?

22. The temperature at which certain thermostats are set to go on is normally distributed with variance σ^2. A random sample is to be drawn and the sample variance S^2 computed. How many observations are required to ensure that

(a) $P\{S^2/\sigma^2 \leq 1.8\} \geq .95$

(b) $P\left\{.85 \leq \dfrac{S^2}{\sigma^2} \leq 1.15\right\} \geq .99$

23. If $X_1, \ldots, X_n$ is a sample from a normal population with mean μ and variance σ^2, prove that $\sqrt{n}\,(\bar{X} - \mu)/S$ has a t-distribution with $n - 1$ degrees of freedom.

Hint: Use Theorem 4.5.1 and the definition of the t-distribution.

24. Consider 2 independent samples—the first of size 10 from a normal population having variance 4 and the second of size 5 from a normal population having variance 2. Compute the probability that the sample variance from the second sample exceeds the one from the first.

Hint: Relate it to the F distribution.

25. A bank executive wishes to determine the average amount of money contained in a savings account in his bank. Describe a method for estimating this quantity without examining each account.

26. A university administrator is interested in the average class size in her university. To estimate this quantity, she randomly chooses, from a list of the 10,000 students at her university, a group of 100 students who are then questioned about the number of students in each of their classes. She then uses the average of all the data obtained as her estimate. (For instance, if each student was taking 5 classes, then there would be 500 data values.) What do you think of this procedure? Is the administrator really estimating the average class size?

27. The workers in a certain industrial firm are considering forming a union. This is strongly opposed by the management, and to get an indication as to their workers' feelings, they hired a pollster to estimate the proportion of workers that favor a union. The pollster telephoned 500 randomly chosen workers. Of these 500, 200 refused to talk to the pollster, 200 were against and 100 were for the union. He then reported that two-thirds of the workers were against the union. What do you think of his conclusion?

28. Verify that Equation 4.6.7 follows from Equation 4.6.6.

C H A P T E R 5

Parameter Estimation

0 INTRODUCTION

Let $X_1, \ldots, X_n$ be a random sample from a distribution F_θ that is specified up to a vector of unknown parameters θ. For instance, the sample could be from a Poisson distribution whose mean value is unknown; or it could be from a normal distribution having an unknown mean and variance. Whereas in probability theory it is usual to suppose that all of the parameters of a distribution are known, the opposite is true in statistics, where a central problem is to use the observed data to make inferences about the unknown parameters.

In Sections 1 and 2, we present two general approaches, called the *method of moments* and the *maximum likelihood* approach for determining estimates of unknown parameters. These estimates are called *point estimates* because they specify a single quantity as an estimate of θ. In Section 3, we consider the problem of obtaining *interval estimates*. In this case, rather than specifying a certain value as our estimate of θ, we specify an interval in which we estimate that θ lies. We also consider the question of how much *confidence* we can attach to a given interval estimate. We then illustrate this approach by showing how to obtain an interval estimate of the unknown mean of a normal distribution when the variance of the distribution is specified. We then consider a variety of interval estimation problems. Specifically, in Section 3.1, we consider an interval estimate of the mean of a normal distribution when the variance is not assumed known. In Section 3.2, we determine an interval estimate for the difference of two normal means both when their variances are assumed known and unknown (though in this latter case we suppose that their unknown variances are equal). In Section 3.3 we consider the problem of obtaining an interval estimate of the variance of a normal population. Finally in Sections 3.4 and 3.5 we present interval estimates of the mean of a Bernoulli random variable and the mean of an exponential random variable.

In Section 4 we return to the problem of obtaining point estimates of unknown parameters, and show how to evaluate an estimator by considering its mean square error. The bias of an estimator is discussed and the relationship between it and the mean square error of the estimator is explored.

160

In Section 5, we consider the problem of determining an estimate of an unknown parameter when there is some prior information available. This is the *Bayesian* approach, which supposes that prior to observing the data there is always information about θ that is available to the decision maker, and which can be expressed in terms of a probability distribution on θ. For such a situation, we show how to compute the *Bayes* estimator—that is, the estimator whose average squared distance from θ is minimal.

1 METHOD OF MOMENTS ESTIMATORS

Any statistic used to estimate the value of an unknown parameter θ is called an *estimator* of θ. The observed value of the estimator is called the *estimate*. For instance, as we shall see, the usual estimator of the mean of a normal population, based on a sample $X_1, \ldots, X_n$ from that population, is the sample average $\overline{X} = \Sigma_i X_i / n$. If a sample of size 3 yields the data $X_1 = 2$, $X_2 = 3$, $X_3 = 4$, then the estimate of the population mean, resulting from the estimator $\overline{X}$ is the value 3.

The earliest general method for determining an estimator of an unknown parameter, introduced by Karl Pearson in 1894, is known as the method of moments. It works as follows: Suppose that $X = (X_1, \ldots, X_n)$ is a sample from a distribution F_θ where θ is an unknown parameter that we wish to estimate. Suppose also that θ can be expressed as a function of the mean of the distribution—that is, say, $\theta = g(E[X])$. The method of moment approach is to use $\overline{X} = \Sigma_{i=1}^n X_i / n$ as an estimate of $E[X]$ and then estimate θ by $g(\overline{X})$.

It follows that the method of moments estimate of a population mean is always the sample mean. Thus, if the sample was from a Poisson distribution whose mean value d was unknown or from a normal distribution whose mean μ was unknown, then in both cases the method of moments estimator would be the sample mean $\overline{X}$.

Suppose now that it were not possible to express the unknown parameter θ as a function of only $E[X]$. For instance, the sample could be from a normal population whose variance σ^2 is unknown, and there is no way to write σ^2 as a function of only $E(X)$. In this situation, the method of moments approach is to proceed as follows: Consider the moments $E[X^k]$, $k \geq 1$, of the distribution and express θ as a function of these moments—for instance, suppose θ depends on the first r moments and so we can express it as

$$\theta = g(E[X], E[X^2], \ldots, E[X^r])$$

Then, letting

$$M_k = \frac{\sum_{i=1}^n X_i^k}{n}, \qquad k = 1, \ldots, r$$

we call M_k the kth *sample moment* and use it to estimate $E[X^k]$. The method of moments estimator of θ is then given by $\hat{\theta}$ where

$$\hat{\theta} = g(M_1, M_2, \ldots, M_r)$$

Example 5.1a Suppose we want to estimate both the mean μ and the variance σ^2 when we have a sample of size n from a normal population. Since

$$\mu = E[X]$$

and

$$\sigma^2 = E[X^2] - E^2[X]$$

it follows that the method of moments estimators of μ and σ^2 are $\hat{\mu}$ and $\hat{\sigma}^2$ given by

$$\hat{\mu} = \frac{\sum_{i=1}^n X_i}{n}$$

$$\hat{\sigma}^2 = \frac{\sum_{i=1}^n X_i^2}{n} - \left(\frac{\sum_{i=1}^n X_i}{n}\right)^2$$

or

$$\hat{\mu} = \overline{X}$$

$$\hat{\sigma}^2 = \frac{\sum_{i=1}^n X_i^2 - n\overline{X}^2}{n} = \frac{\sum_{i=1}^n \left(X_i - \overline{X}\right)^2}{n}$$

where the final equality made use of the useful identity (established in Chapter 4, Section 3):

$$\sum_{i=1}^n \left(X_i - \overline{X}\right)^2 = \sum_{i=1}^n X_i^2 - n\overline{X}^2 \quad \blacksquare$$

REMARK
We do not expect an estimator to always generate an estimate that is equal, or even always close to, the unknown parameter. In general, what we look for in an estimator is that it leads to estimates that are close to the unknown parameter "on the average." A discussion on how one can evaluate estimators will be given in Section 4.

2 MAXIMUM LIKELIHOOD ESTIMATORS

Suppose that the random variables $X_1, \ldots, X_n$, whose joint distribution is assumed given except for an unknown parameter θ, are to be observed. The problem of interest is to use the observed values to estimate θ. For example, the X_i's might be independent, exponential random variables each having the same unknown mean θ. In this case, the joint density function of the random variables would be given by

$$
\begin{aligned}
f(x_1, & x_2, \ldots, x_n) \\
&= f_{X_1}(x_1) f_{X_2}(x_2) \cdots f_{X_n}(x_n) \\
&= \frac{1}{\theta} e^{-x_1/\theta} \frac{1}{\theta} e^{-x_2/\theta} \cdots \frac{1}{\theta} e^{-x_n/\theta} \qquad 0 < x_i < \infty, \ i = 1, \ldots, n \\
&= \frac{1}{\theta^n} \exp\left\{-\sum_1^n x_i/\theta\right\} \qquad 0 < x_i < \infty, \ i = 1, \ldots, n
\end{aligned}
$$

and the object would be to estimate θ from the observed data $X_1, X_2, \ldots, X_n$.

A particular type of estimator, known as the *maximum-likelihood* estimator, is widely used in statistics. It is obtained by reasoning as follows. Let $f(x_1, \ldots, x_n|\theta)$ denote the joint probability mass function of the random variables $X_1, X_2, \ldots, X_n$ when they are discrete, and let it be their joint probability density function when they are jointly continuous random variables. Because θ is assumed unknown, we also write f as a function of θ. Now since $f(x_1, \ldots, x_n|\theta)$ represents the likelihood that the values $x_1, x_2, \ldots, x_n$ will be observed when θ is the true value of the parameter, it would seem that a reasonable estimate of θ would be that value yielding the largest likelihood of the observed values. In other words, the maximum-likelihood estimate $\hat{\theta}$ is defined to be that value of θ maximizing $f(x_1, \ldots, x_n|\theta)$ where $x_1, \ldots, x_n$ are the observed values. The function $f(x_1, \ldots, x_n|\theta)$ is often referred to as the *likelihood* function of θ.

In determining the maximizing value of θ it is often useful to use the fact that $f(x_1, \ldots, x_n|\theta)$ and $\log[f(x_1, \ldots, x_n|\theta)]$ have their maximum at the same value of θ. Hence, we may also obtain $\hat{\theta}$ by maximizing $\log[f(x_1, \ldots, x_n|\theta)]$.

Example 5.2a Maximum Likelihood Estimator of a Bernoulli Parameter Suppose that n independent trials each of which is a success with probability p are performed. What is the maximum likelihood estimator of p?

Solution The data consists of the values of $X_1, \ldots, X_n$ where

$$X_i = \begin{cases} 1 & \text{if trial } i \text{ is a success} \\ 0 & \text{otherwise} \end{cases}$$

Now

$$P\{X_i = 1\} = p = 1 - P\{X_i = 0\}$$

which can be succinctly expressed as

$$P\{X_i = x\} = p^x(1 - p)^{1-x} \qquad x = 0, 1$$

Hence, by the assumed independence of the trials, the likelihood (that is, the joint probability mass function) of the data is given by

$$f(x_1, \ldots, x_n|p) = P\{X_1 = x_1, \ldots, X_n = x_n|p\}$$
$$= p^{x_1}(1 - p)^{1-x_1} \cdots p^{x_n}(1 - p)^{1-x_n}$$
$$= p^{\sum_1^n x_i}(1 - p)^{n - \sum_1^n x_i} \qquad x_i = 0, 1 \qquad i = 1, \ldots, n$$

To determine the value of p that maximizes the likelihood, first take logs to obtain:

$$\log f(x_1, \ldots, x_n|p) = \sum_1^n x_i \log p + \left(n - \sum_1^n x_i\right)\log(1 - p)$$

Differentiation yields

$$\frac{d}{dp}\log f(x_1, \ldots, x_n|p) = \frac{\sum_1^n x_i}{p} - \frac{(n - \sum_1^n x_i)}{1 - p}$$

Upon equating to zero and solving, we obtain that the maximum likelihood estimate $\hat{p}$ satisfies

$$\frac{\sum_1^n x_i}{\hat{p}} = \frac{n - \sum_1^n x_i}{1 - \hat{p}}$$

or

$$\hat{p} = \frac{\sum_{i=1}^n x_i}{n}$$

Hence, the maximum likelihood estimator of the unknown mean of a Bernoulli distribution is given by

$$d(X_1, \ldots, X_n) = \frac{\sum_{i=1}^n X_i}{n}$$

In words, the maximum likelihood estimator of p agrees with the method of moments estimator and is equal to the proportion of the observed trials that result in successes. For an illustration, suppose that each RAM (random access memory) chip produced by a certain manufacturer is, independently, of acceptable quality with probability p. Then if out of a sample of 1000 tested 921 are acceptable, it follows that the maximum likelihood estimate of p is .921. ∎

Example 5.2b Maximum-Likelihood Estimator of a Poisson Parameter Suppose $X_1, \ldots, X_n$ are independent Poisson random variables each having mean λ. Determine the maximum-likelihood estimator of λ.

Solution The likelihood function is given by

$$f(x_1, \ldots, x_n | \lambda) = \frac{e^{-\lambda} \lambda^{x_1}}{x_1!} \cdots \frac{e^{-\lambda} \lambda^{x_n}}{x_n!}$$

$$= \frac{e^{-n\lambda} \lambda^{\sum_1^n x_i}}{x_1! \cdots x_n!}$$

Thus,

$$\log f(x_1, \ldots, x_n | \lambda) = -n\lambda + \sum_1^n x_i \log \lambda - \log c$$

where $c = \prod_{i=1}^n x_i!$ does not depend on λ, and

$$\frac{d}{d\lambda} \log f(x_1, \ldots, x_n | \lambda) = -n + \frac{\sum_1^n x_i}{\lambda}$$

By equating to zero, we obtain that the maximum-likelihood estimate $\hat{\lambda}$ equals

$$\hat{\lambda} = \frac{\sum_1^n x_i}{n}$$

and so the maximum-likelihood estimator is given by

$$d(X_1, \ldots, X_n) = \frac{\sum_{i=1}^n X_i}{n}$$

For example, suppose that the number of people that enter a certain retail establishment in any day is a Poisson random variable having an unknown mean λ, which must be estimated. If after 20 days a total of 857 people have entered the establishment, then the maximum likelihood estimate of λ is $857/20 = 42.85$. That is, we estimate that on average, 42.85 customers will enter the establishment on a given day. ∎

Example 5.2c Maximum-Likelihood Estimator in a Normal Population Suppose $X_1, \ldots, X_n$ are independent, normal random variables each with unknown mean μ and unknown standard deviation σ. The joint density is given by

$$f(x_1, \ldots, x_n | \mu, \sigma) = \prod_{i=1}^{n} \frac{1}{\sqrt{2\pi}\,\sigma} \exp\left[\frac{-(x_i - \mu)^2}{2\sigma^2}\right]$$

$$= \left(\frac{1}{2\pi}\right)^{n/2} \frac{1}{\sigma^n} \exp\left[\frac{-\Sigma_1^n(x_i - \mu)^2}{2\sigma^2}\right]$$

The logarithm of the likelihood is thus given by

$$\log f(x_1, \ldots, x_n | \mu, \sigma) = -\frac{n}{2}\log(2\pi) - n\log\sigma - \frac{\Sigma_1^n(x_i - \mu)^2}{2\sigma^2}$$

In order to find the value of μ and σ maximizing the foregoing, we compute

$$\frac{\partial}{\partial \mu}\log f(x_1, \ldots, x_n | \mu, \sigma) = \frac{\Sigma_{i=1}^n(x_i - \mu)}{\sigma^2}$$

$$\frac{\partial}{\partial \sigma}\log f(x_1, \ldots, x_n | \mu, \sigma) = -\frac{n}{\sigma} + \frac{\Sigma_1^n(x_i - \mu)^2}{\sigma^3}$$

Equating these equations to zero yields that

$$\hat{\mu} = \frac{\Sigma_{i=1}^n x_i}{n}$$

and

$$\hat{\sigma} = \left(\frac{\Sigma_{i=1}^n(x_i - \hat{\mu})^2}{n}\right)^{1/2}$$

Hence, the maximum-likelihood estimators of μ and σ are given respectively by

$$\overline{X} \quad \text{and} \quad \left(\frac{\Sigma_{i=1}^n(X_i - \overline{X})^2}{n}\right)^{1/2} \tag{5.2.1}$$

It should be noted that the maximum likelihood estimator of the standard deviation σ differs from the sample standard deviation

$$S = \left[\Sigma_{i=1}^n(X_i - \overline{X})^2/(n-1)\right]^{1/2}$$

in that the denumerator in Equation 5.2.1 is $\sqrt{n}$ rather than $\sqrt{n-1}$. However, for n of reasonable size, these two estimators of σ will be approximately equal. ∎

In all of the foregoing examples, the maximum likelihood estimator of the population mean turned out to be the sample mean $\overline{X}$. To show that this is not always the situation, consider the following example.

Example 5.2d Estimating the Mean of a Uniform Distribution Suppose $X_1, \ldots, X_n$ constitute a sample from a uniform distribution on $(0, \theta)$ where θ is unknown. Their joint density is thus

$$f(x_1, x_2, \ldots, x_n | \theta) = \begin{cases} \dfrac{1}{\theta^n} & 0 < x_i < \theta, \quad i = 1, \ldots, n \\ 0 & \text{otherwise} \end{cases}$$

This density is maximized by choosing θ as small as possible. Since θ must be at least as large as all of the observed values x_i, it follows that the smallest possible choice of θ is equal to $\max(x_1, x_2, \ldots, x_n)$. Hence, the maximum likelihood estimator of θ is

$$\hat{\theta} = \max(X_1, X_2, \ldots, X_n)$$

It easily follows from the foregoing that the maximum likelihood estimator of $\theta/2$, the mean of the distribution, is $\max(X_1, X_2, \ldots, X_n)/2$.

On the other hand, since $E[X] = \theta/2$, or $\theta = 2E[X]$, the method of moments estimator of θ is

$$d_2 = \frac{2\sum_{i=1}^n X_i}{n}$$

Thus, in the preceding situation, the method of moments estimator differs from the maximum likelihood estimator. How can we determine which one is a better estimator of θ? One way is by taking an empirical approach via a simulation study. That is, suppose we choose a value θ and then simulate random variables from a uniform $(0, \theta)$ distribution. We can then actually compare the two estimated values (namely, the maximum and twice the sample mean) with the true value θ to see which estimator performed better. We can then repeat this procedure (using different random numbers) and see if we can spot a general pattern of the superiority of one of the estimators. Let's see how such an approach works in the case under consideration.

To start, recall that if U is a uniform $(0, 1)$ random variable (that is, U is a random number) then θU is uniform on $(0, \theta)$. Hence, we can choose a value of n, the sample size, and θ, and then have the computer generate random numbers $U_1, U_2, \ldots, U_n$ and set $X_i = \theta U_i$, $i = 1, \ldots, n$. We then compare

$$d_1 = \max(X_1, \ldots, X_n) = \theta \max(U_1, \ldots, U_n)$$

and

$$d_2 = \frac{2\sum_{i=1}^n X_i}{n} = \frac{\theta\, 2\sum_{i=1}^n U_i}{n}$$

with the actual value θ. In fact as we can see, θ can be factored out from all of the foregoing and all comparisons can be done with $\theta = 1$. That is, the approach is to generate n, a chosen number, of random numbers $U_1, \ldots, U_n$ and then compare how closely $\max_i U_i$ and $2\sum_{i=1}^{n} U_i/n$ estimate 1. This should then be repeated until one can come to some conclusion as to which estimator performs better.

To perform the foregoing we have written the following Basic program to generate 12 samples of 15 uniform $(0, 1)$ random variables and then compute the 2 estimators for each sample. The absolute value of the error involved was then determined.

```
10 RANDOMIZE
20 FOR J=1 TO 12
30 M=0
40 T=0
50 FOR I=1 TO 15
60 U=RND
70 IF U>M THEN M=U
80 T=T+U
90 NEXT
100 PRINT "1-MAX = "1-M, "ABS(2*XBAR-1) = "ABS(2*T/15-1)
110 NEXT
120 END
Ok
RUN
Random number seed (-32768 to 32767)? 1324
1-MAX =   .1957145          ABS(2*XBAR-1) =   .1538227
1-MAX =   3.121013E-02      ABS(2*XBAR-1) =   1.537955E-02
1-MAX =   5.913019E-03      ABS(2*XBAR-1) =   .3164175
1-MAX =   .2225734          ABS(2*XBAR-1) =   .2307838
1-MAX =   .1352978          ABS(2*XBAR-1) =   2.284479E-02
1-MAX =   1.971906E-02      ABS(2*XBAR-1) =   5.976451E-02
1-MAX =   .0321312          ABS(2*XBAR-1) =   .1067939
1-MAX =   1.243359E-02      ABS(2*XBAR-1) =   .1624677
1-MAX =   6.085456E-03      ABS(2*XBAR-1) =   .2474407
1-MAX =   7.542491E-02      ABS(2*XBAR-1) =   5.232442E-02
1-MAX =   3.900433E-02      ABS(2*XBAR-1) =   .0588156
1-MAX =   .0743739          ABS(2*XBAR-1) =   .1999309
Ok
```

As the preceding data indicate, for estimating θ in a uniform $(0, \theta)$ population, the maximum likelihood estimator certainly appears to be superior to the method of moments estimator for samples of size 15. In fact, as we shall show by a theoretical argument in Section 4, this will always be the case no matter what the sample size. ■

3 INTERVAL ESTIMATES

Suppose that $X_1, \ldots, X_n$ is a sample from a normal population having unknown mean μ and known variance σ^2. It has been shown that $\overline{X} = \sum_{i=1}^{n} X_i/n$ is the maximum likelihood estimator for μ. However, we don't expect that the sample mean $\overline{X}$ will exactly equal μ, but rather that it will "be close." Hence, rather than a point estimate, it is sometimes more valuable to be able to specify an interval for which we have a certain degree of confidence that μ lies within. To obtain such an interval estimator, we make use of the probability distribution of the point estimator. Let us see how it works for the preceding situation.

In the foregoing, since $\sqrt{n}\,(\overline{X}-\mu)/\sigma$ will have a unit normal distribution, it follows that

$$P\left\{-1.96 < \sqrt{n}\,\frac{(\overline{X}-\mu)}{\sigma} < 1.96\right\} = .95$$

or, equivalently

$$P\left\{\overline{X} - 1.96\frac{\sigma}{\sqrt{n}} < \mu < \overline{X} + 1.96\frac{\sigma}{\sqrt{n}}\right\} = .95$$

That is, 95 percent of the time μ will lie within $1.96\sigma/\sqrt{n}$ units of the sample average. If we now observe the sample and it turns out that $\overline{X} = \bar{x}$, then we say that "with 95 percent confidence"

$$\bar{x} - 1.96\frac{\sigma}{\sqrt{n}} < \mu < \bar{x} + 1.96\frac{\sigma}{\sqrt{n}} \qquad (5.3.1)$$

That is, "with 95 percent confidence" we assert that the true mean lies within $1.96\sigma/\sqrt{n}$ of the observed sample mean. The interval $(\bar{x} - 1.96\sigma/\sqrt{n}, \bar{x} + 1.96\sigma/\sqrt{n})$ is called a 95 percent level confidence interval for μ.

The interval in Equation 5.3.1 is called a *two-sided confidence interval*. Sometimes, however, we are interested in determining a value so that we can assert with, say, 95 percent confidence, that μ is at least as large as that value. To do so we note that, since for a unit normal random variable Z,

$$P\{Z < 1.64\} = .95$$

it follows that

$$P\left\{\sqrt{n}\,\frac{(\overline{X}-\mu)}{\sigma} < 1.64\right\} = .95$$

or

$$P\left\{\overline{X} - 1.64\frac{\sigma}{\sqrt{n}} < \mu\right\} = .95$$

and so a 95 percent *one-sided confidence interval* for μ is

$$\left(\bar{x} - 1.64\frac{\sigma}{\sqrt{n}}, \infty\right)$$

where $\bar{x}$ is the observed sample mean. We can also obtain a one-sided confidence interval that specifies that μ is less than a certain value (see Problem 8).

We can also obtain confidence intervals of any specified level of confidence. To do so recall that z_α is such that

$$P\{Z > z_\alpha\} = \alpha \qquad 0 < \alpha < 1$$

when Z is a unit normal. Hence, for any specified level of confidence $1 - \alpha$, we have (see Figure 5.3.1) that because

$$P\{-z_{\alpha/2} < Z < z_{\alpha/2}\} = 1 - \alpha$$

it follows that

$$P\left\{-z_{\alpha/2} < \sqrt{n}\,\frac{(\overline{X} - \mu)}{\sigma} < z_{\alpha/2}\right\} = 1 - \alpha$$

or

$$P\left\{\overline{X} - z_{\alpha/2}\,\frac{\sigma}{\sqrt{n}} < \mu < \overline{X} + z_{\alpha/2}\,\frac{\sigma}{\sqrt{n}}\right\} = 1 - \alpha$$

Hence a $100(1 - \alpha)$ percent two-sided confidence interval for μ is

$$\left(\overline{x} - z_{\alpha/2}\,\frac{\sigma}{\sqrt{n}},\quad \overline{x} + z_{\alpha/2}\,\frac{\sigma}{\sqrt{n}}\right)$$

where $\overline{x}$ is the observed sample mean. Similarly, we can obtain one-sided confidence intervals for μ at any desired level of confidence.

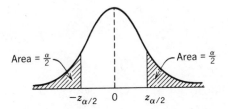

Area $= \frac{\alpha}{2}$ Area $= \frac{\alpha}{2}$

$-z_{\alpha/2}$ 0 $z_{\alpha/2}$

FIGURE 5.3.1 Unit normal density

Example 5.3a Suppose that when a signal having value μ is transmitted from location A the value received at location B is normally distributed with mean μ and variance 4. That is, if μ is sent, then the value received is $\mu + N$ where N, representing noise, is normal with mean 0 and variance 4. To reduce error, suppose the same value is sent 9 times. If the successive values received are $5, 8.5, 12, 15, 7, 9, 7.5, 6.5, 10.5$, let us construct a 95 percent confidence interval for μ.

Since

$$\overline{x} = \frac{81}{9} = 9$$

It follows, under the assumption that the values received are independent, that a 95 percent confidence interval for μ is

$$\left(9 - 1.96\frac{\sigma}{3}, 9 + 1.96\frac{\sigma}{3}\right) = (7.69, 10.31)$$

Hence, we are "95 percent confident" that the true message value lies between 7.69 and 10.31.

If we were interested in a one-sided interval that gave a lower bound on μ, then we would use the fact that

$$P\left\{\frac{3(\overline{X}-\mu)}{2} < z_\alpha\right\} = 1 - \alpha$$

or

$$P\left\{\mu > \overline{X} - \tfrac{2}{3}z_\alpha\right\} = 1 - \alpha$$

For instance, with "95 percent confidence"

$$\mu \in \left(\overline{x} - \tfrac{2}{3}(1.64), \infty\right)$$

or

$$\mu \in (7.91, \infty) \qquad \text{with "95 percent confidence"} \quad \blacksquare$$

Sometimes one is interested in a two-sided confidence interval of a certain level, say $1 - \alpha$, and the problem is to choose the sample size n so that the interval is of a certain size. For instance, suppose that we want to compute an interval of length .1 that we can assert, with 99 percent confidence, contains μ. How large need n be? To solve this, note that as $z_{.005} = 2.58$ it follows that the 99 percent confidence interval for μ from a sample of size n is

$$\left(\overline{x} - 2.58\frac{\sigma}{\sqrt{n}}, \quad \overline{x} + 2.58\frac{\sigma}{\sqrt{n}}\right)$$

Hence, its length is

$$5.16\frac{\sigma}{\sqrt{n}}$$

Thus, to make the length of the interval equal to .1, we must choose

$$5.16\frac{\sigma}{\sqrt{n}} = .1$$

or

$$n = (51.6\sigma)^2$$

REMARK

The interpretation of "a $100(1 - \alpha)$ percent confidence interval" can be confusing. It should be noted that we are *not* asserting that the probability that $\mu \in (\overline{x} - 1.96\sigma/\sqrt{n}, \overline{x} + 1.96\sigma/\sqrt{n})$ is .95, for there are no random variables involved in this assertion and thus nothing is random. What we are asserting is that the technique utilized to obtain this interval is such that 95 percent of the time that it is employed it will result in an interval in which μ lies. In other words, before the data are observed we can assert that with probability .95 the interval that will be obtained will contain μ, whereas after the data are obtained we can only assert that the resultant interval indeed contains μ "with confidence .95."

Example 5.3b Suppose we wish to estimate the average CPU (central processing unit) service time of computer jobs at a given facility, so that we can assert with 95 percent confidence that the estimated value is within $\tfrac{1}{2}$ second of the true value. If past experience leads us to believe that the CPU service times are normally distributed with a variance of $\sigma^2 = 2.25$ (seconds), how large a random sample is needed?

Solution A 95 percent confidence interval for the unknown mean μ, based on a sample of size n, is

$$\mu \in \left(\bar{x} - 1.96 \frac{\sigma}{\sqrt{n}}, \quad \bar{x} + 1.96 \frac{\sigma}{\sqrt{n}} \right)$$

As the length of this interval is $3.92\sigma/\sqrt{n} = 3.92\sqrt{2.25/n}$, we must choose n so that

$$3.92\sqrt{\frac{2.25}{n}} = .5$$

or

$$n = 2.25 \times (7.84)^2 = 138.298$$

Hence a sample of size 139 is required. ∎

3.1 Confidence Interval for a Normal Mean When the Variance Is Unknown

Suppose now that $X_1, \ldots, X_n$ is a sample from a normal distribution with unknown mean μ and unknown variance σ^2 and that we wish to construct a $100(1 - \alpha)$ percent confidence interval for μ. Since σ is unknown, we can no longer base our interval on the fact that $[\sqrt{n}(\bar{X} - \mu)]/\sigma$ has a unit normal random variable. However, by letting $S^2 = \sum_{i=1}^{n}(X_i - \bar{X})^2/(n - 1)$ denote the sample variance, then from Corollary 4.5.2 of Chapter 4, $[\sqrt{n}(\bar{X} - \mu)]/S$ has a t distribution with $n - 1$ degrees of freedom. Hence, for any $\alpha \in (0, \frac{1}{2})$

$$P\left\{ t_{1-\alpha/2, n-1} < \sqrt{n} \frac{(\bar{X} - \mu)}{S} < t_{\alpha/2, n-1} \right\} = 1 - \alpha$$

or, using that $t_{1-\alpha/2, n-1} = -t_{\alpha/2, n-1}$

$$P\left\{ \bar{X} - t_{\alpha/2, n-1} \frac{S}{\sqrt{n}} < \mu < \bar{X} + t_{\alpha/2, n-1} \frac{S}{\sqrt{n}} \right\} = 1 - \alpha$$

Thus, if it is observed that $\bar{X} = \bar{x}$ and $S = s$, then we can say that "with $100(1 - \alpha)$ percent confidence"

$$\mu \in \left(\bar{x} - t_{\alpha/2, n-1} \frac{s}{\sqrt{n}}, \quad \bar{x} + t_{\alpha/2, n-1} \frac{s}{\sqrt{n}} \right)$$

Example 5.3c Let us again consider Example 5.3a but let us now suppose that when the value μ is transmitted at location A then the value received at location B is normal with mean μ and variance σ^2 but with σ^2 being unknown. If 9 successive values, having the same μ, are, as in Example 5.3a, $5, 8.5, 12, 15, 7, 9, 7.5, 6.5, 10.5$, compute a 95 percent confidence interval for μ.

A simple calculation yields that

$$\bar{x} = 9$$

and

$$s^2 = \frac{\Sigma x_i^2 - 9(\bar{x})^2}{8} = 9.5$$

or

$$s = 3.082.$$

Hence, as $t_{.025,8} = 2.306$ (from Program 3-8-2-B or Table A-3-8-2), a 95 percent confidence interval for μ is

$$\left(9 - 2.306\frac{(3.082)}{3}, 9 + 2.306\frac{(3.082)}{3}\right) = (6.63, 11.37)$$

a larger interval than obtained in Example 5.3a. The reason why the interval just obtained is larger than the one in Example 5.3a is twofold. The primary reason is that we have a larger estimated variance than in Example 5.3a. That is, in Example 5.3a we assumed that σ^2 was known to equal 4, whereas in this example we assumed it to be unknown and our estimate of it turned out to be 9.5, which resulted in a larger confidence interval. In fact, the confidence interval would have been larger than in Example 5.3a even if our estimate of σ^2 was again 4 because by having to estimate the variance we need to utilize the t distribution, which has a greater variance and thus a larger spread than the unit normal (which can be used when σ^2 is assumed known). For instance, if it had turned out that $\bar{x} = 9$ and $s^2 = 4$, then our confidence interval would have been

$$\left(9 - 2.306 \cdot \tfrac{2}{3}, 9 + 2.306 \cdot \tfrac{2}{3}\right) = (7.46, 10.54)$$

which is larger than that obtained in Example 5.3a. ∎

REMARKS

(a) It should be noted that the confidence interval for μ when σ is known is based on the fact that $\sqrt{n}(\bar{X} - \mu)/\sigma$ has a unit normal distribution. When σ is unknown, the foregoing approach is to estimate it by S and then use the fact that $\sqrt{n}(\bar{X} - \mu)/S$ has a t distribution with $n - 1$ degrees of freedom.

(b) It should be noted that the length of a $100(1 - \alpha)$ percent confidence interval for μ is not always larger when the variance is unknown. For the length of such an interval is $2z_\alpha\sigma/\sqrt{n}$ when σ is known, whereas it is $2t_{\alpha, n-1}S/\sqrt{n}$ when σ is unknown; and it is certainly possible that the sample standard deviation S can turn out to be much smaller than σ. However, it can be shown that the mean length of the interval is longer when σ is unknown. That is, it can be shown that

$$t_{\alpha, n-1}E[S] \geq z_\alpha\sigma$$

Indeed, $E[S]$ is evaluated in Section 2 of Chapter 11, and it is shown, for instance that

$$E[S] = \begin{cases} .94\sigma & \text{when } n = 5 \\ .97\sigma & \text{when } n = 9 \end{cases}$$

Since

$$z_{.025} = 1.96 \qquad t_{.025,4} = 2.78 \qquad t_{.025,8} = 2.31$$

the length of a 95 percent interval from a sample of size 5 is $2 \times 1.96\sigma/\sqrt{5} = 1.75\sigma$ when σ is known, whereas its expected length is $2 \times 2.78 \times .94\sigma/\sqrt{5} = 2.34\sigma$ when σ is unknown—an increase of 33.7 percent. If the sample is of size 9, then the two values to compare are 1.31σ and 1.49σ—a gain of 13.7 percent.

A one-sided upper confidence interval can be obtained by noting that

$$P\left\{\sqrt{n}\,\frac{(\overline{X} - \mu)}{S} < t_{\alpha, n-1}\right\} = 1 - \alpha$$

or

$$P\left\{\overline{X} - \mu < \frac{S}{\sqrt{n}} t_{\alpha, n-1}\right\} = 1 - \alpha$$

or

$$P\left\{\mu > \overline{X} - \frac{S}{\sqrt{n}} t_{\alpha, n-1}\right\} = 1 - \alpha$$

Hence, if it is observed that $\overline{X} = \bar{x}$, $S = s$, then we can assert "with $100(1 - \alpha)$ percent confidence" that

$$\mu \in \left(\bar{x} - \frac{s}{\sqrt{n}} t_{\alpha, n-1}, \infty\right)$$

Similarly, a $100(1 - \alpha)$ lower confidence interval would be

$$\mu \in \left(-\infty, \bar{x} + \frac{s}{\sqrt{n}} t_{\alpha, n-1}\right)$$

Program 5-3-1 will compute both one- and two-sided confidence intervals for the mean of a normal distribution when the variance is unknown. It uses Program 3-8-2-B as a subroutine to compute the necessary percentile value of the t-statistic.

Example 5.3d Determine a 95 percent confidence interval for the average resting pulse of the members of a health club if a random selection of 15 members of the club yielded the data: 54, 63, 58, 72, 49, 92, 70, 73, 69, 104, 48, 66, 80, 64, 77. Also determine a 95 percent lower confidence interval for this mean.

Solution We run Program 5-3-1 to obtain:

```
RUN
THIS PROGRAM COMPUTES A 100(1-a)% CONFIDENCE INTERVAL FOR THE MEAN OF
A NORMAL POPULATION WHEN THE VARIANCE IS UNKNOWN
ENTER THE SAMPLE SIZE
? 15
ENTER THE DATA VALUES ONE AT A TIME
? 54
? 63
? 58
? 72
? 49
? 92
? 70
? 73
? 69
? 104
? 48
? 66
? 80
? 64
? 77
ENTER THE VALUE OF a
? .05
IS A TWO-SIDED INTERVAL DESIRED? ENTER 1 IF THE ANSWER IS YES AND 0
IF NO
? 1
THE 95 % CONFIDENCE INTERVAL FOR THE MEAN IS
(60.86499 , 77.66835 )
IS ANOTHER CONFIDENCE INTERVAL DESIRED? IF YES ENTER 1.
IF NO ENTER 0.
? 1
ENTER THE VALUE OF a
? .05
IS A TWO-SIDED INTERVAL DESIRED? ENTER 1 IF THE ANSWER IS YES AND 0
IF NO
? 0
IS THE ONE-SIDED CONFIDENCE INTERVAL TO BE UPPER OR LOWER? ENTER 1
FOR UPPER AND 0 FOR LOWER
? 0
THE 95
% LOWER CONFIDENCE INTERVAL FOR THE MEAN IS                    (-INFINITY,
76.16618 )
IS ANOTHER CONFIDENCE INTERVAL DESIRED? IF YES ENTER 1.
IF NO ENTER 0.
? 0
Ok
```

3.2   Estimating the Difference in Means of Two Normal Populations

Let $X_1, X_2, \ldots, X_n$ be a sample of size n from a normal population having mean μ_1 and variance σ_1^2 and let $Y_1, \ldots, Y_m$ be a sample of size m from a different normal population having mean μ_2 and variance σ_2^2, and suppose that the two samples are independent of each other. We are interested in estimating $\mu_1 - \mu_2$.

Since $\overline{X} = \sum_{i=1}^{n} X_i/n$ and $\overline{Y} = \sum_{i=1}^{m} Y_i/m$ are the maximum likelihood estimators of μ_1 and μ_2 it seems intuitive (and can be proven) that $\overline{X} - \overline{Y}$ is the maximum likelihood estimator of $\mu_1 - \mu_2$.

To obtain a confidence interval estimator, we need the distribution of $\overline{X} - \overline{Y}$. As

$$\overline{X} \sim \mathcal{N}\left(\mu_1, \sigma_1^2/n\right)$$
$$\overline{Y} \sim \mathcal{N}\left(\mu_2, \sigma_2^2/m\right)$$

it follows from the fact that the sum of independent normal random variables is also normal, that

$$\overline{X} - \overline{Y} \sim \mathcal{N}\left(\mu_1 - \mu_2, \frac{\sigma_1^2}{n} + \frac{\sigma_2^2}{m}\right)$$

Hence, assuming σ_1^2 and σ_2^2 are known, we have that

$$\frac{\overline{X} - \overline{Y} - (\mu_1 - \mu_2)}{\sqrt{\dfrac{\sigma_1^2}{n} + \dfrac{\sigma_2^2}{m}}} \sim \mathcal{N}(0,1) \qquad (5.3.2)$$

and so

$$P\left\{ -z_{\alpha/2} < \frac{\overline{X} - \overline{Y} - (\mu_1 - \mu_2)}{\sqrt{\dfrac{\sigma_1^2}{n} + \dfrac{\sigma_2^2}{m}}} < z_{\alpha/2} \right\} = 1 - \alpha$$

or, equivalently

$$P\left\{ \overline{X} - \overline{Y} - z_{\alpha/2}\sqrt{\frac{\sigma_1^2}{n} + \frac{\sigma_2^2}{m}} < \mu_1 - \mu_2 \right.$$

$$\left. < \overline{X} - \overline{Y} + z_{\alpha/2}\sqrt{\frac{\sigma_1^2}{n} + \frac{\sigma_2^2}{m}} \right\} = 1 - \alpha$$

Hence, if $\overline{X}$ and $\overline{Y}$ are observed to equal $\bar{x}$ and $\bar{y}$ respectively, a $100(1 - \alpha)$ percent two-sided confidence interval for $\mu_1 - \mu_2$ is

$$\left(\bar{x} - \bar{y} - z_{\alpha/2}\sqrt{\frac{\sigma_1^2}{n} + \frac{\sigma_2^2}{m}}, \bar{x} - \bar{y} + z_{\alpha/2}\sqrt{\frac{\sigma_1^2}{n} + \frac{\sigma_2^2}{m}} \right)$$

One-sided confidence intervals for $\mu_1 - \mu_2$ are obtained in a similar fashion, and we leave it for the reader to verify that a $100(1 - \alpha)$ percent one-sided interval is given by

$$\mu_1 - \mu_2 \in \left(-\infty, \bar{x} - \bar{y} + z_\alpha\sqrt{\sigma_1^2/n + \sigma_2^2/m} \right)$$

Program 5-3-2-A will compute both one- and two-sided confidence intervals for $\mu_1 - \mu_2$.

Example 5.3e Two different types of electrical cable insulation have recently been tested to determine the voltage level at which failures tend to occur. When specimens were subjected to an increasing voltage stress in a laboratory experiment, failures for the 2 types of cable insulation occurred at the following voltages:

Type A		Type B	
36	54	52	60
44	52	64	44
41	37	38	48
53	51	68	46
38	44	66	70
36	35	52	62
34	44		

Suppose that it is known that the amount of voltage that cables having type A insulation can withstand is normally distributed with unknown mean μ_A and known variance $\sigma_A^2 = 40$, whereas the corresponding distribution for type B insulation is normal with unknown mean μ_B and known variance $\sigma_B^2 = 100$. Determine a 95 percent confidence interval for $\mu_A - \mu_B$. Determine a value that we can assert, with 95 percent confidence, exceeds $\mu_A - \mu_B$.

Solution We run Program 5-3-2-A:

```
RUN
THIS PROGRAM COMPUTES A 100(1-a)% CONFIDENCE INTERVAL FOR THE         DIFFERENCE
 OF MEANS IN TWO NORMAL POPULATIONS HAVING KNOWN VARIANCES
ENTER THE SIZE OF SAMPLE 1
? 14
ENTER THE SAMPLE 1 DATA VALUES ONE AT A TIME
? 36? 44? 41? 53? 38? 36? 34? 54? 52? 37? 51? 44? 35? 44
ENTER THE POPULATION VARIANCE OF SAMPLE 1
? 40
ENTER THE SIZE OF SAMPLE 2
? 12
ENTER THE SAMPLE 2 DATA VALUES ONE AT A TIME
? 52? 64? 38? 68? 66? 52? 60? 44? 48? 46? 70? 62
ENTER THE POPULATION VARIANCE OF SAMPLE 2
? 100
ENTER THE VALUE OF a
? .05
IS A TWO-SIDED INTERVAL DESIRED? ENTER 1 IF THE ANSWER IS YES AND 0  IF NO
? 1
THE 95 % CONFIDENCE INTERVAL  IS (-19.60556 ,-6.489673 )
IS ANOTHER CONFIDENCE INTERVAL DESIRED? IF YES ENTER 1. IF NO         ENTER 0.
? 1
ENTER THE VALUE OF a
? .05
IS A TWO-SIDED INTERVAL DESIRED? ENTER 1 IF THE ANSWER IS YES AND 0  IF NO
? 0
IS THE ONE-SIDED CONFIDENCE INTERVAL TO BE UPPER OR LOWER? ENTER 1   FOR UPPER
AND 0 FOR LOWER
? 0
THE 95 % LOWER CONFIDENCE INTERVAL  IS (-INFINITY,-7.54403 )
IS ANOTHER CONFIDENCE INTERVAL DESIRED? IF YES ENTER 1. IF NO         ENTER 0.
? 0
Ok  ■
```

Let us suppose now that we again desire an interval estimator of $\mu_1 - \mu_2$ but that the population variances σ_1^2 and σ_2^2 are unknown. In this case, it is natural to try to replace σ_1^2 and σ_2^2 in Equation 5.3.2 by the sample variances

$$S_1^2 = \sum_{i=1}^{n} \frac{(X_i - \bar{X})^2}{n-1}$$

$$S_2^2 = \sum_{i=1}^{m} \frac{(Y_i - \bar{Y})^2}{m-1}$$

That is, it is natural to somehow base our interval estimate on something like

$$\frac{\bar{X} - \bar{Y} - (\mu_1 - \mu_2)}{\sqrt{S_1^2/n + S_2^2/m}}$$

However, in order to utilize the foregoing to obtain a confidence interval, we need its distribution and it must not depend on any of the unknown parameters σ_1^2 and σ_2^2. Unfortunately, this distribution is both complicated and does indeed depend on the unknown parameters σ_1^2 and σ_2^2. In fact, it is only in the special case when $\sigma_1^2 = \sigma_2^2$ that we will be able to obtain an interval estimator. So let us suppose that the population variances, though unknown, are equal and let σ^2 denote their common value. Now, from Theorem 4.5.1 of Chapter 4, it follows that

$$(n-1)\frac{S_1^2}{\sigma^2} \sim \chi_{n-1}^2$$

and

$$(m-1)\frac{S_2^2}{\sigma^2} \sim \chi_{m-1}^2$$

Also, because the samples are independent, it follows that these two chi-square random variables are independent. Hence, from the additive property of chi-square random variables, which states that the sum of independent chi-square random variables is also chi-square with a degree of freedom equal to the sum of the degrees of freedom, it follows that

$$(n-1)\frac{S_1^2}{\sigma^2} + (m-1)\frac{S_2^2}{\sigma^2} \sim \chi_{n+m-2}^2 \qquad (5.3.3)$$

Also, since

$$\bar{X} - \bar{Y} \sim \mathcal{N}\left(\mu_1 - \mu_2, \frac{\sigma^2}{n} + \frac{\sigma^2}{m}\right)$$

we see that

$$\frac{\bar{X} - \bar{Y} - (\mu_1 - \mu_2)}{\sqrt{\frac{\sigma^2}{n} + \frac{\sigma^2}{m}}} \sim \mathcal{N}(0,1) \qquad (5.3.4)$$

Now it follows from the fundamental result that in normal sampling $\bar{X}$ and S^2 are independent (Theorem 4.5.1 of Chapter 4), that $\bar{X}_1, S_1^2, \bar{X}_2, S_2^2$ are independent random variables. Hence, using the definition of a t-random variable (as the ratio of two independent random variables, the numerator being a unit normal and the denominator being the square root of a chi-square random variable divided by its degree of freedom parameter), it follows from Equations 5.3.3 and 5.3.4 that

$$\frac{\bar{X} - \bar{Y} - (\mu_1 - \mu_2)}{\sqrt{\dfrac{1}{n} + \dfrac{1}{m}}} \left(\frac{n + m - 2}{(n-1)S_1^2 + (m-1)S_2^2} \right)^{1/2} \sim t_{n+m-2}$$

That is

$$\frac{\bar{X} - \bar{Y} - (\mu_1 - \mu_2)}{\sqrt{\dfrac{1}{n} + \dfrac{1}{m}} \sqrt{\dfrac{(n-1)S_1^2 + (m-1)S_2^2}{n + m - 2}}}$$

is t distributed with $n + m - 2$ degrees of freedom. Hence,

$$P\left\{ -t_{\alpha/2, n+m-2} \leq \frac{\bar{X} - \bar{Y} - (\mu_1 - \mu_2)}{\sqrt{\left(\dfrac{1}{n} + \dfrac{1}{m}\right)\left[\dfrac{(n-1)S_1^2 + (m-1)S_2^2}{n + m - 2}\right]}} \leq t_{\alpha/2, n+m-2} \right\}$$

$$= 1 - \alpha$$

Thus, when the data results in $\bar{X} = \bar{x}$, $\bar{Y} = \bar{y}$, $S_1 = s_1$, $S_2 = s_2$, we obtain the following $100(1 - \alpha)$ percent confidence interval for $\mu_1 - \mu_2$

$$\left(\bar{x} - \bar{y} - t_{\alpha/2, n+m-2} \sqrt{\left(\frac{1}{n} + \frac{1}{m}\right)\left(\frac{(n-1)s_1^2 + (m-1)s_2^2}{n + m - 2}\right)}, \right.$$

$$\left. \bar{x} - \bar{y} + t_{\alpha/2, n+m-2} \sqrt{\left(\frac{1}{n} + \frac{1}{m}\right)\left(\frac{(n-1)s_1^2 + (m-1)s_2^2}{n + m - 2}\right)} \right)$$

One-sided confidence intervals are similarly obtained.

Program 5-3-2-B can be used to obtain both one- and two-sided confidence intervals for the difference in means in two normal populations having unknown but equal variances.

Example 5.3f There are two different techniques a given manufacturer can employ to produce batteries. A random selection of 12 batteries produced by technique I and of 14 produced by technique II resulted in the following capacities (in ampere hours):

Technique I		Technique II	
140	132	144	134
136	142	132	130
138	150	136	146
150	154	140	128
152	136	128	131
144	142	150	137
		130	135

Determine a 90 percent level two-sided confidence interval for the difference in means, assuming a common variance. Also determine a 95 percent upper confidence interval for $\mu_I - \mu_{II}$.

Solution We run Program 5-3-2-B to obtain:

```
RUN
THIS PROGRAM COMPUTES A 100(1-a)% CONFIDENCE INTERVAL FOR THE DIFFERENCE OF
  MEANS IN TWO NORMAL POPULATIONS HAVING UNKNOWN BUT EQUAL VARIANCES
ENTER THE SIZE OF SAMPLE NUMBER 1
? 12
ENTER THE SAMPLE 1 DATA VALUES ONE AT A TIME
? 140
? 136
? 138
? 150
? 152
? 144
? 132
? 142
? 150
? 154
? 136
? 142
ENTER THE SIZE OF SAMPLE NUMBER 2
? 14
ENTER THE SAMPLE 2 DATA VALUES ONE AT A TIME
? 144
? 132
? 136
? 140
? 128
? 150
? 130
? 134
? 130
? 146
? 128
? 131
? 137
? 135
ENTER THE VALUE OF a
? .1
IS A TWO-SIDED INTERVAL DESIRED? ENTER 1 IF THE ANSWER IS YES AND O IF NO
? 1
THE 90 % CONFIDENCE INTERVAL  IS ( 2.497077 , 11.93148 )
IS ANOTHER CONFIDENCE INTERVAL DESIRED? IF YES ENTER 1. IF NO ENTER O.
? 1
ENTER THE VALUE OF a
? .05
IS A TWO-SIDED INTERVAL DESIRED? ENTER 1 IF THE ANSWER IS YES AND O IF NO
? 0
IS THE ONE-SIDED CONFIDENCE INTERVAL TO BE UPPER OR LOWER? ENTER 1 FOR UPPER AND
  O FOR LOWER
? 1
THE 95 % UPPER CONFIDENCE INTERVAL  IS ( 2.497077 ,INFINITY)
IS ANOTHER CONFIDENCE INTERVAL DESIRED? IF YES ENTER 1. IF NO ENTER O.
? 0
Ok
```

3.3 Confidence Intervals for the Variance of a Normal Distribution

If $X_1, \ldots, X_n$ is a sample from a normal distribution having unknown parameters μ and σ^2, then we can construct a confidence interval for σ^2 by using the fact that

$$(n-1)\frac{S^2}{\sigma^2} \sim \chi^2_{n-1}$$

Hence,

$$P\left\{\chi^2_{1-\alpha/2,\,n-1} \le (n-1)\frac{S^2}{\sigma^2} \le \chi^2_{\alpha/2,\,n-1}\right\} = 1 - \alpha$$

or, equivalently

$$P\left\{\frac{(n-1)S^2}{\chi^2_{\alpha/2,\,n-1}} \le \sigma^2 \le \frac{(n-1)S^2}{\chi^2_{1-\alpha/2,\,n-1}}\right\} = 1 - \alpha$$

Hence, a $100(1 - \alpha)$ percent confidence interval for σ^2 is, when $S^2 = s^2$

$$\sigma^2 \in \left(\frac{(n-1)S^2}{\chi^2_{\alpha/2,\,n-1}}, \frac{(n-1)S^2}{\chi^2_{1-\alpha/2,\,n-1}}\right) \qquad \text{with confidence } 1 - \alpha$$

Example 5.3g A standardized procedure is expected to produce washers with very small deviation in their thicknesses. Suppose that 10 such washers were chosen and measured. If the thickness of these washers were, in inches,

0.123	0.133
0.124	0.125
0.126	0.128
0.120	0.124
0.130	0.126

what is a 90 percent confidence interval for the standard deviation of the thickness of a washer produced by this procedure?

Running Program 4-3 yields that

$$S^2 = 1.366 \times 10^{-5}$$

Since $\chi^2_{.05,\,9} = 16.917$ and $\chi^2_{.95,\,9} = 3.334$ we obtain that

$$\sigma^2 \in \left(9 \times 1.366 \times 10^{-5}/16.917, 9 \times 1.366 \times 10^{-5}/3.334\right)$$

$$\text{with confidence .90}$$

or

$$\sigma^2 \in \left(7.267 \times 10^{-6}, 36.875 \times 10^{-6}\right) \qquad \text{with confidence .90}$$

or, upon taking square roots

$$\sigma \in \left(2.696 \times 10^{-3}, 6.072 \times 10^{-3}\right) \qquad \text{with confidence .90} \quad \blacksquare$$

3.4 Approximate Confidence Interval for the Mean of a Bernoulli Random Variable

Consider a population of items each of which independently meets a certain standard with some unknown probability p. If n of these items are tested to determine whether they meet the standards, how can we use the resulting data to obtain a confidence interval for p?

If we let X denote the number of the n items that meet the standards, then X is a binomial random variable with parameters n and p. Thus, when n is large, it follows by the normal approximation to the binomial that X is approximately normally distributed with mean np and variance $np(1-p)$. Hence,

$$\frac{X - np}{\sqrt{np(1-p)}} \doteq \mathcal{N}(0,1) \qquad (5.3.5)$$

which $\doteq$ means "is approximately distributed as." Therefore, for any $\alpha \in (0,1)$

$$P\left\{ -z_{\alpha/2} < \frac{X - np}{\sqrt{np(1-p)}} < z_{\alpha/2} \right\} \approx 1 - \alpha$$

and so if X is observed to equal x, then an approximate $100(1-\alpha)$ percent confidence *region* for p is

$$\left\{ p : -z_{\alpha/2} < \frac{x - np}{\sqrt{np(1-p)}} < z_{\alpha/2} \right\}$$

However, the foregoing region is not an interval. To obtain a confidence *interval* for p, we use the fact that X/n, the fraction of items that meet the standards, is the maximum likelihood estimator of p. Hence it follows that $\sqrt{X[1 - (X/n)]}$ will approximately equal $\sqrt{np(1-p)}$ and thus from Equation 5.3.5 we see that

$$\frac{X - np}{\sqrt{X\left(1 - \dfrac{X}{n}\right)}} \doteq \mathcal{N}(0,1)$$

Therefore, for any $\alpha \in (0,1)$

$$P\left\{ -z_{\alpha/2} < \frac{X - np}{\sqrt{X\left(1 - \dfrac{X}{n}\right)}} < z_{\alpha/2} \right\} \approx 1 - \alpha$$

or, equivalently

$$P\left\{ \frac{X}{n} - \frac{\sqrt{X\left(1 - \dfrac{X}{n}\right)}}{n} z_{\alpha/2} < p < \frac{X}{n} + \frac{\sqrt{X\left(1 - \dfrac{X}{n}\right)}}{n} z_{\alpha/2} \right\} \approx 1 - \alpha$$

which yields the desired $100(1-\alpha)$ percent confidence interval for p.

Example 5.3h Suppose from a large batch of transistor tubes, 100 are randomly selected and are tested to see if they meet the current standards. If 80 of the 100 meet the standards, then a 95 percent confidence interval for p, the fraction of all transistors that meet the standards, is given by

$$\left(.8 - \frac{\sqrt{80(.2)}}{100}(1.96), .8 + \frac{\sqrt{80(.2)}}{100}(1.96) \right)$$

That is, with "95 percent confidence," p is between .7216 and .8784. ∎

It often occurs that one is interested in obtaining a $100(1 - \alpha)$ percent confidence interval for p of some given length—say l. The problem then is to determine the appropriate sample size n to obtain such an interval. Now the length of a $100(1 - \alpha)$ percent confidence interval for p from a sample of size n is

$$\frac{2}{n}\sqrt{X\left(1 - \frac{X}{n}\right)}\, z_{\alpha/2} = \frac{2z_{\alpha/2}}{\sqrt{n}}\sqrt{\frac{X}{n}\left(1 - \frac{X}{n}\right)}$$

which, since X/n is approximately p, will be approximately equal to $2z_{\alpha/2}\sqrt{p(1 - p)/n}$. That is,

$$\frac{2z_{\alpha/2}}{\sqrt{n}}\sqrt{p(1 - p)} \approx \begin{array}{l}\text{length of } 100(1 - \alpha) \text{ percent}\\ \text{confidence interval from sample of size } n\end{array}$$

Unfortunately, p is not known in advance (if it were, there would be no need to try to estimate it) and so we cannot just equate $2z_{\alpha/2}\sqrt{p(1 - p)/n}$ to l to determine n. What we can do, however, is to first take a preliminary sample of say—size 30, which will be used to initially estimate p; and this preliminary estimate will then be used to determine n. That is, if Y equals the number of successes in the preliminary sample of size 30, then we can initially estimate that p will be approximately $Y/30$. Hence, to determine a $100(1 - \alpha)$ percent confidence interval of length l for p, we would approximately need a sample of size n where

$$\frac{2z_{\alpha/2}}{\sqrt{n}}\sqrt{\frac{Y}{30}\left(1 - \frac{Y}{30}\right)} = l$$

or, upon squaring both sides,

$$\frac{(2z_{\alpha/2})^2}{n}\frac{Y}{30}\left(1 - \frac{Y}{30}\right) = l^2$$

or

$$n = \frac{(2z_{\alpha/2})^2}{l^2}\left[\frac{Y}{30}\left(1 - \frac{Y}{30}\right)\right]$$

Therefore, to complete our sample, we would take an additional sample of size $n - 30$ (if $n \leq 30$, we need not take an additional sample).

Example 5.3i Each computer chip produced by a certain manufacturer is either acceptable or unacceptable. A large batch of such chips is produced and it is supposed that each chip in this batch will, independently, be acceptable with some unknown probability p. To obtain a 99 percent confidence interval for p, which is to be of length approximately .05, a sample of size 30 is initially taken. If 26 of the 30 chips are then deemed acceptable then the preliminary estimate of p is 26/30. Hence, to meet the requirements of a 99 percent confidence interval of length approximately .05 would require approximately a sample of size

$$n = \frac{4(z_{.005})^2}{(.05)^2} \frac{26}{30}\left(1 - \frac{26}{30}\right) = \frac{4(2.58)^2}{(.05)^2} \frac{26}{30} \frac{4}{30} = 1231$$

Hence, we should now sample an additional 1201 chips and if, for instance, 1040 of them are acceptable, then the final 99 percent confidence interval for p is

$$\left(\frac{1066}{1231} - \sqrt{1066\left(1 - \frac{1066}{1231}\right)} \frac{z_{.005}}{1231}, \frac{1066}{1231} + \sqrt{1066\left(1 - \frac{1066}{1231}\right)} \frac{z_{.005}}{1231}\right)$$

or

$$p \in (.84091, .89101) \quad \blacksquare$$

REMARK
As shown, a $100(1 - \alpha)$ percent confidence interval for p will be of approximate length l when the sample size is

$$n = \frac{(2z_{\alpha/2})^2}{l^2} p(1 - p)$$

Now it is easily shown that the function $g(p) = p(1 - p)$ attains its maximum value of $\frac{1}{4}$, in the interval $0 \le p \le 1$, when $p = \frac{1}{2}$. Thus an upper bound on n is

$$n \le \frac{(z_{\alpha/2})^2}{l^2}$$

and so by choosing a sample whose size is at least as large as $(z_{\alpha/2})^2/l^2$, one can be assured of obtaining a confidence interval of length no greater than l without any need of any additional sampling.

3.5 Confidence Interval of the Mean of the Exponential Distribution

If $X_1, \ldots, X_n$ are independent exponential random variables each having mean θ then, as will be shown in Example 5.4d, the maximum likelihood estimator of θ is $\sum_{i=1}^n X_i/n$. To obtain a confidence interval estimator for θ, we recall, from Section 7 of Chapter 3, that $\sum_{i=1}^n X_i$ has a gamma distribution with parameters $n, 1/\theta$. This in turn implies (from the relationship between the gamma and chi-square distribution shown in Section 8.1 of Chapter 3) that

$$\frac{2}{\theta} \sum_{i=1}^n X_i \sim \chi^2_{2n}$$

Hence, for any $\alpha \in (0,1)$

$$P\left\{ \chi^2_{1-\alpha/2,2n} < \frac{2}{\theta} \sum_{i=1}^{n} X_i < \chi^2_{\alpha/2,2n} \right\} = 1 - \alpha$$

or, equivalently

$$P\left\{ \frac{2\sum_{i=1}^{n} X_i}{\chi^2_{\alpha/2,2n}} < \theta < \frac{2\sum_{i=1}^{n} X_i}{\chi^2_{1-\alpha/2,2n}} \right\} = 1 - \alpha$$

Hence, a $100(1 - \alpha)$ percent confidence interval for θ is

$$\theta \in \left(\frac{2\sum_{i=1}^{n} X_i}{\chi^2_{\alpha/2,2n}}, \frac{2\sum_{i=1}^{n} X_i}{\chi^2_{1-\alpha/2,2n}} \right)$$

Example 5.3j The successive items produced by a certain manufacturer are assumed to have useful lives that (in hours) are independent with a common density function

$$f(x) = \frac{1}{\theta} e^{-x/\theta} \qquad 0 < x < \infty$$

If the sum of the lives of the first 10 items is equal to 1740, what is a 95 percent confidence interval for the population mean θ?

From Program 3-8-1-B (or Table A-3-8-1) we see that

$$\chi^2_{.025,20} = 34.169, \qquad \chi^2_{.975,20} = 9.661$$

and so we can conclude, with 95 percent confidence, that

$$\theta \in \left(\frac{2 x 1740}{34.169}, \frac{2 x 1740}{9.661} \right)$$

or, equivalently

$$\theta \in (101.847, 360.211) \qquad \blacksquare$$

4 EVALUATING A POINT ESTIMATOR

Let $\mathbf{X} = (X_1, \ldots, X_n)$ be a sample from a population whose distribution is specified up to an unknown parameter θ; and let $d = d(\mathbf{X})$ be an estimator of θ. How are we to determine its worth as an estimator of θ? One way is to consider the square of the difference between $d(\mathbf{X})$ and θ. However, since $(d(\mathbf{X}) - \theta)^2$ is a random variable, let us agree to consider $r(d, \theta)$, the *mean square error* of the estimator d, which is defined by

$$r(d, \theta) = E\left[(d(\mathbf{X}) - \theta)^2 \right]$$

as an indication of the worth of d as an estimator of θ.

It would be nice if there was a single estimator d that minimized $r(d, \theta)$ for all possible values of θ. However, except in trivial situations, this will never be the case. For example, consider the estimator $d*$ defined by

$$d*(X_1, \ldots, X_n) = 4$$

That is, no matter what the outcome of the sample data, the estimator $d*$ chooses 4 as its estimate of θ. While this seems like a silly estimator (since it makes no use of the data), it is, however, true that when θ actually equals 4, the mean square error of this estimator is 0. Thus, the mean square error of any estimator different than $d*$ must, in most situations, be larger than the mean square error of $d*$ when $\theta = 4$.

Although minimum mean square estimators rarely exist, it is sometimes possible to find an estimator having the smallest mean square error among all estimators that satisfy a certain property. One such property is that of unbiasedness.

Definition 5.4.1 Let $d = d(\mathbf{X})$ be an estimator of the parameter θ. Then

$$b_\theta(d) = E[d(\mathbf{X})] - \theta$$

is called the *bias* of d as an estimator of θ. If $b_\theta(d) = 0$ for all θ, then we say that d is an *unbiased* estimator of θ. In other words, an estimator is unbiased if its expected value always equals the value of the parameter it is attempting to estimate.

Example 5.4a Let $X_1, X_2, \ldots, X_n$ be a random sample from a distribution having unknown mean θ. Then

$$d_1(X_1, X_2, \ldots, X_n) = X_1$$

and

$$d_2(X_1, X_2, \ldots, X_n) = \frac{X_1 + X_2 + \cdots + X_n}{n}$$

are both unbiased estimators of θ since

$$E[X_1] = E\left[\frac{X_1 + X_2 + \cdots + X_n}{n}\right] = \theta$$

More generally, $d_3(X_1, X_2, \ldots, X_n) = \sum_{i=1}^{n} \lambda_i X_i$ is an unbiased estimator of θ whenever $\sum_{i=1}^{n} \lambda_i = 1$. This follows since

$$E\left[\sum_{i=1}^{n} \lambda_i X_i\right] = \sum_{i=1}^{n} E[\lambda_i X_i]$$

$$= \sum_{i=1}^{n} \lambda_i E(X_i)$$

$$= \theta \sum_{i=1}^{n} \lambda_i$$

$$= \theta \quad \blacksquare$$

If $d(X_1, \ldots, X_n)$ is an unbiased estimator, then its mean square error is given by

$$r(d, \theta) = E\left[(d(\mathbf{X}) - \theta)^2\right]$$

$$= E\left[(d(\mathbf{X}) - E[d(\mathbf{X})])^2\right] \quad \text{since } d \text{ is unbiased}$$

$$= \text{Var}(d(\mathbf{X}))$$

Thus the mean square error of an unbiased estimator is equal to its variance.

Example 5.4b Combining Independent Unbiased Estimators Let d_1 and d_2 denote independent unbiased estimators of θ, having known variances σ_1^2 and σ_2^2. That is, for $i = 1, 2$,

$$E[d_i] = \theta \quad \text{Var}(d_i) = \sigma_i^2$$

Any estimator of the form

$$d = \lambda d_1 + (1 - \lambda)d_2$$

will also be unbiased. To determine the value of λ that results in d having the smallest possible mean square error, note that

$$r(d, \theta) = \text{Var}(d)$$

$$= \lambda^2 \text{Var}(d_1) + (1 - \lambda)^2 \text{Var}(d_2)$$

$$\text{by the independence of } d_1 \text{ and } d_2$$

$$= \lambda^2 \sigma_1^2 + (1 - \lambda)^2 \sigma_2^2$$

Differentiation yields that

$$\frac{d}{d\lambda}r(d, \theta) = 2\lambda\sigma_1^2 - 2(1 - \lambda)\sigma_2^2$$

To determine the value of λ that minimizes $r(d, \theta)$—call it $\hat{\lambda}$—set this equal to 0 and solve for λ to obtain

$$2\hat{\lambda}\sigma_1^2 = 2(1 - \hat{\lambda})\sigma_2^2$$

or

$$\hat{\lambda} = \frac{\sigma_2^2}{\sigma_1^2 + \sigma_2^2} = \frac{1/\sigma_1^2}{1/\sigma_1^2 + 1/\sigma_2^2}$$

In words, the optimal weight to give an estimator is inversely proportional to its variance (when all the estimators are unbiased and independent).

For an application of the foregoing, suppose that a conservation organization wants to determine the acidity content of a certain lake. To determine this quantity, they draw some water from the lake and then send samples of this water to n different laboratories. These laboratories will then, independently, test for acidity content by using their respective titration equipment, which is of differing precision. Specifically, suppose that d_i, the result of a titration test at laboratory i, is a random variable having mean θ, the true acidity of the sample water, and variance σ_i^2, $i = 1, \ldots, n$. If the quantities σ_i^2, $i = 1, \ldots, n$ are known to the conservation organization, then they should estimate the acidity of the sampled water from the lake by

$$d = \frac{\sum_{i=1}^{n} d_i / \sigma_i^2}{\sum_{i=1}^{n} 1 / \sigma_i^2}$$

The mean square error of d is as follows:

$$r(d, \theta) = \text{Var}(d) \qquad \text{since } d \text{ is unbiased}$$

$$= \left(\frac{1}{\sum_{i=1}^{n} 1/\sigma_i^2} \right)^2 \sum_{i=1}^{n} \left(\frac{1}{\sigma_i^2} \right)^2 \sigma_i^2$$

$$= \frac{1}{\sum_{i=1}^{n} 1/\sigma_i^2} \quad \blacksquare$$

A generalization of the result that the mean square error of an unbiased estimator is equal to its variance is that the mean square error of any estimator is equal to its variance plus the square of its bias. This follows since

$$r(d, \theta) = E\left[(d(\mathbf{X}) - \theta)^2 \right]$$

$$= E\left[(d - E[d] + E[d] - \theta)^2 \right]$$

$$= E\left[(d - E[d])^2 + (E[d] - \theta)^2 + 2(E[d] - \theta)(d - E[d]) \right]$$

$$= E\left[(d - E[d])^2 \right] + E\left[(E[d] - \theta)^2 \right]$$

$$\quad + 2E\left[(E[d] - \theta)(d - E[d]) \right]$$

$$= E\left[(d - E[d])^2 \right] + (E[d] - \theta)^2 + 2(E[d] - \theta)E[d - E[d]]$$

$$\text{since } E[d] - \theta \text{ is constant}$$

$$= E\left[(d - E[d])^2 \right] + (E[d] - \theta)^2$$

The last equality follows since

$$E[d - E[d]] = 0$$

Hence

$$r(d, \theta) = \text{Var}(d) + b_\theta^2(d)$$

Example 5.4c Let $X_1, \ldots, X_n$ denote a sample from a uniform $(0, \theta)$ distribution, where θ is assumed unknown. Since

$$E[X_i] = \frac{\theta}{2}$$

a "natural" estimator to consider is the unbiased estimator

$$d_1 = d_1(\mathbf{X}) = \frac{2\sum_{i=1}^{n} X_i}{n}$$

Since $E[d_1] = \theta$, it follows that

$$r(d_1, \theta) = \text{Var}(d_1)$$

$$= \frac{4}{n} \text{Var}(X_i)$$

$$= \frac{4}{n} \frac{\theta^2}{12} \qquad \text{since Var}(X_i) = \frac{\theta^2}{12}$$

$$= \frac{\theta^2}{3n}$$

A second possible estimator of θ is the maximum likelihood estimator, which, as shown in Example 5.2d, is given by

$$d_2 = d_2(\mathbf{X}) = \max_i X_i$$

To compute the mean square error of d_2 as an estimator of θ, we need first compute its mean (so as to determine its bias) and variance. To do so, note that the distribution function of d_2 is as follows:

$$F_2(x) \equiv P\{d_2(\mathbf{X}) \le x\}$$

$$= P\left\{ \max_i X_i \le x \right\}$$

$$= P\{X_i \le x \qquad \text{for all } i = 1, \ldots, n\}$$

$$= \prod_{i=1}^{n} P\{X_i \le x\} \qquad \text{by independence}$$

$$= \left(\frac{x}{\theta}\right)^n \qquad x \le \theta$$

Hence, upon differentiating, we obtain that the density function of d_2 is

$$f_2(x) = \frac{nx^{n-1}}{\theta^n} \qquad x \le \theta$$

Therefore,

$$E[d_2] = \int_0^{\theta} x \frac{nx^{n-1}}{\theta^n} \, dx = \frac{n}{n+1} \theta \qquad (5.4.1)$$

Also

$$E[d_2^2] = \int_0^{\theta} x^2 \frac{nx^{n-1}}{\theta^n} \, dx = \frac{n}{n+2} \theta^2$$

and so

$$\text{Var}(d_2) = \frac{n}{n+2}\theta^2 - \left(\frac{n}{n+1}\theta\right)^2 \tag{5.4.2}$$

$$= n\theta^2\left[\frac{1}{n+2} - \frac{n}{(n+1)^2}\right] = \frac{n\theta^2}{(n+2)(n+1)^2}$$

Hence

$$r(d_2,\theta) = \left(E(d_2) - \theta\right)^2 + \text{Var}(d_2) \tag{5.4.3}$$

$$= \frac{\theta^2}{(n+1)^2} + \frac{n\theta^2}{(n+2)(n+1)^2}$$

$$= \frac{\theta^2}{(n+1)^2}\left[1 + \frac{n}{n+2}\right]$$

$$= \frac{2\theta^2}{(n+1)(n+2)}$$

Since

$$\frac{2\theta^2}{(n+1)(n+2)} \le \frac{\theta^2}{3n} \qquad n = 1,2,\ldots$$

it follows that d_2 is a superior estimator of θ than is d_1.

Equation 5.4.1 suggests the use of even another estimator—namely, the unbiased estimator $(1 + 1/n)d_2(\mathbf{X}) = (1 + 1/n)\max_i X_i$. However, rather than considering this estimator directly, let us consider all estimators of the form

$$d_c(\mathbf{X}) = c \max_i X_i = cd_2(\mathbf{X})$$

where c is a given constant. The mean square error of this estimator is

$$r(d_c(\mathbf{X}),\theta) = \text{Var}(d_c(\mathbf{X})) + \left(E[d_c(\mathbf{X})] - \theta\right)^2 \tag{5.4.4}$$

$$= c^2 \text{Var}(d_2(\mathbf{X})) + \left(cE[d_2(\mathbf{X})] - \theta\right)^2$$

$$= \frac{c^2 n\theta^2}{(n+2)(n+1)^2} + \theta^2\left(\frac{cn}{n+1} - 1\right)^2$$

by Equations 5.4.2 and 5.4.1

To determine the constant c resulting in minimal mean square error, we differentiate to obtain

$$\frac{d}{dc}r(d_c(\mathbf{X}),\theta) = \frac{2cn\theta^2}{(n+2)(n+1)^2} + \frac{2\theta^2 n}{n+1}\left(\frac{cn}{n+1} - 1\right)$$

Equating this to 0 shows that the best constant c—call it c^*—is such that

$$\frac{c^*}{n+2} + c^*n - (n+1) = 0$$

or

$$c^* = \frac{(n+1)(n+2)}{n^2+2n+1} = \frac{n+2}{n+1}$$

Substituting this value of c into Equation 5.4.4 yields that

$$r\left(\frac{n+2}{n+1}\max_i X_i, \theta\right) = \frac{(n+2)n\theta^2}{(n+1)^4} + \theta^2\left(\frac{n(n+2)}{(n+1)^2} - 1\right)^2$$

$$= \frac{(n+2)n\theta^2}{(n+1)^4} + \frac{\theta^2}{(n+1)^4}$$

$$= \frac{\theta^2}{(n+1)^2}$$

A comparison with Equation 5.4.3 shows that the (biased) estimator $(n+2)/(n+1)\max_i X_i$ has about half the mean square error of the maximum likelihood estimator $\max_i X_i$. ∎

The methods of moments and maximum likelihood are two general approaches for obtaining estimators. Whereas the rationale for both is heuristic, time and experience have shown that both (particularly the maximum likelihood estimates) tend to perform well in a variety of practical situations. In addition, the maximum likelihood estimators are generally considered to be more reliable (though counterexamples are certainly known) than method of moment estimators. In fact, it can be shown that the maximum likelihood estimator will have an asymptotically smaller mean square error than any other estimator as the sample size increases to ∞. That is, it can be shown, subject to certain regularity conditions, that if d_n represents the maximum likelihood estimator of θ from a sample of size n and d_n^* is any other estimator of θ based on a sample of size n then

$$\lim_{n\to\infty} \frac{r(d_n,\theta)}{r(d_n^*,\theta)} \leq 1$$

(The regularity conditions preclude the uniform $(0,\theta)$ distribution which, as we have seen in the previous example, provides a counterexample to the above.)[†]

However, in smaller samples, even when the regularity conditions hold, it should be noted that one is sometimes able to improve upon maximum likelihood estimators. For instance, consider the following example, which considers the estimation of the unknown mean of an exponential distribution.

[†]Part of the regularity conditions is that, in the continuous case, the density $f_\theta(x)$ should be differentiable in θ for fixed x.

Example 5.4d Suppose that $X_1, \ldots, X_n$ is a sample from an exponential distribution having an unknown mean θ. The likelihood is thus

$$f(x_1, \ldots, x_n|\theta) = \frac{1}{\theta^n} \exp\left\{ -\sum_{i=1}^{n} x_i/\theta \right\}$$

To obtain the maximum likelihood estimator, first take logs to obtain

$$\log f(x_1, \ldots, x_n|\theta) = -n \log \theta - \frac{1}{\theta} \sum_{i=1}^{n} x_i$$

Differentiation yields that

$$\frac{d}{d\theta} \log f(x_1, \ldots, x_n|\theta) = -\frac{n}{\theta} + \frac{\sum_{i=1}^{n} x_i}{\theta^2}$$

On equating this to 0 and solving for θ, we see that the maximum likelihood estimate is given by

$$\hat{\theta} = \sum_{i=1}^{n} \frac{x_i}{n}$$

Hence, in this case, both the maximum likelihood and the method of moments estimator of θ is the sample mean $\bar{X}$. Since $\bar{X}$ is an unbiased estimator, it follows that its mean square error is

$$r(\bar{X}, \theta) = \text{Var}(\bar{X})$$
$$= \frac{\text{Var}(X_i)}{n}$$
$$= \frac{\theta^2}{n}$$

where the last equality follows since the variance of an exponential is equal to the square of its mean.

However, let us now consider the estimator $c\bar{X}$, where c is a fixed constant. The mean square error of this estimator is

$$r(c\bar{X}, \theta) = \text{Var}(c\bar{X}) + (E[c\bar{X}] - \theta)^2 \qquad (5.4.5)$$
$$= c^2 \text{Var}(\bar{X}) + (c\theta - \theta)^2$$
$$= c^2 \frac{\theta^2}{n} + (c - 1)^2 \theta^2$$
$$= \theta^2 \left[\frac{c^2}{n} + (c - 1)^2 \right]$$

To choose the value of c that minimizes this quantity, we differentiate to obtain

$$\frac{d}{dc} r(c\bar{X}, \theta) = \theta^2 \left[\frac{2c}{n} + 2(c - 1) \right]$$

Setting equal to 0 shows that the minimizing c is given by

$$c^* = \frac{n}{n + 1}$$

That is, the estimator

$$\frac{n}{n+1}\overline{X} = \frac{\sum_{i=1}^{n}X_i}{n+1}$$

has a uniformly smaller mean square error than $\overline{X}$. In fact, letting $c = n/(n+1)$ in Equation 5.4.5 shows that

$$r\left(\frac{n}{n+1}\overline{X}, \theta\right) = \theta^2\left[\frac{n}{(n+1)^2} + \frac{1}{(n+1)^2}\right] = \frac{\theta^2}{n+1}$$

while

$$r(\overline{X}, \theta) = \frac{\theta^2}{n}$$

Whereas this difference becomes unimportant in large samples, it can be significant when one must estimate with a small sample size. For instance, when $n = 10$, the mean square error of $n/(n+1)\overline{X}$ is 9 percent less than that of $\overline{X}$. ∎

5 BAYES ESTIMATORS

In certain situations it seems reasonable to regard an unknown parameter θ as being the value of a random variable from a given probability distribution. This usually arises when, prior to the observance of the outcomes of the data $X_1, \ldots, X_n$, we have some information about the value of θ and this information is expressible in terms of a probability distribution (called appropriately the *prior* distribution of θ). For instance, suppose that from past experience we know that θ is equally likely to be near any value in the interval $(0, 1)$. Hence, we could reasonably assume that θ is chosen from a uniform distribution on $(0, 1)$.

Suppose now that our prior feelings about θ are that it can be regarded as being the value of a continuous random variable having probability density function $p(\theta)$; and suppose that we are about to observe the value of a sample whose distribution depends on θ. Specifically, suppose that $f(x|\theta)$ represents the likelihood—that is, it is the probability mass function in the discrete case or the probability density function in the continuous case—that a data value is equal to x when θ is the value of the parameter. If the observed data values are $X_i = x_i$, $i = 1, \ldots, n$, then the updated, or conditional, probability density function of θ is as follows:

$$f(\theta|x_1, \ldots, x_n) = \frac{f(\theta, x_1, \ldots, x_n)}{f(x_1, \ldots, x_n)}$$

$$= \frac{p(\theta)f(x_1, \ldots, x_n|\theta)}{\int f(x_1, \ldots, x_n|\theta)p(\theta)\, d\theta}$$

The conditional density function $f(\theta|x_1, \ldots, x_n)$ is called the *posterior* density function. (Thus, before observing the data, one's feelings about θ are expressed in terms of the prior distribution, whereas once the data are observed, this prior distribution is updated to yield the posterior distribution.)

Now we have shown in Chapter 4 that whenever we are given the probability distribution of a random variable, the best estimate of the value of that random variable, in the sense of minimizing the expected squared error, is its mean. Therefore, it follows that the best estimate of θ, given the data values $X_i = x_i$, $i = 1, \ldots, n$, is the mean of the posterior distribution $f(\theta|x_1, \ldots, x_n)$. This estimator, called the *Bayes estimator*, is written as $E[\theta|X_1, \ldots, X_n]$. That is, if $X_i = x_i$, $i = 1, \ldots, n$, then the value of the Bayes estimator is

$$E[\theta|X_1 = x_1, \ldots, X_n = x_n] = \int \theta f(\theta|x_1, \ldots, x_n)\, d\theta$$

Example 5.5a Suppose that $X_1, \ldots, X_n$ are independent Bernoulli random variables, each having probability mass function given by

$$f(x|\theta) = \theta^x(1 - \theta)^{1-x} \qquad x = 0, 1$$

where θ is unknown. Further, suppose that θ is chosen from a uniform distribution on $(0, 1)$. Compute the Bayes estimator of θ.

Solution We must compute $E[\theta|X_1, \ldots, X_n]$. Since the prior density of θ is the uniform density

$$p(\theta) = 1 \qquad 0 < \theta < 1$$

we have that the conditional density of θ given $X_1, \ldots, X_n$ is given by

$$
\begin{aligned}
f(\theta|x_1, \ldots, x_n) &= \frac{f(x_1, \ldots, x_n, \theta)}{f(x_1, \ldots, x_n)} \\
&= \frac{f(x_1, \ldots, x_n|\theta)p(\theta)}{\int_0^1 f(x_1, \ldots, x_n|\theta)p(\theta)\, d\theta} \\
&= \frac{\theta^{\Sigma_1^n x_i}(1 - \theta)^{n - \Sigma_1^n x_i}}{\int_0^1 \theta^{\Sigma_1^n x_i}(1 - \theta)^{n - \Sigma_1^n x_i}\, d\theta}
\end{aligned}
$$

Now it can be shown that for integral values m and r

$$\int_0^1 \theta^m(1 - \theta)^r\, d\theta = \frac{m!r!}{(m + r + 1)!} \qquad (5.5.1)$$

Hence, upon letting $x = \Sigma_{i=1}^n x_i$

$$f(\theta|x_1, \ldots, x_n) = \frac{(n + 1)!\theta^x(1 - \theta)^{n-x}}{x!(n - x)!} \qquad (5.5.2)$$

Therefore,

$$
\begin{aligned}
E[\theta|x_1, \ldots, x_n] &= \frac{(n + 1)!}{x!(n - x)!}\int_0^1 \theta^{1+x}(1 - \theta)^{n-x}\, d\theta \\
&= \frac{(n + 1)!}{x!(n - x)!}\frac{(1 + x)!(n - x)!}{(n + 2)!} \qquad \text{from Equation 5.5.1} \\
&= \frac{x + 1}{n + 2}
\end{aligned}
$$

Thus, the Bayes estimator is given by

$$E[\theta|X_1,\ldots,X_n] = \frac{\sum_{i=1}^n X_i + 1}{n + 2}$$

As an illustration, if 10 independent trials, each of which results in a success with probability θ, result in 6 successes, then assuming a uniform $(0,1)$ prior distribution on θ, the Bayes estimate of θ is $7/11$ (as opposed, for instance, to the maximum likelihood estimate of $6/10$). ∎

REMARK

The conditional distribution of θ given that $X_i = x_i$, $i = 1,\ldots,n$, whose density function is given by Equation (5.5.2), is called the beta distribution with parameters $\sum_{i=1}^n x_i + 1$, $n - \sum_{i=1}^n x_i + 1$.

Example 5.5b Suppose $X_1,\ldots,X_n$ are independent normal random variables, each having unknown mean θ and known variance σ_0^2. If θ is itself selected from a normal population having known mean μ and known variance σ^2, what is the Bayes estimator of θ?

In order to determine $E[\theta|X_1,\ldots,X_n]$, the Bayes estimator, we need first determine the conditional density of θ given the values of $X_1,\ldots,X_n$. Now

$$f(\theta|x_1,\ldots,x_n) = \frac{f(x_1,\ldots,x_n|\theta)p(\theta)}{f(x_1,\ldots,x_n)}$$

where

$$f(x_1,\ldots,x_n|\theta) = \frac{1}{(2\pi)^{n/2}\sigma_0^n} \exp\left\{-\sum_{i=1}^n (x_i - \theta)^2/2\sigma_0^2\right\}$$

$$p(\theta) = \frac{1}{\sqrt{2\pi}\,\sigma} \exp\left\{-(\theta - \mu)^2/2\sigma^2\right\}$$

and

$$f(x_1,\ldots,x_n) = \int_{-\infty}^{\infty} f(x_1,\ldots,x_n|\theta)p(\theta)\,d\theta$$

With the help of a little algebra, it can now be shown that this conditional density is a *normal* density with mean

$$E[\theta|X_1,\ldots,X_n] = \frac{n\sigma^2}{n\sigma^2 + \sigma_0^2}\overline{X} + \frac{\sigma_0^2}{n\sigma^2 + \sigma_0^2}\mu \qquad (5.5.3)$$

$$= \frac{\frac{n}{\sigma_0^2}}{\frac{n}{\sigma_0^2} + \frac{1}{\sigma^2}}\overline{X} + \frac{\frac{1}{\sigma^2}}{\frac{n}{\sigma_0^2} + \frac{1}{\sigma^2}}\mu$$

and variance

$$\mathrm{Var}\left(\theta|X_1,\ldots,X_n\right) = \frac{\sigma_0^2\sigma^2}{n\sigma^2 + \sigma_0^2}$$

Writing the Bayes estimator as we did in Equation 5.5.3 is informative, for it shows that it is a weighted average of $\overline{X}$, the sample mean, and μ, the a priori mean. In fact, the weights given to these two quantities are in proportion to the inverses of σ_0^2/n (the conditional variance of the sample mean $\overline{X}$ given θ) and σ^2 (the variance of the prior distribution). ■

REMARK. ON CHOOSING A NORMAL PRIOR

As illustrated by Example 5.5b, it is computationally very convenient to choose a normal prior for the unknown mean θ of a normal distribution—for then the Bayes estimator is simply given by Equation 5.5.3. This raises the question of how one should go about determining whether there is a normal prior that reasonably represents one's prior feelings about the unknown mean.

To begin, it seems reasonable to determine the value—call it μ—that you a priori feel is most likely to be near θ. That is, we start with the mode (which equals the mean when the distribution is normal) of the prior distribution. We should then try to ascertain whether or not we believe that the prior distribution is symmetric about μ. That is, for each $a > 0$ do we believe that it is just as likely that θ will lie between $\mu - a$ and μ as it is that it will be between μ and $\mu + a$? If the answer is positive, then we accept, as a working hypothesis, that our prior feelings about θ can be expressed in terms of a prior distribution that is normal with mean μ. To determine σ, the standard deviation of the normal prior, think of an interval centered about μ that you a priori feel is 90 percent certain to contain θ. For instance, suppose you feel 90 percent (no more and no less) certain that θ will lie between $\mu - a$ and $\mu + a$. Then, since a normal random variable θ with mean μ and variance σ^2 is such that

$$P\left\{ -1.64 < \frac{\theta - \mu}{\sigma} < 1.64 \right\} = .90$$

or

$$P\{\mu - 1.64\sigma < \theta < \mu + 1.64\sigma\} = .90$$

it seems reasonable to take

$$1.64\sigma = a \quad \text{or} \quad \sigma = \frac{a}{1.64}$$

Thus, if your prior feelings can indeed reasonably be described by a normal distribution, then that distribution would have mean μ and standard deviation $\sigma = a/1.64$. As a test of whether this distribution indeed fits your prior feelings you might ask yourself such questions as whether you are 95 percent certain that θ will fall between $\mu - 1.96\sigma$ and $\mu + 1.96\sigma$, or whether you are 99 percent certain that θ will fall between $\mu - 2.58\sigma$ and $\mu + 2.58\sigma$, where these intervals are determined by the equalities

$$P\left\{ -1.96 < \frac{\theta - \mu}{\sigma} < 1.96 \right\} = .95$$

$$P\left\{ -2.58 < \frac{\theta - \mu}{\sigma} < 2.58 \right\} = .99$$

which hold when θ is normal with mean μ and variance σ^2.

Example 5.5c Consider the likelihood function $f(x_1, \ldots, x_n | \theta)$ and suppose that θ is uniformly distributed over some interval (a, b). The posterior density of θ given $X_1, \ldots, X_n$ equals

$$f(\theta | x_1, \ldots, x_n) = \frac{f(x_1, \ldots, x_n | \theta) p(\theta)}{\int_a^b f(x_1, \ldots, x_n | \theta) p(\theta) \, d\theta}$$

$$= \frac{f(x_1, \ldots, x_n | \theta)}{\int_a^b f(x_1, \ldots, x_n | \theta) \, d\theta} \qquad a < \theta < b$$

Now the *mode* of a density $f(\theta)$ was defined to be that value of θ that maximizes $f(\theta)$. By the foregoing, it follows that the mode of the density $f(\theta | x_1, \ldots, x_n)$ is that value of θ maximizing $f(x_1, \ldots, x_n | \theta)$, that is, it is just the maximum likelihood estimate of θ [when it is constrained to be in (a, b)]. In other words, the maximum likelihood estimate equals the mode of the posterior distribution when a uniform prior distribution is assumed. ∎

If, rather than a point estimate, we desire an interval in which θ lies with a specified probability—say $1 - \alpha$—we can accomplish this by choosing values a and b such that

$$\int_a^b f(\theta | x_1, \ldots, x_n) \, d\theta = 1 - \alpha$$

Example 5.5d Suppose that if a signal of value s is sent from location A, then the signal value received at location B is normally distributed with mean s and variance 60. Suppose also that the value of a signal sent at location A is, a priori, known to be normally distributed with mean 50 and variance 100. If the value received at location B is equal to 40, determine an interval that will contain the actual value sent with probability .90.

Solution It follows from Example 5.5b that the conditional distribution of S, the signal value sent, given that 40 is the value received, is normal with mean and variance given by

$$E[S | \text{data}] = \frac{1/60}{1/60 + 1/100} 40 + \frac{1/100}{1/60 + 1/100} 50 = 43.75$$

$$\text{Var}(S | \text{data}) = \frac{1}{1/60 + 1/100} = 37.5$$

Hence, given that the value received is 40, $(S - 43.75)/\sqrt{37.5}$ has a unit normal distribution and so

$$P\left\{ -1.64 < \frac{S - 43.75}{\sqrt{37.5}} < 1.64 | \text{data} \right\} = .95$$

or

$$P\left\{ 43.75 - 1.64\sqrt{37.5} < S < 43.75 + 1.64\sqrt{37.5} | \text{data} \right\} = .95$$

That is, with *probability* .95, the true signal sent lies within the interval (33.71, 53.79). ■

PROBLEMS

1. Let $X_1, \ldots, X_n$ be a sample from the distribution whose density function is

$$f(x) = \begin{cases} e^{-(x-\theta)} & x \geq \theta \\ 0 & \text{otherwise} \end{cases}$$

(a) Determine the method of moments estimator of θ.
(b) Determine the maximum likelihood estimate of θ.

2. (a) Show that $-\log U$ is an exponential random variable with parameter 1, when U is uniform on $(0, 1)$.
(b) If X is exponential with parameter 1, show that $X + \theta$ has the density function of Problem 1.
(c) Using parts (a) and (b), fix a value of θ—say $\theta = 0.5$—and write a program to simulate 10 random variables from the density in Problem 1. Then compare the method of moments estimator and the maximum likelihood estimator with the value 0.5. Repeat this 20 times, using different random numbers. Which of the 2 estimators do you think is superior in this case?

3. Let $X_1, \ldots, X_n$ be a sample from the distribution whose density function is

$$f(x) = \begin{cases} \lambda e^{-\lambda(x-1)} & x \geq 1 \\ 0 & \text{otherwise} \end{cases}$$

Show that the method of moments and maximum likelihood estimators for λ are equal.

4. Determine the maximum likelihood estimator of θ when $X_1, \ldots, X_n$ is a sample with density function

$$f(x) = \tfrac{1}{2} e^{-|x-\theta|} \qquad -\infty < x < \infty$$

5. Let $X_1, \ldots, X_n$ be a sample from a normal μ, σ^2 population. Determine the maximum likelihood estimator of σ^2 when μ is known. What is the expected value of this estimator?

6. Let $X_1, \ldots, X_{2n+1}$ be a sample from a normal population with unknown mean μ and known variance $\sigma^2 = 1$. Two possible estimators of μ are (a) the sample mean $\bar{X}$, and (b) the sample median $X_{(n+1)}$, which is defined to be the $(n + 1)$st largest of the $2n + 1$ values $X_1, \ldots, X_{2n+1}$. To empirically determine which estimator is preferable, simulate 9 unit normal random variables and compute the square of the sample average and the square of the sample median. Repeat this 10 times so as to be able to estimate the expected square of the error for both estimators. (Approximately, what should the average square error of the sample mean be?) What conclusions can you draw?

7. In determining a confidence interval for the mean of a normal population whose variance is assumed known, how large a sample is needed to make the confidence interval one-third as large as it is when the sample size is n?

8. If $X_1, \ldots, X_n$ is a sample from a normal population whose mean μ is unknown but whose variance σ^2 is known, show that $[-\infty, \overline{X} + z_\alpha \sigma / \sqrt{n}]$ is a $100(1 - \alpha)$ percent lower confidence interval for μ.

9. A sample of 20 cigarettes is tested to determine nicotine content and the average value observed was 1.2 mg. Compute a 99 percent two-sided confidence interval for the mean nicotine content of a cigarette if it is known that the standard deviation of a cigarette's nicotine content is $\sigma = 0.2$ mg.

10. In Problem 9, suppose that the population variance is not known in advance of the experiment. If the sample variance is 0.04, compute a 99 percent two-sided confidence interval for the mean nicotine content.

11. In Problem 10, compute a value c, for which we can assert "with 99 percent confidence" that c is larger than the mean nicotine content of a cigarette.

12. Suppose that when sampling from a normal population having an unknown mean μ and unknown variance σ^2, we wish to determine a sample size n so as to guarantee that the resulting $100(1 - \alpha)$ percent confidence interval for μ will be of size no greater than A, for given values α and A. Explain how we can approximately do this by a double sampling scheme that first takes a subsample of size 30 and then chooses the total sample size by using the results of the first subsample.

13. The following data resulted from 24 independent measurements of the melting point of lead

330°C	322°C	345°C
328.6°C	331°C	342°C
342.4°C	340.4°C	329.7°C
334°C	326.5°C	325.8°C
337.5°C	327.3°C	322.6°C
341°C	340°C	333°C
343.3°C	331°C	341°C
329.5°C	332.3°C	340°C

$\overline{X} = 334.08$

$\hat{S} = 6.93$

Assuming that the measurements can be regarded as constituting a normal sample whose mean is the true melting point of lead, determine a 95 percent two-sided confidence interval for this value. Also determine a 99 percent two-sided confidence interval.

14. If a sample of size 16 results in a sample mean of 42 and a sample standard deviation of 4.6, determine a 95 percent confidence interval for the mean. Assume a normal population.

15. Repeat Problem 13 if it is known that the standard deviation of the sample values is $4°C$.

16. The range of a new type of mortar shell is being investigated. The observed ranges, in meters, of 20 such shells are as follows:

2100	1984	2072	1898
1950	1992	2096	2103
2043	2218	2244	2206
2210	2152	1962	2007
2018	2106	1938	1956

Assuming that a shell's range is normally distributed, construct (a) a 95 percent and (b) 99 percent two-sided confidence interval for the mean range of a shell. (c) Determine the largest value v which, "with 95 percent confidence," will be less than the mean range.

17. Studies were conducted in Los Angeles to determine the carbon monoxide concentration near freeways. The basic technique used was to capture air samples in special bags and to then determine the carbon monoxide concentration by using a spectrophotometer. The measurements in ppm (parts per million) over a sampled period during the year were 102.2, 98.4, 104.1, 101, 102.2, 100.4, 98.6, 88.2, 78.8, 83, 84.7, 94.8, 105.1, 106.2, 111.2, 108.3, 105.2, 103.2, 99, 98.8. Compute a 95 percent two-sided confidence interval for the mean carbon monoxide concentration.

18. Let $X_1, \ldots, X_n, X_{n+1}$ denote a sample from a normal population whose mean and variance are unknown. We are interested in using the observed values of $X_1, \ldots, X_n$ to determine an interval, called a *prediction* interval, which we predict will contain X_{n+1} "with 95 percent confidence."

   (a) Determine such an interval.
      Hint: What is the distribution of $X_{n+1} - \sum_{i=1}^{n} X_i/n$?
   (b) Explain the meaning of the phrase "with 95 percent confidence."

19. The daily dissolved oxygen concentration for a water stream has been recorded over 30 days. If the sample average of the 30 values is 2.5 mg/liter and the sample standard deviation is 2.12 mg/liter, determine a value which, with 90 percent confidence, exceeds the mean daily concentration.

20. A civil engineer wishes to measure the compressive strength of 2 different types of concrete. A random sample of 10 specimens of the first type yielded the following data (in psi)

Type 1: 3250, 3268, 4302, 3184, 3266
         3297, 3332, 3502, 3064, 3116

whereas a sample of 10 specimens of the second yielded the data

Type 2: 3094, 3106, 3004, 3066, 2984,
          3124, 3316, 3212, 3380, 3018

If we assume that the samples are normal with a common variance, determine

(a) a 95 percent two-sided confidence interval for $\mu_1 - \mu_2$, the difference in means,

(b) a 95 percent one-sided upper confidence interval for $\mu_1 - \mu_2$,

(c) a 95 percent one-sided lower confidence interval for $\mu_1 - \mu_2$

21. Independent random samples are taken from the output of two machines on a production line. The weight of each item is of interest. From the first machine, a sample of size 36 is taken, with sample mean weight of 120 grams and a sample variance of 4. From the second machine, a sample of size 64 is taken, with sample mean weight of 130 grams and a sample variance of 5. It is assumed that the weights of items from the first machine are normally distributed with mean μ_1 and variance σ^2, and that the weights of items from the second machine are normally distributed with mean μ_2 and variance σ^2 (that is, the variances are assumed to be equal). From a 99 percent confidence interval for $\mu_1 - \mu_2$, the difference in population means.

22. Do Problem 21 when it is known in advance that the sample variances are 4 and 5.

23. The following are the burning times in seconds of floating smoke pots of two different types:

Type I		Type II	
481	572	526	537
506	561	511	582
527	501	556	605
661	487	542	558
501	524	491	578

Find a 99 percent confidence interval for the mean difference in burning times assuming normality with unknown but equal variances.

24. The capacities (in ampere-hours) of 10 batteries were recorded as follows:

140, 136, 150, 144, 148, 152, 138, 141, 143, 151

(a) Estimate the population variance σ^2.

(b) Compute a 99 percent two-sided confidence interval for σ^2.

(c) Compute a value v which enables us to state, with 90 percent confidence, that σ^2 is less than v.

25. Find a 95 percent two-sided confidence interval for the variance of the diameter of the rivet based on the data below:

6.68	6.66	6.62	6.72
6.76	6.67	6.70	6.72
6.78	6.66	6.76	6.72
6.76	6.70	6.76	6.76
6.74	6.74	6.81	6.66
6.64	6.79	6.72	6.82
6.81	6.77	6.60	6.72
6.74	6.70	6.64	6.78
6.70	6.70	6.75	6.79

Assume a normal population.

26. Ten units of rocket powder were tested and their burning times (in seconds) were recorded as follows:

50.6	69.8
54.8	53.6
54.4	66.1
44.9	48
42.1	37.8

Compute a 90 percent two-sided confidence interval for the variance of the burning time. Assume that the underlying population is normal.

27. If $X_1, \ldots, X_n$ is a sample from a normal population, explain how to obtain a $100(1 - \alpha)$ percent confidence interval for the population variance σ^2 when the population mean μ is known. Explain in what sense knowledge of μ improves the interval estimator compared with the case that it is unknown.

28. Repeat Problem 26 if it is known that the mean burning time is 53.6 seconds.

29. If $X_1, \ldots, X_n$ is a sample from a normal population having known mean μ and unknown variance σ_1^2 and $Y_1, \ldots, Y_m$ is an independent sample from a normal population having known mean μ_2 and unknown variance σ_2^2, determine a $100(1 - \alpha)$ percent confidence interval for σ_1^2/σ_2^2.
Hint: Use the definition of an F-random variable along with the fact that $(n - 1)S_x^2/\sigma_1^2$ and $(m - 1)S_y^2/\sigma_2^2$ are independent chi-square random variables.

30. Two analysts took repeated readings on the hardness of city water. Assuming that the readings of analyst i constitute a sample from a normal population having variance σ_i^2, $i = 1, 2$, compute a 95 percent two-sided confidence interval for σ_1^2/σ_2^2 when the data are as follows:

Coded Measures of Hardness	
Analyst 1	*Analyst 2*
0.46	0.82
0.62	0.61
0.37	0.89
0.40	0.51
0.44	0.33
0.58	0.48
0.48	0.23
0.53	0.25
	0.67
	0.88

31. In a recent study, 79 of 140 meteorites were observed to enter the atmosphere with a velocity of less than 25 miles per second. If we take $\hat{p} = 79/140$ as an estimate of the probability that an arbitrary meteorite that enters the atmosphere will have a speed less than 25 miles per second, what can we say, with 99 percent confidence, about the maximum error of our estimate?

32. A random sample of 100 items from a production line revealed 17 of them to be defective. Compute a 95 percent two-sided confidence interval for the probability that an item produced is defective. Determine also a 99 percent upper confidence interval for this value. What assumptions are you making?

33. Of 100 randomly detected cases of individuals having lung cancer, 67 died within 5 years of detection.

(a) Estimate the probability that a person contracting lung cancer will die within 5 years.

(b) How large an additional sample would be required to be 95 percent confident that the error in estimating the probability in (a) is less than .02?

34. Suppose the lifetimes of batteries are exponentially distributed with mean θ. If the average of a sample of 10 batteries is 36 hours, determine a 95 percent two-sided confidence interval for θ.

35. Determine $100(1 - \alpha)$ percent one-sided upper and lower confidence intervals for θ in Problem 34.

36. Compute $r(d_i, \theta)$ for the estimator d_1, d_2, and d_3 in Example 5.4a.

37. Let $X_1, X_2, \ldots, X_n$ denote a sample from a population whose mean value θ is unknown. Use the results of Example 5.4b to argue that among all unbiased estimators of θ of the form $\sum_{i=1}^{n}\lambda_i X_i, \sum_{i=1}^{n}\lambda_i = 1$ the one with minimal mean square error has $\lambda_i \equiv 1/n$, $i = 1, \ldots, n$.

38. Consider two independent samples from normal populations having the same variance σ^2, of respective sizes n and m. That is, $X_1, \ldots, X_n$ and

$Y_1, \ldots, Y_m$ are independent samples from normal populations each having variance σ^2. Let S_x^2 and S_y^2 denote the respective sample variances. Thus both S_x^2 and S_y^2 are unbiased estimators of σ^2. Show by using the results of Example 5.5b along with the fact that

$$\text{Var}\left(\chi_k^2\right) = 2k$$

where χ_k^2 is chi-square with k degrees of freedom, that the minimum mean square estimator of σ^2 of the form $\lambda S_x^2 + (1 - \lambda)S_y^2$ is

$$\frac{(n - 1)S_x^2 + (m - 1)S_y^2}{n + m - 2}$$

This is called the *pooled estimator* of σ^2.

39. Consider two estimators d_1 and d_2 of a parameter θ. If $E[d_1] = \theta$, $\text{Var}(d_1) = 6$ and $E[d_2] = \theta + 2$, $\text{Var}(d_2) = 2$, which estimator should be preferred?

40. Suppose that the number of accidents occurring daily in a certain plant has a Poisson distribution with an unknown mean λ. Based on previous experience in similar industrial plants, suppose that a statistician's initial feelings about the possible value of λ can be expressed by an exponential distribution with parameter 1. That is, the prior density is

$$p(\lambda) = e^{-\lambda} \qquad 0 < \lambda < \infty$$

Determine the Bayes estimate of λ if there are a total of 83 accidents over the next 10 days. What is the maximum likelihood estimate?

41. The functional lifetimes in hours of computer chips produced by a certain semiconductor firm are exponentially distributed with mean $1/\lambda$. Suppose that the prior distribution on λ is the gamma distribution with density function

$$g(x) = \frac{e^{-x}x^2}{2} \qquad 0 < x < \infty$$

If the average life of the first 20 chips tested is 4.6 hours, compute the Bayes estimate of λ.

42. Each item produced will, independently, be defective with probability p. If the prior distribution on p is uniform on $(0, 1)$, compute the posterior probability that p is less than .2 given

(a) a total of 2 defectives out of a sample of size 10,
(b) a total of 1 defective out of a sample of size 10,
(c) a total of 10 defectives out of a sample of size 10.

43. The breaking strength of a certain type of cloth is to be measured for 10 specimens. The underlying distribution is normal with unknown mean θ but with a standard deviation equal to 3 psi. Suppose also that based on previous experience we feel that the unknown mean has a prior distribution that is normally distributed with mean 200 and standard deviation 2. If the average breaking strength of a sample of 20 specimens is 182 psi, determine a region that contains θ with probability .95.

Hypothesis Testing

1 INTRODUCTION

As in the previous chapter, let us suppose that a random sample from a population distribution, specified except for a vector of unknown parameters, is to be observed. However, rather than wishing to explicitly estimate the unknown parameters, let us now suppose that we are primarily concerned with using the resulting sample to test some particular hypothesis concerning them. For an illustration, suppose that a construction firm has just purchased a large supply of cables that have been guaranteed to have an average breaking strength of at least 7000 psi. To verify this claim, the firm has decided to take a random sample of 10 of these cables to determine their breaking strengths. They will then use the result of this experiment to ascertain whether or not they accept the cable manufacturer's hypothesis that the population mean is at least 7000 pounds per square inch.

A statistical hypothesis is usually a statement about a set of parameters of a population distribution. It is called a hypothesis because it is not known whether or not it is true. A primary problem is to develop a procedure for determining whether or not the values of a random sample from this population are consistent with the hypothesis. For instance, consider a particular normally distributed population having an unknown mean value θ and known variance 1. The statement "θ is less than 1" is a statistical hypothesis that we could try to test by observing a random sample from this population. If the random sample is deemed to be consistent with the hypothesis under consideration, we say that the hypothesis has been "accepted"; otherwise we say that it has been "rejected."

It should be noted that in accepting a given hypothesis we are not actually claiming that it is true but rather we are saying that the resulting data appear to be consistent with it. For instance, in the case of a normal $(\theta, 1)$ population, if a resulting sample of size 10 has an average value of 1.25, then although such a result cannot be regarded as being evidence in favor of the hypothesis "$\theta < 1$," it is not inconsistent with this hypothesis, which would thus be accepted. On the other hand, if the sample of size 10 has an average value of 3, then even though a sample value that large is possible when $\theta < 1$, it is so unlikely that it seems inconsistent with this hypothesis, which would thus be rejected.

2 SIGNIFICANCE LEVELS

Consider a population having distribution F_θ, where θ is unknown, and suppose we want to test a specific hypothesis about θ. We shall denote this hypothesis by H_0 and call it the *null hypothesis*. For example, if F_θ is a normal distribution function with mean θ and variance equal to 1, then two possible null hypothesis about θ are

    (a) $H_0 : \theta = 1$
    (b) $H_0 : \theta \leq 1$

Thus the first of these hypotheses states that the population is normal with mean 1 and variance 1, whereas the second states that it is normal with variance 1 and a mean less than or equal to 1. It should be noted that the null hypothesis in (a), when true, completely specifies the population distribution; whereas the null hypothesis in (b) does not. A hypothesis that, when true, completely specifies the population distribution is called a *simple* hypothesis; one that does not is called a *composite* hypothesis.

Suppose now that in order to test a specific null hypothesis H_0 a population sample of size n—say $X_1, \ldots, X_n$—is to be observed. Based on these n values, we must decide whether or not to accept H_0. A test for H_0 can be specified by defining a region C in n-dimensional space with the proviso that the hypothesis is to be rejected if the random sample $X_1, \ldots, X_n$ turns out to lie in C and accepted otherwise. The region C is called the *critical region*. In other words, the statistical test determined by the critical region C is the one that

$$\text{accepts}\quad H_0 \quad \text{if}\quad (X_1, X_2, \ldots, X_n) \notin C$$

and

$$\text{rejects}\quad H_0 \quad \text{if}\quad (X_1, \ldots, X_n) \in C$$

For instance, a common test of the hypothesis that θ, the mean of a normal population with variance 1, is equal to 1 has a critical region given by

$$C = \left\{ (X_1, \ldots, X_n) : \left| \frac{\sum_{i=1}^n X_i}{n} - 1 \right| > \frac{1.96}{\sqrt{n}} \right\} \tag{6.2.1}$$

Thus, this test calls for rejection of the null hypothesis that $\theta = 1$ when the sample average differs from 1 by more than 1.96 divided by the square root of the sample size.

It is important to note when developing a procedure for testing a given null hypothesis H_0 that, in any test, there are two different types of errors that can result. The first of these, called a *type I error*, is said to result if the test incorrectly calls for rejecting H_0 when it is indeed correct. The second, called a *type II error*, results if the test calls for accepting H_0 when it is false. Now, as was previously mentioned, the objective of a statistical test of H_0 is not to determine whether or not H_0 is true but rather to determine if its validity is consistent with the resultant data. Hence, with this objective it seems reasonable that H_0 should only be rejected if the resultant data are very unlikely when H_0 is true. The classical way of accomplishing this is to specify a value α and then require the test to have the property that whenever H_0 is true its probability of being rejected is never greater than α. The value α, called the *level of significance of the test*, is usually set in advance, with commonly chosen values being $\alpha = 0.1$, 0.05, 0.005. In other words, the classical approach to testing H_0 is to fix a significance level α and then require that the test have the property that the probability of a type I error occurring can never be greater than α.

Suppose now that we are interested in testing a certain hypothesis concerning θ, an unknown parameter of the population. Specifically, for a given set of parameter values w, suppose we are interested in testing

$$H_0 : \theta \in w$$

A common approach to developing a test of H_0, say at level of significance α, is to start by determining a point estimator of θ—say $d(\mathbf{X})$. The hypothesis is then rejected if $d(\mathbf{X})$ is "far away" from the region w. However, in order to determine how "far away" it need be to justify rejection of H_0, we need to determine the probability distribution of $d(\mathbf{X})$ when H_0 is true since this will usually enable us to determine the appropriate critical region so as to make the test have the required significance level α. For example, the test of the hypothesis that the mean of a normal $(\theta, 1)$ population is equal to 1, given by Equation 6.2.1, calls for rejection when the point estimate of θ—that is, the sample average—is further than $1.96/\sqrt{n}$ away from 1. As we will see in the next section, the value $1.96/\sqrt{n}$ was chosen to meet a level of significance of $\alpha = 0.05$.

3 TESTS CONCERNING THE MEAN OF A NORMAL POPULATION

3.1 Case of Known Variance

Suppose that $X_1, \ldots, X_n$ is a sample of size n from a normal distribution having an unknown mean μ and a known variance σ^2 and suppose we are interested in testing the null hypothesis

$$H_0 : \mu = \mu_0$$

against the alternative hypothesis

$$H_1 : \mu \neq \mu_0$$

where μ_0 is some specified constant.

Since $\overline{X} = \sum_{i=1}^{n} X_i/n$ is a natural point estimator of μ, it seems reasonable to accept H_0 if $\overline{X}$ is not too far from μ_0. That is, the critical region of the test would be of the form

$$C = \left\{ X_1, \ldots, X_n : |\overline{X} - \mu_0| > c \right\} \tag{6.3.1}$$

for some suitably chosen value c.

If we desire that the test has significance level α, then we must determine the critical value c in Equation 6.3.1 that will make the type I error equal to α. That is, c must be such that

$$P_{\mu_0}\left\{ |\overline{X} - \mu_0| > c \right\} = \alpha \tag{6.3.2}$$

where we write P_{μ_0} to mean that the preceding probability is to be computed under the assumption that $\mu = \mu_0$. However, when $\mu = \mu_0$, $\overline{X}$ will be normally distributed with mean μ_0 and variance σ^2/n and so Z, defined by

$$Z \equiv \frac{\overline{X} - \mu_0}{\sigma/\sqrt{n}}$$

will have a unit normal distribution. Now Equation 6.3.2 is equivalent to

$$P\left\{ |Z| > \frac{c\sqrt{n}}{\sigma} \right\} = \alpha$$

or equivalently

$$2P\left\{ Z > \frac{c\sqrt{n}}{\sigma} \right\} = \alpha$$

where Z is a unit normal random variable. However, we know that

$$P\left\{ Z > z_{\alpha/2} \right\} = \alpha/2$$

and so

$$\frac{c\sqrt{n}}{\sigma} = z_{\alpha/2}$$

or

$$c = \frac{z_{\alpha/2}\sigma}{\sqrt{n}}$$

Thus, the significance level α test is to reject H_0 if $|\overline{X} - \mu_0| > z_{\alpha/2}\sigma/\sqrt{n}$ and accept otherwise; or, equivalently, to

$$\text{reject} \quad H_0 \quad \text{if} \quad \frac{\sqrt{n}}{\sigma}|\overline{X} - \mu_0| > z_{\alpha/2} \tag{6.3.3}$$

$$\text{accept} \quad H_0 \quad \text{if} \quad \frac{\sqrt{n}}{\sigma}|\overline{X} - \mu| \leq z_{\alpha/2}$$

This can be pictorially represented as follows:

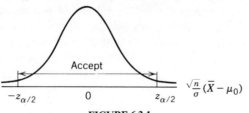

FIGURE 6.3.1

where we have superimposed the unit normal density function [which is the density of the test statistic $\sqrt{n}\,(\overline{X}-\mu_0)/\sigma$ when H_0 is true].

Example 6.3a It is known that if a signal of value μ is sent from location A, then the value received at location B is normally distributed with mean μ and standard deviation 2. That is, the random noise added to the signal is a $N(0,4)$ random variable. There is reason for the people at location B to suspect that the signal value $\mu = 8$ will be sent today. Test this hypothesis if the same signal value is independently sent 5 times and the average value received at location B is $\overline{X} = 9.5$.

Solution Suppose we are testing at the 5 percent level of significance. To begin, we compute the test statistic

$$\frac{\sqrt{n}}{\sigma}|\overline{X}-\mu_0| = \frac{\sqrt{5}}{2}(1.5) = 1.68$$

Since this value is less than $z_{.025} = 1.96$, the hypothesis is accepted. In other words, the data are not inconsistent with the null hypothesis in the sense that a sample average as far from the value 8 as observed would be expected, when the true mean is 8, over 5 percent of the time. It should be noted, however, that if a less stringent significance level were chosen—say $\alpha = .1$—then the null hypothesis would have been rejected. This follows since $z_{.05} = 1.64$, which is less than 1.68. Hence, if we would have chosen a test that had a 10 percent chance of rejecting H_0 when H_0 was true, then the null hypothesis would have been rejected.

The "correct" level of significance to use in a given situation depends on the individual circumstances involved in that situation. For instance, if rejecting a null hypothesis H_0 would result in large costs that would thus be wasted if H_0 were indeed true, then we might elect to be quite conservative and so choose a significance level of 0.05 or 0.01. Also, if we initially feel strongly that H_0 was correct, then we would require very stringent data evidence to the contrary for us to reject H_0. (That is, we would set a very low significance level in this situation.) ■

The test given by Equation 6.3.3 can be described as follows: For any observed value of the test statistic $(\sqrt{n}/\sigma)|\overline{X}-\mu_0|$, call that value v, the test calls for rejection of the null hypothesis if the probability that the test statistic would be as large as v when H_0 is true is less than or equal to the significance level α. From this, it follows that we can determine whether or not to accept the null hypothesis by computing, first, the value of the test statistic and, second, the probability that a unit normal would (in absolute value) exceed that quantity. This probability—called the *p-value* of the test—gives the critical significance level in the sense that H_0 will be accepted if the significance level α is less than the *p*-value and rejected if it is greater than or equal.

In practice, the significance level is often not set in advance but rather the data are looked at to determine the resultant *p*-value. Sometimes, this critical significance level is clearly much higher than we would use and so the null hypothesis can be readily accepted.

Example 6.3b In Example 6.3a, suppose that the average of the 5 values received is $\overline{X} = 8.5$. In this case,

$$\frac{\sqrt{n}}{\sigma}|\overline{X} - \mu_0| = \frac{\sqrt{5}}{4} = .559$$

Since

$$P\{|Z| > .559\} = 2P\{Z > .559\}$$
$$= 2 \times .288 = .576$$

from Table A-3-5-1-A or Program 3-5-1-A

it follows that the p-value is .576 and thus the null hypothesis H_0 that the signal sent has value 8 would be accepted at any significance level $\alpha < .576$. Since we would clearly never want to test a null hypothesis using a significance level as large as .576, H_0 would be accepted. ∎

We have not yet talked about the probability of a type II error—that is, the probability of accepting the null hypothesis when the true mean μ is unequal to μ_0. This probability will depend on the value of μ, and so let us define $\beta(\mu)$ by

$$\beta(\mu) = P_\mu\{\text{acceptance of } H_0\}$$

$$= P_\mu\left\{\left|\frac{\overline{X} - \mu_0}{\sigma/\sqrt{n}}\right| \leq z_{\alpha/2}\right\}$$

$$= P_\mu\left\{-z_{\alpha/2} \leq \frac{\overline{X} - \mu_0}{\sigma/\sqrt{n}} \leq z_{\alpha/2}\right\}$$

The function $\beta(\mu)$ is called the *operating characteristic* (or OC) *curve* and represents the probability that H_0 will be accepted when the true mean is μ.

To compute this probability, we use the fact that $\overline{X}$ is normal with mean μ and variance σ^2/n and so

$$Z \equiv \frac{\overline{X} - \mu}{\sigma/\sqrt{n}} \sim \mathcal{N}(0, 1)$$

Hence,

$$\beta(\mu) = P_\mu\left\{-z_{\alpha/2} \leq \frac{\overline{X} - \mu_0}{\sigma/\sqrt{n}} \leq z_{\alpha/2}\right\} \tag{6.3.4}$$

$$= P_\mu\left\{-z_{\alpha/2} - \frac{\mu}{\sigma/\sqrt{n}} \leq \frac{\overline{X} - \mu_0 - \mu}{\sigma/\sqrt{n}} \leq z_{\alpha/2} - \frac{\mu}{\sigma/\sqrt{n}}\right\}$$

$$= P\left\{\frac{\mu_0 - \mu}{\sigma/\sqrt{n}} - z_{\alpha/2} \leq Z \leq \frac{\mu_0 - \mu}{\sigma/\sqrt{n}} + z_{\alpha/2}\right\}$$

$$= \Phi\left(\frac{\mu_0 - \mu}{\sigma/\sqrt{n}} + z_{\alpha/2}\right) - \Phi\left(\frac{\mu_0 - \mu}{\sigma/\sqrt{n}} - z_{\alpha/2}\right)$$

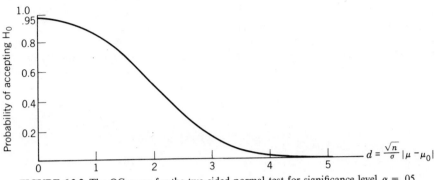

FIGURE 6.3.2 The OC curve for the two-sided normal test for significance level $\alpha = .05$

For a fixed significance level α the OC curve given by Equation 6.3.4 is symmetric about μ_0 and indeed will depend on μ only through $(\sqrt{n}/\sigma)|\mu - \mu_0|$. This curve with the abscissa changed from μ to $d = (\sqrt{n}/\sigma)|\mu - \mu_0|$ is presented in Figure 6.3.2 when $\alpha = .05$.

Example 6.3c For the problem presented in Example 6.3a, let us determine the probability of accepting the null hypothesis that $\mu = 8$ when the actual value sent is 10. To do so, we compute

$$\frac{\sqrt{n}}{\sigma}(\mu_0 - \mu) = -\frac{\sqrt{5}}{2} \times 2 = -\sqrt{5}$$

As $z_{.025} = 1.96$, the desired probability is, from Equation 6.3.4,

$$\Phi(-\sqrt{5} + 1.96) - \Phi(-\sqrt{5} - 1.96)$$

$$= 1 - \Phi(\sqrt{5} - 1.96) - \left[1 - \Phi(\sqrt{5} + 1.96)\right]$$

$$= \Phi(4.196) - \Phi(.276)$$

$$= .392. \quad \blacksquare$$

REMARK
The function $1 - \beta(\mu)$ is called the *power-function* of the test. Thus, for a given value μ, the power of the test is equal to the probability of rejection when μ is the true value.

The operating characteristic function is useful in determining how large the random sample need be to meet certain specifications concerning type II errors. For instance, suppose that we desire to determine the sample size n necessary to ensure that the probability of accepting $H_0 : \mu = \mu_0$ when the true mean is actually μ_1 is no greater than β. That is, we want n to be such that

$$\beta(\mu_1) = \beta$$

But from Equation (6.3.4) this is equivalent to

$$\Phi\left(\frac{\sqrt{n}\,(\mu_0 - \mu_1)}{\sigma} + z_{\alpha/2}\right) - \Phi\left(\frac{\sqrt{n}\,(\mu_0 - \mu_1)}{\sigma} - z_{\alpha/2}\right) = \beta \qquad (6.3.5)$$

Although the foregoing cannot be analytically solved for n, a solution can be obtained by using the standard normal distribution table. In addition, an approximation for n can be derived from Equation 6.3.5 as follows. To start, suppose that $\mu_1 > \mu_0$. Then, as this implies that

$$\frac{\mu_0 - \mu_1}{\sigma/\sqrt{n}} - z_{\alpha/2} \leq -z_{\alpha/2}$$

it follows, since Φ is an increasing function, that

$$\Phi\left(\frac{\mu_0 - \mu_1}{\sigma/\sqrt{n}} - z_{\alpha/2}\right) \leq \Phi(-z_{\alpha/2}) = P\{Z \leq -z_{\alpha/2}\} = P\{Z \geq z_{\alpha/2}\} = \alpha/2$$

Hence, we can take

$$\Phi\left(\frac{\mu_0 - \mu_1}{\sigma/\sqrt{n}} - z_{\alpha/2}\right) \approx 0$$

and so from Equation 6.3.5

$$\beta \approx \Phi\left(\frac{\mu_0 - \mu_1}{\sigma/\sqrt{n}} + z_{\alpha/2}\right) \qquad (6.3.6)$$

or, since

$$\beta = P\{Z > z_\beta\} = P\{Z < -z_\beta\} = \Phi(-z_\beta)$$

we obtain from Equation 6.3.6 that

$$-z_\beta \approx (\mu_0 - \mu_1)\frac{\sqrt{n}}{\sigma} + z_{\alpha/2}$$

or

$$n \approx \frac{(z_{\alpha/2} + z_\beta)^2 \sigma^2}{(\mu_1 - \mu_0)^2} \qquad (6.3.7)$$

In fact, the same approximation would result when $\mu_1 < \mu_0$ (the details are left as an exercise) and so Equation 6.3.7 is in all cases a reasonable approximation to the sample size necessary to ensure that the type II error at the value $\mu = \mu_1$ is approximately equal to β.

Example 6.3d For the problem of Example 6.3a, how many signals need be sent so that the 0.05 level test of $H_0 : \mu = 8$ has at least a 75 percent probability of rejection when $\mu = 9.2$?

Solution Since $z_{.025} = 1.96$, $z_{.25} = .67$ the approximation 6.3.7 yields

$$n \approx \frac{(1.96 + .67)^2}{(1.2)^2} 4 = 19.21$$

Hence a sample at size 20 is needed. From Equation 6.3.4 we see that with $n = 20$

$$\beta(9.2) = \Phi\left(-\frac{1.2\sqrt{20}}{2} + 1.96\right) - \Phi\left(-\frac{1.2\sqrt{20}}{2} - 1.96\right)$$

$$= \Phi(-.723) - \Phi(-4.643)$$

$$= 1 - \Phi(.723)$$

$$= .235. \quad \blacksquare$$

3.1.1 One-Sided Tests In testing the null hypothesis that $\mu = \mu_0$, we have chosen a test that calls for rejection when $\overline{X}$ is far from μ_0. That is, a very small value of $\overline{X}$ or a very large value appear to make it unlikely that μ (which $\overline{X}$ is estimating) could equal μ_0. However, what happens when the only alternative to μ being equal to μ_0 is for μ to be greater than μ_0? That is, what happens when the alternative hypothesis to $H_0 : \mu = \mu_0$ is $H_1 : \mu > \mu_0$? Clearly in this latter case we would not want to reject H_0 when $\overline{X}$ is small (since a small $\overline{X}$ is more likely when H_0 is true than when H_1 is true). Thus, in testing

$$H_0 : \mu = \mu_0 \qquad \text{versus} \qquad H_1 : \mu > \mu_0 \qquad (6.3.8)$$

we should reject H_0 when $\overline{X}$, the point estimate of μ_0, is much greater than μ_0. That is, the critical region should be of the following form:

$$C = \left\{ (X_1, \ldots, X_n) : \overline{X} - \mu_0 > c \right\}$$

Since the probability of rejection should equal α when H_0 is true (that is, when $\mu = \mu_0$), we require that c be such that

$$P_{\mu_0}\left\{ \overline{X} - \mu_0 > c \right\} = \alpha \qquad (6.3.9)$$

But since

$$Z = \frac{\overline{X} - \mu_0}{\sigma/\sqrt{n}}$$

has a unit normal distribution when H_0 is true, Equation 6.3.9 is equivalent to

$$P\left\{ Z > \frac{c\sqrt{n}}{\sigma} \right\} = \alpha$$

when Z is a unit normal. But since

$$P\{Z \geq z_\alpha\} = \alpha$$

we see that

$$c = \frac{z_\alpha \sigma}{\sqrt{n}}$$

Hence, the test of the hypothesis 6.3.8 is to reject H_0 if $\overline{X} - \mu_0 > z_\alpha \sigma/\sqrt{n}$, and accept otherwise; or equivalently to

$$\text{accept} \quad H_0 \quad \text{if} \quad \frac{\sqrt{n}}{\sigma}\left(\overline{X} - \mu_0\right) \leq z_\alpha \qquad (6.3.10)$$

$$\text{reject} \quad H_0 \quad \text{if} \quad \frac{\sqrt{n}}{\sigma}\left(\overline{X} - \mu_0\right) > z_\alpha$$

This is called a *one-sided* critical region (since it calls for rejection only when $\overline{X}$ is large). Correspondingly, the hypothesis testing problem

$$H_0 : \mu = \mu_0$$
$$H_1 : \mu > \mu_0$$

is called a one-sided testing problem (in contrast to the *two-sided* problem that results when the alternative hypothesis is $H_1 : \mu \neq \mu_0$).

To compute the *p*-value in the one-sided test Equation 6.3.10, we first use the data to determine the value of the statistic $\sqrt{n}\,(\overline{X} - \mu_0)/\sigma$. The *p*-value is then equal to the probability that a unit normal would be at least as large as this value.

Example 6.3e Suppose in Example 6.3a that it is known in advance that the signal value is at least as large as 8. What can be concluded in this case?

To see if the data are consistent with the hypothesis that the mean is 8 we test

$$H_0 : \mu = 8$$

against the one-sided alternative

$$H_1 : \mu > 8$$

The value of the test statistic is $\sqrt{n}\,(\overline{X} - \mu_0)/\sigma = \sqrt{5}\,(9.5 - 8)/2 = 1.68$, and the *p*-value is the probability that a unit normal would exceed 1.68, namely

$$p\text{-value} = 1 - \Phi(1.68) = .0465$$

Since the test would call for rejection at all significance levels greater than or equal to .0465, it would, for instance, reject the null hypothesis at the $\alpha = .05$ level of significance. (Intuitively, why is it that the null hypothesis $H_0 : \mu = 8$ is now being rejected at $\alpha = .05$ level of significance when it was originally, in Example 6.3a, accepted at that level?) ∎

The operating characteristic function of the one-sided test Equation 6.3.10

$$\beta(\mu) = P_\mu\{\text{accepting } H_0\}$$

can be obtained as follows:

$$\beta(\mu) = P_\mu\left\{ \overline{X} \leq \mu_0 + z_\alpha \frac{\sigma}{\sqrt{n}} \right\}$$

$$= P\left\{ \frac{\overline{X} - \mu}{\sigma/\sqrt{n}} \leq \frac{\mu_0 - \mu}{\sigma/\sqrt{n}} + z_\alpha \right\}$$

$$= P\left\{ Z \leq \frac{\mu_0 - \mu}{\sigma/\sqrt{n}} + z_\alpha \right\} \qquad Z \sim \mathcal{N}(0,1)$$

where the last equation follows since $\sqrt{n}\,(\overline{X} - \mu)/\sigma$ has a unit normal distribution. Hence we can write

$$\beta(\mu) = \Phi\left(\frac{\mu_0 - \mu}{\sigma/\sqrt{n}} + z_\alpha \right)$$

Since Φ, being a distribution function, is increasing in its argument, it follows that $\beta(\mu)$ decreases in μ; which is intuitively pleasing since it certainly seems reasonable that the larger the true mean μ the less likely it should be to conclude that $\mu \leq \mu_0$. Also since $\Phi(z_\alpha) = 1 - \alpha$, it follows that

$$\beta(\mu_0) = 1 - \alpha$$

The test given by Equation 6.3.10, which was designed to test $H_0: \mu = \mu_0$ versus $H_1: \mu > \mu_0$, can also be used to test, at level of significance α, the one-sided hypothesis

$$H_0: \mu \leq \mu_0$$

versus

$$H_1: \mu > \mu_0$$

To verify that it remains a level α test, we need show that the probability of rejection is never greater than α when H_0 is true. That is, we must verify that

$$P_\mu\left\{\overline{X} > \mu_0 + z_\alpha \frac{\sigma}{\sqrt{n}}\right\} \leq \alpha \qquad \text{for all } \mu \leq \mu_0$$

or, equivalently

$$P_\mu\left\{\overline{X} \leq \mu_0 + z_\alpha \frac{\sigma}{\sqrt{n}}\right\} \geq 1 - \alpha \qquad \text{for all } \mu \leq \mu_0$$

or

$$\beta(\mu) \geq 1 - \alpha \qquad \text{for all } \mu \leq \mu_0$$

But it has previously been shown that for the test given by Equation 6.3.10, $\beta(\mu)$ decreases in μ and $\beta(\mu_0) = 1 - \alpha$. This gives that

$$\beta(\mu) \geq \beta(\mu_0) = 1 - \alpha \qquad \text{for all } \mu \leq \mu_0$$

which shows that the test given by Equation 6.3.10 remains a level α test for $H_0: \mu \leq \mu_0$ against the alternative hypothesis $H_1: \mu > \mu_0$.

REMARK
We can also test the one-sided hypothesis

$$H_0: \mu = \mu_0 \ (\text{or } \mu \geq \mu_0) \qquad \text{versus} \qquad H_1: \mu < \mu_0$$

at significance level α by

$$\text{accepting} \quad H_0 \quad \text{if} \quad \frac{\sqrt{n}}{\sigma}(\overline{X} - \mu_0) > -z_\alpha$$

$$\text{rejecting} \quad H_0 \quad \text{otherwise}$$

This test can alternatively be performed by first computing the value of the test statistic $\sqrt{n}\,(\overline{X} - \mu_0)/\sigma$. The p-value would then equal the probability that a unit normal would be less than this value, and the hypothesis would be rejected at any significance level greater than or equal to this p-value.

Example 6.3f All cigarettes presently on the market have an average nicotine content of at least 1.6 mg per cigarette. A firm that produces cigarettes claims that it has discovered a new way to cure tobacco leaves that will result in the average nicotine content of a cigarette being less than 1.6 mg. To test this claim, a sample of 20 of the firm's cigarettes were analyzed. If it is known that the standard deviation of a cigarette's nicotine content is 0.8 mg, what conclusions can be drawn, at the 5 percent level of significance, if the average nicotine content of the 20 cigarettes is 1.54?

Note: The above raises the question of how we would know in advance that the standard deviation is 0.8. One possibility is that the variation in a cigarette's nicotine content is due to variability in the amount of tobacco in each cigarette and not on the method of curing that is used. Hence, the standard deviation can be known from previous experience.

Solution We must first decide on the appropriate null hypothesis. As was previously noted, our approach to testing is not symmetric with respect to the null and the alternative hypotheses since we consider only tests having the property that their probability of rejecting the null hypothesis when it is true will never exceed the significance level α. Thus, whereas rejection of the null hypothesis is a strong statement about the data not being consistent with this hypothesis, an analogous statement cannot be made when the null hypothesis is accepted. Hence, since in the preceding example we would like to endorse the producer's claims only when there is substantial evidence for it, we should take this claim as the alternative hypothesis. That is, we should test

$$H_0 : \mu \geq 1.6 \qquad \text{versus} \qquad H_1 : \mu < 1.6$$

Now, the value of the test statistic is

$$\sqrt{n}\,(\overline{X} - \mu_0)/\sigma = \sqrt{20}\,(1.54 - 1.6)/.8 = -.336$$

and so the p-value is given by

$$p\text{-value} = P\{Z < -.336\} \qquad Z \sim N(0,1)$$
$$= .368$$

Since this value is greater than .05, the foregoing data do not enable us to reject, at the .05 percent level of significance, the hypothesis that the mean nicotine content exceeds 1.6 mg. In other words, the evidence, although supporting the cigarette producer's claim, is not strong enough to prove that claim. ∎

REMARKS

(a) There is a direct analogy between confidence interval estimation and hypothesis testing. For instance, for a normal population having mean μ and known variance σ^2 we have shown in Section 3 of Chapter 5 that a $100(1 - \alpha)$ percent confidence interval for μ is given by

$$\mu \in \left(\overline{x} - z_{\alpha/2} \frac{\sigma}{\sqrt{n}}, \overline{x} + z_{\alpha/2} \frac{\sigma}{\sqrt{n}} \right)$$

where $\bar{x}$ is the observed sample mean. More formally, the preceding confidence interval statement is equivalent to

$$P\left\{\mu \in \left(\bar{X} - z_{\alpha/2}\frac{\sigma}{\sqrt{n}}, \bar{X} + z_{\alpha/2}\frac{\sigma}{\sqrt{n}} \right) \right\} = 1 - \alpha$$

Hence, if $\mu = \mu_0$, then the probability that μ_0 will fall in the interval

$$\left(\bar{X} - z_{\alpha/2}\frac{\sigma}{\sqrt{n}}, \bar{X} + z_{\alpha/2}\frac{\sigma}{\sqrt{n}} \right)$$

is $1 - \alpha$, implying that a significance level α test of $H_0 : \mu = \mu_0$ versus $H_1 : \mu \neq \mu_0$ is to reject when

$$\mu_0 \notin \left(\bar{X} - z_{\alpha/2}\frac{\sigma}{\sqrt{n}}, \bar{X} + z_{\alpha/2}\frac{\sigma}{\sqrt{n}} \right)$$

Similarly, since a $100(1 - \alpha)$ percent one-sided confidence interval for μ is given by

$$\mu \in \left(\bar{X} - z_{\alpha}\frac{\sigma}{\sqrt{n}}, \infty \right)$$

it follows that an α-level significance test of $H_0 : \mu \leq \mu_0$ versus $H_1 : \mu > \mu_0$ is to reject H_0 when $\mu_0 \notin (\bar{X} - z_{\alpha}\sigma/\sqrt{n}, \infty)$—that is, when $\mu_0 < \bar{X} - z_{\alpha}\sigma/\sqrt{n}$.

(b) *A Remark on Robustness*

A test that performs well even when the underlying assumptions on which it is based are violated is said to be *robust*. For instance, the tests of Sections 3.1 and 3.1.1 were derived under the assumption that the underlying population distribution is normal with known variance σ^2. However, in deriving these tests, this assumption was used only to conclude that $\bar{X}$ also has a normal distribution. But, by the central limit theorem, it follows that for a reasonably large sample size, $\bar{X}$ will approximately have a normal distribution no matter what the underlying distribution. Thus we can conclude that these tests will be relatively robust for any population distribution with variance σ^2.

3.2   Unknown Variance: The *t*-Test

Up to now we have supposed that the only unknown parameter of the normal population distribution is its mean. However, the more common situation is one where the mean μ and variance σ^2 are both unknown. Let us suppose this to be the case and again consider a test of the hypothesis that the mean is equal to some specified value μ_0. That is, consider a test of

$$H_0 : \mu = \mu_0$$

versus the alternative

$$H_1 : \mu \neq \mu_0$$

It should be noted that the null hypothesis is not a simple hypothesis since it does not specify the value of σ^2.

As before, it seems reasonable to reject H_0 when the sample mean $\overline{X}$ is far from μ_0. However, how far away it need be to justify rejection will depend on the variance σ^2. For recall that when the value of σ^2 was known, the test called for rejecting H_0 when $|\overline{X} - \mu_0|$ exceeded $z_{\alpha/2}\sigma/\sqrt{n}$ or equivalently when

$$\left| \frac{\overline{X} - \mu_0}{\sigma/\sqrt{n}} \right| > z_{\alpha/2}$$

Now when σ^2 is no longer known, it seems reasonable to estimate it by

$$S^2 = \frac{\sum_{i=1}^{n}(X_i - \overline{X})^2}{n-1}$$

and then to reject H_0 when

$$\left| \frac{\overline{X} - \mu_0}{S/\sqrt{n}} \right|$$

is large.

To determine how large a value of the statistic

$$\left| \frac{\sqrt{n}\,(\overline{X} - \mu_0)}{S} \right|$$

to require for rejection, in order that the resulting test have significance level α, we must determine the probability distribution of this statistic when H_0 is true. However, as shown in Section 5 of Chapter 4 the statistic T, defined by

$$T = \frac{\sqrt{n}\,(\overline{X} - \mu_0)}{S}$$

has, when $\mu = \mu_0$, a t-distribution with $n-1$ degrees of freedom. Hence,

$$P_{\mu_0}\left\{ -t_{\alpha/2,\,n-1} \le \frac{\sqrt{n}\,(\overline{X} - \mu_0)}{S} \le t_{\alpha/2,\,n-1} \right\} = 1 - \alpha \qquad (6.3.11)$$

where $t_{\alpha/2,\,n-1}$ is the 100 $\alpha/2$ upper percentile value of the t-distribution with $n-1$ degrees of freedom. (That is $P\{T_{n-1} \ge t_{\alpha/2,\,n-1}\} = P\{T_{n-1} \le -t_{\alpha/2,\,n-1}\} = \alpha/2$ when T_{n-1} has a t-distribution with $n-1$ degrees of freedom). From Equation 6.3.11 we see that the appropriate significance level α test of

$$H_0 : \mu = \mu_0 \qquad \text{versus} \qquad H_1 : \mu \ne \mu_0$$

is, when σ^2 is unknown, to

$$\text{accept} \quad H_0 \quad \text{if} \quad \left| \frac{\sqrt{n}\,(\overline{X} - \mu_0)}{S} \right| \le t_{\alpha/2,\,n-1} \qquad (6.3.12)$$

$$\text{reject} \quad H_0 \quad \text{if} \quad \left| \frac{\sqrt{n}\,(\overline{X} - \mu_0)}{S} \right| > t_{\alpha/2,\,n-1}$$

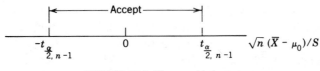

FIGURE 6.3.3 The two-sided t-test

The test defined by Equation 6.3.12 is called a *two-sided t-test*. It is pictorially illustrated in Figure 6.3.3.

If we let t denote the observed value of the test statistic $T = \sqrt{n}(\overline{X} - \mu_0)/S$, then the *p*-value of the test is the probability that $|T|$ would exceed $|t|$ when H_0 is true. That is, the *p*-value is the probability that the absolute value of a t-random variable with $n - 1$ degrees of freedom would exceed $|t|$. The test then calls for rejection at all significance levels higher than the *p*-value and acceptance at all lower significance levels.

Program 6-3-2 computes the value of the test statistic and the corresponding *p*-value. It can be applied both for one and two-sided tests. (The one-sided material will be presented shortly.)

Example 6.3g Among a clinic's patients having blood cholesterol levels ranging in the medium to high range (at least 220 milliliters per deciliter of serum), volunteers were recruited to test a new drug designed to reduce blood cholesterol. A group of 50 volunteers were given the drug for one month and the changes in their blood cholesterol levels were noted. If the average change was a reduction of 14.8 with a sample standard deviation of 6.4, what conclusions can be drawn?

Let us start by testing the hypothesis that the change could be due solely to chance—that is, that the 50 changes constitute a normal sample with mean 0. Because the value of the t-statistic used to test the hypothesis that a normal mean is equal to 0 is

$$T = \sqrt{n}\,\overline{X}/S = \sqrt{50}\,14.8/6.4 = 16.352$$

it is clear that we should reject the hypothesis that the changes were solely due to chance. Unfortunately, however, we are not justified at this point in concluding that the changes were due to the specific drug used and not to some other possibility. For instance, it is well known that any medication received by a patient (whether or not this medication is directly relevant to the patient's suffering) often leads to an improvement in the patient's condition—the so-called placebo effect. Also, another possibility that may need to be taken into account would be the weather conditions during the month of testing, for it is certainly conceivable that this affects blood cholesterol level. Indeed, it must be concluded that the foregoing was a very poorly designed experiment, for in order to test whether a specific treatment has an effect on a disease that may be affected by many things, we should try to design the experiment so as to neutralize all other possible causes. The accepted approach for accomplishing this is to divide the volunteers at random into two groups—one group to receive the drug and the other to receive a placebo (that is, a tablet that looks and tastes like the actual drug but has no physiological effect). The volunteers should not be told whether they are in the actual or control group, and indeed it is best if even the clinicians do not have this information (the so-called double-blind test) so as not to allow their own biases to play a role. Since the two groups are chosen at random from among the volunteers, we can now hope that on average all factors affecting the two groups will be the same except that one received the actual drug and the other a placebo. Hence, any difference in performance between the groups can be attributed to the drug. ■

Example 6.3h A public health official claims that the mean home water use is 350 gallons a day. To verify this claim, a study of 20 randomly selected homes was instigated with the result that the average daily water uses of these 20 homes were as follows:

340	344	362	375
356	386	354	364
332	402	340	355
362	322	372	324
318	360	338	370

Do the data contradict the official's claim?

Solution To test $H_0 : \mu = 350$ against $H_1 : \mu \neq 350$ we run Program 6-3-2.

```
RUN
THIS PROGRAM COMPUTES THE p-value WHEN TESTING THAT A NORMAL POPULATION WHOSE VA
RIANCE IS UNKNOWN HAS MEAN EQUAL TO MU-ZERO.
ENTER THE VALUE OF MU-ZERO
? 350
ENTER THE SAMPLE SIZE
? 20
ENTER THE DATA VALUES ONE AT A TIME
? 340? 356? 332? 362? 318? 344? 386? 402? 322? 360? 362? 354? 340? 372? 338? 37
5
? 364? 355? 324? 370THE VALUE OF THE t-STATISTIC IS .7778385
IS THE ALTERNATIVE HYPOTHESIS TWO-SIDED? ENTER 1 IF YES AND 0 IF NO
? 1
THE p-value IS .4461928
Ok
```

Such a large *p*-value shows that the data are not inconsistent with the claim of the health official. ■

We can use a one-sided *t*-test to test the hypothesis

$$H_0 : \mu = \mu_0 \quad (\text{or } H_0 : \mu \leq \mu_0)$$

against the one-sided alternative

$$H_1 : \mu > \mu_0$$

The significance level α test is to

$$\text{accept} \quad H_0 \quad \text{if} \quad \frac{\sqrt{n}\,(\overline{X} - \mu_0)}{S} \leq t_{\alpha, n-1} \qquad (6.3.13)$$

$$\text{reject} \quad H_0 \quad \text{if} \quad \frac{\sqrt{n}\,(\overline{X} - \mu_0)}{S} > t_{\alpha, n-1}$$

If $\sqrt{n}\,(\overline{X} - \mu_0)/S = t$ then the p-value of the test is the probability that a t-random variable with $n - 1$ degrees of freedom would exceed t.

The significance level α test of

$$H_0 : \mu = \mu_0 \qquad (\text{or } H_0 : \mu \geq \mu_0)$$

versus the alternative

$$H_1 : \mu < \mu_0$$

is to

accept H_0 if $\quad \dfrac{\sqrt{n}\,(\overline{X} - \mu_0)}{S} \geq -t_{\alpha,\,n-1}$

reject H_0 if $\quad \dfrac{\sqrt{n}\,(\overline{X} - \mu_0)}{S} < -t_{\alpha,\,n-1}$

The p-value of this test is the probability that a t-random variable with $n - 1$ degrees of freedom would be less than the observed value of $\sqrt{n}\,(\overline{X} - \mu_0)/S$.

Example 6.3i The manufacturer of a new fiberglass tire claims that its average life will be at least 40,000 miles. To verify this claim a sample of 12 tires is tested, with their lifetimes being as follows:

Tire	1	2	3	4	5	6	7	8	9	10	11	12
Life	36.1	40.2	33.8	38.5	42	35.8	37	41	36.8	37.2	33	36

(in 1000 miles)

Test the manufacturer's claim at the 5 percent level of significance.

Solution To determine whether the foregoing data are consistent with the hypothesis that the mean life is at least 40,000 miles, we will test

$$H_0 : \mu \geq 40{,}000 \qquad \text{versus} \qquad H_1 : \mu < 40{,}000$$

To do this test we run Program 6-3-2:

```
RUN
THIS PROGRAM COMPUTES THE p-value WHEN TESTING THAT A NORMAL POPULATION WHOSE VA
RIANCE IS UNKNOWN HAS MEAN EQUAL TO MU-ZERO.
ENTER THE VALUE OF MU-ZERO
? 40
ENTER THE SAMPLE SIZE
? 12
ENTER THE DATA VALUES ONE AT A TIME
? 36.1? 40/.2? 33.8? 38.5? 42? 35.8? 37? 41? 36.8? 37.2? 33? 36
THE VALUE OF THE t-STATISTIC IS-3.444766
IS THE ALTERNATIVE HYPOTHESIS TWO-SIDED? ENTER 1 IF YES AND O IF NO
? 0
IS THE ALTERNATIVE THAT THE MEAN EXCEEDS MU-ZERO OR THAT IT IS LESS?  ENTER 1 IN
  THE FORMER CASE AND O IN THE LATTER
? 0
THE p-value IS 2.762199E-03
```

Since the p-value is less than $\alpha = .05$, the manufacturer's claim is rejected. ∎

4   TESTING THE EQUALITY OF MEANS OF TWO NORMAL POPULATIONS

A common situation faced by a practicing engineer is one in which he must decide whether two different approaches lead to the same solution. Often such a situation can be modeled as a test of the hypothesis that two normal populations have the same mean value.

4.1   Case of Known Variances

Suppose that $X_1, \ldots, X_n$ and $Y_1, \ldots, Y_m$ are independent samples from normal populations having unknown means μ_x and μ_y but known variances σ_x^2 and σ_y^2. Let us consider the problem of testing the hypothesis

$$H_0 : \mu_x = \mu_y$$

versus the alternative

$$H_1 : \mu_x \neq \mu_y$$

Since $\overline{X}$ is an estimate of μ_x and $\overline{Y}$ of μ_y, it follows that $\overline{X} - \overline{Y}$ can be used to estimate $\mu_x - \mu_y$. Hence, as the null hypothesis can be written as $H_0 : \mu_x - \mu_y = 0$, it seems reasonable to reject it when $\overline{X} - \overline{Y}$ is far from zero. That is, the form of the test should be to

$$\text{reject} \quad H_0 \quad \text{if} \quad |\overline{X} - \overline{Y}| > c \tag{6.4.1}$$

$$\text{accept} \quad H_0 \quad \text{if} \quad |\overline{X} - \overline{Y}| \leq c$$

for some suitably chosen value c.

To determine that value of c that would result in the test in Equations 6.4.1 having a significance level α, we need determine the distribution of $\overline{X} - \overline{Y}$ when H_0 is true. However, as was shown in Section 3.2 of Chapter 5

$$\overline{X} - \overline{Y} \sim \mathcal{N}\left(\mu_x - \mu_y, \frac{\sigma_x^2}{n} + \frac{\sigma_y^2}{m}\right)$$

which implies that

$$\frac{\overline{X} - \overline{Y} - (\mu_x - \mu_y)}{\sqrt{\dfrac{\sigma_x^2}{n} + \dfrac{\sigma_y^2}{m}}} \sim \mathcal{N}(0, 1) \tag{6.4.2}$$

Hence, when H_0 is true (and so $\mu_x - \mu_y = 0$), it follows that

$$(\overline{X} - \overline{Y}) \Big/ \sqrt{\frac{\sigma_x^2}{n} + \frac{\sigma_y^2}{m}}$$

has a standard normal distribution; and thus

$$P_{H_0}\left\{-z_{\alpha/2} < \frac{\overline{X} - \overline{Y}}{\sqrt{\dfrac{\sigma_x^2}{n} + \dfrac{\sigma_y^2}{m}}} < z_{\alpha/2}\right\} = 1 - \alpha \qquad (6.4.3)$$

From Equation 6.4.3 we obtain the significance level α test of $H_0 : \mu_x = \mu_y$ versus $H_1 : \mu_x \neq \mu_y$ is

$$\text{accept} \quad H_0 \quad \text{if} \quad \frac{|\overline{X} - \overline{Y}|}{\sqrt{\sigma_x^2/n + \sigma_y^2/m}} < z_{\alpha/2}$$

$$\text{reject} \quad H_0 \quad \text{if} \quad \frac{|\overline{X} - \overline{Y}|}{\sqrt{\sigma_x^2/n + \sigma_y^2/m}} > z_{\alpha/2}$$

Program 6-4-1 will compute the value of the test statistic $(\overline{X} - \overline{Y})/\sqrt{\sigma_x^2/n + \sigma_y^2/m}$.

Example 6.4a Two new methods for producing a tire have been proposed. To ascertain which is superior, a tire manufacturer produces a sample of 10 tires using the first method and a sample of 8 using the second. The first set are to be road tested at location A and the second at location B. It is known from past experience that the lifetime of a tire that is road tested at one of these locations is normally distributed with a mean life due to the tire but with a variance due (for the most part) to the location. Specifically, it is known that the lifetimes of tires tested at location A are normal with standard deviation equal to 4000 kilometers, whereas those tested at location B are normal with $\sigma = 6000$ kilometers. If the manufacturer is interested in testing the hypothesis that there is no appreciable difference in the mean life of tires produced by either method, what conclusion should be drawn at the 5 percent level of significance if the resulting data are as given in Table 6.4.1?

TABLE 6.4.1

Tire Lives in Units of 100 Kilometers

Tires tested at A	Tires tested at B
61.1	62.2
58.2	56.6
62.3	66.4
64	56.2
59.7	57.4
66.2	58.4
57.8	57.6
61.4	65.4
62.2	
63.6	

Solution We run Program 6-4-1:

```
RUN
THIS PROGRAM COMPUTES THE VALUE OF THE TEST STATISTIC IN TESTING THAT TWO NORMAL
  MEANS ARE EQUAL WHEN THE VARIANCES ARE KNOWN
ENTER THE SAMPLE SIZES
? 10,8
ENTER THE SAMPLE VARIANCES
? 1600,3600
ENTER THE FIRST SAMPLE ONE AT A TIME
? 61.1? 58.2? 62.3? 64? 59.7? 66.2? 57.8? 61.4? 62.2? 63.6
ENTER THE SECOND SAMPLE ONE AT A TIME
? 62.2? 56.6? 66.4? 56.2? 57.4? 58.4? 57.6? 65.4
THE VALUE OF THE TEST STATISTIC IS        6.579402E-02
```

For such a small value of the test statistic (which has a unit normal distribution when H_0 is true), it is clear that the null hypothesis is accepted. ∎

It follows from Equation 6.4.1 that a test of the hypothesis $H_0 : \mu_x = \mu_y$ (or $H_0 : \mu_x \le \mu_y$) against the one-sided alternative $H_1 : \mu_x > \mu_y$ would be to

$$\text{accept} \quad H_0 \quad \text{if} \quad \overline{X} - \overline{Y} < z_\alpha \sqrt{\frac{\sigma_x^2}{n} + \frac{\sigma_y^2}{m}}$$

$$\text{reject} \quad H_0 \quad \text{if} \quad \overline{X} - \overline{Y} > z_\alpha \sqrt{\frac{\sigma_x^2}{n} + \frac{\sigma_y^2}{m}}$$

4.2 Case of Unknown Variances

Suppose again that $X_1, \ldots, X_n$ and $Y_1, \ldots, Y_m$ are independent samples from normal populations having respective parameters (μ_x, σ_x^2) and (μ_y, σ_y^2); but now suppose that all four parameters are unknown. We will once again consider a test of

$$H_0 : \mu_x = \mu_y \qquad \text{versus} \qquad H_1 : \mu_x \neq \mu_y$$

To determine a significance level α test of H_0 we will need to make the additional assumption that the unknown variances σ_x^2 and σ_y^2 are equal. Let σ^2 denote their value—that is,

$$\sigma^2 = \sigma_x^2 = \sigma_y^2$$

As before, we would like to reject H_0 when $\overline{X} - \overline{Y}$ is "far" from zero. To determine how far from zero it need be, let

$$S_x^2 = \frac{\sum_{i=1}^{n}(X_i - \overline{X})^2}{n - 1}$$

$$S_y^2 = \frac{\sum_{i=1}^{m}(Y_i - \overline{Y})^2}{m - 1}$$

denote the sample variances of the two samples. Then, as was shown in

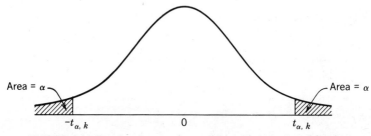

FIGURE 6.4.1 Density of a t-random variable with k degrees of freedom

Section 3.2 of Chapter 5

$$\frac{\overline{X} - \overline{Y} - (\mu_x - \mu_y)}{\sqrt{\dfrac{(n-1)S_x^2 + (m-1)S_y^2}{n+m-2}}\sqrt{1/n + 1/m}} \sim t_{n+m-2}$$

Hence, when H_0 is true, and so $\mu_x - \mu_y = 0$, we see that the statistic T, defined by

$$T \equiv \frac{\overline{X} - \overline{Y}}{\sqrt{\dfrac{(n-1)S_x^2 + (m-1)S_y^2}{n+m-2}}\sqrt{\dfrac{1}{n} + \dfrac{1}{m}}}$$

has a t-distribution with $n + m - 2$ degrees of freedom. From this, it follows that we can test the hypothesis that $\mu_x = \mu_y$ as follows:

$$\text{accept} \quad H_0 \quad \text{if} \quad |T| < t_{\alpha/2,\,n+m-2}$$

$$\text{reject} \quad H_0 \quad \text{if} \quad |T| > t_{\alpha/2,\,n+m-2}$$

where $t_{\alpha/2,\,n+m-2}$ is the 100 $\alpha/2$ percentile point of a t-random variable with $n + m - 2$ degrees of freedom (see Figure 6.4.1).

If the observed absolute value of T is $|T| = |t|$, then the p-value of the test data is the probability that the absolute value of a t-random variable with $n + m - 2$ degrees of freedom exceeds $|t|$. That is,

$$p\text{-value} = P\{|T_{n+m-2}| \geq |t|\} = 2P\{T_{n+m-2} \geq |t|\}$$

A one-sided test—say of

$$H_0 : \mu_x \leq \mu_y \qquad \text{versus} \qquad H_1 : \mu_x > \mu_y$$

is as follows:

$$\text{accept} \quad H_0 \quad \text{if} \quad T \leq t_{\alpha,\,n+m-2}$$

$$\text{reject} \quad H_0 \quad \text{otherwise}$$

If $T = t$, the p-value is the probability that a t-random variable with $n + m - 2$ degrees of freedom exceeds t.

Program 6-4-2 computes both the value of the test statistic and the corresponding p-value.

Example 6.4b Reconsider Example 6.4a but suppose now that the population variances are unknown but equal.

Solution Run Program 6-4-2:

```
RUN
THIS PROGRAM COMPUTES THE p-value WHEN TESTING THAT TWO NORMAL POPULATIONS HAVIN
G EQUAL BUT UNKNOWN VARIANCES HAVE A COMMON MEAN
ENTER THE SIZE OF SAMPLE 1
? 10
ENTER SAMPLE 1 ONE AT A TIME
? 61.1
? 58.2
? 62.3
? 64
? 59.7
? 66.2
? 57.8
? 61.4
? 62.2
? 63.6
ENTER THE SIZE OF SAMPLE 2
? 8
ENTER SAMPLE 2 ONE AT A TIME
? 62.2
? 56.6
? 66.4
? 56.2
? 57.4
? 58.4
? 57.6
? 65.4
THE VALUE OF THE t-STATISTIC IS 1.028025
IS THE ALTERNATIVE HYPOTHESIS TWO-SIDED? ENTER 1 IF YES AND 0 IF NO
? 1
THE p-value IS .3191708
Ok
```

Thus the null hypothesis is accepted at any significance level less than .3191708. ■

4.3   Unknown and Unequal Variances

Let us now suppose that the population variances σ_x^2 and σ_y^2 are not only unknown but also cannot be considered to be equal. In this situation, since S_x^2 is the natural estimator of σ_x^2 and S_y^2 of σ_y^2, it would seem reasonable to base our test of

$$H_0 : \mu_x = \mu_y \qquad \text{versus} \qquad H_1 : \mu_x \neq \mu_y$$

on the test statistic

$$\frac{\overline{X} - \overline{Y}}{\sqrt{\dfrac{S_x^2}{n} + \dfrac{S_y^2}{m}}} \tag{6.4.4}$$

However, the foregoing has a complicated distribution, which, even when H_0 is true, depends on the unknown parameters, and thus cannot be generally employed. The one situation in which we can utilize the statistic of Equation

6.4.4 is when n and m are both large. For in this case, it can be shown that when H_0 is true Equation 6.4.4 will have *approximately* a unit normal distribution. Hence, when n and m are large an *approximate* level α test of $H_0 : \mu_x = \mu_y$ versus $H_1 : \mu_x \neq \mu_y$ is to

$$\text{accept} \quad H_0 \quad \text{if} \quad -z_{\alpha/2} < \frac{\bar{X} - \bar{Y}}{\sqrt{\dfrac{S_x^2}{n} + \dfrac{S_y^2}{m}}} < z_{\alpha/2}$$

reject   otherwise

The problem of determining an exact level α test of the hypothesis that the means of two normal populations, having unknown and not necessarily equal variances, are equal is known as the Behrens-Fisher problem. There is no completely satisfactory solution known.

4.4   The Paired t-Test

Suppose we are interested in determining whether the installation of a certain antipollution device will affect a car's mileage. To test this, a collection of n cars that do not have this device are gathered. Each car's mileage per gallon is then determined both before and after the device is installed. How can we test the hypothesis that the antipollution control has no effect on gas consumption?

The data can be described by the n pairs (X_i, Y_i), $i = 1, \ldots, n$, where X_i is the gas consumption of the ith car before installation of the pollution control device, and Y_i of the same car after installation. It is important to note that, since each of the n cars will be inherently different, we cannot treat $X_1, \ldots, X_n$ and $Y_1, \ldots, Y_n$ as being independent samples. For example, if we know that X_1 is large (say 40 miles per gallon), we would certainly expect that Y_1 would also probably be large. Thus, we cannot employ the earlier methods presented in this section.

One way in which we can test the hypothesis that the antipollution device does not affect gas mileage is to let the data consist of each car's difference in gas mileage. That is, let $W_i = X_i - Y_i$, $i = 1, \ldots, n$. Now, if there is no affect from the device, it should follow that the W_i would have mean 0. Hence, we can test the hypothesis of no effect by testing

$$H_0 : \mu_w = 0 \qquad \text{versus} \qquad H_1 : \mu_w \neq 0$$

where $W_1, \ldots, W_n$ are assumed to be a sample from a normal population having unknown mean μ_w and unknown variance σ_w^2. But the t-test described in Section 3.2 shows that this can be tested by

$$\text{accepting} \quad H_0 \quad \text{if} \quad -t_{\alpha/2, n-1} < \sqrt{n}\,\frac{\bar{W}}{S_w} < t_{\alpha/2, n-1}$$

rejecting H_0 otherwise

Example 6.4c An industrial safety program was recently instituted in the computer chip industry. The average weekly loss (averaged over one month) in man-hours due to accidents in 10 similar plants both before and after the program are as follows:

Plant	Before	After
1	30.5	23
2	18.5	21
3	24.5	22
4	32	28.5
5	16	14.5
6	15	15.5
7	23.5	24.5
8	25.5	21
9	28	23.5
10	18	16.5

Determine, at the 5 percent level of significance, whether the safety program has been proven to be effective.

Solution To determine this we will test

$$H_0 : \mu_A - \mu_B \geq 0 \qquad \text{versus} \qquad H_1 : \mu_A - \mu_B < 0$$

for this will enable us to see whether the null hypothesis that the safety program has not had a beneficial effect is a reasonable possibility. To test this we run Program 6-3-2:

```
RUN
THIS PROGRAM COMPUTES THE p-value WHEN TESTING THAT A NORMAL POPULATION WHOSE VA
RIANCE IS UNKNOWN HAS MEAN EQUAL TO MU-ZERO.
ENTER THE VALUE OF MU-ZERO
? 0
ENTER THE SAMPLE SIZE
? 10
ENTER THE DATA VALUES ONE AT A TIME
? 7.5? -2.5? 2.5? 3.5? 1.5? -.5? -1? 4.5? 4.5? 1.5
THE VALUE OF THE t-STATISTIC IS 2.265949
IS THE ALTERNATIVE HYPOTHESIS TWO-SIDED? ENTER 1 IF YES AND 0 IF NO
? 0
IS THE ALTERNATIVE THAT THE MEAN EXCEEDS MU-ZERO OR THAT IT IS LESS?  ENTER 1 IN
 THE FORMER CASE AND 0 IN THE LATTER
? 1
THE p-value IS 2.489293E-02
Ok
```

Since the *p*-value is less than .05, the hypothesis that the safety program has not been effective is rejected and so we can conclude that its effectiveness has been established (at least for any significance level greater than .025). ∎

It should be noted that the paired-sample *t*-test can be used even though the samples are not independent and the population variances are unequal. Its disadvantage is, of course, that the test is being performed as if there are only *n* rather than $2n$ data points.

5 HYPOTHESIS TESTS CONCERNING THE VARIANCE OF A NORMAL POPULATION

Let $X_1, \ldots, X_n$ denote a sample from a normal population having unknown mean μ and unknown variance σ^2, and suppose we desire to test the hypothesis

$$H_0 : \sigma^2 = \sigma_0^2$$

versus the alternative

$$H_1 : \sigma^2 \neq \sigma_0^2$$

for some specified value σ_0^2.

To obtain a test, recall (as was shown in Section 5 of Chapter 4) that $(n-1)S^2/\sigma^2$ has a chi-square distribution with $n-1$ degrees of freedom. Hence, when H_0 is true

$$\frac{(n-1)S^2}{\sigma_0^2} \sim \chi_{n-1}^2$$

and so

$$P_{H_0}\left\{ \chi_{1-\alpha/2, \, n-1}^2 < \frac{(n-1)S^2}{\sigma_0^2} < \chi_{\alpha/2, \, n-1}^2 \right\} = 1 - \alpha$$

Therefore, a significance level α test is to

$$\text{accept} \quad H_0 \quad \text{if} \quad \chi_{1-\alpha/2, \, n-1}^2 < \frac{(n-1)S^2}{\sigma_0^2} < \chi_{\alpha/2, \, n-1}^2$$

$$\text{reject} \quad H_0 \quad \text{otherwise}$$

The preceding test can be implemented by first computing the value of the test statistic $(n-1)S^2/\sigma_0^2$—call it c. Then compute the probability that a chi-square random variable with $n-1$ degrees of freedom would be (a) less than and (b) greater than c. If either of these probabilities is less than $\alpha/2$, then the hypothesis is rejected. In other words, the p-value of the test data is

$$p\text{-value} = 2\min\left(P\{\chi_{n-1}^2 < c\}, 1 - P\{\chi_{n-1}^2 < c\} \right)$$

The quantity $P\{\chi_{n-1}^2 < c\}$ can be obtained from Program 3-8-1-A. The p-value for a one-sided test is similarly obtained.

Example 6.5a A machine that automatically controls the amount of ribbon on a tape has recently been installed. This machine will be judged to be effective if the standard deviation σ of the amount of ribbon on a tape is less than 0.15 cm. If a sample of 20 tapes yields a sample variance of $S^2 = .025$ cm^2, are we justified in concluding that the machine is ineffective?

Solution We will test the hypothesis that the machine is effective, since a rejection of this hypothesis will then enable us to conclude that it is ineffective.

Since we are thus interested in testing

$$H_0 : \sigma^2 \leq .0225 \qquad \text{versus} \qquad H_1 : \sigma^2 > .0225$$

it follows that we would want to reject H_0 when S^2 is large. Hence, the p-value of the preceding test data is the probability that a chi-square random variable with 19 degrees of freedom would exceed the observed value of $19S^2/.0225 = 19 \times .025/.0225 = 21.111$. That is,

$$p\text{-value} = P\{\chi^2_{19} > 21.111\}$$
$$= 1 - .6693 = .3307 \qquad \text{from Program 3-8-1-A}$$

Therefore, we must conclude that the observed value of $S^2 = .025$ is not large enough to reasonably preclude the possibility that $\sigma^2 \leq .0225$, and so the null hypothesis is accepted. ■

5.1   Testing for the Equality of Variances of Two Normal Populations

Let $X_1, \ldots, X_n$ and $Y_1, \ldots, Y_m$ denote independent samples from two normal populations having respective (unknown) parameters μ_x, σ_x^2 and μ_y, σ_y^2 and consider a test of

$$H_0 : \sigma_x^2 = \sigma_y^2 \qquad \text{versus} \qquad H_1 : \sigma_x^2 \neq \sigma_y^2$$

If we let

$$S_x^2 = \frac{\sum_{i=1}^n (X_i - \overline{X})^2}{n - 1}$$

$$S_y^2 = \frac{\sum_{i=1}^m (Y_i - \overline{Y})^2}{m - 1}$$

denote the sample variances, then as shown in Section 5 of Chapter 4, $(n - 1)S_x^2/\sigma_x^2$ and $(m - 1)S_y^2/\sigma_y^2$ are independent chi-squared random variables with $n - 1$ and $m - 1$ degrees of freedom respectively. Therefore, $(S_x^2/\sigma_x^2)/(S_y^2/\sigma_y^2)$ has an F distribution with parameters $n - 1$ and $m - 1$. Hence, when H_0 is true

$$S_x^2/S_y^2 \sim F_{n-1, m-1}$$

and so

$$P_{H_0}\{F_{1-\alpha/2, n-1, m-1} < S_x^2/S_y^2 < F_{\alpha/2, n-1, m-1}\} = 1 - \alpha$$

Thus, a significance level α test of H_0 against H_1 is to

accept H_0 if $F_{1-\alpha/2, n-1, m-1} < S_x^2/S_y^2 < F_{\alpha/2, n-1, m-1}$

reject H_0 otherwise

The preceding test can be effected by first determining the value of the test statistic S_x^2/S_y^2, say its value is v, and then computing (using Program 3-8-3) $P\{F_{n-1, m-1} < v\}$ where $F_{n-1, m-1}$ is an F random variable with parameters $n - 1, m - 1$. If this probability is either less than $\alpha/2$ (which occurs when S_x^2

is significantly less than S_y^2) or greater than $1 - \alpha/2$ (which occurs when S_x^2 is significantly greater than S_y^2) then the hypothesis is rejected. In other words, the p-value of the test data is

$$p\text{-value} = 2 \min \left(P\{ F_{n-1, m-1} < v \}, 1 - P\{ F_{n-1, m-1} < v \} \right)$$

The test now calls for rejection whenever the significance level α is at least as large as the p-value.

Example 6.5b There are two different choices of a catalyst to stimulate a certain chemical process. To test whether the variance of the yield is the same no matter which catalyst is used, a sample of 10 batches is produced using the first catalyst, and 12 using the second. If the resulting data is $S_1^2 = .14$ and $S_2^2 = .28$, can we reject, at the 5 percent level, the hypothesis of equal variance?

Solution Utilizing Program 3-8-3-A, which computes the F cumulative distribution function, yields that

$$P\{ F_{9,11} \le .5 \} = .1539$$

Hence,

$$p\text{-value} = 2 \min \{ .1539, .8461 \}$$
$$= .3074$$

and so the hypothesis of equal variance is accepted. ■

6   HYPOTHESIS TESTS IN BERNOULLI POPULATIONS

The binomial distribution is frequently encountered in engineering problems. For a typical example, consider a production process that manufactures items that can be classified in one of two ways—either as acceptable or as defective. An assumption often made is that each item produced will, independently, be defective with probability p; and so the number of defects in a sample of n items will thus have a binomial distribution with parameters (n, p). We will now consider a test of

$$H_0 : p \le p_0 \qquad \text{versus} \qquad H_1 : p > p_0$$

where p_0 is some specified value.

If we let X denote the number of defects in the sample of size n, then it is clear that we wish to reject H_0 when X is large. To see how large it need be to justify rejection at the α level of significance, note that

$$P\{ X \ge k \} = \sum_{i=k}^{n} P\{ X = i \} = \sum_{i=k}^{n} \binom{n}{i} p^i (1 - p)^{n-i}$$

Now it is certainly intuitive (and can be proven) that $P\{ X \ge k \}$ is an increasing function of p—that is, the probability that the sample will contain at least k errors increases in the defect probability p. Using this, we see that when H_0 is true (and so $p \leq p_0$)

$$P\{X \geq k\} \leq \sum_{i=k}^{n} \binom{n}{i} p_0^i (1 - p_0)^{n-i}$$

Hence, a significance level α test of $H_0 : p \leq p_0$ versus $H_1 : p > p_0$ is to reject H_0 when

$$X \geq k*$$

where $k*$ is the smallest value of k for which $\sum_{i=k}^{n} \binom{n}{i} p_0^i (1 - p_0)^{n-i} \leq \alpha$. That is,

$$k* = \min \left\{ k : \sum_{i=k}^{n} \binom{n}{i} p_0^i (1 - p_0)^{n-i} \leq \alpha \right\}$$

This test can best be performed by first determining the value of the test statistic—say $X = x$—and then computing the p-value given by

$$p\text{-value} = P\{ B(n, p_0) \geq x \}$$

$$= \sum_{i=x}^{n} \binom{n}{i} p_0^i (1 - p_0)^{n-i}$$

To compute this we can employ Program 3-1.

Example 6.6a A computer chip manufacturer claims that no more than 2 percent of the chips it sends out are defective. An electronics company, impressed with this claim, has purchased a large quantity of such chips. To determine if the manufacturer's claim can be taken literally, the company has decided to test a sample of 300 of these chips. If 10 of these 300 chips are found to be defective, should the manufacturer's claim be rejected?

Solution Let us test the claim at the 5 percent level of significance. To see if rejection is called for, we need to compute the probability that the sample of size 300 would have resulted in 10 or more defectives when p is equal to .02. (That is, we compute the p-value.) If this probability is less than or equal to 0.05, then the manufacturer's claim should be rejected. Now

$$P_{.02}\{ X \geq 10 \} = 1 - P_{.02}\{ X < 10 \}$$

$$= 1 - \sum_{i=0}^{9} \binom{300}{i} (.02)^i (.98)^{300-i}$$

$$= 0.0818 \qquad \text{from Program 3-1}$$

and so the manufacturer's claim cannot be rejected at the 5 percent level of significance. ∎

When the sample size n is large, we can derive an *approximate* significance level α test of $H_0 : p \leq p_0$ versus $H_1 : p > p_0$, which will significantly reduce the computational effort needed, by using the normal approximation to the binomial. It works as follows: since X will approximately have, when n is large, a normal distribution with mean and variance

$$E[X] = np, \qquad \text{Var}(X) = np(1 - p)$$

it follows that

$$\frac{X - np}{\sqrt{np(1 - p)}}$$

will approximately have a unit normal distribution. Therefore, an approximate significance level α test would be to reject H_0 if

$$\frac{X - np_0}{\sqrt{np_0(1 - p_0)}} \geq z_\alpha$$

Example 6.6b For the data of Example 6.6a the value of the test statistic $(X - np_0)/\sqrt{np_0(1 - p_0)}$ is $(10 - 300 \times .02)/\sqrt{300 \times .02 \times .98} = 1.6496$. Hence, using the normal approximation, it follows that H_0 should be rejected at any significance level greater than or equal to the p-value given by

$$p\text{-value} = P\{Z \geq 1.6496\}$$

$$= .0495 \qquad \text{from Program 3-5-1-A}$$

Thus, for instance, H_0 would be rejected at the 5 percent level of significance, which is contrary to the result of the exact α-level test of Example 6.6a. This result indicates the danger in using the approximate test—namely, that if the sample size is not sufficiently large, it can lead to a different conclusion than the exact test. A general rule of thumb is that the p-value given by the approximate test will be quite close to the p-value given by the exact test when the sample size n is large enough so that $np_0 \geq 20$. Thus, in the present case, since $np_0 = 6$, it is not too surprising that the approximate test might lead to a different conclusion than the exact test. ■

6.1   Testing the Equality of Parameters in Two Bernoulli Populations

Suppose there are two distinct methods for producing a certain type of transistor; and suppose that transistors produced by the first method will, independently, be defective with probability p_1, with the corresponding probability being p_2 for those produced by the second method. To test the hypothesis that $p_1 = p_2$, a sample of n_1 transistors is produced using method 1 and n_2 using method 2.

Let X_1 denote the number of defective transistors obtained from the first sample and X_2 for the second. Thus, X_1 and X_2 are independent binomial random variables with respective parameters (n_1, p_1) and (n_2, p_2). Suppose that $X_1 + X_2 = k$ and so there have been a total of k defectives. Now, if H_0 is true, then each of the $n_1 + n_2$ transistors produced will have the same probability of being defective, and so the determination of the k defectives will have the same distribution as a random selection of a sample of size k from a population of $n_1 + n_2$ items of which n_1 are white and n_2 are black. In other words, given a total of k defectives, the conditional distribution of the number of defective transistors obtained from method 1 will, when H_0 is true, have the following hypergeometric distribution[†]:

$$P_{H_0}\{ X_1 = i | X_1 + X_2 = k \} = \frac{\binom{n_1}{i}\binom{n_2}{k-i}}{\binom{n_1 + n_2}{k}} \qquad i = 0, 1, \ldots, k \quad (6.6.1)$$

Now, in testing

$$H_0 : p_1 = p_2 \qquad \text{versus} \qquad H_1 : p_1 \neq p_2$$

it seems reasonable to reject the null hypothesis when the proportion of defective transistors produced by method 1 is much different than the proportion of defectives obtained under method 2. Therefore, if there is a total of k defectives, then we would expect, when H_0 is true, that X_1/n_1 (the proportion of defective transistors produced by method 1) would be close to $(k - X_1)/n_2$ (the proportion of defective transistors produced by method 2). Because X_1/n_1 and $(k - X_1)/n_2$ will be furthest apart when X_1 is either very small or very large, it thus seems that a reasonable significance level α test of Equation 6.6.1 is as follows. If $X_1 + X_2 = k$ and $X_i = x_i$, then one should

reject H_0 if either $P\{ X \le x_1 \} \le \alpha/2$ or $P\{ X \ge x_1 \} \le \alpha/2$

accept H_0 otherwise

where X is a hypergeometric random variable with probability mass function

$$P\{ X = i \} = \frac{\binom{n_1}{i}\binom{n_2}{k-i}}{\binom{n_1 + n_2}{k}} \qquad i = 0, 1, \ldots, k \qquad (6.6.2)$$

In other words, this test will call for rejection if the significance level is at least as large as the p-value given by

$$p\text{-value} = 2 \min \left(P\{ X \le x_1 \}, P\{ X \ge x_1 \} \right) \qquad (6.6.3)$$

This is called the *Fisher-Irwin test*.

6.1.1 Computations for the Fisher-Irwin Test   To utilize the Fisher-Irwin test, we need to be able to compute the hypergeometric distribution function. To do so, note that with X having mass function Equation 6.6.2

$$\frac{P\{ X = i + 1 \}}{P\{ X = i \}} = \frac{\binom{n_1}{i+1}\binom{n_2}{k-i-1}}{\binom{n_1}{i}\binom{n_2}{k-i}} \qquad (6.6.4)$$

$$= \frac{(n_1 - i)(k - i)}{(i + 1)(n_2 - k + i + 1)} \qquad (6.6.5)$$

where the verification of the final equality is left as an exercise.

[†]See Example 3.3b of Chapter 3 for a formal verification of Equation 6.1.

To determine the p-value, we start by computing

$$P\{X = x_1\} = \binom{n_1}{x_1}\binom{n_2}{k - x_1}\bigg/\binom{n_1 + n_2}{k}$$

This is accomplished by first taking logs and computing $\log P\{X = x_1\}$ and then setting

$$P\{X = x_1\} = \exp\{\log P\{X = x_i\}\}$$

We now either recursively compute, using Equation 6.6.4, $P\{X = x_1 + 1\}$, $P\{X = x_1 + 2\}, \ldots, P\{X = \min(k, n)\}$ or $P\{X = x_1 - 1\}$, $P\{X = x_1 - 2\}$, $\ldots, P\{X = \max(0, k - n_2)\}$ depending on which sequence is of smaller size. By adding together these additional probabilities—call their sum S—we would have determined both $P\{X = x_1\}$ and S, which is either $P\{X > x_1\}$ or $P\{X < x_1\}$. Now, the pair of values $1 - S$ and $S + P\{X = x_1\}$ is the same as the pair $P\{X \leq x_1\}$ and $P\{X \geq x_1\}$, since if $S = P\{X > x_i\}$, then $1 - S = P\{X \leq x_i\}$ and $S + P\{X = x_i\} = P\{X \geq x_i\}$, whereas if $S = P\{X < x_i\}$, then $1 - S = P\{X \geq x_i\}$ and $S + P\{X = x_i\} = P\{X \leq x_i\}$. Therefore, from Equation 6.6.3a, we have that

$$p\text{-value} = 2\min(1 - S, S + P\{X = x_i\})$$

Program 6-6-1 uses the preceding analysis to compute the p-value of the data for the Fisher-Irwin test of the equality of two Bernoulli probabilities. The program will work best if the Bernoulli outcome that is called unsuccessful (or defective) is the one whose probability is less than .5. For instance, if over half the items produced are defective, then rather than testing that the defect probability is the same in both samples, one should test that the probability of producing an acceptable item is the same in both samples.

Example 6.6c Suppose that method 1 resulted in 20 unacceptable transistors out of 100 produced; whereas method 2 resulted in 12 unacceptable transistors out of 100 produced. Can we conclude from this, at the 10 percent level of significance, that the two methods are equivalent?

Upon running Program 6-6-1 we obtain:

```
RUN
THIS PROGRAM COMPUTES THE p-value FOR THE TEST DATA IN THE FISHER-IRWIN TEST
ENTER THE SIZE OF THE FIRST SAMPLE
? 100
ENTER THE SIZE OF THE SECOND SAMPLE
? 100
ENTER THE TOTAL NUMBER OF FAILURES
? 32
ENTER THE NUMBER OF FAILURES IN THE FIRST SAMPLE
? 20
THE p-value IS .1763395
```

Hence, the hypothesis that the two methods are equivalent is accepted. ∎

When n_1 and n_2 are large, an approximate level α test of $H_0: p_1 = p_2$, based on the normal approximation to the binomial, is outlined in Problem 31.

7 TESTS CONCERNING THE MEAN OF A POISSON DISTRIBUTION

Let X denote a Poisson random variable having mean λ and consider a test of

$$H_0 : \lambda = \lambda_0 \qquad \text{versus} \qquad H_1 : \lambda \neq \lambda_0$$

If the observed value of X is $X = x$, then an α-level test would reject H_0 if either

$$P_{\lambda_0}\{ X \geq x \} \leq \alpha/2 \qquad \text{or} \qquad P_{\lambda_0}\{ X \leq x \} \leq \alpha/2 \qquad (6.7.1)$$

where P_{λ_0} means that the probability is computed under the assumption that the Poisson mean is λ_0. It follows from Equation 6.7.1 that the p-value is given by

$$p\text{-value} = 2 \min \left(P_{\lambda_0}\{ X \geq x \}, P_{\lambda_0}\{ X \leq x \} \right)$$

The calculation of the preceding probabilities that a Poisson random variable with mean λ_0 is greater (less) than or equal to x can be obtained by using Program 3-2.

Example 6.7a Management's claim that the mean number of defective computer chips produced daily is not greater than 25 is in dispute. Test this hypothesis, at the 5 percent level of significance, if a sample of 5 days revealed $28, 34, 32, 38, 22$ defective chips.

As each individual computer chip has a very small chance of being defective, it is probably reasonable to suppose that the daily number of defective chips is approximately a Poisson random variable, with mean, say, λ. To see whether or not the manufacturer's claim is credible we shall test the hypothesis

$$H_0 : \lambda \leq 25 \qquad \text{versus} \qquad H_1 : \lambda > 25$$

Now, under H_0, the total number of defective chips produced over a 5-day period is Poisson distributed (since the sum of independent Poisson random variables is Poisson) with a mean no greater than 125. Since this number is equal to 154, it follows that the p-value of the data is given by

$$p\text{-value} = P_{125}\{ X \geq 154 \}$$
$$= 1 - P_{125}\{ X \leq 153 \}$$
$$= .0066 \qquad \text{from Program 3-2}$$

Therefore, the manufacturer's claim is rejected at the 5 percent (as it would be even at the 1 percent) level of significance. ∎

REMARK

If the mean of the relevant Poisson probability calculation is too large to effectively employ Program 3-2, one can use the fact that a Poisson random variable with mean λ is, for large λ, approximately normally distributed with a mean and variance equal to λ.

7.1 Testing the Relationship Between Two Poisson Parameters

Let X_1 and X_2 be independent Poisson random variables with respective means λ_1 and λ_2, and consider a test of

$$H_0 : \lambda_2 = c\lambda_1 \qquad \text{versus} \qquad H_1 : \lambda_2 \neq c\lambda_1$$

for a given constant c. Our test of this is a conditional test (similar in spirit to the Fisher-Irwin test of Section 6.1), which is based on the fact that the conditional distribution of X_1 given the sum of X_1 and X_2 is binomial. More specifically we have the following proposition.

Proposition 6.7.1

$$P\{ X_1 = k \mid X_1 + X_2 = n \} = \binom{n}{k} [\lambda_1/(\lambda_1 + \lambda_2)]^k [\lambda_2/(\lambda_1 + \lambda_2)]^{n-k}$$

Proof

$$P\{ X_1 = k \mid X_1 + X_2 = n \}$$

$$= \frac{P\{ X_1 = k, X_1 + X_2 = n \}}{P\{ X_1 + X_2 = n \}}$$

$$= \frac{P\{ X_1 = k, X_2 = n - k \}}{P\{ X_1 + X_2 = n \}}$$

$$= \frac{P\{ X_1 = k \} P\{ X_2 = n - k \}}{P\{ X_1 + X_2 = n \}} \qquad \text{by independence}$$

$$= \frac{\exp\{-\lambda_1\}\lambda_1^k/k! \exp\{-\lambda_2\}\lambda_2^{n-k}/(n-k)!}{\exp\{-(\lambda_1 + \lambda_2)\}(\lambda_1 + \lambda_2)^n/n!}$$

$$= \frac{n!}{(n-k)!k!} [\lambda_1/(\lambda_1 + \lambda_2)]^k [\lambda_2/(\lambda_1 + \lambda_2)]^{n-k} \qquad \blacksquare$$

It follows from Proposition 6.7.1 that, if H_0 is true, then the conditional distribution of X_1 given that $X_1 + X_2 = n$ is the binomial distribution with parameters n and $p = 1/(1 + c)$. From this we can conclude that if $X_1 + X_2 = n$, then H_0 should be rejected if the observed value of X_1, call it x_1, is such that either

$$P\{ \text{Bin}(n, 1/(1 + c)) \geq x_1 \} \leq \alpha/2$$

or

$$P\{ \text{Bin}(n, 1/(1 + c)) \leq x_1 \} \leq \alpha/2$$

Example 6.7b An industrial concern runs two large plants. If the number of accidents over the last 8 weeks at plant 1 were $16, 18, 9, 22, 17, 19, 24, 8$ while the number of accidents over the last 6 weeks at plant 2 were $22, 18, 26, 30, 25, 28$, can we conclude, at the 5 percent level of significance, that the safety conditions differ from plant to plant?

Since there is a small probability of an industrial accident in any given minute, it would seem that the weekly number of such accidents should approximately have a Poisson distribution. If we let X_1 denote the total number of accidents over an 8-week period at plant 1, and let X_2 be the number over a 6-week period at plant 2, then if the safety conditions did not differ at the two plants we would have that

$$\lambda_2 = \tfrac{3}{4}\lambda_1$$

where $\lambda_i \equiv E[X_i]$, $i = 1, 2$. Hence, as $X_1 = 133$, $X_2 = 149$ it follows that the p-value of the test of

$$H_0 : \lambda_2 = \tfrac{3}{4}\lambda_1 \qquad \text{versus} \qquad H_1 : \lambda_2 \neq \tfrac{3}{4}\lambda_1$$

is given by

$$p\text{-value} = 2 \min \left(P\{\text{Bin}\,(282, \tfrac{4}{7}) \geq 133\}, \, P\{\text{Bin}\,(282, \tfrac{4}{7}) \leq 133\} \right)$$

$$= 9.408 \times 10^{-4} \qquad \text{by Program 3-1}$$

Thus, the hypothesis that the safety conditions at the two plants are equivalent is rejected. ∎

PROBLEMS

1. In a certain chemical process, it is very important that a particular solution that is to be used as a reactant have a pH of exactly 8.20. A method for determining pH that is available for solutions of this type is known to give measurements that are normally distributed with a mean equal to the actual pH and with a standard deviation of 0.02. Suppose 10 independent measurements yielded the following pH values

8.18	8.17
8.16	8.15
8.17	8.21
8.22	8.16
8.19	8.18

   (a) What conclusion can be drawn at the $\alpha = .10$ level of significance?
   (b) What about at the $\alpha = .05$ level of significance?

2. The mean breaking strength of a certain type of fiber is required to be at least 200 psi. Past experience indicates that the standard deviation of breaking strength is 5 psi. If a sample of 8 pieces of fiber yielded breakage at the following pressures

210	198
195	202
197.4	196
199	195.5

   would you conclude, at the 5 percent level of significance, that the fiber is acceptable? What about at the 10 percent level of significance?

3. It is known that the average height of a male residing in the United States is 5 feet 10 inches and the standard deviation is 3 inches. To test the hypothesis that males in your city are "average," a sample of 20 men has

been chosen. The heights of the men in the sample are

Man	Height in inches	Man
1	72 70.4	11
2	68.1 76	12
3	69.2 72.5	13
4	72.8 74	14
5	71.2 71.8	15
6	72.2 69.6	16
7	70.8 75.6	17
8	74 70.6	18
9	66 76.2	19
10	70.3 77	20

What do you conclude? Explain what assumptions you are making.

4. Suppose in Problem 1 that we wished to design a test so that if the pH were really equal to 8.20 this conclusion will be reached with probability equal to 0.95. On the other hand, if the pH differs from 8.20 by 0.03 (in either direction), we want the probability of picking up such a difference to exceed 0.95.

   (a) What is the test procedure that should be used?
   (b) What is the required sample size?
   (c) If $\bar{x} = 8.31$, what is your conclusion?
   (d) If the actual pH is 8.32, what is the probability of concluding that the pH is not 8.20, using the foregoing procedure?

5. Verify that the approximation in Equation 6.3.7 remains valid even when $\mu_1 < \mu_0$.

6. A car is advertised as having a gas mileage rating of at least 30 miles/gallon in highway driving. If the miles per gallon obtained in 10 independent experiments are $26, 24, 20, 25, 27, 25, 28, 30, 26, 33$, should you believe the advertisement? What assumptions are you making?

7. A producer specifies that the mean lifetime of a certain type of battery is at least 240 hours. A sample of 18 such batteries yielded the following data:

237	242	232
242	248	230
244	243	254
262	234	220
225	236	232
218	228	240

Assuming that the life of the batteries is approximately normally distributed, do the data indicate that the specifications are not being met?

8. In the manufacture of cylindrical rods with a circular cross section that fit into a circular socket, the lathe is set so that the mean diameter is 5.00 cm. It was necessary to test the accuracy of the machine setting. A sample of

size 10 was taken with the results as follows:

5.036	5.031
5.085	5.064
4.991	4.942
4.935	5.051
4.999	5.011

Test the hypothesis that the setting is correct at the 5 percent level of significance. What assumptions are you making?

9. A company supplies plastic sheets for industrial use. A new type of plastic has been produced and the company would like to claim that the average stress resistance of this new product is at least 30.0, where stress resistance is measured in pounds per square inch (psi) necessary to crack the sheet. The following random sample was drawn off the production line. Based on this sample, what conclusion can you draw?

30.1	32.7	22.5	27.5
27.7	29.8	28.9	31.4
31.2	24.3	26.4	22.8
29.1	33.4	32.5	21.7

Assume normality.

10. A sample of 10 fish were caught at lake A and their PCB concentrations were measured using a certain technique. The resulting data in parts per million being

Lake A: 11.5, 10.8, 11.6, 9.4, 12.4, 11.4, 12.2, 11, 10.6, 10.8

In addition a sample of 8 fish were caught at lake B and their levels of PCB were measured by a different technique than used at lake A. The resultant data were

Lake B: 11.8, 12.6, 12.2, 12.5, 11.7, 12.1, 10.4, 12.6

If it is known that the measuring technique used at lake A has a variance of 0.09 whereas the one used at lake B has a variance of 0.16, could you reject (at the 5 percent level of significance) a claim that the two lakes are equally contaminated?

11. Consider two different producers of a type of rocket powder. A sample of 10 units was selected from each producer and their burning times in seconds were recorded. This resulted in the following data:

Producer 1	Producer 2
50.7	60.3
54.8	58.8
48.6	56.2
36.9	48.6
52.4	40
51.6	42.8
53	58
38	44.3
42.2	55
50.3	48.6

Test the hypothesis that the average burning times are equal. What assumption are you making?

12. The data below give the lifetimes in hundreds of hours of samples of two types of electronic tubes. Past lifetime data of such tubes has shown that they can often be modeled as arising from a lognormal distribution. That is, the logarithms of the data are normally distributed. Assuming that variance of the logarithms is equal for the two populations, test, at the 5 percent level of significance, the hypothesis that the two population distributions are identical

| Type 1 | 32, 84, 37, 42, 78, 62, 59, 74 |
| Type 2 | 39, 111, 55, 106, 90, 87, 85 |

13. The viscosity of two different brands of car oil are measured and the following data resulted:

| Brand 1 | 10.62, 10.58, 10.33, 10.72, 10.44, 10.74 |
| Brand 2 | 10.50, 10.52, 10.58, 10.62, 10.55, 10.51, 10.53 |

Test the hypothesis that the mean viscosity of the two brands is equal, assuming that the populations have normal distributions with equal variances.

14. It is argued that the resistance of wire A is greater than the resistance of wire B. You make tests on each wire with the following results:

Wire A	Wire B
0.140 ohm	0.135 ohm
0.138	0.140
0.143	0.136
0.142	0.142
0.144	0.138
0.137	0.140

What conclusion can you draw at the 10 percent significance level? Explain what assumptions you are making.

15. In a certain experimental laboratory, a method A for producing gasoline from a crude oil is being investigated. Before completing experimentation, a new method B is proposed. All other things being equal, it was decided to abandon A in favor of B only if the average yield of the latter was substantially greater. The yield of both processes is assumed to be normally distributed. However, there has been insufficient time to ascertain their true standard deviations, although there appears to be no reason why they cannot be assumed equal. Cost considerations impose size limits on the size of samples that can be obtained. If a 1 percent significance level is all that is allowed, what would be your recommendation based on the following random samples? The numbers represent percent yield of crude oil.

| A | 23.2, 26.6, 24.4, 23.5, 22.6, 25.7, 25.5 |
| B | 25.7, 27.7, 26.2, 27.9, 25.0, 21.4, 26.1 |

16. Ten pregnant women were given an injection of pitocin to induce labor. Their systolic blood pressures immediately before and after the injection were:

Patient	Before	After
1	134	140
2	122	130
3	132	135
4	130	126
5	128	134
6	140	138
7	118	124
8	127	126
9	125	132
10	142	144

Do the data indicate that injection of this drug changes blood pressure?

17. A question of medical importance is whether jogging leads to a reduction in one's pulse rate. To test this hypothesis, eight nonjogging volunteers have agreed to begin a one-month jogging program. After the month their pulse rates were determined and compared with their earlier values. If the data are as follows, can we conclude that jogging has had an effect on the pulse rates?

Subject	1	2	3	4	5	6	7	8
Pulse Rate Before	74	86	98	102	78	84	79	70
Pulse Rate After	70	85	90	110	71	80	69	74

18. If $X_1, \ldots, X_n$ is a sample from a normal population having unknown parameters μ and σ^2, devise a significance level α test of

$$H_0 = \sigma^2 \leq \sigma_0^2$$

against the alternative

$$H_1 = \sigma^2 > \sigma_0^2$$

for a given positive value σ_0^2.

19. In Problem 18, explain how the test would be modified if the population mean μ were known in advance.

20. A gun-like apparatus has recently been designed to replace needles in administering vaccines. The apparatus can be set to inject different amounts of the serum, but because of random fluctuations the actual amount injected is normally distributed with a mean equal to the setting and with an unknown variance σ^2. It has been decided that the apparatus would be too dangerous to use if σ exceeds .10. If a random sample of 50 injections resulted in a sample standard deviation of .08, should use of the new apparatus be discontinued? Suppose the level of significance is $\alpha = .10$. Comment on the appropriate choice of a significance level for this problem, as well as the appropriate choice of the null hypothesis.

21. A pharmaceutical house produces a certain drug item whose weight has a standard deviation of 0.5 milligrams. The company's research team has proposed a new method of producing the drug. However, this entails some costs and will be adopted only if there is strong evidence that the standard deviation of the weight of the items drops to below 0.4 milligrams. If a sample of 10 items is produced and has the following weights, should the new method be adopted?

5.728	5.731
5.722	5.719
5.727	5.724
5.718	5.726
5.723	5.722

22. The production of large electrical transformers and capacitators requires the use of Polychlorinated Biphenyls (PCBs), which are extremely hazardous when released into the environment. Two methods have been suggested to monitor the levels of PCB in fish near a large plant. It is believed that each method will result in a normal random variable that depends on the method. Test the hypothesis at the $\alpha = .10$ level of significance that both methods have the same variance if a given fish is checked 8 times by each method with the following data (in parts per million) recorded

Method 1	6.2, 5.8, 5.7, 6.3, 5.9, 6.1, 6.2, 5.7
Method 2	6.3, 5.7, 5.9, 6.4, 5.8, 6.2, 6.3, 5.5

23. In Problem 13, test the hypothesis that the populations have the same variances.

24. If $X_1, \ldots, X_n$ is a sample from a normal population with variance σ_x^2 and $Y_1, \ldots, Y_n$ is an independent sample from normal population with variance σ_y^2, develop a significance level α test of

$$H_0 : \sigma_x^2 < \sigma_y^2 \qquad \text{versus} \qquad H_1 : \sigma_x^2 > \sigma_y^2$$

25. The amount of surface wax on each side of waxed paper bags is believed to be normally distributed. However, there is reason to believe that there is greater variation in the amount on the inner side of the paper than on the outside. A sample of 75 observations of the amount of wax on each side of these bags is obtained and the following data recorded:

Wax in pounds per unit area of sample

Outside surface	Inside surface
$\bar{x} = 0.948$	$\bar{y} = 0.652$
$\Sigma x_i^2 = 91$	$\Sigma y_i^2 = 82$

Conduct a test to determine whether or not the variability of the amount of wax on the inner surface is greater than the amount on the outer surface ($\alpha = 0.05$).

26. A standard drug is known to be effective in 75 percent of the cases in which it is used to treat a certain infection. A new drug has been developed and has been found to be effective in 42 cases out of 50. Based on this, would you accept, at the 5 percent level of significance, the hypothesis that the two drugs are of equal effectiveness? Use an exact test. What is the p-value?

27. Do Problem 26 by using the test based on the normal approximation to the binomial.

28. In a recently conducted poll, 54 out of 200 people surveyed claimed to have a firearm in their homes. In a similar survey done earlier, 30 out of 150 people made that claim. Does this prove that more people now have (or at least claim to have) firearms in their homes, or is it possible that the proportion of the population having firearms has not changed and the foregoing is due to the inherent randomness in sampling?

29. Let X_1 denote a binomial random variable with parameters (n_1, p_1) and X_2 an independent binomial random variable with parameters (n_2, p_2). Develop a test, using the same approach as in the Fisher-Irwin test, of

$$H_0: p_1 \leq p_2$$

versus the alternative

$$H_1: p_1 > p_2$$

30. Verify that Equation 6.6.5 follows from Equation 6.6.4.

31. Let X_1 and X_2 be binomial random variables with respective parameters n_1, p_1 and n_2, p_2. Show that when n_1 and n_2 are large, an approximate level α test of $H_0: p_1 = p_2$ versus $H_1: p_1 \neq p_2$ is as follows:

reject $\quad H_0 \quad$ if $\quad \dfrac{|X_1/n_1 - X_2/n_2|}{\sqrt{\dfrac{X_1 + X_2}{n_1 + n_2}\left(1 - \dfrac{X_1 + X_2}{n_1 + n_2}\right)\left(\dfrac{1}{n_1} + \dfrac{1}{n_2}\right)}} > z_{\alpha/2}$

Hint: **(a)** Argue first that when n_1 and n_2 are large

$$\dfrac{\dfrac{X_1}{n_1} - \dfrac{X_2}{n_2} - (p_1 - p_2)}{\sqrt{\dfrac{p_1(1 - p_1)}{n_1} + \dfrac{p_2(1 - p_2)}{n_2}}} \stackrel{.}{\sim} N(0, 1)$$

where $\stackrel{.}{\sim}$ means "approximately has the distribution."
(b) Now argue that when H_0 is true and so $p_1 = p_2$, their common value can be best estimated by $(X_1 + X_2)/(n_1 + n_2)$.

32. Use the approximate test given in Problem 31 on the data of Problem 28.

33. Test the hypothesis, at the .05 level of significance, that the yearly number of earthquakes felt on a certain island has mean 52 if the readings for the

last 8 years are $46, 62, 60, 58, 47, 50, 59, 49$. Assume an underlying Poisson distribution and give an explanation to justify this assumption.

34. For the following data, sample 1 is from a Poisson distribution with mean λ_1 and sample 2 is from a Poisson distribution with mean λ_2. Test the hypothesis that $\lambda_1 = \lambda_2$.

Sample 1	24, 32, 29, 33, 40, 28, 34, 36
Sample 2	42, 36, 41

CHAPTER 7

Regression

1 INTRODUCTION

Many engineering and scientific problems are concerned with determining a relationship between a set of variables. For instance, in a chemical process, we might be interested in the relationship between the output of the process, the temperature at which it occurs, and the amount of catalyst employed. Knowledge of such a relationship would enable us to predict the output for various values of temperature and amount of catalyst.

In many situations, there is a single *response* variable Y, also called the *dependent* variable, which depends on the value of a set of *input*, also called *independent*, variables $x_1, \ldots, x_r$. The simplest type of relationship between the dependent variable Y and the input variables $x_1, \ldots, x_r$ is a linear relationship. That is, for some constants $\beta_0, \beta_1, \ldots, \beta_r$ the equation

$$Y = \beta_0 + \beta_1 x_1 + \cdots + \beta_r x_r \qquad (7.1.1)$$

would hold. If this was the relationship between Y and the x_i, $i = 1, \ldots, r$, then, it would be possible (once the β_i were learned) to exactly predict the response for any set of input values. However, in practice, such precision is almost never attainable, and the most that one can expect is that Equation 7.1.1 would be valid *subject to random error*. By this we mean that the explicit relationship is

$$Y = \beta_0 + \beta_1 x_1 + \cdots + \beta_r x_r + e \qquad (7.1.2)$$

where e, representing the random error, is assumed to be a random variable having mean 0. Indeed, another way of expressing Equation 7.1.2 is as follows:

$$E[Y|\mathbf{x}] = \beta_0 + \beta_1 x_1 + \cdots + \beta_r x_r$$

where $\mathbf{x} = (x_1, \ldots, x_r)$ is the set of independent variables, and $E[Y|\mathbf{x}]$ is the expected response given the inputs $\mathbf{x}$.

Equation 7.1.2 is called a linear regression equation. We say that it describes the regression of Y on the set of independent variables $x_1, \ldots, x_r$. The quantities $\beta_0, \beta_1, \ldots, \beta_r$ are called the *regression coefficients*, and must usually be estimated from a set of data. A regression equation containing a single independent variable—that is, one in which $r = 1$—is called a *simple regression equation*, whereas one containing many independent variables is called a *multiple regression equation*.

Thus, a simple linear regression model supposes a linear relationship between the mean response and the value of a single independent variable. It can be expressed as

$$Y = \alpha + \beta x + e$$

where x is the value of the independent variable, also called the input level, Y is the response, and e, representing the random error, is a random variable having mean 0.

Example 7.1a Consider the following 10 data pairs (x_i, y_i), $i = 1, \ldots, 10$, relating y, the percent yield of a laboratory experiment, to x, the temperature at which the experiment was run.

i	x_i	y_i	i	x_i	y_i
1	100	45	6	150	68
2	110	52	7	160	75
3	120	54	8	170	76
4	130	63	9	180	92
5	140	62	10	190	88

A plot of y_i versus x_i—called a *scatter diagram*—is given in Figure 7.1.1. As this scatter diagram appears to reflect, subject to random error, a linear relation between y and x, it seems that a simple linear regression model would be appropriate. ∎

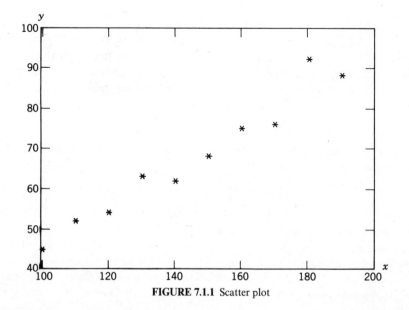

FIGURE 7.1.1 Scatter plot

2 LEAST SQUARES ESTIMATORS OF THE REGRESSION PARAMETERS

Suppose that the responses Y_i corresponding to the input values x_i, $i = 1, \ldots, n$ are to be observed and used to estimate α and β in a simple linear regression model. To determine estimators of α and β we reason as follows: If A is the estimator of α and B of β, then the estimator of the response corresponding to the input variable x_i would be $A + Bx_i$. Since the actual response is Y_i, the squared difference is $(Y_i - A - Bx_i)^2$, and so if A and B are the estimators of α and β, then the sum of the squared differences between the estimated responses and the actual response values—call it SS—is given by

$$SS = \sum_{i=1}^{n} (Y_i - A - Bx_i)^2$$

The method of least squares chooses as estimators of α and β the values of A and B that minimize SS. To determine these estimators we differentiate SS first with respect to A and then to B as follows:

$$\frac{\partial SS}{\partial A} = -2 \sum_{i=1}^{n} (Y_i - A - Bx_i)$$

$$\frac{\partial SS}{\partial B} = -2 \sum_{i=1}^{n} x_i(Y_i - A - Bx_i)$$

Setting these partial derivatives equal to zero yields the following equations for the minimizing values A and B:

$$\sum_{i=1}^{n} Y_i = nA + B \sum_{i=1}^{n} x_i \qquad (7.2.1)$$

$$\sum_{i=1}^{n} x_iY_i = A \sum_{i=1}^{n} x_i + B \sum_{i=1}^{n} x_i^2$$

The Equations 7.2.1 are known as the *normal equations*. If we let

$$\overline{Y} = \sum_i Y_i/n, \qquad \overline{x} = \sum_i x_i/n$$

then we can write the first normal equation as

$$A = \overline{Y} - B\overline{x} \qquad (7.2.2)$$

Substituting this value of A into the second normal equation yields

$$\sum_i x_iY_i = (\overline{Y} - B\overline{x})n\overline{x} + B \sum_i x_i^2$$

or

$$B\left(\sum_i x_i^2 - n\overline{x}^2\right) = \sum_i x_iY_i - n\overline{x}\,\overline{Y}$$

or

$$B = \frac{\Sigma_i x_i Y_i - n\bar{x}\bar{Y}}{\Sigma_i x_i^2 - n\bar{x}^2}$$

Hence, using Equation 7.2.2 and the fact that $n\bar{Y} = \Sigma_{i=1}^n Y_i$, we have proven the following proposition.

Proposition 7.2.1

The least squares estimators of β and α corresponding to the data set x_i, Y_i, $i = 1, \ldots, n$ are, respectively

$$B = \frac{\Sigma_{i=1}^n x_i Y_i - \bar{x}\Sigma_{i=1}^n Y_i}{\Sigma_{i=1}^n x_i^2 - n\bar{x}^2}$$

$$A = \bar{Y} - B\bar{x}$$

The straight line $A + Bx$ is called the estimated regression line.

Program 7-2 computes the least squares estimators A and B. It also gives the user the option of computing some other statistics whose values will be needed in the following sections.

Example 7.2a The raw material used in the production of a certain synthetic fiber is stored in a location without a humidity control. Measurements of the relative humidity in the storage location and the moisture content of a sample

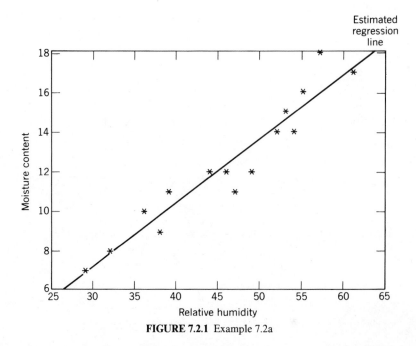

FIGURE 7.2.1 Example 7.2a

of the raw material were taken over 15 days with the following data (in percentages) resulting.

Relative humidity	46 53 29 61 36 39 47 49 52 38 55 32 57 54 44
Moisture content	12 15 7 17 10 11 11 12 14 9 16 8 18 14 12

These data are plotted in Figure 7.2.1.

To compute the least squares estimator and the estimated regression line we run Program 7-2.

```
RUN
THIS PROGRAM COMPUTES THE LEAST SQUARES ESTIMATORS AND RELATED STATISTICS IN SIM
PLE LINEAR REGRESSION MODELS
ENTER THE NUMBER OF DATA PAIRS n
? 15
ENTER THE n  SUCCESSIVE PAIRS x,Y ONE PAIR AT A TIME
? 46,12
? 53,15
? 29,7
? 61,17
? 36,10
? 39,11
? 47,11
? 49,12
? 52,14
? 38,9
? 55,16
? 32,8
? 57,18
? 54,14
? 44,12
THE LEAST SQUARES ESTIMATORS ARE   AS FOLLOWS
A = -2.510452
B =  .3232035
THE ESTIMATED REGRESSION LINE IS  Y = -2.510452  +  .3232035  x
DO YOU WANT OTHER COMPUTED VALUES? ENTER 1 IF YES AND 0 IF NO.
? 0
Ok
```

3 DISTRIBUTION OF THE ESTIMATORS

To specify the distribution of the estimators A and B it is necessary to make additional assumptions about the random errors aside from just assuming that their mean is 0. The usual approach is to assume that the random errors are independent normal random variables having mean 0 and variance σ^2. That is, we suppose that if Y_i is the response corresponding to the input value x_i, then $Y_1, \ldots, Y_n$ are independent and

$$Y_i \sim \mathcal{N}\left(\alpha + \beta x_i, \sigma^2\right)$$

It should be noted that the foregoing supposes that the variance of the random error does not depend on the input value but rather is a constant. This value σ^2 is not assumed to be known but rather must be estimated from the data.

Since the least squares estimator B of β can be expressed as

$$B = \frac{\Sigma_i(x_i - \bar{x})Y_i}{\Sigma_i x_i^2 - n\bar{x}^2} \tag{7.3.1}$$

we see that it is a linear combination of the independent normal random variables Y_i, $i = 1, \ldots, n$ and so is itself normally distributed. Using Equation 7.3.1 the mean and variance of B are computed as follows:

$$\begin{aligned} E[B] &= \frac{\Sigma_i(x_i - \bar{x})E[Y_i]}{\Sigma_i x_i^2 - n\bar{x}^2} \\[2mm] &= \frac{\Sigma_i(x_i - \bar{x})(\alpha + \beta x_i)}{\Sigma_i x_i^2 - n\bar{x}^2} \\[2mm] &= \frac{\alpha\Sigma_i(x_i - \bar{x}) + \beta\Sigma_i x_i(x_i - \bar{x})}{\Sigma_i x_i^2 - n\bar{x}^2} \\[2mm] &= \beta\frac{[\Sigma_i x_i^2 - \bar{x}\Sigma_i x_i]}{\Sigma_i x_i^2 - n\bar{x}^2} \qquad \text{since } \sum_i(x_i - \bar{x}) = 0 \\[2mm] &= \beta \end{aligned}$$

Thus $E[B] = \beta$ and so B is an unbiased estimator of β. We will now compute the variance of B.

$$\begin{aligned} \text{Var}(B) &= \frac{\text{Var}\left(\Sigma_{i=1}^n(x_i - \bar{x})Y_i\right)}{\left(\Sigma_{i=1}^n x_i^2 - n\bar{x}^2\right)^2} \tag{7.3.2} \\[2mm] &= \frac{\Sigma_{i=1}^n(x_i - \bar{x})^2\,\text{Var}(Y_i)}{\left(\Sigma_{i=1}^n x_i^2 - n\bar{x}^2\right)^2} \qquad \text{by independence} \\[2mm] &= \frac{\sigma^2\Sigma_{i=1}^n(x_i - \bar{x})^2}{\left(\Sigma_{i=1}^n x_i^2 - n\bar{x}^2\right)^2} \\[2mm] &= \frac{\sigma^2}{\Sigma_{i=1}^n x_i^2 - n\bar{x}^2} \end{aligned}$$

where the final equality results from the use of the identity

$$\sum_{i=1}^n(x_i - \bar{x})^2 = \sum_{i=1}^n x_i^2 - n\bar{x}^2$$

Using Equation 7.3.1 along with the relationship

$$A = \sum_{i=1}^n \frac{Y_i}{n} - B\bar{x}$$

shows that A can also be expressed as a linear combination of the independent normal random variables Y_i, $i = 1, \ldots, n$; and is thus also normally distributed. Its mean is obtained from

$$E[A] = \sum_{i=1}^{n} \frac{E[Y_i]}{n} - \bar{x}E[B]$$

$$= \sum_{i=1}^{n} \frac{(\alpha + \beta x_i)}{n} - \bar{x}\beta$$

$$= \alpha + \beta\bar{x} - \bar{x}\beta$$

$$= \alpha$$

Thus A is also an unbiased estimator. The variance of A is computed by first expressing A as a linear combination of the Y_i. The result (whose details are left as an exercise) is that

$$\text{Var}(A) = \frac{\sigma^2 \sum_{i=1}^{n} x_i^2}{n\left(\sum_{i=1}^{n} x_i^2 - n\bar{x}^2\right)} \tag{7.3.3}$$

The quantities $Y_i - A - Bx_i$, $i = 1, \ldots, n$, which represent the differences between the actual responses (that is, the Y_i) and their least squares estimators (that is, $A + Bx_i$) are called the *residuals*. The sum of squares of the residuals

$$SS_R = \sum_{i=1}^{n} (Y_i - A - Bx_i)^2$$

can be utilized to estimate the unknown error variance σ^2. Indeed, it can be shown that

$$\frac{SS_R}{\sigma^2} \sim \chi_{n-2}^2$$

That is, SS_R/σ^2 has a chi-square distribution with $n - 2$ degrees of freedom, which implies that

$$E\left[\frac{SS_R}{\sigma^2}\right] = n - 2$$

or

$$E\left[\frac{SS_R}{(n-2)}\right] = \sigma^2$$

Thus $SS_R/(n-2)$ is an unbiased estimator of σ^2. In addition it can be shown that SS_R is independent of the pair A and B.

REMARK

A plausibility argument as to why SS_R/σ^2 might have a chi-square distribution with $n - 2$ degrees of freedom and be independent of A and B runs as follows. Since the Y_i are independent normal random variables, it follows that $(Y_i - E[Y_i])/\sqrt{\text{Var}(Y_i)}$, $i = 1, \ldots, n$ are independent unit normals and so

$$\sum_{i=1}^{n} \frac{(Y_i - E[Y_i])^2}{\text{Var}(Y_i)} = \sum_{i=1}^{n} \frac{(Y_i - \alpha - \beta x_i)^2}{\sigma^2} \sim \chi_n^2$$

Now if we substitute the estimators A and B for α and β, then 2 degrees of freedom are lost, and so it is not an altogether surprising result that SS_R/σ^2 has a chi-square distribution with $n - 2$ degrees of freedom.

The fact that SS_R is independent of A and B is quite similar to the fundamental result that in normal sampling $\overline{X}$ and S^2 are independent. Indeed this latter result states that if $Y_1,\ldots,Y_n$ is a normal sample with population mean μ and variance σ^2, then if in the sum of squares $\sum_{i=1}^{n}(Y_i - \mu)^2/\sigma^2$, which has a chi-square distribution with n degrees of freedom, one substitutes the estimator $\overline{Y}$ for μ to obtain the new sum of squares $\sum_i(Y_i - \overline{Y})^2/\sigma^2$, then this quantity [equal to $(n - 1)S^2/\sigma^2$] will be independent of $\overline{Y}$ and will have a chi-square distribution with $n - 1$ degrees of freedom. Since SS_R/σ^2 is obtained by substituting the estimators A and B for α and β in the sum of squares $\sum_{i=1}^{n}(Y_i - \alpha - \beta x_i)^2/\sigma^2$, it is not unreasonable to expect that this quantity might be independent of A and B.

Notation   If we let

$$S_{xY} = \sum_{i=1}^{n}(x_i - \overline{x})(Y_i - \overline{Y}) = \sum_{i=1}^{n}x_iY_i - n\overline{x}\,\overline{Y}$$

$$S_{xx} = \sum_{i=1}^{n}(x_i - \overline{x})^2 = \sum_{i=1}^{n}x_i^2 - n\overline{x}^2$$

$$S_{YY} = \sum_{i=1}^{n}(Y_i - \overline{Y})^2 = \sum_{i=1}^{n}Y_i^2 - n\overline{Y}^2$$

then the least squares estimators can be expressed as

$$B = \frac{S_{xY}}{S_{xx}}$$

$$A = \overline{Y} - B\overline{x}$$

The following computational identity for SS_R, the sum of squares of the residuals, can be established.

Computational Identity for SS_R

$$SS_R = \frac{S_{xx}S_{YY} - S_{xY}^2}{S_{xx}} \tag{7.3.4}$$

The following proposition sums up the results of this section.

Proposition 7.3.1

Suppose that the responses Y_i, $i = 1,\ldots,n$ are independent normal random variables with means $\alpha + \beta x_i$ and common variance σ^2. The least squares estimators of β and α are respectively

$$B = \frac{S_{xY}}{S_{xx}}, \qquad A = \overline{Y} - B\overline{x}$$

Furthermore, the distributions of A and B are as follows:

$$A \sim \mathcal{N}\left(\alpha, \frac{\sigma^2 \Sigma_i x_i^2}{n S_{xx}}\right)$$

$$B \sim \mathcal{N}\left(\beta, \sigma^2 / S_{xx}\right)$$

In addition, if we let

$$SS_R = \sum_i (Y_i - A - Bx_i)^2$$

denote the sum of squares of the residuals, then

$$\frac{SS_R}{\sigma^2} \sim \chi_{n-2}^2$$

and SS_R is independent of the least squares estimators A and B. Also, SS_R can be computed from

$$SS_R = \frac{S_{xx} S_{YY} - (S_{xY})^2}{S_{xx}}$$

The Basic program 7-2 will compute the least squares estimators A and B as well as $\bar{x}$, $\Sigma_i x_i^2$, S_{xx}, S_{xY}, S_{YY}, and SS_R.

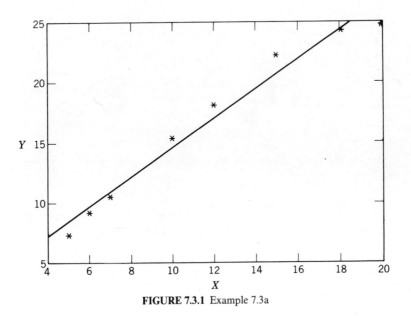

FIGURE 7.3.1 Example 7.3a

Example 7.3a The following data relate x, the moisture of a wet mix of a certain product, to Y, the density of the finished product

x_i	Y_i
5	7.4
6	9.3
7	10.6
10	15.4
12	18.1
15	22.2
18	24.1
20	24.8

Fit a linear curve to these data. Also determine SS_R.

To solve the foregoing, run Program 7-2:

```
RUN
THIS PROGRAM COMPUTES THE LEAST SQUARES ESTIMATORS AND RELATED STATISTICS IN SIM
PLE LINEAR REGRESSION MODELS
ENTER THE NUMBER OF DATA PAIRS n
? 8
ENTER THE n  SUCCESSIVE PAIRS x,Y ONE PAIR AT A TIME
? 5,7.4
? 6,9.3
? 7,10.6
? 10,15.4
? 12,18.1
? 15,22.2
? 18,24.1
? 20,24.8
THE LEAST SQUARES ESTIMATORS ARE  AS FOLLOWS
A =  2.463487
B =  1.206367
THE ESTIMATED REGRESSION LINE IS  Y =  2.463487  +  1.206367  x
DO YOU WANT OTHER COMPUTED VALUES? ENTER 1 IF YES AND 0 IF NO.
? 1
S(x,Y) = 267.6626
S(x,x) = 221.875
S(Y,Y) = 332.3692
SSR = 9.46993
THE AVERAGE x VALUE IS 11.625
THE SUM OF THE SQUARES OF THE x VALUES IS 1303
Ok  ■
```

A plot of the data points and the estimated regression line is presented in Figure 7.3.1.

4 STATISTICAL INFERENCES ABOUT THE REGRESSION PARAMETERS

Using Proposition 7.3.1 it is a simple matter to devise hypothesis tests and confidence intervals for the regression parameters.

4.1 Inferences Concerning β

An important hypothesis to consider regarding the simple linear regression model

$$Y = \alpha + \beta x + e$$

is the hypothesis that $\beta = 0$. Its importance derives from fact that it is equivalent to stating that the mean response does not depend on the input, or equivalently, that there is no regression on the input variable. To test

$$H_0 : \beta = 0 \qquad \text{against} \qquad H_1 : \beta \neq 0$$

note that, from Proposition 7.3.1

$$\frac{B - \beta}{\sqrt{\sigma^2/S_{xx}}} = \sqrt{S_{xx}}\,\frac{(B - \beta)}{\sigma} \sim \mathcal{N}(0,1) \tag{7.4.1}$$

and is independent of

$$\frac{SS_R}{\sigma^2} \sim \chi^2_{n-2}$$

Hence, from the definition of a t-random variable it follows that

$$\frac{\sqrt{S_{xx}}\,\dfrac{(B - \beta)}{\sigma}}{\sqrt{\dfrac{SS_R}{\sigma^2(n-2)}}} \sim t_{n-2} \tag{7.4.2}$$

That is, $\sqrt{(n-2)S_{xx}/SS_R}\,(B - \beta)$ has a t-distribution with $n-2$ degrees of freedom. Therefore, if H_0 is true (and so $\beta = 0$), then

$$\sqrt{\frac{(n-2)S_{xx}}{SS_R}}\,B \sim t_{n-2}$$

which gives rise to the following test of H_0.

Hypothesis Test of $H_0 : \beta = 0$ A significance level γ test of H_0 is to

$$\text{reject} \quad H_0 \quad \text{if} \quad \sqrt{\frac{(n-2)S_{xx}}{SS_R}}\,|B| > t_{\gamma/2,\,n-2}$$

$$\text{accept} \quad H_0 \quad \text{otherwise}$$

This test can be performed by first computing the value of the test statistic $\sqrt{(n-2)S_{xx}/SS_R}\,|B|$—call its value v—and then rejecting H_0 if the desired significance level is at least as large as

$$p\text{-value} = P\{|T_{n-2}| > v\}$$

$$= 2P\{T_{n-2} > v\}$$

where T_{n-2} is a t-random variable with $n-2$ degrees of freedom. This latter probability can be obtained by using Program 3-8-2-A.

Example 7.4a The term *regression* was originally employed by Francis Galton while describing the laws of inheritance. Galton believed that the laws of inheritance caused population extremes to "regress toward the mean." By this he meant that the children of individuals having extreme values of a certain characteristic would tend to have less extreme values of this characteristic than their parent. To illustrate this, the British statistician Karl Pearson plotted the heights of 10 randomly chosen sons versus that of their fathers. The resulting data (in inches) were as follows.

Fathers' height	60	62	64	65	66	67	68	70	72	74
Sons' height	63.6	65.2	66	65.5	66.9	67.1	67.4	68.3	70.1	70

It should be noted that whereas the data appear to indicate that taller fathers tend to have taller sons, it also appears to indicate that the sons of fathers that are either extremely short or extremely tall tend to be more "average" than their fathers—that is, there is a "regression toward the mean."

If indeed there is a regression toward the mean, then the response Y (that is, the height of a son) would tend to be larger than the input value x (equal to the height of the father) when x is small, and would tend to be smaller than x when x is large. Since this would imply that the slope of the regression line would be less than 1 (see Figure 7.4.1), we will determine whether the preceding data strongly indicate a regression toward the mean by testing the alternative hypothesis that the slope of the regression line is not less than 1. That is, let us test

$$H_0 : \beta \geq 1 \qquad \text{versus} \qquad H_1 : \beta < 1$$

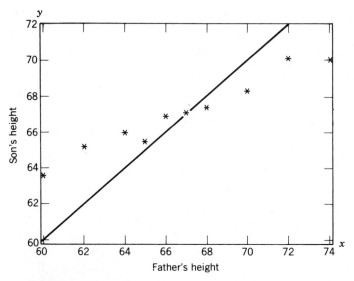

FIGURE 7.4.1 Example 7.4a. For x small, $y > x$. For x large, $y < x$

To test H_0, run Program 7-2:

```
RUN
THIS PROGRAM COMPUTES THE LEAST SQUARES ESTIMATORS AND RELATED STATISTICS IN SIM
PLE LINEAR REGRESSION MODELS
ENTER THE NUMBER OF DATA PAIRS n
? 10
ENTER THE n  SUCCESSIVE PAIRS x,Y ONE PAIR AT A TIME
? 60,63.6
? 62,65.2
? 64,66
? 65,65.5
? 66,66.9
? 67,67.1
? 68,67.4
? 70,68.3
? 72,70.1
? 74,70
THE LEAST SQUARES ESTIMATORS ARE   AS FOLLOWS
A =  35.97757
B =  .4645573
THE ESTIMATED REGRESSION LINE IS  Y =  35.97757  +  .4645573  x
DO YOU WANT OTHER COMPUTED VALUES? ENTER 1 IF YES AND 0 IF NO.
? 1
S(x,Y) = 79.71875
S(x,x) = 171.6016
S(Y,Y) = 38.53125
SSR = 1.497325
THE AVERAGE x VALUE IS 66.8
THE SUM OF THE SQUARES OF THE x VALUES IS 44794
Ok
```

Since the observed value of $\sqrt{(n-2)S_{xx}/SS_R}$ is $\sqrt{8 \times 171.6/1.49721} = 30.28045$, we note that if H_0 were true, and so $\beta \geq 1$, then we would have that

$$\sqrt{\frac{(n-2)S_{xx}}{SS_R}} (B - \beta) \leq (.4645797 - 1)30.28045 = -16.212721$$

Hence, the p-value of the data is given by

$$p\text{-value} = P\{T_8 \leq -16.212721\}$$

$$< 10^{-4} \qquad \text{by Program 3-8-2-A}$$

Thus the hypothesis that $\beta \geq 1$ would be rejected, even at significance level $\alpha = 10^{-4}$, strongly indicating a regression toward the mean. ∎

A confidence interval estimator for β is easily obtained from Equation 7.4.2. Indeed, it follows from Equation 7.4.2 that for any a, $0 < a < 1$,

$$P\left\{-t_{a/2, n-2} < \sqrt{\frac{(n-2)S_{xx}}{SS_R}} (B - \beta) < t_{a/2, n-2}\right\} = 1 - a$$

or, equivalently

$$P\left\{B - \sqrt{\frac{SS_R}{(n-2)S_{xx}}} t_{a/2, n-2} < \beta < B + \sqrt{\frac{SS_R}{(n-2)S_{xx}}} t_{a/2, n-2}\right\} = 1 - a$$

which yields the following

Confidence Interval for β   A $100(1 - a)$ percent confidence interval estimator of $β$ is

$$\left(B - \sqrt{\frac{SS_R}{(n-2)S_{xx}}}\, t_{a/2,\,n-2}, \quad B + \sqrt{\frac{SS_R}{(n-2)S_{xx}}}\, t_{a/2,\,n-2} \right)$$

REMARK
The result that

$$\frac{B - β}{\sqrt{σ^2/S_{xx}}} \sim \mathcal{N}(0,1)$$

cannot be immediately applied to make inferences about $β$ since it involves the unknown parameter $σ^2$. Instead, what we do is use the preceding statistic with $σ^2$ replaced by its estimator $SS_R/(n-2)$, which has the effect of changing the distribution of the statistic from the unit normal to the t-distribution with $n - 2$ degrees of freedom.

4.2   Inferences Concerning α

The determination of confidence intervals and hypotheses tests for $α$ is accomplished in exactly the same manner as was done for $β$. Specifically, Proposition 7.3.1 can be used to show that

$$\sqrt{\frac{n(n-2)S_{xx}}{\Sigma_i x_i^2 SS_R}}\,(A - α) \sim t_{n-2} \tag{7.4.3}$$

which leads to the following confidence interval estimator of $α$.

Confidence Interval Estimator of α   The $100(1 - a)$ percent confidence interval for $α$ is the interval

$$A \pm \sqrt{\frac{\Sigma_i x_i^2 SS_R}{n(n-2)S_{xx}}}\, t_{a/2,\,n-2}$$

Hypotheses tests concerning $α$ are easily obtained from Equation 7.4.3, and their development is left as an exercise.

4.3   Inferences Concerning the Mean Response α + βx

It is often of interest to use the data pairs (x_i, Y_i), $i = 1, \ldots, n$ to estimate $α + βx_0$, the mean response for a given input level x_0. If it is a point estimator that is desired, then the natural estimator is $A + Bx_0$, which is an unbiased estimator since

$$E[A + Bx_0] = E[A] + x_0 E[B] = α + βx_0$$

However, if we desire a confidence interval, or are interested in testing some hypothesis about this mean response, then it is necessary to first determine the probability distribution of the estimator $A + Bx_0$. We now do so.

Using the expression for B given by Equation 7.3.1 yields that

$$B = c \sum_{i=1}^{n} (x_i - \bar{x}) Y_i$$

where

$$c = \frac{1}{\sum_{i=1}^{n} x_i^2 - n\bar{x}^2} = \frac{1}{S_{xx}}$$

Since

$$A = \bar{Y} - B\bar{x}$$

we see that

$$A + Bx_0 = \frac{\sum_{i=1}^{n} Y_i}{n} - B(\bar{x} - x_0)$$

$$= \sum_{i=1}^{n} Y_i \left[\frac{1}{n} - c(x_i - \bar{x})(\bar{x} - x_0) \right]$$

Since the Y_i are independent normal random variables, the foregoing equation shows that $A + Bx_0$ can be expressed as a linear combination of independent normal random variables, and is thus itself normally distributed. As we already know its mean, we need only compute its variance, which is accomplished as follows:

$$\text{Var}(A + Bx_0)$$

$$= \sum_{i=1}^{n} \left[\frac{1}{n} - c(x_i - \bar{x})(\bar{x} - x_0) \right]^2 \text{Var}(Y_i)$$

$$= \sigma^2 \sum_{i=1}^{n} \left[\frac{1}{n^2} + c^2(\bar{x} - x_0)^2 (x_i - \bar{x})^2 - 2c(x_i - \bar{x}) \frac{(\bar{x} - x_0)}{n} \right]$$

$$= \sigma^2 \left[\frac{1}{n} + c^2(\bar{x} - x_0)^2 \sum_{i=1}^{n} (x_i - \bar{x})^2 - 2c(\bar{x} - x_0) \sum_{i=1}^{n} \frac{(x_i - \bar{x})}{n} \right]$$

$$= \sigma^2 \left[\frac{1}{n} + \frac{(\bar{x} - x_0)^2}{S_{xx}} \right]$$

where the last equality followed from

$$\sum_{i=1}^{n} (x_i - \bar{x})^2 = \sum_{i=1}^{n} x_i^2 - n\bar{x}^2 = 1/c = S_{xx}, \qquad \sum_{i=1}^{n} (x_i - \bar{x}) = 0$$

Hence, we have shown that

$$A + Bx_0 \sim \mathcal{N}\left(\alpha + \beta x_0, \sigma^2 \left[\frac{1}{n} + \frac{(x_0 - \bar{x})^2}{S_{xx}} \right] \right) \qquad (7.4.4)$$

In addition, as $A + Bx_0$ is independent of

$$SS_R/\sigma^2 \sim \chi^2_{n-2}$$

it follows that

$$\frac{A + Bx_0 - (\alpha + \beta x_0)}{\sqrt{\dfrac{1}{n} + \dfrac{(x_0 - \bar{x})^2}{S_{xx}}}\sqrt{\dfrac{SS_R}{(n-2)}}} \sim t_{n-2} \qquad (7.4.5)$$

Equation 7.4.5 can now be used to obtain the following confidence interval estimator of $\alpha + \beta x_0$.

Confidence Interval Estimator of $\alpha + \beta x_0$ With $100(1 - a)$ percent confidence, $\alpha + \beta x_0$ will lie within

$$A + Bx_0 \pm \sqrt{\frac{1}{n} + \frac{(x_0 - \bar{x})^2}{S_{xx}}}\sqrt{\frac{SS_R}{(n-2)}}\, t_{a/2,\, n-2}$$

Example 7.4b Using the data of Example 4a determine a 95 percent confidence interval for the average height of all males whose fathers are 68 inches tall.

Solution Since the observed values are

$$n = 10, \qquad x_0 = 68, \qquad \bar{x} = 66.8, \qquad S_{xx} = 171.6, \qquad SS_R = 1.49721$$

we see that

$$\sqrt{\frac{1}{n} + \frac{(x_0 - \bar{x})^2}{S_{xx}}}\sqrt{\frac{SS_R}{(n-2)}} = .1424276$$

Also, as

$$t_{.025,\, 8} = 2.306 \qquad A + Bx_0 = 67.56751$$

we obtain the following 95 percent confidence interval

$$\alpha + \beta x_0 \in (67.239, 67.896). \quad \blacksquare$$

4.4 Prediction Interval of a Future Response

It is often the case that it is more important to estimate the actual value of a future response rather than its mean value. For instance, if an experiment is to be performed at temperature level x_0, then we would probably be more interested in predicting $Y(x_0)$, the yield from this experiment, than we would be in estimating the expected yield—$E[Y(x_0)] = \alpha + \beta x_0$. (On the other hand, if a series of experiments were to be performed at input level x_0, then we would probably want to estimate $\alpha + \beta x_0$, the mean yield.)

Suppose first that we are interested in a single value (as opposed to an interval) to use as a predictor of $Y(x_0)$, the response at level x_0. Now, it is clear that the best predictor of $Y(x_0)$, is its mean value $\alpha + \beta x_0$. [Actually this is not so immediately obvious since one could argue that the best predictor of a random variable is (1) its mean—which minimizes the expected square of the difference between the predictor and the actual value; or (2) its median—which minimizes the expected absolute difference between the predictor and the actual value; or (3) its mode—which is the most likely value to occur. However, as the mean, median, and mode of a normal random variable are all equal—and the response is, by assumption, normally distributed—there is no doubt in this situation.] Since α and β are not known, it seems reasonable to use their estimators A and B and thus use $A + Bx_0$ as the predictor of a new response at input level x_0.

Let us now suppose that rather than being concerned with determining a single value to predict a response, we are interested in finding a prediction interval that, with a given degree of confidence, will contain the response. To obtain such an interval let Y denote the future response whose input level is x_0 and consider the probability distribution of the response minus its predicted value—that is, the distribution of $Y - A - Bx_0$. Now,

$$Y \sim \mathcal{N}\left(\alpha + \beta x_0, \sigma^2\right)$$

and, as was shown in Section 4.3

$$A + Bx_0 \sim \mathcal{N}\left(\alpha + \beta x_0, \sigma^2\left[\frac{1}{n} + \frac{(x_0 - \bar{x})^2}{S_{xx}}\right]\right)$$

Hence, because Y is independent of the earlier data values $Y_1, Y_2, \ldots, Y_n$ that were used to determine A and B, it follows that Y is independent of $A + Bx_0$ and so

$$Y - A - Bx_0 \sim \mathcal{N}\left(0, \sigma^2\left[1 + \frac{1}{n} + \frac{(x_0 - \bar{x})^2}{S_{xx}}\right]\right)$$

or, equivalently

$$\frac{Y - A - Bx_0}{\sigma\sqrt{\dfrac{(n+1)}{n} + \dfrac{(x_0 - \bar{x})^2}{S_{xx}}}} \sim \mathcal{N}(0,1) \tag{7.4.6}$$

Now, using once again the result that SS_R is independent of A and B (and also of Y) and

$$\frac{SS_R}{\sigma^2} \sim \chi_{n-2}^2$$

we obtain, by the usual argument, upon replacing σ^2 in Equation 7.4.6 by its estimator $SS_R/(n-2)$ that

$$\frac{Y - A - Bx_0}{\sqrt{\dfrac{(n+1)}{n} + \dfrac{(x_0 - \bar{x})^2}{S_{xx}}}\sqrt{\dfrac{SS_R}{(n-2)}}} \sim t_{n-2}$$

and so, for any value of a, $0 < a < 1$

$$P\left\{-t_{a/2, n-2} < \frac{Y - A - Bx_0}{\sqrt{\dfrac{(n+1)}{n} + \dfrac{(x_0 - \bar{x})^2}{S_{xx}}}\sqrt{\dfrac{SS_R}{(n-2)}}} < t_{a/2, n-2}\right\} = 1 - a$$

That is, we have just established the following

Prediction Interval for a Response at the Input Level x_0 Based on the response values Y_i corresponding to the input values x_i, $i = 1, 2, \ldots, n$: With $100(1 - a)$ percent confidence the response Y at the input level x_0 will be contained in the interval

$$A + Bx_0 \pm t_{a/2, n-2} \sqrt{\left[\frac{(n + 1)}{n} + \frac{(x_0 - \bar{x})^2}{S_{xx}} \right] \frac{SS_R}{(n - 2)}}$$

REMARK

There is often some confusion about the difference between a confidence and a prediction interval. A confidence interval is an interval that does contain, with a given degree of confidence, a fixed parameter of interest. A prediction interval, on the other hand, is an interval that will contain, again with a given degree of confidence, a random variable of interest.

Example 7.4c In Example 7.4a suppose we want an interval that we can "be 95 percent certain" will contain the height of a given male whose father is 68 inches tall. A simple computation now yields the following prediction interval:

$$Y(68) \in 67.568 \pm 1.050$$

or, with 95 percent confidence, the person's height will be between 66.518 and 68.618.   ∎

4.5   Summary of Distributional Results

We now summarize the distributional results of this section.

$$\text{Model:} \quad Y = \alpha + \beta x + e \qquad e \sim \mathcal{N}(0, \sigma^2)$$
$$\text{Data:} \quad (x_i, Y_i) \qquad i = 1, 2, \ldots, n$$

Inferences About	*Use the Distributional Result*
β	$\sqrt{\dfrac{(n - 2)S_{xx}}{SS_R}} (B - \beta) \sim t_{n-2}$
α	$\sqrt{\dfrac{n(n - 2)S_{xx}}{\Sigma_i x_i^2 SS_R}} (A - \alpha) \sim t_{n-2}$
$\alpha + \beta x_0$	$\dfrac{A + Bx_0 - \alpha - \beta x_0}{\sqrt{\left(\dfrac{1}{n} + \dfrac{(x_0 - \bar{x})^2}{S_{xx}} \right) \left(\dfrac{SS_R}{(n - 2)} \right)}} \sim t_{n-2}$
$Y(x_0)$	$\dfrac{Y(x_0) - A - Bx_0}{\sqrt{\left(1 + \dfrac{1}{n} + \dfrac{(x_0 - \bar{x})^2}{S_{xx}} \right) \left(\dfrac{SS_R}{(n - 2)} \right)}} \sim t_{n-2}$

5 INDEX OF FIT

If we consider the least squares fit of the data pairs (x_i, Y_i), $i = 1, \ldots, n$ to the model

$$Y = \alpha + e \qquad (7.5.1)$$

then it is easy to check that the least squares estimator of α is $\overline{Y}$; and so the minimal possible sum of squares of the residuals based on the model in Equation 7.5.1 is $\Sigma_i(Y_i - \overline{Y})^2$. On the other hand, since the linear regression model

$$Y = \alpha + \beta x + e$$

allows for greater flexibility than does Equation 7.5.1 it follows that it will have a value of SS_R that is less than or equal to $\Sigma_i(Y_i - \overline{Y})^2$—indeed, it will be strictly less, unless the least squares estimate of β is 0. The quantity R^2, defined by

$$R^2 = 1 - \frac{SS_R}{\Sigma_i (Y_i - \overline{Y})^2}$$

$$= 1 - \frac{SS_R}{S_{YY}}$$

$$= \frac{S_{xY}^2}{S_{xx}S_{YY}} \qquad \text{by } (7.3.4)$$

is called the *coefficient of determination*. It will have a value between 0 (which it would assume if $B = 0$) and 1 (which it would assume if there was a perfect fit and thus $SS_R = 0$). The value of R^2 is sometimes interpreted as being the proportion of the variation of the responses Y that is explained by the input levels x.

The quantity $R = \sqrt{R^2}$ is called the *index of fit*, and is often used as an indicator of how well the regression model fits the data. However, one must be cautioned that a high value of R does not necessarily mean that the linear regression model is correct, nor does a small value necessarily mean that it is incorrect, for in fact, the value of R is only an indication of the improvement in fit when using the regression model $Y = \alpha + \beta x + e$ over the model $Y = \alpha + e$.

Example 7.5a From the readout from Example 7.4a we see that

$$SS_R = 1.497 \qquad S_{YY} = 38.5344$$

and thus

$$R^2 = .9612 \qquad R = .9804.$$

Such a large value of R indicates that the regression model results in a much better fit of the data than we would have obtained by assuming that there was no relation between the height of a father and the subsequent height of his son. ∎

The index of fit R is also referred to as the *sample correlation coefficient*. This nomenclature stems from the fact that whereas we have supposed that the successive input levels are given values, it often turns out that rather than being fixed they are actually random variables. In such a situation it is often reasonable to suppose that the pairs $X_i, Y_i,\ i = 1, \ldots, n$, are independent and have a common joint distribution, where X_i and Y_i denote the ith input and response values. Suppose now that we are interested in estimating ρ, the correlation between X_i and Y_i. That is.

$$\rho = \frac{E[(X - E[X])(Y - E[Y])]}{\sqrt{\operatorname{Var}(X)\operatorname{Var}(Y)}}$$

where X and Y represent any of the pairs X_i, Y_i. A natural estimator of ρ is obtained by

estimating $\quad E[(X - E[X])(Y - E[Y])] \quad$ by $\quad \displaystyle\sum_{i=1}^{n} \frac{[(X_i - \bar{X})(Y_i - \bar{Y})]}{n}$

estimating $\quad \operatorname{Var}(X) \quad$ by $\quad \displaystyle\sum_{i=1}^{n} \frac{(X_i - \bar{X})^2}{n}$

estimating $\quad \operatorname{Var}(Y) \quad$ by $\quad \displaystyle\sum_{i=1}^{n} \frac{(Y_i - \bar{Y})^2}{n}$

That is, a natural estimator of ρ is

$$\hat{\rho} = \frac{\Sigma(X_i - \bar{X})(Y_i - \bar{Y})}{\sqrt{\Sigma(X_i - \bar{X})^2 \Sigma(Y_i - \bar{Y})^2}}$$

$$= \frac{S_{XY}}{\sqrt{S_{XX} S_{YY}}} = R$$

In other words, the index of fit R is the estimate of the correlation between input and response.

6　ANALYSIS OF RESIDUALS: ASSESSING THE MODEL

The initial step for ascertaining whether or not the simple linear regression model

$$Y = \alpha + \beta x + e, \qquad e \sim \mathcal{N}(0, \sigma^2)$$

is appropriate in a given situation is to investigate the scatter diagram. Indeed, this is often sufficient to convince one that the regression model is or is not correct. When the scatter diagram does not by itself rule out the preceding model, then the least square estimators A and B should be computed and the residuals $Y_i - (A + Bx_i),\ i = 1, \ldots, n$ analyzed. The analysis begins by normalizing, or standardizing, the residuals, by dividing them by $\sqrt{SS_R/(n-2)}$, the estimate of the standard deviation of the Y_i. The resulting quantities

$$\frac{Y_i - (A + Bx_i)}{\sqrt{SS_R/(n-2)}} \qquad i = 1, \ldots, n$$

are called the *standardized residuals*.

When the simple linear regression model is correct, the standardized residuals are approximately, independent unit normal random variables, and,

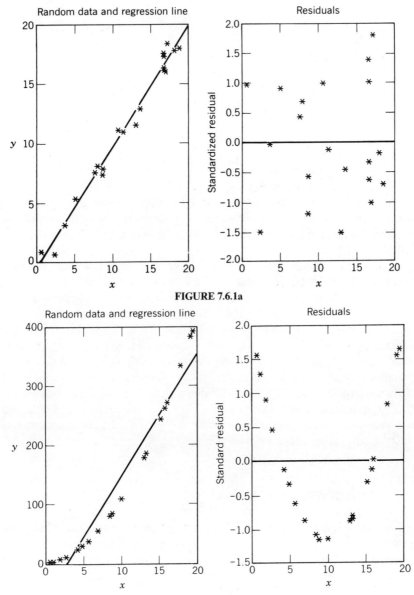

FIGURE 7.6.1a

FIGURE 7.6.1b

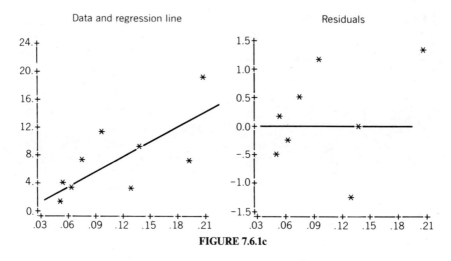

FIGURE 7.6.1c

as such, they should be randomly distributed about 0 with about 95 percent of their values being between -2 and $+2$ (since $P\{-1.96 < Z < 1.96\} = .95$). In addition, a plot of the standardized residuals should not indicate any distinct pattern. Indeed, any indication of a distinctive pattern should make one suspicious about the validity of the assumed simple linear regression model.

Figure 7.6.1 presents three different scatter diagrams and their associated standardized residuals. The first of these, as indicated both by its scatter diagram and the random nature of its standardized residuals, appears to fit the straight-line model quite well. The second residual plot shows a discernible pattern, in that the residuals appear to be first decreasing and then increasing as the input level increases. This often means that higher order (than just linear) terms are needed to describe the relationship between the input and response. Indeed, this is also indicated by the scatter diagram in this case. The third standardized residual plot also shows a pattern, in that the absolute value of the residuals, and thus their squares, appear to be increasing, as the input level increases. This often indicates that the variance of the response is not constant but, rather, increases with the input level.

7 TRANSFORMING TO LINEARITY

In many situations, it is clear that the mean response is not a linear function of the input level. In such cases, if the form of the relationship can be determined, it is sometimes possible, by a change of variables, to transform it into a linear form. For instance, in certain applications it is known that $W(t)$, the amplitude of a signal a time t after its origination, is approximately related to t by the functional form

$$W(t) \approx ce^{-dt}$$

On taking logarithms, this can be expressed as

$$\log W(t) \approx \log c - dt$$

If we now let

$$Y = \log W(t)$$
$$\alpha = \log c$$
$$\beta = -d$$

then, the foregoing can be modeled as a regression of the form

$$Y = \alpha + \beta t + e$$

The regression parameters α and β would then be estimated by the usual least squares approach and the original functional relationships can be predicted from

$$W(t) \approx e^{A+Bt}$$

Example 7.7a Studies have shown that the probability that a 40-year-old smoker, who has been smoking for the past 10 years, will contract lung cancer within the next 20 years (assuming that he continues smoking at the same level) is a function of the average number of cigarettes he consumes. The following are the results of an extensive study (performed on mice and extrapolated to humans).

Number of cigarettes smoked per day	Probability of lung cancer
5	.061
10	.113
20	.192
30	.259
40	.339
50	.401
60	.461
80	.551

Using the preceding data we would like to estimate the probability that an individual whose daily intake is 35 cigarettes will contract lung cancer.

Let $P(i)$ denote the probability of contracting lung cancer within the next 20 years if one continues to smoke i cigarettes daily. Even though a plot of $P(i)$ looks roughly linear (see Figure 7.7.1), we can improve upon the fit by considering a nonlinear functional form. To obtain such a functional relationship between $P(i)$ and i, we reason as follows: Let us suppose that each cigarette smoked has a fixed probability of causing lung cancer (possibly by damaging the DNA within a lung cell). Hence, if someone smoked i cigarettes

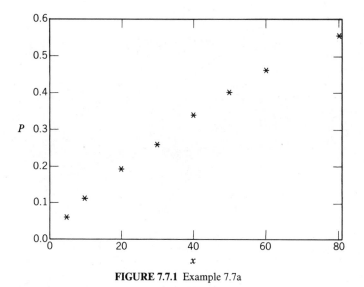

FIGURE 7.7.1 Example 7.7a

daily, then the probability of not having a lung cancer caused by smoking is the product of the probabilities that each of these i cigarettes do not cause a cancer. Since there is also a chance of contracting lung cancer from other means, it would seem that

$$1 - P(i) = P\{\text{no lung cancer if smoke } i \text{ daily}\}$$

$$\simeq c(P\{\text{no lung cancer if smoke 1 daily}\})^i$$

where $1 - c$ is the probability of getting lung cancer from a nonsmoking cause. This relationship can be written as

$$1 - P \simeq c(1 - d)^x$$

or

$$\log(1 - P) \simeq \log c + x \log(1 - d)$$

Thus, setting

$$Y = -\log(1 - P)$$
$$\alpha = -\log c$$
$$\beta = -\log(1 - d)$$

we obtain the usual regression equation

$$Y = \alpha + \beta x + e .$$

To see whether the data support this model, we can plot $-\log(1 - P)$ versus x. The transformed data are presented in Table 7.7.1 and the graph in Figure 7.7.2.

Running Program 7.2 yields that the least square estimates of α and β are

$$A = .0154$$
$$B = .0099$$

TABLE 7.7.1

Number of cigarettes	$-\log(1-P)$
5	.063
10	.120
20	.213
30	.300
40	.414
50	.512
60	.618
80	.801

Transforming this back into the original variable gives that the estimates of c and d are

$$\hat{c} = e^{-A} = .9847$$

$$1 - \hat{d} = e^{-B} = .9901$$

and so the estimated functional relationship is

$$\hat{P} = 1 - .9847\,(.9901)^{x}$$

The residuals $P - \hat{P}$ are presented in Table 7.7.2.

REMARK

If P denotes the proportion of a population whose exposure level is x that contract a certain disease, then Example 7.7a uses the model

$$-\log(1-P) = \alpha + \beta x + e$$

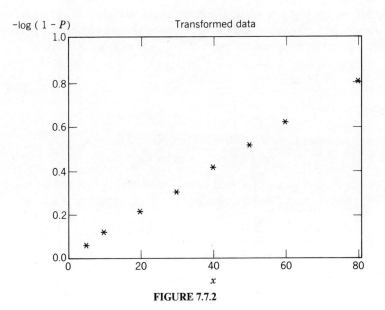

FIGURE 7.7.2

TABLE 7.7.2

x	P	$\hat{P}$	$P - \hat{P}$
5	.061	.063	−.002
10	.113	.109	.040
20	.192	.193	−.001
30	.259	.269	−.010
40	.339	.339	.000
50	.401	.401	.000
60	.461	.458	.003
80	.551	.556	−.005

Another model often employed and called the logistic (or logit) uses the relationship

$$\log\left(\frac{P}{1-P}\right) = \alpha + \beta x + e$$

The quantity $P/(1-P)$ is called odds-ratio. For instance, if an event has probability $P = 3/4$ of occurring, then the odds-ratio is $P/(1-P) = 3/1$, which represents the "odds" in favor of the event.

8 WEIGHTED LEAST SQUARES

In the regression model

$$Y = \alpha + \beta x + e$$

it often turns out that the variance of a response is not constant but rather depends on its input level. If these variances are known—at least up to a proportionality constant—then the regression parameters α and β should be estimated by minimizing a weighted sum of squares. Specifically, if

$$\text{Var}(Y_i) = \frac{\sigma^2}{w_i}$$

then the estimators A and B should be chosen to minimize

$$\sum_i \frac{[Y_i - (A + Bx_i)]^2}{\text{Var}(Y_i)} = \frac{1}{\sigma^2}\sum_i w_i(Y_i - A - Bx_i)^2$$

On taking partial derivatives with respect to A and B and setting them equal to 0, we obtain the following equations for the minimizing A and B.

$$\sum_i w_i Y_i = A\sum_i w_i + B\sum_i w_i x_i \qquad (7.8.1)$$

$$\sum_i w_i x_i Y_i = A\sum_i w_i x_i + B\sum_i w_i x_i^2$$

These equations are easily solved to yield the least squares estimators.

Example 7.8a To develop a feel as to why the estimators should be obtained by minimizing the weighted sum of squares rather than the ordinary sum of squares, consider the following situation. Suppose that $X_1, \ldots, X_n$ are independent normal random variables each having mean μ and variance σ^2. Suppose further that the X_i are not directly observable but rather only Y_1 and Y_2, defined by

$$Y_1 = X_1 + \cdots + X_k \qquad Y_2 = X_{k+1} + \cdots + X_n \qquad k < n$$

are directly observable. Based on Y_1 and Y_2 how should we estimate μ?

Whereas the best estimator of μ is clearly $\overline{X} = \sum_{i=1}^n X_i / n = (Y_1 + Y_2)/n$, let us see what the ordinary least squares estimator would be. Since

$$E[Y_1] = k\mu \qquad E[Y_2] = (n - k)\mu$$

the least squares estimator of μ would be that value of μ that minimizes

$$(Y_1 - k\mu)^2 + (Y_2 - [n - k]\mu)^2$$

On differentiating and setting equal to zero, we see that the least squares estimator of μ—call it $\hat{\mu}$—is such that

$$-2k(Y_1 - k\hat{\mu}) - 2(n - k)[Y_2 - (n - k)\hat{\mu}] = 0$$

or

$$\left[k^2 + (n - k)^2\right]\hat{\mu} = kY_1 + (n - k)Y_2$$

or

$$\hat{\mu} = \frac{kY_1 + (n - k)Y_2}{k^2 + (n - k)^2}$$

Thus we see that while the ordinary least squares estimator is an unbiased estimator of μ—since

$$E[\hat{\mu}] = \frac{kE[Y_1] + (n - k)E[Y_2]}{k^2 + (n - k)^2} = \frac{k^2\mu + (n - k)^2\mu}{k^2 + (n - k)^2} = \mu$$

it is not the best estimator $\overline{X}$.

Now let us determine the estimator produced by minimizing the weighted sum of squares. That is, let us determine the value of μ—call it μ_w—that minimizes

$$\frac{(Y_1 - k\mu)^2}{\text{Var}(Y_1)} + \frac{[Y_2 - (n - k)\mu]^2}{\text{Var}(Y_2)}$$

Since

$$\text{Var}(Y_1) = k\sigma^2 \qquad \text{Var}(Y_2) = (n - k)\sigma^2$$

this is equivalent to choosing μ to minimize

$$\frac{(Y_1 - k\mu)^2}{k} + \frac{[Y_2 - (n - k)\mu]^2}{n - k}$$

Upon differentiating and then equating to 0, we see that μ_w, the minimizing value, satisfies

$$\frac{-2k(Y_1 - k\mu_w)}{k} - \frac{2(n-k)[Y_2 - (n-k)\mu_w]}{n-k} = 0$$

or

$$Y_1 + Y_2 = n\mu_w$$

or

$$\mu_w = \frac{Y_1 + Y_2}{n}$$

That is, the weighted least squares estimator is indeed the preferred estimator $(Y_1 + Y_2)/n = \overline{X}$. ∎

REMARKS

(a) The weighted sum of squares can also be seen as the relevant quantity to be minimized by multiplying the regression equation

$$Y = \alpha + \beta x + e$$

by $\sqrt{w}$. This results in the equation

$$Y\sqrt{w} = \alpha\sqrt{w} + \beta x\sqrt{w} + e\sqrt{w}$$

Now, in this latter equation the error term $e\sqrt{w}$ has mean 0 and constant variance. Hence, the natural least squares estimators of α and β would be the values of A and B that minimize

$$\sum_i \left(Y_i\sqrt{w_i} - A\sqrt{w_i} - Bx_i\sqrt{w_i} \right)^2 = \sum_i w_i(Y_i - A - Bx_i)^2$$

(b) The weighted least squares approach puts the greatest emphasis on those data pairs having the greatest weights (and thus the smallest variance in their error term).

At this point it might appear that the weighted least squares approach is not particularly useful since it requires a knowledge, up to a constant, of the variance of a response at an arbitrary input level. However, by analyzing the model that generates the data, it is often possible to determine these values. This will be indicated by the following two examples.

Example 7.8b The following data represent travel times in a downtown area of a certain city. The independent, or input, variable is the distance to be traveled.

Distance (miles)	.5	1	1.5	2	3	4	5	6	8	10
Travel time (minutes)	15.0	15.1	16.5	19.9	27.7	29.7	26.7	35.9	42	49.4

Assuming a linear relationship of the form

$$Y = \alpha + \beta x + e$$

between Y, the travel time, and x, the distance, how should we estimate α and

β? To utilize the weighted least squares approach we need to know, up to a multiplicative constant, the variance of Y as a function of x. We will now present an argument that $\text{Var}(Y)$ should be proportional to x.

Let d denote the length of a city block. Thus a trip of distance x will consist of x/d blocks. If we let Y_i, $i = 1, \ldots, x/d$, denote the time it takes to traverse block i, then the total travel time can be expressed as

$$Y = Y_1 + Y_2 + \cdots + Y_{x/d}$$

Now in many applications it is probably reasonable to suppose that the Y_i are independent random variables with a common variance, and thus,

$$\begin{aligned}
\text{Var}(Y) &= \text{Var}(Y_1) + \cdots + \text{Var}(Y_{x/d}) \\
&= (x/d)\,\text{Var}(Y_1) \qquad \text{since } \text{Var}(Y_i) = \text{Var}(Y_1) \\
&= x\sigma^2 \qquad\qquad\quad \text{where } \sigma^2 = \text{Var}(Y_1)/d
\end{aligned}$$

Thus, it would seem that the estimators A and B should be chosen so as to minimize

$$\sum_i \frac{(Y_i - A - Bx_i)^2}{x_i}$$

Using the preceding data with the weights $w_i = 1/x_i$, the least squares Equations 7.8.1 are

$$104.22 = 5.34A + 10B$$
$$277.9 = 10A + 41B$$

which yield the solution

$$A = 12.561 \qquad B = 3.714$$

A graph of the estimated regression line $12.561 + 3.714x$ along with the data points is presented in Figure 7.8.1. As a qualitative check of our solution, note that the regression line fits the data pairs best when the input levels are small, which is as it should be since the weights are inversely proportional to the inputs.

Example 7.8c Consider the relationship between Y, the number of accidents on a heavily traveled highway and x, the number of cars traveling on the highway. After a little thought it would probably seem to most that the linear model

$$Y = \alpha + \beta x + e$$

would be appropriate. However, as there does not appear to be any a priori reason why $\text{Var}(Y)$ should not depend on the input level x, it is not clear that we would be justified in using the ordinary least squares approach to estimate α and β. Indeed, we will now argue that a weighted least squares approach with weights $1/x$ should be employed—that is, we should choose A and B to minimize

$$\sum_i \frac{(Y_i - A - Bx_i)^2}{x_i}$$

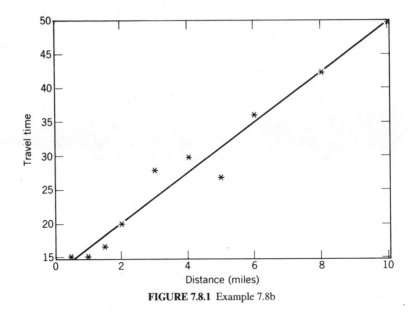

FIGURE 7.8.1 Example 7.8b

The rationale behind this claim is that it seems reasonable to suppose that Y approximately has a Poisson distribution. This is so since we can imagine that each of the x cars will have a small probability of causing an accident and so, for large x, the number of accidents should approximately be a Poisson random variable. Since the variance of a Poisson random variable is equal to its mean, we see that

$$\text{Var}(Y) \simeq E[Y] \quad \text{since } Y \text{ is approximately Poisson}$$
$$= \alpha + \beta x$$
$$\simeq \beta x \quad \text{for large } x$$

REMARKS

(a) Another technique that is often employed when the variance of the response depends on the input level is to attempt to stabilize the variance by an appropriate transformation. For example, if Y is a Poisson random variable with mean λ, then it can be shown [see Remark (b)] that $\sqrt{Y}$ has approximate variance .25 no matter what the value of λ. Based on this fact one might try to model $E[\sqrt{Y}]$ as a linear function of the input. That is, one might consider the model

$$\sqrt{Y} = \alpha + \beta x + e$$

The problem with such an approach, however, is that in situations where it is reasonable to suppose that the mean response is approximately a linear function of the input, it is not at all clear why the mean square root of the response should also have such a relationship with the input level. It is for this reason that the author prefers the weighted least squares approach.

(b) Proof that Var $(\sqrt{Y}) \approx .25$ when Y is Poisson with mean λ. Consider the Taylor series expansion of $g(y) = \sqrt{y}$ about the value λ. By ignoring all terms beyond the second derivative term, we obtain that

$$g(y) \approx g(\lambda) + g'(\lambda)(y - \lambda) + \frac{g''(\lambda)(y - \lambda)^2}{2} \qquad (7.8.2)$$

Since

$$g'(\lambda) = \frac{1}{2}\lambda^{-1/2} \qquad g''(\lambda) = -\frac{1}{4}\lambda^{-3/2}$$

we obtain, on evaluating Equation 7.8.2 at $y = Y$, that

$$\sqrt{Y} \approx \sqrt{\lambda} + \frac{1}{2}\lambda^{-1/2}(Y - \lambda) - \frac{1}{8}\lambda^{-3/2}(Y - \lambda)^2$$

Taking expectations, and using the results that

$$E[Y - \lambda] = 0 \qquad E[(Y - \lambda)^2] = \text{Var}(Y) = \lambda$$

yields that

$$E[\sqrt{Y}] \approx \sqrt{\lambda} - \frac{1}{8\sqrt{\lambda}}$$

Hence

$$(E[\sqrt{Y}])^2 \approx \lambda + \frac{1}{64\lambda} - \frac{1}{4}$$

$$\approx \lambda - \frac{1}{4}$$

and so

$$\text{Var}(\sqrt{Y}) = E[Y] - (E[\sqrt{Y}])^2$$

$$\approx \lambda - \left(\lambda - \frac{1}{4}\right)$$

$$= \frac{1}{4}.$$

9 POLYNOMIAL REGRESSION

In situations where the functional relationship between the response Y and the independent variable x cannot be adequately approximated by a linear relationship, it is sometimes possible to obtain a reasonable fit by considering a polynomial relationship. That is, we might try to fit to the data set a functional relationship of the form

$$Y = \beta_0 + \beta_1 x + \beta_2 x^2 + \cdots + \beta_r x^r + e$$

where $\beta_0, \beta_1, \ldots, \beta_r$ are regression coefficients that would have to be estimated. If the data set consists of the n pairs (x_i, Y_i), $i = 1, \ldots, n$, then the least square estimators of $\beta_0, \ldots, \beta_r$—call them $B_0, \ldots, B_r$—are those values that minimize

$$\sum_{i=1}^{n} (Y_i - B_0 - B_1 x - B_2 x^2 - \cdots - B_r x^r)^2$$

To determine these estimators, we take partial derivatives with respect to $B_0 \cdots B_r$ of the foregoing sum of squares, and then set these equal to 0 so as to determine the minimizing values. On doing so, and then rearranging the resulting equations, we obtain that the least square estimators $B_0, B_1, \ldots, B_r$ satisfy the following set of $r + 1$ linear equations called the normal equations.

$$\sum_{i=1}^{n} Y_i = B_0 n + B_1 \sum_{i=1}^{n} x_i + B_2 \sum_{i=1}^{n} x_i^2 + \cdots + B_r \sum_{i=1}^{n} x_i^r$$

$$\sum_{i=1}^{n} x_i Y_i = B_0 \sum_{i=1}^{n} x_i + B_1 \sum_{i=1}^{n} x_i^2 + B_2 \sum_{i=1}^{n} x_i^3 + \cdots + B_r \sum_{i=1}^{n} x_i^{r+1}$$

$$\sum_{i=1}^{n} x_i^2 Y_i = B_0 \sum_{i=1}^{n} x_i^2 + B_1 \sum_{i=1}^{n} x_i^3 + \cdots + B_r \sum_{i=1}^{n} x_i^{r+2}$$

$$\vdots \qquad \vdots \qquad \qquad \vdots$$

$$\sum_{i=1}^{n} x_i^r Y_i = B_0 \sum_{i=1}^{n} x_i^r + B_1 \sum_{i=1}^{n} x_i^{r+1} + \cdots + B_r \sum_{i=1}^{n} x_i^{2r}$$

In fitting a polynomial to a set of data pairs, it is often possible to determine the necessary degree of the polynomial by a study of the scatter diagram. It should be emphasized that one should always use the lowest possible degree that appears to adequately describe the data. [Thus, for instance, whereas it is almost always possible to find a polynomial of degree n, that passes through all the n pairs (x_i, Y_i), $i = 1, \ldots, n$, it would be hard to ascribe much confidence to such a fit.]

Even more so than in linear regression, it is extremely risky to use a polynomial fit to predict the value of a response at an input level x_0 that is far away from the input levels x_i, $i = 1, \ldots, n$ used in finding the polynomial fit. (For one thing, the polynomial fit may be valid only in a region around the x_i, $i = 1, \ldots, n$ and not including x_0).

Example 7.9a Fit a polynomial to the following data.

x	Y
1	20.6
2	30.8
3	55
4	71.4
5	97.3
6	131.8
7	156.3
8	197.3
9	238.7
10	291.7

A plot of these data (see Figure 7.9.1) indicates that a quadratic relationship

$$Y = \beta_0 + \beta_1 x + \beta_2 x^2 + e$$

might hold. Since

$$\sum_i x_i = 55, \qquad \sum_i x_i^2 = 385, \qquad \sum_i x_i^3 = 3025, \qquad \sum_i x_i^4 = 25333$$

$$\sum_i Y_i = 1291.1, \qquad \sum_i x_i Y_i = 9549.3, \qquad \sum_i x_i^2 Y_i = 77758.9$$

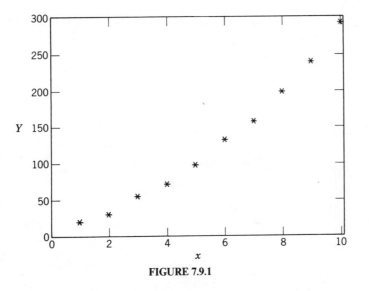

FIGURE 7.9.1

the least squares estimates are the solution of the following set of equations.

$$1291.1 = 10B_0 + 55B_1 + 385B_2 \qquad (7.9.1)$$

$$9549.3 = 55B_0 + 385B_1 + 3025B_2$$

$$77758.9 = 385B_0 + 3025B_1 + 25333B_2$$

Solving these equations (see the remark following this example) yields that the least squares estimates are

$$B_0 = 12.59326 \qquad B_1 = 6.326172 \qquad B_2 = 2.122818$$

Thus, the estimated quadratic regression equation is

$$Y = 12.59 + 6.33x + 2.12x^2$$

This equation, along with the data, is plotted in Figure 7.9.2. ■

REMARK

In matrix notation Equation 7.9.1 can be written as

$$\begin{bmatrix} 1291.1 \\ 9549.3 \\ 77758.9 \end{bmatrix} = \begin{bmatrix} 10 & 55 & 385 \\ 55 & 385 & 3025 \\ 385 & 3025 & 25333 \end{bmatrix} \begin{bmatrix} B_0 \\ B_1 \\ B_2 \end{bmatrix}$$

which has the solution

$$\begin{bmatrix} B_0 \\ B_1 \\ B_2 \end{bmatrix} = \begin{bmatrix} 10 & 55 & 385 \\ 55 & 385 & 3025 \\ 385 & 3025 & 25333 \end{bmatrix}^{-1} \begin{bmatrix} 1291.1 \\ 9549.3 \\ 77758.9 \end{bmatrix}$$

Program Inv in the appendix can be used to invert a matrix.

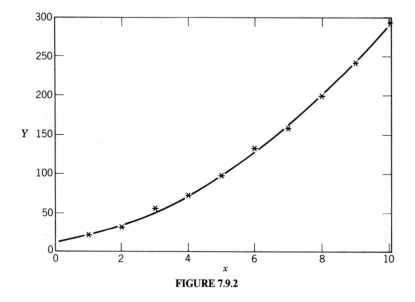

FIGURE 7.9.2

10 MULTIPLE LINEAR REGRESSION

In the majority of applications the response of an experiment can more adequately be predicted not on the basis of a single independent input variable but on a collection of such variables. Indeed, a typical situation is one in which there are a set of, say k, input variables and the response Y is related to them by the relation

$$Y = \beta_0 + \beta_1 x_1 + \cdots + \beta_k x_k + e$$

where x_j, $j = 1, \ldots, k$ is the level of the jth input variable and e is a random error that we shall assume is normally distributed with mean 0 and (constant) variance σ^2. The parameters $\beta_0, \beta_1, \ldots, \beta_k$ and σ^2 are assumed to be unknown and must be estimated from the data, which we shall suppose will consist of the values of $Y_1, \ldots, Y_n$ where Y_i is the response level corresponding to the k input levels $x_{i1}, x_{i2}, \ldots, x_{ik}$. That is, the Y_i are related to these input levels through

$$E[Y_i] = \beta_0 + \beta_1 x_{i1} + \beta_2 x_{i2} + \cdots + \beta_k x_{ik}$$

If we let $B_0, B_1, \ldots, B_k$ denote estimators of $\beta_0, \ldots, \beta_k$, then the sum of the squared differences between the Y_i and their estimated expected values is

$$\sum_{i=1}^{n} \left(Y_i - B_0 - B_1 x_{i1} - B_2 x_{i2} - \cdots - B_k x_{ik} \right)^2$$

The least squares estimators are those values of $B_0, B_1, \ldots, B_k$ that minimize the foregoing.

To determine the least squares estimators we repeatedly take partial derivatives of the preceding sum of squares first with respect to B_0, then to $B_1, \ldots,$

then to B_k. On equating these $k + 1$ equations to 0, we obtain the following set of equations:

$$\sum_{i=1}^{n} (Y_i - B_0 - B_1 x_{i1} - B_2 x_{i2} \cdots - B_k x_{ik}) = 0$$

$$\sum_{i=1}^{n} x_{i1}(Y_i - B_0 - B_1 x_{i1} - \cdots - B_k x_{ik}) = 0$$

$$\sum_{i=1}^{n} x_{i2}(Y_i - B_0 - B_1 x_{i1} - \cdots - B_k x_{ik}) = 0$$

$$\vdots$$

$$\sum_{i=1}^{n} x_{ik}(Y_i - B_0 - B_1 x_{i1} - \cdots - B_i x_{ik}) = 0$$

Rewriting these equations yields that the least squares estimators $B_0, B_1, \ldots,$ B_k satisfy the following set of linear equations, called the *normal equations*:

$$\sum_{i=1}^{n} Y_i = nB_0 + B_1 \sum_{i=1}^{n} x_{i1} + B_2 \sum_{i=1}^{n} x_{i2} + \cdots + B_k \sum_{i=1}^{n} x_{ik} \qquad (7.10.1)$$

$$\sum_{i=1}^{n} x_{i1}Y_i = B_0 \sum_{i=1}^{n} x_{i1} + B_1 \sum_{i=1}^{n} x_{i1}^2 + B_2 \sum_{i=1}^{n} x_{i1}x_{i2} + \cdots + B_k \sum_{i=1}^{n} x_{i1}x_{ik}$$

$$\vdots$$

$$\sum_{i=1}^{k} x_{ik}Y_i = B_0 \sum_{i=1}^{n} x_{ik} + B_1 \sum_{i=1}^{n} x_{ik}x_{i1} + B_2 \sum_{i=1}^{n} x_{ik}x_{i2} + \cdots + B_k \sum_{i=1}^{n} x_{ik}^2$$

Before solving the normal equations it is convenient to introduce matrix notation. If we let

$$\mathbf{Y} = \begin{bmatrix} Y_1 \\ Y_2 \\ \vdots \\ Y_n \end{bmatrix} \qquad \mathbf{X} = \begin{bmatrix} 1 & x_{11} & x_{12} & \cdots & x_{1k} \\ 1 & x_{21} & x_{22} & \cdots & x_{2k} \\ \vdots & \vdots & \vdots & & \vdots \\ 1 & x_{n1} & x_{n2} & \cdots & x_{nk} \end{bmatrix}$$

$$\boldsymbol{\beta} = \begin{bmatrix} \beta_0 \\ \beta_1 \\ \vdots \\ \beta_k \end{bmatrix} \qquad \mathbf{e} = \begin{bmatrix} e_1 \\ e_2 \\ \vdots \\ e_n \end{bmatrix}$$

then $\mathbf{Y}$ is an $n \times 1$, $\mathbf{X}$ an $n \times p$, $\boldsymbol{\beta}$ a $p \times 1$, and $\mathbf{e}$ an $n \times 1$ matrix where $p \equiv k + 1$.

The multiple regression model can now be written as

$$\mathbf{Y} = \mathbf{X}\boldsymbol{\beta} + \mathbf{e}$$

In addition, if we let

$$\mathbf{B} = \begin{bmatrix} B_0 \\ B_1 \\ \vdots \\ B_k \end{bmatrix}$$

be the matrix of least squares estimators, then the normal Equations 7.10.1 can be written as

$$\mathbf{X'XB} = \mathbf{X'Y} \qquad (7.10.2)$$

where $\mathbf{X'}$ is the transpose of $\mathbf{X}$.

To see that Equation 7.10.2 is equivalent to the normal Equations 7.10.1 note that

$$\mathbf{X'X} = \begin{bmatrix} 1 & 1 & \cdots & 1 \\ x_{11} & x_{21} & \cdots & x_{n1} \\ x_{12} & x_{22} & \cdots & x_{n2} \\ \vdots & & & \\ x_{1k} & x_{2k} & \cdots & x_{nk} \end{bmatrix} \begin{bmatrix} 1 & x_{11} & x_{12} & \cdots & x_{1k} \\ 1 & x_{21} & x_{22} & \cdots & x_{2k} \\ \vdots & \vdots & & & \\ 1 & x_{n1} & x_{n2} & \cdots & x_{nk} \end{bmatrix}$$

$$= \begin{bmatrix} n & \sum_i x_{i1} & \sum_i x_{i2} & \cdots & \sum_i x_{ik} \\ \sum_i x_{i1} & \sum_i x_{i1}^2 & \sum_i x_{i1} x_{i2} & \cdots & \sum_i x_{i1} x_{ik} \\ \vdots & & & & \\ \sum_i x_{ik} & \sum_i x_{ik} x_{i1} & \sum_i x_{ik} x_{i2} & \cdots & \sum_i x_{ik}^2 \end{bmatrix}$$

and

$$\mathbf{X'Y} = \begin{bmatrix} \sum_i Y_i \\ \sum_i x_{i1} Y_i \\ \vdots \\ \sum_i x_{ik} Y_i \end{bmatrix}$$

It is now easy to see that the matrix equation

$$\mathbf{X'XB} = \mathbf{X'Y}$$

is equivalent to the set of normal Equations 7.10.1. Assuming that $(\mathbf{X'X})^{-1}$ exists, which is almost always the case, we obtain, upon multiplying it by both sides of the foregoing, that the least squares estimators are given by

$$\mathbf{B} = (\mathbf{X'X})^{-1}\mathbf{X'Y} \qquad (7.10.3)$$

Program 7-10 computes the least squares estimates, the inverse matrix $(\mathbf{X'X})^{-1}$, and SS_R.

Example 7.10a The following data relate the suicide rate to the population size and the divorce rate at eight different locations.

Location	Population in thousands	Divorce rate per 100,000	Suicide rate per 100,000
Akron, Ohio	679	30.4	11.6
Anaheim, Ca.	1420	34.1	16.1
Buffalo, N.Y.	1349	17.2	9.3
Austin, Texas	296	26.8	9.1
Chicago, Ill.	6975	29.1	8.4
Columbia, S.C.	323	18.7	7.7
Detroit, Mich.	4200	32.6	11.3
Gary, Indiana	633	32.5	8.4

Fit a multiple linear regression model to these data. That is, fit a model of the form

$$Y = \beta_0 + \beta_1 x_1 + \beta_2 x_2 + e$$

where Y is the suicide rate, x_1 is the population, and x_2 is the divorce rate. We run Program 7-10:

```
RUN
THIS PROGRAM COMPUTES THE LEAST SQUARES ESTIMATES OF THE COEFFICIENTS AND THE
 SUM OF SQUARES OF THE RESIDUALS IN MULTIPLE LINEAR REGRESSION
IT BEGINS BY COMPUTING THE INVERSE OF THE X-TRANSPOSE*X MATRIX
ENTER THE NUMBER OF ROWS OF THE X-MATRIX
? 8
ENTER THE NUMBER OF COLUMNS OF THE X-MATRIX
? 3
ENTER ROW 1 ONE AT A TIME
? 1? 679? 30.4
ENTER ROW 2 ONE AT A TIME
? 1? 1420? 34.1
ENTER ROW 3 ONE AT A TIME
? 1? 1349? 17.2
ENTER ROW 4 ONE AT A TIME
? 1? 296? 26.8
ENTER ROW 5 ONE AT A TIME
? 1? 6975? 29.1
ENTER ROW 6 ONE AT A TIME
? 1? 323? 18.7
ENTER ROW 7 ONE AT A TIME
? 1? 4200? 32.6
ENTER ROW 8 ONE AT A TIME
? 1? 633? 32.5
THE INVERSE MATRIX IS AS FOLLOWS
 2.783111   1.707031E-05  -9.727136E-02

 1.707031E-05   2.69611E-08  -2.55E-06

-9.727136E-02  -2.55E-06   3.697616E-03

ENTER THE RESPONSE VALUES ONE AT A TIME
? 11.6? 16.1? 9.3? 9.1? 8.4? 7.7? 11.3? 8.4
THE ESTIMATES OF THE REGRESSION COEFFICIENTS ARE AS FOLLOWS

           B( 0 )= 3.507385
           B( 1 )=-2.47709E-04
           B( 2 )= .2609463
THE SUM OF SQUARES OF THE RESIDUALS IS   SS(R) = 34.1192
Ok
```

Thus the estimated regression line is

$$Y = 3.5073 - .0002x_1 + .2609x_2$$

The value of B_1 indicates that the population does not play a major role in predicting the suicide rate (at least when the divorce rate is also given). Perhaps the population density, rather than the actual population, would have been more useful. ∎

It follows from Equation 7.10.3 that the least squares estimators $B_0, B_1, \ldots, B_k$—the elements of the matrix **B**—are all linear combinations of the independent normal random variables $Y_1, \ldots, Y_n$ and so will also be normally distributed. Indeed in such a situation—namely when each member of a set of random variables can be expressed as a linear combination of independent normal random variables—we say that the set of random variables has a joint *multivariate normal distribution*.

The least squares estimators turn out to be unbiased. This can be shown as follows:

$$E[\mathbf{B}] = E\left[(\mathbf{X'X})^{-1}\mathbf{X'Y}\right]$$

$$= E\left[(\mathbf{X'X})^{-1}\mathbf{X'}(\mathbf{X}\boldsymbol{\beta} + \mathbf{e})\right] \qquad \text{since } \mathbf{Y} = \mathbf{X}\boldsymbol{\beta} + \mathbf{e}$$

$$= E\left[(\mathbf{X'X})^{-1}\mathbf{X'X}\boldsymbol{\beta} + (\mathbf{X'X})^{-1}\mathbf{X'e}\right]$$

$$= E\left[\boldsymbol{\beta} + (\mathbf{X'X})^{-1}\mathbf{X'e}\right]$$

$$= \boldsymbol{\beta} + (\mathbf{X'X})^{-1}\mathbf{X'}E[\mathbf{e}]$$

$$= \boldsymbol{\beta}$$

The variances of the least squares estimators can be obtained from the matrix $(\mathbf{X'X})^{-1}$. Indeed, the values of this matrix are related to the covariances of the B_i's. Specifically the element in the $(i + 1)$st row, $(j + 1)$st column of $(\mathbf{X'X})^{-1}$ is equal to $\text{Cov}(B_i, B_j)/\sigma^2$.

To verify the preceding statement concerning $\text{Cov}(B_i, B_j)$ let

$$\mathbf{C} = (\mathbf{X'X})^{-1}\mathbf{X'}$$

Since **X** is an $n \times p$ matrix and **X′** a $p \times n$ matrix, it follows that **X′X** is $p \times p$, as is $(\mathbf{X'X})^{-1}$, and so **C** will be a $p \times n$ matrix. Let C_{ij} denote the element in row i, column j of this matrix. Now

$$\begin{bmatrix} B_0 \\ \vdots \\ B_{i-1} \\ \vdots \\ B_k \end{bmatrix} = \mathbf{B} = \mathbf{CY} = \begin{bmatrix} C_{11} & \cdots & C_{1n} \\ & & \\ C_{i1} & \cdots & C_{in} \\ & & \\ C_{p1} & \cdots & C_{pn} \end{bmatrix} \begin{bmatrix} Y_1 \\ \vdots \\ \\ \vdots \\ Y_n \end{bmatrix}$$

and so

$$B_{i-1} = \sum_{l=1}^{n} C_{il}Y_l$$

$$B_{j-1} = \sum_{r=1}^{n} C_{jr}Y_r$$

Hence

$$\text{Cov}\left(B_{i-1}, B_{j-1}\right) = \text{Cov}\left(\sum_{l=1}^{n} C_{il}Y_l, \sum_{r=1}^{n} C_{jr}Y_r\right)$$

$$= \sum_{r=1}^{n} \sum_{l=1}^{n} C_{il}C_{jr} \text{Cov}\left(Y_l, Y_r\right)$$

Now Y_l and Y_r are independent when $l \neq r$ and so

$$\text{Cov}\left(Y_l, Y_r\right) = \begin{cases} 0 & \text{if } l \neq r \\ \text{Var}\left(Y_r\right) & \text{if } l = r \end{cases}$$

Since $\text{Var}\left(Y_r\right) = \sigma^2$, we see that

$$\text{Cov}\left(B_{i-1}, B_{j-1}\right) = \sigma^2 \sum_{r=1}^{n} C_{ir}C_{jr} \qquad (7.10.4)$$

$$= \sigma^2 (\mathbf{CC'})_{ij}$$

where $(\mathbf{CC'})_{ij}$ is the element in row i, column j of $\mathbf{CC'}$.
  If we now let $\text{Cov}(\mathbf{B})$ denote the matrix of covariances—that is,

$$\text{Cov}(\mathbf{B}) = \begin{bmatrix} \text{Cov}\left(B_0, B_0\right) & \cdots & \text{Cov}\left(B_0, B_k\right) \\ \vdots & & \\ \text{Cov}\left(B_k, B_0\right) & \cdots & \text{Cov}\left(B_k, B_k\right) \end{bmatrix}$$

then it follows from Equation 7.10.4 that

$$\text{Cov}(\mathbf{B}) = \sigma^2 \mathbf{CC'} \qquad (7.10.5)$$

Now

$$\mathbf{C'} = \left((\mathbf{X'X})^{-1}\mathbf{X'}\right)'$$

$$= \mathbf{X}\left((\mathbf{X'X})^{-1}\right)'$$

$$= \mathbf{X}(\mathbf{X'X})^{-1}$$

where the last equality follows since $(\mathbf{X'X})^{-1}$ is symmetric (since $\mathbf{X'X}$ is) and so is equal to its transpose. Hence

$$\mathbf{CC'} = (\mathbf{X'X})^{-1}\mathbf{X'X}(\mathbf{X'X})^{-1}$$

$$= (\mathbf{X'X})^{-1}$$

and so we can conclude from Equation 7.10.5 that

$$\text{Cov}(\mathbf{B}) = \sigma^2(\mathbf{X'X})^{-1} \tag{7.10.6}$$

Since $\text{Cov}(B_i, B_i) = \text{Var}(B_i)$, it follows that the variances of the least squares estimators are given by σ^2 multiplied by the diagonal elements of $(\mathbf{X'X})^{-1}$.

The quantity σ^2 can be estimated by using the sum of squares of the residuals. That is, if we let

$$SS_R = \sum_{i=1}^{n} (Y_i - B_0 - B_1 x_{i1} - B_2 x_{i2} - \cdots - B_k x_{ik})^2$$

then it can be shown that

$$\frac{SS_R}{\sigma^2} \sim \chi^2_{n-(k+1)}$$

and so

$$E\left[\frac{SS_R}{\sigma^2}\right] = n - k - 1$$

or

$$E[SS_R/(n - k - 1)] = \sigma^2$$

That is, $SS_R/(n - k - 1)$ is an unbiased estimator of σ^2. In addition, as in the case of simple linear regression, SS_R will be independent of the least squares estimators $B_0, B_1, \ldots, B_k$.

REMARK

If we let r_i denote the ith residual

$$r_i = Y_i - B_0 - B_1 x_{i1} - \cdots - B_k x_{ik}, \qquad i = 1, \ldots, n$$

then

$$\mathbf{r} = \mathbf{Y} - \mathbf{XB}$$

where

$$\mathbf{r} = \begin{bmatrix} r_1 \\ r_2 \\ \vdots \\ r_n \end{bmatrix}$$

Hence, we may write

$$
\begin{aligned}
SS_R &= \sum_{i=1}^{n} r_i^2 \\
&= \mathbf{r'r} \\
&= (\mathbf{Y} - \mathbf{XB})'(\mathbf{Y} - \mathbf{XB}) \\
&= [\mathbf{Y'} - (\mathbf{XB})'](\mathbf{Y} - \mathbf{XB}) \\
&= (\mathbf{Y'} - \mathbf{B'X'})(\mathbf{Y} - \mathbf{XB}) \\
&= \mathbf{Y'Y} - \mathbf{Y'XB} - \mathbf{B'X'Y} + \mathbf{B'X'XB} \\
&= \mathbf{Y'Y} - \mathbf{Y'XB}
\end{aligned}
\tag{7.10.7}
$$

where the last equality follows from the normal equations

$$X'XB = X'Y$$

Now as Y' is $1 \times n$, X is $n \times p$ and B is $p \times 1$, it follows that $Y'XB$ is a 1×1 matrix. That is, $Y'XB$ is a scalar and thus is equal to its transpose, which shows that

$$Y'XB = (Y'XB)'$$
$$= B'X'Y$$

Hence, using Equation 7.10.7 we have proven the following identity:

$$SS_R = Y'Y - B'X'Y$$

The foregoing is a useful computational formula for SS_R (though one must be careful of possible round-off error when using it).

Example 7.10b For the data of Example 7.10a we computed that $SS_R = 34.12$. Since $n = 8$, $k = 2$, the estimate of σ^2 is $34.12/5 = 6.824$. ■

Example 7.10c The diameter of a tree at its breast height is influenced by many factors. The following data relate the diameter of a particular type of Eucalyptus tree to its age, average rainfall at its site, site's elevation, and the wood's mean specific gravity. (The data come from R. G. Skolmen, 1975, "Shrinkage and Specific Gravity Variation in Robusta Eucalyptus Wood Grown in Hawaii," USDA Forest Service PSW-298.)

	Age (years)	Elevation (1000 ft)	Rainfall (inches)	Specific gravity	Diameter at breast height (inches)
1	44	1.3	250	.63	18.1
2	33	2.2	115	.59	19.6
3	33	2.2	75	.56	16.6
4	32	2.6	85	.55	16.4
5	34	2.0	100	.54	16.9
6	31	1.8	75	.59	17.0
7	33	2.2	85	.56	20.0
8	30	3.6	75	.46	16.6
9	34	1.6	225	.63	16.2
10	34	1.5	250	.60	18.5
11	33	2.2	255	.63	18.7
12	36	1.7	175	.58	19.4
13	33	2.2	75	.55	17.6
14	34	1.3	85	.57	18.3
15	37	2.6	90	.62	18.8

Assuming a linear regression model of the form

$$Y = \beta_0 + \beta_1 x_1 + \beta_2 x_2 + \beta_3 x_3 + \beta_4 x_4 + e$$

where x_1 is the age, x_2 is the elevation, x_3 is the rainfall, x_4 is the specific gravity, and Y is the tree's diameter, test the hypothesis that $\beta_2 = 0$. That is, test the hypothesis that, given the other three factors, the elevation of the tree does not affect its diameter.

To test this hypothesis we begin by running Program 7-10, which yields, among other things, the following:

$$(\mathbf{X'X})_{3,3}^{-1} = .379 \qquad SS_R = 19.262 \qquad B_2 = .075$$

It now follows from Equation 7.10.6 that

$$\text{Var}(B_2) = .379\sigma^2$$

Since B_2 is normal and

$$E[B_2] = \beta_2$$

we see that

$$\frac{B_2 - \beta_2}{.616\sigma} \sim N(0,1)$$

Replacing σ by its estimator $SS_R/10$ transforms the foregoing unit normal distribution into a t-distribution with 10 ($= n - k - 1$) degrees of freedom. That is,

$$\frac{B_2 - \beta_2}{.616\sqrt{SS_R/10}} \sim t_{10}$$

Hence, if $\beta_2 = 0$ then

$$\frac{\sqrt{10/SS_R}\, B_2}{.616} \sim t_{10}$$

Since the value of the preceding statistic is $(\sqrt{10/19.262})\,(.075)/.616 = .088$ the p-value of the test of the hypothesis that $\beta_2 = 0$ is

$$p\text{-value} = P\{|T_{10}| > .088\}$$
$$= 2P\{T_{10} > .088\}$$
$$= .9316 \qquad \text{by Program 3-8-2-}A$$

Hence, the hypothesis is accepted (and, in fact, would be accepted at any significance level less than .9316). ∎

REMARK
The quantity

$$R^2 = 1 - \frac{SS_R}{\Sigma_i(Y_i - \overline{Y})^2}$$

which measures the amount of reduction in the sum of squares of the residuals when using the model

$$Y = \beta_0 + \beta_1 x_1 + \cdots + \beta_n x_n + e$$

as opposed to the model

$$Y = \beta_0 + e$$

is called the *coefficient of multiple determination*. The value

$$R = \sqrt{R^2}$$

is called the *multiple correlation coefficient* between Y and the input values.

10.1 Predicting Future Responses

Let us now suppose that a series of experiments is to be performed using the input levels $x_1, \ldots, x_k$. Based on our data, consisting of the prior responses $Y_1, \ldots, Y_n$, suppose we would like to estimate the mean response. Since the mean response is

$$E[Y|x] = \beta_0 + \beta_1 x_1 + \cdots + \beta_k x_k$$

a point estimate of it is simply $\sum_{i=0}^{k} B_i x_i$ where $x_0 \equiv 1$.

In order to determine a confidence interval estimator we need the distribution of $\sum_{i=0}^{k} B_i x_i$. Because it can be expressed as a linear combination of the independent normal random variables Y_i, $i = 1, \ldots, n$, it follows that it is also normally distributed. Its mean and variance are obtained as follows:

$$E\left[\sum_{i=0}^{k} x_i B_i\right] = \sum_{i=0}^{k} x_i E[B_i] \qquad (7.10.8)$$

$$= \sum_{i=0}^{k} x_i \beta_i \qquad \text{since } E[B_i] = \beta_i$$

That is, it is an unbiased estimator. Also, using the fact that the variance of a random variable is equal to the covariance between that random variable and itself, we see that

$$\text{Var}\left(\sum_{i=0}^{k} x_i B_i\right) = \text{Cov}\left(\sum_{i=0}^{k} x_i B_i, \sum_{j=0}^{k} x_j B_j\right) \qquad (7.10.9)$$

$$= \sum_{i=0}^{k} \sum_{j=0}^{k} x_i x_j \, \text{Cov}\left(B_i, B_j\right)$$

If we let $\mathbf{x}$ denote the matrix

$$\mathbf{x} = \begin{bmatrix} x_0 \\ x_1 \\ \vdots \\ x_k \end{bmatrix}$$

then, recalling that $\text{Cov}(B_i, B_j)/\sigma^2$ is the element in the $(i + 1)$st row and $(j + 1)$st column of $(\mathbf{X}'\mathbf{X})^{-1}$, we can express Equation 7.10.9 as

$$\text{Var}\left(\sum_{i=0}^{k} x_i B_i\right) = \mathbf{x}'(\mathbf{X}'\mathbf{X})^{-1}\mathbf{x}\sigma^2 \qquad (7.10.10)$$

Using Equations 7.10.8 and 7.10.10 we see that

$$\frac{\sum_{i=0}^{k} x_i B_i - \sum_{i=0}^{k} x_i \beta_i}{\sigma\sqrt{\mathbf{x}'(\mathbf{X}'\mathbf{X})^{-1}\mathbf{x}}} \sim N(0, 1)$$

If we now replace σ by its estimator $\sqrt{SS_R/(n - k - 1)}$ we obtain, by the usual argument, that

$$\frac{\sum_{i=0}^{k}x_{i}B_{i} - \sum_{i=0}^{k}x_{i}\beta_{i}}{\sqrt{\dfrac{SS_{R}}{(n-k-1)}}\sqrt{\mathbf{x}'(\mathbf{X}'\mathbf{X})^{-1}\mathbf{x}}} \sim t_{n-k-1}$$

which gives rise to the following confidence interval estimator of $\sum_{i=0}^{k}x_{i}\beta_{i}$.

Confidence Interval Estimate of $E[Y|\mathbf{x}] = \sum_{i=0}^{k}x_{i}\beta_{i}, (x_0 \equiv 1)$ A $100(1-a)$
percent confidence interval estimate of $\sum_{i=0}^{k}x_{i}\beta_{i}$ is given by

$$\sum_{i=0}^{k}x_{i}b_{i} \pm \sqrt{\frac{ss_{r}}{(n-k-1)}}\sqrt{\mathbf{x}'(\mathbf{X}'\mathbf{X})^{-1}\mathbf{x}} \quad t_{a/2,\,n-k-1}$$

where $b_0, \ldots, b_k$ are the values of the least squares estimators $B_0, B_1, \ldots, B_k$, and ss_r is the value of SS_R.

Example 7.10d A steel company is planning to produce cold reduced sheet steel consisting of .15 percent copper at an annealing temperature of 1150 (degrees F), and are interested in estimating the average (Rockwell 30-T) hardness of a sheet. To determine this, they have collected the following data on 10 different specimens of sheet steel having different copper contents and annealing temperatures:

Hardness	Copper Content	Annealing Temperature (units of 1000 F)
79.2	.02	1.05
64.0	.03	1.20
55.7	.03	1.25
56.3	.04	1.30
58.6	.10	1.30
84.3	.15	1.00
70.4	.15	1.10
61.3	.09	1.20
51.3	.13	1.40
49.8	.09	1.40

Estimate the average hardness and determine an interval in which it will lie with 95 percent confidence.

To solve this we first run Program 7-10, which gives the following result:

```
THE INVERSE MATRIX IS AS FOLLOWS
  9.427627  -5.223003  -7.290261

 -5.223003  43.74858   1.304812

 -7.290261   1.304812   5.886853

THE ESTIMATES OF THE REGRESSION COEFFICIENTS ARE AS FOLLOWS
            B( 0 )= 160.292
            B( 1 )= 16.65271
            B( 2 )=-80.80713
THE SUM OF SQUARES OF THE RESIDUALS IS   SS(R) = 67.01172
```

Now $\mathbf{x}' = (1, .15, 1.15)$ and so, using the value of the inverse matrix given before, a simple computation yields that

$$\sqrt{\mathbf{x}'(\mathbf{X}'\mathbf{X})^{-1}\mathbf{x}} = .5607 \qquad \sum_i x_i B_i = 69.862 \qquad \sqrt{\frac{SS_R}{7}} = 3.094$$

Hence, a point estimate of the expected hardness of sheets containing .15 percent copper at an annealing temperature of 1150 is 69.862. In addition, since $t_{.025,7} = 2.365$, a 95 percent confidence interval for this value is

$$69.862 \pm 4.130 \quad \blacksquare$$

When it is only a single experiment that is going to be performed at the input levels $x_1, \ldots, x_k$ we are usually more concerned with predicting the actual response rather than its mean value. That is, we are interested in utilizing our data set $Y_1, \ldots, Y_n$ to predict

$$Y(\mathbf{x}) = \sum_{i=0}^{k} \beta_i x_i + e \qquad \text{where } x_0 = 1$$

A point prediction is given by $\sum_{i=0}^{k} B_i x_i$ where B_i is the least squares estimator of β_i based on the set of prior responses $Y_1, \ldots, Y_n$, $i = 1, \ldots, k$.

To determine a prediction interval for $Y(\mathbf{x})$ note first that since $B_0, \ldots, B_k$ are based on prior responses, it follows that they are independent of $Y(\mathbf{x})$. Hence it follows that $Y(\mathbf{x}) - \sum_{i=0}^{k} B_i x_i$ is normal with mean 0 and variance given by

$$\text{Var}\left[Y(\mathbf{x}) - \sum_{i=0}^{k} B_i x_i\right] = \text{Var}\left[Y(\mathbf{x})\right] + \text{Var}\left(\sum_{i=0}^{k} B_i x_i\right) \qquad \text{by independence}$$

$$= \sigma^2 + \sigma^2 \mathbf{x}'(\mathbf{X}'\mathbf{X})^{-1}\mathbf{x} \qquad \text{from Equation 7.10.10}$$

and so

$$\frac{Y(\mathbf{x}) - \sum_{i=0}^{k} B_i x_i}{\sigma\sqrt{1 + \mathbf{x}'(\mathbf{X}'\mathbf{X})^{-1}\mathbf{x}}} \sim N(0,1)$$

which yields, upon replacing σ by its estimator, that

$$\frac{Y(\mathbf{x}) - \sum_{i=0}^{k} B_i x_i}{\sqrt{\frac{SS_R}{(n-k-1)}}\sqrt{1 + \mathbf{x}'(\mathbf{X}'\mathbf{X})^{-1}\mathbf{x}}} \sim t_{n-k-1}$$

We thus have:

Prediction Interval for $Y(\mathbf{x})$ With $100(1 - a)$ percent confidence $Y(\mathbf{x})$ will lie between

$$\sum_{i=0}^{k} x_i b_i \pm \sqrt{\frac{ss_r}{(n-k-1)}}\sqrt{1 + \mathbf{x}'(\mathbf{X}'\mathbf{X})^{-1}\mathbf{x}} \qquad t_{a/2, n-k-1}$$

where $b_0, \ldots, b_k$ are the values of the least squares estimators $B_0, B_1, \ldots, B_k$, and ss_r is the value of SS_R.

Example 7.10e If in Example 7.10d we were interested in determining an interval in which a single steel sheet, produced with a carbon content of .15 percent and at an annealing temperature of 1150 F, would lie, then the midpoint of the prediction interval would be as given before. However, the half-length of this prediction interval would differ from the confidence interval for the mean value by the factor $\sqrt{1.3144} / \sqrt{.3144}$. That is, the 95 percent prediction interval is

$$69.862 \pm 8.389 \quad \blacksquare$$

PROBLEMS

1. The following data relate x, the moisture of a wet mix of a certain product, to Y, the density of the finished product.

x_i	Y_i
5	7.4
6	9.3
7	10.6
10	15.4
12	18.1
15	22.2
18	24.1
20	24.8

   (a) Draw a scatter diagram.
   (b) Fit a linear curve to the data.

2. The following data relate the number of units of a good that were ordered as a function of the price of the good at 6 different locations.

Number ordered	88	112	123	136	158	172
Price	50	40	35	30	20	15

   How many units do you think would be ordered if the price was 25?

3. The corrosion of a certain metallic substance has been studied in dry oxygen at 500 degrees Centigrade. In this experiment the gain in weight after various periods of exposure was used as a measure of the amount of oxygen that had reacted with the sample. The table presents the data:

Hours	Percent gain
1.0	.02
2.0	.03
2.5	.035
3.0	.042
3.5	.05
4.0	.054

(a) Plot a scatter diagram.
(b) Fit a linear relation.
(c) Predict the percent weight gain when the metal is exposed for 3.2
hours.

4. The following data indicate the relationship between x, the specific gravity
of a wood sample, and Y, its maximum crushing strength in compression
parallel to the grain.

x_i	y_i (psi)
0.41	1850
0.46	2620
0.44	2340
0.47	2690
0.42	2160
0.39	1760
0.41	2500
0.44	2750
0.43	2730
0.44	3120

(a) Plot a scatter diagram. Does a linear relation seem reasonable?
(b) Estimate the regression coefficients.
(c) Predict the maximum crushing strength of a wood sample whose
specific gravity is 0.43.

5. The following data indicate the gain in reading speed versus the number of
weeks in the program of 10 students in a speed-reading program.

Number of weeks	Speed gain (wds / min)
2	21
3	42
8	102
11	130
4	52
5	57
9	105
7	85
5	62
7	90

(a) Plot a scatter diagram to see if a linear relationship is indicated.
(b) Find the least squares estimated of the regression coefficients.
(c) Estimate the expected gain of a student that plans to take the program
for 7 weeks.

6. Verify Equation 7.3.3, which states that

$$\text{Var}\,(A) = \frac{\sigma^2 \Sigma_{i=1}^n x_i^2}{n \Sigma_{i=1}^n (x_i - \bar{x})^2}$$

7. In Problem 4

(a) Estimate the variance of an individual response.
(b) Determine a 90 percent confidence interval for the variance.

8. Verify that

$$SS_R = \frac{S_{xx} S_{YY} - S_{xY}^2}{S_{xx}}$$

9. The following table relates the number of sunspots that appeared each year from 1970 to 1983 to the number of auto accident deaths during that year. Test the hypothesis that the number of auto deaths is not affected by the number of sunspots. (The sunspot data is from Jastrow and Thompson *Fundamentals and Frontiers of Astronomy*, and the auto death data is from *General Statistics of the U.S. 1985*.)

Year	Sunspots	Auto accidents deaths (1000s)
70	165	54.6
71	89	53.3
72	55	56.3
73	34	49.6
74	9	47.1
75	30	45.9
76	59	48.5
77	83	50.1
78	109	52.4
79	127	52.5
80	153	53.2
81	112	51.4
82	80	46
83	45	44.6

Also compute a 95 percent confidence interval estimate for α.

10. Verify Equation 7.4.3, which states that

$$\sqrt{\frac{n(n-2)S_{xx}}{\Sigma_i x_i^2 SS_R}}\,(A - \alpha) \sim t_{n-2}$$

11. The following data represent the relationship between the number of alignment errors and the number of missing rivets for 10 different aircrafts.

Number of missing rivets $= x$	Number of alignment errors $= y$
13	7
15	7
10	5
22	12
30	15
7	2
25	13
16	9
20	11
15	8

(a) Plot a scatter diagram.
(b) Estimate the regression coefficients.
(c) Test the hypothesis that $\alpha = 1$.
(d) Estimate the expected number of alignment errors of a plane having 24 missing rivets.
(e) Compute a 90 percent confidence interval estimate for the quantity in (d).

Problems 12 through 16 refer to the following data, which are based in part on 1960 records from the 43 states indicated and the District of Columbia, all of which charged a cigarette tax.

Cigarette Smoking and Cancer Death Rates

State	Cigarettes per person[a]	Deaths per year per 100,000 population			
		Bladder cancer	Lung cancer	Kidney cancer	Leukemia
Alabama	1,820	2.90	17.05	1.59	6.15
Arizona	2,582	3.52	19.80	2.75	6.61
Arkansas	1,824	2.99	15.98	2.02	6.94
California	2,860	4.46	22.07	2.66	7.06
Connecticut	3,110	5.11	22.83	3.35	7.20
Delaware	3,360	4.78	24.55	3.36	6.45
District of Columbia	4,046	5.60	27.27	3.13	7.08
Florida	2,827	4.46	23.57	2.41	6.07
Idaho	2,010	3.08	13.58	2.46	6.62
Illinois	2,791	4.75	22.80	2.95	7.27
Indiana	2,618	4.09	20.30	2.81	7.00
Iowa	2,212	4.23	16.59	2.90	7.69
Kansas	2,184	2.91	16.84	2.88	7.42
Kentucky	2,344	2.86	17.71	2.13	6.41
Louisiana	2,158	4.65	25.45	2.30	6.71
Maine	2,892	4.79	20.94	3.22	6.24
Maryland	2,591	5.21	26.48	2.85	6.81
Massachusetts	2,692	4.69	22.04	3.03	6.89
Michigan	2,496	5.27	22.72	2.97	6.91
Minnesota	2,206	3.72	14.20	3.54	8.28

Cigarette Smoking and Cancer Death Rates (*Continued*)

State	Cigarettes per person[a]	Deaths per year per 100,000 population			
		Bladder cancer	Lung cancer	Kidney cancer	Leukemia
Mississippi	1,608	3.06	15.60	1.77	6.08
Missouri	2,756	4.04	20.98	2.55	6.82
Montana	2,375	3.95	19.50	3.43	6.90
Nebraska	2,332	3.72	16.70	2.92	7.80
Nevada	4,240	6.54	23.03	2.85	6.67
New Jersey	2,864	5.98	25.95	3.12	7.12
New Mexico	2,116	2.90	14.59	2.52	5.95
New York	2,914	5.30	25.02	3.10	7.23
North Dakota	1,996	2.89	12.12	3.62	6.99
Ohio	2,638	4.47	21.89	2.95	7.38
Oklahoma	2,344	2.93	19.45	2.45	7.46
Pennsylvania	2,378	4.89	22.11	2.75	6.83
Rhode Island	2,918	4.99	23.68	2.84	6.35
South Carolina	1,806	3.25	17.45	2.05	5.82
South Dakota	2,094	3.64	14.11	3.11	8.15
Tennessee	2,008	2.94	17.60	2.18	6.59
Texas	2,257	3.21	20.74	2.69	7.02
Utah	1,400	3.31	12.01	2.20	6.71
Vermont	2,589	4.63	21.22	3.17	6.56
Washington	2,117	4.04	20.34	2.78	7.48
West Virginia	2,125	3.14	20.55	2.34	6.73
Wisconsin	2,286	4.78	15.53	3.28	7.38
Wyoming	2,804	3.20	15.92	2.66	5.78
Alaska	3,034	3.46	25.88	4.32	4.90

[a] Estimated from cigarette tax revenues.

12. (a) Draw a scatter diagram of cigarette consumption versus death rate from bladder cancer.
    (b) Does the diagram indicate the possibility of a linear relationship?
    (c) Find the best linear fit.
    (d) If next year's average cigarette consumption is 2500, what is your prediction of the death rate from bladder cancer?

13. (a) Draw a scatter diagram relating cigarette use and death rates from lung cancer.
    (b) Estimate the regression parameters α and β.
    (c) Test at the .05 level of significance the hypothesis that cigarette consumption does not affect the death rate from lung cancer.
    (d) What is the p-value of the test in (c)?

14. (a) Draw a scatter diagram of cigarette use versus death rate from kidney cancer.
    (b) Estimate the regression line.
    (c) What is the p-value in the test that the slope of the regression line is 0?
    (d) Determine a 90 percent confidence interval for the mean death rate from kidney cancer in a state whose citizens smoke an average of 3400 cigarettes per year.

15. (a) Draw a scatter diagram of cigarettes smoked versus death rate from leukemia.
    (b) Estimate the regression coefficients.
    (c) Test the hypothesis that there is no regression of the death rate from leukemia on the number of cigarettes used. That is, test that $\beta = 0$.
    (d) Determine a 90 percent prediction interval for the leukemia death rate in a state whose citizens smoke an average of 2500 cigarettes.

16. (a) Estimate the variances in Problems 12 through 15.
    (b) Determine a 95 percent confidence interval for the variance in the data relating to lung cancer.
    (c) For the lung cancer data break up the data into two parts—the first corresponding to states whose average cigarette consumption is less than 2300 and the second, above. Assume a linear regression model for both sets of data. How would you test the hypothesis that the variance of a response is the same for both sets?
    (d) Do the test in (c) at the .05 level of significance.

17. Plot the standardized residuals from the data of Problem 1. What does the plot indicate about the assumptions of the linear regression model?

18. The determination of the shear strength of spot welds is relatively difficult, whereas measuring the weld diameter of spot welds is relatively simple. As a result, it would be advantageous if shear strength could be predicted from a measurement of weld diameter. The data are as follows:

Shear strength (psi)	Weld diameter (0.0001 in.)
370	400
780	800
1210	1250
1560	1600
1980	2000
2450	2500
3070	3100
3550	3600
3940	4000
3950	4000

(a) Draw a scatter diagram.
(b) Find the least squares estimates of the regression coefficients.
(c) Test the hypothesis that the slope of the regression line is equal to 1 at the .05 level significance.
(d) Estimate the expected value of shear strength when the weld diameter is 0.2500.
(e) Find a prediction interval such that, with 95 percent confidence, the value of shear strength corresponding to a weld diameter of 0.2250 in. will be contained in it.
(f) Plot the standardized residuals.
(g) Does the plot in (f) support the assumptions of the model?

19. A screw manufacturer is interested in giving out data to his customers on the relation between nominal and actual lengths. The following results (in inches) were observed:

Nominal x		Actual y	
$\frac{1}{4}$	0.262	0.262	0.245
$\frac{1}{2}$	0.496	0.512	0.490
$\frac{3}{4}$	0.743	0.744	0.751
1	0.976	1.010	1.004
$1\frac{1}{4}$	1.265	1.254	1.252
$1\frac{1}{2}$	1.498	1.518	1.504
$1\frac{3}{4}$	1.738	1.759	1.750
2	2.005	1.992	1.992

(a) Estimate the regression coefficients.
(b) Estimate the variance involved in manufacturing a screw.
(c) For a large set of nominal one-inch screws, find a 90 percent confidence interval for the average length.
(d) For a nominal one-inch screw, find a 90 percent prediction interval for its actual length.
(e) Plot the standardized residuals.
(f) Do the residuals in (e) indicate any flaw in the regression model?
(g) Determine the index of fit.

20. Glass plays a key role in criminal investigations, because criminal activity often results in the breakage of windows and other glass objects. Since glass fragments often lodge in the clothing of the criminal, it is of great importance to be able to identify such fragments as originating at the scene of the crime. Two physical properties of glass that are useful for identification purposes are its refractive index, which is relatively easy to measure, and its density, which is much more difficult to measure. The exact measurement of density is, however, greatly facilitated if one has a good estimate of this value before setting up the laboratory experiment needed to determine it exactly. Thus, it would be quite useful if one could use the refractive index of a glass fragment to estimate its density.

The following data relate the refractive index to the density for 18 pieces of glass.

Refractive index	Density	Refractive index	Density
1.5139	2.4801	1.5161	2.4843
1.5153	2.4819	1.5165	2.4858
1.5155	2.4791	1.5178	2.4950
1.5155	2.4796	1.5181	2.4922
1.5156	2.4773	1.5191	2.5035
1.5157	2.4811	1.5227	2.5086
1.5158	2.4765	1.5227	2.5117
1.5159	2.4781	1.5232	2.5146
1.5160	2.4909	1.5253	2.5187

(a) Predict the density of a piece of glass with a refractive index 1.52.

(b) Determine an interval that, with 95 percent confidence, will contain the density of the glass in (a).

21. The regression model

$$Y = \beta x + e, \qquad e \sim N(0, \sigma^2)$$

is called regression through the origin since it presupposes that the expected response corresponding to the input level $x = 0$ is equal to 0. Suppose that (x_i, Y_i), $i = 1, \ldots, n$ is a data set from this model.

(a) Determine the least squares estimator B of β.

(b) What is the distribution of B?

(c) Define SS_R and give its distribution.

(d) Derive a test of $H_0 : \beta = \beta_0$ versus $H_1 : \beta \neq \beta_0$.

(e) Determine a $100(1 - a)$ percent prediction interval for $Y(x_0)$, the response at input level x_0.

22. Prove the identity

$$R^2 = \frac{S_{xY}^2}{S_{xx}S_{YY}}$$

23. The weight and systolic blood pressure of randomly selected males in age-group 25 to 30 are shown in the following table.

Subject	Weight	Systolic BP	Subject	Weight	Systolic BP
1	165	130	11	172	153
2	167	133	12	159	128
3	180	150	13	168	132
4	155	128	14	174	149
5	212	151	15	183	158
6	175	146	16	215	150
7	190	150	17	195	163
8	210	140	18	180	156
9	200	148	19	143	124
10	149	125	20	240	170

(a) Estimate the regression coefficients.

(b) Do the data support the claim that systolic blood pressure does not depend on an individual's weight?

(c) If a large number of males weighing 182 pounds have their blood pressures taken, determine an interval that, with 95 percent confidence, will contain their average blood pressure.

(d) Analyze the standardized residuals.

(e) Determine the index of fit.

24. It has been determined that the relation between stress (S) and the number of cycles to failure (N) for a particular type alloy is given by

$$S = \frac{A}{N^m}$$

where A and m are unknown constants. An experiment is run yielding the following data:

Stress (thousand psi)	N (million cycles to failure)
55.0	0.223
50.5	0.925
43.5	6.75
42.5	18.1
42.0	29.1
41.0	50.5
35.7	126
34.5	215
33.0	445
32.0	420

Estimate A and m.

25. In 1957 the Dutch Industrial Engineer J. R. DeJong proposed the following model for the time it takes to perform a simple manual task as a function of the number of times the task has been practiced:

$$T \approx ts^{-n}$$

where T is the time, n is the number of times the task has been practiced, and t and s are parameters depending on the task and individual. Estimate t and s for the following data set:

T	22.4	21.3	19.7	15.6	15.2	13.9	13.7
n	0	1	2	3	4	5	6

26. The chlorine residual in a swimming pool at various times after being cleaned is as given:

Time (in hr)	Chlorine residual (pt/million)
2	1.8
4	1.5
6	1.45
8	1.42
10	1.38
12	1.36

Fit a curve of the form

$$Y \approx ae^{-bx}$$

What would you predict for the chlorine residue 15 hours after a cleaning?

27. The proportion of a given heat rise that has dissipated a time t after the source is cut off is of the form

$$P = 1 - e^{-\alpha t}$$

for some unknown constant α. Given the data

P	0.07	0.21	0.32	0.38	0.40	0.45	0.51
t	0.1	0.2	0.3	0.4	0.5	0.6	0.7

estimate the value of α. Estimate the value of t at which half of the heat rise is dissipated.

28. The following data represent the bacterial count of 5 individuals at different times after being inoculated by a vaccine consisting of the bacteria.

Days since inoculation	Bacterial count
3	121,000
6	134,000
7	147,000
8	210,000
9	330,000

(a) Fit a curve.
(b) Estimate the bacteria count of a new patient after 8 days.

29. The following data yield the amount of hydrogen present (in parts per million) in core drillings of fixed size at the following distances (in feet) from the base of a vacuum-cast ingot.

Distance	1	2	3	4	5	6	7	8	9	10
Amount	1.28	1.50	1.12	0.94	0.82	0.75	0.60	0.72	0.95	1.20

(a) Draw a scatter diagram.
(b) Fit a curve of the form

$$Y = \alpha + \beta x + \gamma x^2 + e$$

to the data.

30. A new drug was tested on mice to determine its effectiveness in reducing cancerous tumors. Tests were run on 10 mice, each having a tumor of size 4 grams, by varying the amount of the drug used and then determining the resulting reduction in the weight of the tumor. The data were as follows:

Coded amount of drug	Tumor weight reduction
1	.50
2	.90
3	1.20
4	1.35
5	1.50
6	1.60
7	1.53
8	1.38
9	1.21
10	.65

Estimate the maximum expected tumor reduction and the amount of the drug that attains it by fitting a quadratic regression equation of the form

$$Y = \beta_0 + \beta_1 x + \beta_2 x^2 + e$$

31. The following data represent the relation between the number of cans damaged in a boxcar shipment of cans and the speed of the boxcar at impact.

Speed	Number of cans damaged
3	54
3	62
3	65
5	94
5	122
5	84
6	142
7	139
7	184
8	254

(a) Analyze as a simple linear regression model.
(b) Plot the standardized residuals.
(c) Do the results of (b) indicate any flaw in the model?
(d) If the answer to part (c) is yes, suggest a better model and estimate all resulting parameters.

32. Redo Problem 5 under the assumption that the variance of the gain in reading speed in proportional to the number of weeks in the program.

33. The following data were generated from the model

$$Y = 20 + 4x + e$$

where e is normal with mean variance and variance $15/(5 + x)$:

x_i	y_i
1	23.9
2	27.9
3	31.0
4	36.8
5	41.8
6	43.6
7	48.0
8	49.9
9	56.0
10	59.7

(a) Plot the data.
(b) Fit the data by a straight line using the ordinary least squares approach.

(c) Fit the data by using a weighted least squares approach.

(d) Plot the lines obtained in (c) and (d) along with the data.

34. The following data set refers to Example 7.8c.

Number of cars (daily)	Number of accidents (monthly)
2000	15
2300	27
2500	20
2600	21
2800	31
3000	16
3100	22
3400	23
3700	40
3800	39
4000	27
4600	43
4800	53

(a) Estimate the number of accidents in a month when the number of cars using the highway is 3500.

(b) Use the model

$$\sqrt{Y} = \alpha + \beta x + e$$

and redo part (a).

35. The peak discharge of a river is an important parameter for many engineering design problems. Estimates of this parameter can be obtained by relating it to the watershed area (x_1) and watershed slope (x_2). Estimate the relationship based on the following data:

x_1 (mi²)	x_2 (ft/ft)	Peak discharge (ft³/sec)
36	0.005	50
37	0.040	40
45	0.004	45
87	0.002	110
450	0.004	490
550	0.001	400
1200	0.002	650
4000	0.0005	1550

36. The sediment load in a stream is related to the size of the contributing drainage area (x_1) and the average stream discharge (x_2). Estimate this

relationship using the following data:

Area $(\times 10^3 \ mi^2)$	Discharge (ft^3/sec)	Sediment yield (millions of tons/yr)
8	65	1.8
19	625	6.4
31	1,450	3.3
16	2,400	1.4
41	6,700	10.8
24	8,500	15.0
3	1,550	1.7
3	3,500	0.8
3	4,300	0.4
7	12,100	1.6

37. Fit a multiple linear regression equation to the following data set:

x_1	x_2	x_3	x_4	y
1	11	16	4	275
2	10	9	3	183
3	9	4	2	140
4	8	1	1	82
5	7	2	1	97
6	6	1	−1	122
7	5	4	−2	146
8	4	9	−3	246
9	3	16	−4	359
10	2	25	−5	482

38. The following data refer to Stanford heart transplants. It relates the survival time of patients that have received heart transplants to their age when the transplant occurred and to a so-called mismatch score that is supposed to be an indicator of how well the transplanted heart should fit the recipient.

Survival time (in days)	Mismatch score	Age
624	1.32	51.0
46	0.61	42.5
64	1.89	54.6
1350	0.87	54.1
280	1.12	49.5
10	2.76	55.3
1024	1.13	43.4
39	1.38	42.8
730	0.96	58.4
136	1.62	52.0
836	1.58	45.0
60	0.69	64.5

(a) Letting the dependent variable be the logarithm of the survival time, fit a regression on the independent variables mismatch score and age.

(b) Estimate the variance of the error term.

39. (a) Fit a multiple linear regression equation to the following data set:

x_1	x_2	x_3	y
7.1	.68	4	41.53
9.9	.64	1	63.75
3.6	.58	1	16.38
9.3	.21	3	45.54
2.3	.89	5	15.52
4.6	.00	8	28.55
.2	.37	5	5.65
5.4	.11	3	25.02
8.2	.87	4	52.49
7.1	.00	6	38.05
4.7	.76	0	30.76
5.4	.87	8	39.69
1.7	.52	1	17.59
1.9	.31	3	13.22
9.2	.19	5	50.98

(b) Test the hypothesis that $\beta_0 = 0$.
(c) Test the hypothesis that $\beta_3 = 0$.
(d) Test the hypothesis that the mean response at the input levels $x_1 = x_2 = x_3 = 1$ is 8.5.

40. The tensile strength of a certain synthetic fiber is thought to be related to x_1, the percentage of cotton in the fiber, and x_2, the drying time of the fiber. A test of 10 pieces of fiber produced under different conditions yielded the following results:

Y = tensile strength	x_1 = percentage of cotton	x_2 = drying time
213	13	2.1
220	15	2.3
216	14	2.2
225	18	2.5
235	19	3.2
218	20	2.4
239	22	3.4
243	17	4.1
233	16	4.0
240	18	4.3

(a) Fit a multiple regression equation.
(b) Determine a 90 percent confidence interval for the mean tensile strength of a synthetic fiber having 21 percent cotton whose drying time is 3.6.

41. The time to failure of a machine component is related to the operating voltage (x_1), the motor speed in revolutions per minute (x_2), and the operating temperature (x_2). A designed experiment is run in the research and development laboratory, and the following data are obtained:

y	x_1	x_2	x_3
2145	110	750	140
2155	110	850	180
2220	110	1000	140
2225	110	1100	180
2260	120	750	140
2266	120	850	180
2334	120	1000	140
2340	130	1000	180
2212	115	840	150
2180	115	880	150

where y is the time to failure in minutes.

(a) Fit a multiple regression model to these data.
(b) Estimate the error variance.
(c) Determine a 95 percent confidence interval for the mean time to failure when the operating voltage is 125, the motor speed is 900, and the operating temperature is 160.

42. Explain why, for the same data, a prediction interval for a future response always contains the corresponding confidence interval for the mean response.

43. Consider the following data set:

x_1	x_2	y
5.1	2	55.42
5.4	8	100.21
5.9	-2	27.07
6.6	12	169.95
7.5	-6	-17.93
8.6	16	197.77
9.9	-10	-25.66
11.4	20	264.18
13.1	-14	-53.88
15	24	317.84
17.1	-18	-72.53
19.4	28	385.53

(a) Fit a linear relationship between y and the x_i.
(b) Determine the variance of the error term.
(c) Determine an interval that, with 95 percent confidence, will contain the response when the inputs are $x_1 = 10.2$ and $x_2 = 17$.

44. The cost of producing power per kilowatt hour is a function of the load factor and the cost of coal in cents per million Btu. The following data

was obtained from 12 mills

Load factor (in percent)	Cost of coal	Power cost
84	14	4.1
81	16	4.4
73	22	5.6
74	24	5.1
67	20	5.0
87	29	5.3
77	26	5.4
76	15	4.8
69	29	6.1
82	24	5.5
90	25	4.7
88	13	3.9

(a) Estimate the relationship.
(b) Test the hypothesis that the coefficient of the load factor is equal to 0.
(c) Determine a 95 percent prediction interval for the power cost when the load factor is 85 and the coal cost is 20.

45. The following data relate the systolic blood pressure to the age (x_1) and weight (x_2) of a set of individuals of similar body type and life style.

Age	Weight	Blood pressure
25	162	112
25	184	144
42	166	138
55	150	145
30	192	152
40	155	110
66	184	118
60	202	160
38	174	108

(a) Test the hypothesis that, when an individual's weight is known, age gives no additional information in predicting blood pressure.
(b) Determine an interval that, with 95 percent confidence, will contain the average blood pressure of all individuals of the preceding type that are 45 years old and weigh 180 pounds.
(c) Determine an interval that, with 95 percent confidence, will contain the blood pressure of a given individual of the preceding type who is 45 years old and weighs 180 pounds.

C H A P T E R   8

Analysis of Variance

1 INTRODUCTION

A large company is considering purchasing, in quantity, one of four different computer packages designed to teach a new programming language. Some influential people within this company have claimed that these packages are basically interchangeable in that the one chosen will have little effect on the final competence of its user. To test this hypothesis the company has decided to choose 160 of its engineers, and divide them into 4 groups of size 40. Each member in group i will then be given teaching package i, $i = 1, 2, 3, 4$, to learn the new language. When all the engineers complete their study, a comprehensive exam will be given. The company then wants to use the results of this examination to determine whether the computer teaching packages are really interchangeable or not. How can they do this?

Before answering this question, let us note that we clearly desire to be able to conclude that the teaching packages are indeed interchangeable when the average test scores in all the groups are similar and to conclude that the packages are essentially different when there is a large variation among these average test scores. However, to be able to reach such a conclusion, we should note that the method of division of the 160 engineers into 4 groups is of vital importance. For suppose that the members of the first group score significantly higher than those of the other groups. What can we conclude from this? Specifically, is this result due to teaching package 1 being a superior teaching package, or is it due to the fact that the engineers in group 1 are just better learners? To be able to conclude the former, it is essential that we divide the 160 engineers into the 4 groups in such a way so as to make it extremely unlikely that one of these groups is inherently superior. The time-tested method for doing this is to divide the engineers into 4 groups in a completely random fashion. That is, we should do it in such a way so that all possible divisions are equally likely; for in this case, it would be very unlikely that any one group would be significantly superior to any other group. So let us suppose that the division of the engineers was indeed done "at random." (Whereas it is not at all obvious how this can be accomplished, one efficient procedure is to start by arbitrarily numbering the 160 engineers. Then generate a random permutation of the integers $1, 2, \ldots, 160$—see Example 12.1a of Chapter 12, for the details of this generation—and put the engineers whose numbers are among the first 40 of the permutation into group 1, those whose numbers are among the 41st through the 80th of the permutation into group 2, and so on.)

It is now probably reasonable to suppose that the test score of a given individual should be approximately a normal random variable having parameters that depend on the package from which he was taught. Also, it is probably reasonable to suppose that whereas the average test score of an engineer will depend on the teaching package he was exposed to, the variability in the test score will result from the inherent variation of 160 different people and not from the particular package used. Thus, if we let X_{ij}, $i = 1, \ldots, 4$, $j = 1, \ldots, 40$, denote the test score of the jth engineer in group i, a reasonable model might be to suppose that the X_{ij} are independent random variables with X_{ij} having a normal distribution with unknown mean μ_i and unknown variance σ^2. The hypothesis that the teaching packages are interchangeable is then equivalent to the hypothesis that $\mu_1 = \mu_2 = \mu_3 = \mu_4$.

We now show how such an hypothesis can be tested.

2   ONE-WAY ANALYSIS OF VARIANCE

Consider m independent samples, each of size n, where the members of the ith sample—$X_{i1}, X_{i2}, \ldots, X_{in}$—are normal random variables with unknown mean μ_i and unknown variance σ^2. That is,

$$X_{ij} \sim N(\mu_i, \sigma^2), \qquad i = 1, \ldots, m, \qquad j = 1, \ldots, n$$

Now suppose that we are interested in testing the hypothesis H_0 that all the means are equal—that is

$$H_0 : \mu_1 = \mu_2 = \cdots = \mu_m$$

One way of thinking about this is to imagine that we have m different treatments where the result of applying treatment i on an item is a normal random variable with mean μ_i and variance σ^2. We are then interested in testing the hypothesis that all treatments have the same effect, by applying each treatment to a (different) sample of n items and then analyzing the result.

Our approach to testing the foregoing hypothesis will be to derive two independent estimators of σ^2—the first of which will be a valid estimator whether H_0 is true or not, whereas the second will only be a valid estimator of σ^2 when H_0 is true. In addition, we will show that our second estimator will tend to overestimate σ^2 when H_0 is not true. We will then compare these two estimates and reject the hypothesis H_0 when the second is significantly larger than the first.

To obtain our initial estimate of σ^2, let

$$\overline{X}_{i\cdot} = \sum_{j=1}^{n} \frac{X_{ij}}{n}, \qquad i = 1, \ldots, m$$

denote the average of the ith sample, and let

$$S_i^2 = \sum_{j=1}^{n} \frac{\left(X_{ij} - \overline{X}_{i\cdot}\right)^2}{n-1}, \qquad i = 1, \ldots, m$$

denote the sample variance of the ith sample. Now by the fundamental result concerning the joint distribution of the sample mean and sample variance of a normal population (see Section 5 of Chapter 4), it follows that $\overline{X}_{i\cdot}$ and S_i^2 are independent and

$$(n-1)\frac{S_i^2}{\sigma^2} \sim \chi_{n-1}^2, \qquad i = 1, \ldots, m$$

Also, as the sum of independent chi-square random variables remains a chi-square random variable (with a degree of freedom equal to the sum of the respective degrees of freedom) it follows that

$$\frac{n-1}{\sigma^2} \sum_{i=1}^{m} S_i^2 \sim \chi_{m(n-1)}^2 \qquad (8.2.1)$$

Hence,

$$E\left[\frac{n-1}{\sigma^2} \sum_{i=1}^{m} S_i^2\right] = E\left[\chi_{m(n-1)}^2\right]$$

$$= m(n-1)$$

or

$$E\left[\sum_{i=1}^{m} \frac{S_i^2}{m}\right] = \sigma^2$$

We thus take $\sum_{i=1}^{m} S_i^2/m$ as our initial estimator of σ^2—also, it should be noted that it will be an unbiased estimator of σ^2 even when H_0 is not true. Let

$$SS_w = (n-1) \sum_{i=1}^{m} S_i^2 = \sum_{i=1}^{m} \sum_{j=1}^{n} \left(X_{ij} - X_{i\cdot}\right)^2$$

and call SS_w the within sample sum of squares. Note that from Equation 8.2.1

$$\frac{SS_w}{\sigma^2} \sim \chi_{m(n-1)}^2 \qquad (8.2.2)$$

To obtain our second estimator of σ^2 (which will be a valid estimator of σ^2 only when H_0 is true and which will tend to overestimate σ^2 when H_0 is not true), we will consider the variation between samples. To begin, note that the sample means $\overline{X}_{i\cdot}$, $i = 1, \ldots, m$ are independent normals with mean μ_i and variance σ^2/n. That is,

$$\overline{X}_{i\cdot} = \sum_{j=1}^{n} \frac{X_{ij}}{n} \sim N\left(\mu_i, \sigma^2/n\right)$$

Hence, if H_0 is true, then the sequence of sample averages $\overline{X}_{1\cdot}, \overline{X}_{2\cdot}, \ldots, \overline{X}_{m\cdot}$ constitute a sample from a normal population with mean μ and variance σ^2/n. The sample variance for this sample is

$$\sum_{i=1}^{m} \frac{\left(\overline{X}_{i\cdot} - \overline{X}_{\cdot\cdot}\right)^2}{m-1}$$

where

$$\overline{X}_{\cdot\cdot} = \frac{\overline{X}_{1\cdot} + \cdots + \overline{X}_{m\cdot}}{m} = \frac{\sum_{i=1}^{m}\sum_{j=1}^{n} X_{ij}}{nm}$$

is the average of all the nm data points. Again, from the fundamental result concerning normal samples, it follows that *when H_0 is true*

$$\sum_{i=1}^{m} \frac{\left(\overline{X}_{i\cdot} - \overline{X}_{\cdot\cdot}\right)^2}{\sigma^2/n} \sim \chi_{m-1}^2 \qquad (8.2.3)$$

Also, since S_i^2 is independent of $\overline{X}_{i\cdot}$, it follows that the preceding statistic will be independent of $SS_w = (n-1)\sum_{i=1}^{m} S_i^2$. Let

$$SS_b = n\sum_{i=1}^{m}\left(\overline{X}_{i\cdot} - \overline{X}_{\cdot\cdot}\right)^2 = \sum_{j=1}^{n}\sum_{i=1}^{m}\left(\overline{X}_{i\cdot} - \overline{X}_{\cdot\cdot}\right)^2$$

and call SS_b the between sample sum of squares. Hence, from Equation 8.2.3, we see that *when H_0 is true*

$$\frac{SS_b}{\sigma^2} \sim \chi_{m-1}^2 \qquad (8.2.4)$$

and so

$$E\left[\frac{SS_b}{\sigma^2}\right] = E\left[\chi_{m-1}^2\right] = m-1$$

or

$$E\left[\frac{SS_b}{m-1}\right] = \sigma^2$$

We will now show that when H_0 is not true, the average value of $SS_b/(m-1)$ exceeds σ^2.

Proposition 8.2.1

$$E\left[\frac{SS_b}{m-1}\right] = \sigma^2 + \frac{n}{m-1}\sum_{i=1}^{m}(\mu_i - \bar{\mu})^2$$

where $\bar{\mu} = \sum_{i=1}^{m}\mu_i/m$ is the average of the μ_i's.

Proof Using the identity

$$\sum_{i=1}^{m} (y_i - \bar{y})^2 = \sum_{i=1}^{m} y_i^2 - m\bar{y}^2 \tag{8.2.5}$$

where $\bar{y} = \sum_{i=1}^{m} y_i/m$, we obtain that

$$\frac{SS_b}{n} = \sum_{i=1}^{m} (\bar{X}_{i.} - \bar{X}_{..})^2 \tag{8.2.6}$$

$$= \sum_{i=1}^{m} \bar{X}_{i.}^2 - m\bar{X}_{..}^2$$

with the last equality following from Equation 8.2.5 since $\bar{X}_{..} = \sum_{i=1}^{m} \bar{X}_{i.}/m$. Now as $\bar{X}_{i.}$ is the sample mean of a sample of size n from a normal population having mean μ_i and variance σ^2, it follows that

$$E[\bar{X}_{i.}] = \mu_i, \qquad \text{Var}(\bar{X}_{i.}) = \frac{\sigma^2}{n}$$

and so

$$E[\bar{X}_{i.}^2] = E^2[\bar{X}_{i.}] + \text{Var}(\bar{X}_{i.}) = \mu_i^2 + \frac{\sigma^2}{n} \tag{8.2.7}$$

Also, because

$$\bar{X}_{..} = \sum_{i=1}^{m} \frac{X_{i.}}{m}$$

it follows that

$$E[\bar{X}_{..}] = \sum_{i=1}^{m} \frac{E[X_{i.}]}{m} = \frac{\sum_{i=1}^{m} \mu_i}{m} = \bar{\mu}$$

and

$$\text{Var}(\bar{X}_{..}) = \frac{\sum_{i=1}^{m} \text{Var}(\bar{X}_{i.})}{m^2} \qquad \text{by the independence of } \bar{X}_{1.}, \bar{X}_{2.}, \ldots, \bar{X}_{m.}.$$

$$= \frac{m\sigma^2}{nm^2} = \frac{\sigma^2}{nm}$$

Hence,

$$E[\bar{X}_{..}^2] = \bar{\mu}^2 + \frac{\sigma^2}{nm} \tag{8.2.8}$$

Hence, from Equations 8.2.6, 8.2.7, and 8.2.8, we see that

$$E\left[\frac{SS_b}{n}\right] = \sum_{i=1}^{m} \mu_i^2 + \frac{m\sigma^2}{n} - m\bar{\mu}^2 - \frac{\sigma^2}{n}$$

$$= (m-1)\frac{\sigma^2}{n} + \sum_{i=1}^{m} \mu_i^2 - m\bar{\mu}^2$$

$$= (m-1)\frac{\sigma^2}{n} + \sum_{i=1}^{m} (\mu_i - \bar{\mu})^2 \qquad ∎$$

Thus, we have shown that SS_w and SS_b are independent with

$$\frac{SS_w}{\sigma^2} \sim \chi^2_{m(n-1)}$$

and, *when H_0 is true*

$$\frac{SS_b}{\sigma^2} \sim \chi^2_{m-1}$$

Moreover, from Proposition 8.2.1, we see that when H_0 is not true, SS_b tends to be larger than it is when H_0 is true.

Now let us recall that if χ^2_k and χ^2_l are independent chi-square random variables with k and l degrees of freedom respectively, then

$$\frac{\chi^2_k/k}{\chi^2_l/l} \sim F_{k,l}$$

Hence, when H_0 is true, we see that

$$\frac{\dfrac{SS_b}{\sigma^2}\bigg/(m-1)}{\dfrac{SS_w}{\sigma^2}\bigg/m(n-1)} \sim F_{m-1,\,m(n-1)}$$

or

$$\frac{\dfrac{SS_b}{(m-1)}}{\dfrac{SS_w}{m(n-1)}} \sim F_{m-1,\,m(n-1)}$$

Thus, as SS_b tends to be larger when H_0 is not true, a natural test of the hypothesis that $\mu_1 = \mu_2 = \cdots = \mu_m$ at the α level of significance is to reject the hypothesis if

$$\frac{\dfrac{SS_b}{(m-1)}}{\dfrac{SS_w}{m(n-1)}} > F_{\alpha,\,m-1,\,m(n-1)}$$

where $F_{\alpha,\,m-1,\,m(n-1)}$ is, we recall, the value such that

$$P\left\{ F_{m-1,\,m(n-1)} > F_{\alpha,\,m-1,\,m(n-1)} \right\} = \alpha$$

when $F_{m-1,\,m(n-1)}$ has an F distribution with parameters $m-1$ and $m(n-1)$.

The above is summarized in the ANOVA (analysis of variance) table at the top of page 312.

A useful identity can be obtained by writing each value X_{ij} as follows:

$$X_{ij} = \underbrace{\overline{X}_{..}}_{\substack{\text{grand}\\\text{mean}}} + \underbrace{\overline{X}_{i.} - \overline{X}_{..}}_{\substack{\text{deviation due}\\\text{to row}}} + \underbrace{X_{ij} - \overline{X}_{i.}}_{\substack{\text{error or}\\\text{residual term}}}$$

TABLE 8.2.1

One-way ANOVA Table

Source of variation	Sum of squares	Degrees of freedom	Value of F-statistic
Between treatments	$SS_b = n \sum_{i=1}^{m} (\overline{X}_{i.} - \overline{X}_{..})^2$	$m - 1$	
			$F_{m-1,\, m(n-1)} = \dfrac{SS_b/(m-1)}{SS_w/m(n-1)}$
Within treatments	$SS_w = \sum_{i=1}^{m} \sum_{j=1}^{n} (X_{ij} - \overline{X}_{i.})^2$	$m(n-1)$	

Hence

$$\sum_{i=1}^{m} \sum_{j=1}^{n} X_{ij}^2 = \sum_{i=1}^{m} \sum_{j=1}^{n} \left[\overline{X}_{..} + (\overline{X}_{i.} - \overline{X}_{..}) + (X_{ij} - \overline{X}_{i.}) \right]^2 \quad (8.2.9)$$

On expanding the right side of Equation 8.2.9, all of the cross-product terms will equal 0 since

$$\sum_i \sum_j \overline{X}_{..}(\overline{X}_{i.} - \overline{X}_{..}) = \sum_j \overline{X}_{..} \sum_i (\overline{X}_{i.} - \overline{X}_{..})$$

$$= 0 \quad \text{since } \sum_i (\overline{X}_{i.} - \overline{X}_{..}) = 0$$

$$\sum_i \sum_j \overline{X}_{..}(X_{ij} - \overline{X}_{i.}) = \sum_i \overline{X}_{..} \sum_j (X_{ij} - \overline{X}_{i.})$$

$$= 0 \quad \text{since } \sum_j (X_{ij} - \overline{X}_{i.}) = 0$$

$$\sum_i \sum_j (\overline{X}_{i.} - \overline{X}_{..})(X_{ij} - \overline{X}_{i.}) = \sum_i (\overline{X}_{i.} - \overline{X}_{..}) \sum_j (X_{ij} - \overline{X}_{i.})$$

$$= 0 \quad \text{since } \sum_j (X_{ij} - \overline{X}_{i.}) = 0$$

Therefore, on expanding Equation 8.2.9 we obtain the following.

The Sum of Squares Identity

$$\sum_{i=1}^{m} \sum_{j=1}^{n} X_{ij}^2 = nm\overline{X}_{..}^2 + n \sum_{i=1}^{m} (\overline{X}_{i.} - \overline{X}_{..})^2 + \sum_{i=1}^{m} \sum_{j=1}^{n} (X_{ij} - \overline{X}_{i.})^2 \quad (8.2.10)$$

REMARK

The sum of squares identity is often written as

$$\sum_{i=1}^{m} \sum_{j=1}^{n} (X_{ij} - \overline{X}_{..})^2 = n \sum_{i=1}^{m} (\overline{X}_{i.} - \overline{X}_{..})^2 + \sum_{i=1}^{m} \sum_{j=1}^{n} (X_{ij} - \overline{X}_{i.})^2 \quad (8.2.11)$$

Why is this equivalent to Equation 8.2.10?

Computing in One-way Analysis of Variance The computationally most convenient way to compute SS_b and SS_w when computing by hand is as follows:

$$SS_b = n \sum_{i=1}^{m} \left(\overline{X}_{i\cdot} - \overline{X}_{\cdot\cdot} \right)^2 \qquad (8.2.12)$$

$$= n \sum_{i=1}^{m} \overline{X}_{i\cdot}^2 - nm\overline{X}_{\cdot\cdot}^2$$

where the equality follows from the identity

$$\sum_{i=1}^{k} \left(y_i - \bar{y} \right)^2 = \sum y_i^2 - k\bar{y}^2, \qquad \text{where } \bar{y} = \sum_{1}^{k} y_i / k \qquad (8.2.13)$$

SS_b should be computed using Equation 8.2.12 and then SS_w should be obtained from the sum of squares identity as follows:

$$SS_w \equiv \sum_{i=1}^{m} \sum_{j=1}^{n} \left(X_{ij} - \overline{X}_{i\cdot} \right)^2 \qquad (8.2.14)$$

$$= \sum_{i=1}^{m} \sum_{j=1}^{n} X_{ij}^2 - nm\overline{X}_{\cdot\cdot}^2 - SS_b$$

The formulas for SS_b and SS_w given by Equations 8.2.12 and 8.2.14 are obtained by using the identity in Equation 8.2.13. However, as was noted in Section 3.1 of Chapter 4 although Equation 8.2.13 is a time-saving identity when computing by hand it can, due to computer round-off error, lead to incorrect answers in large problems when one is solving by machine. Program 8-2 will compute the quantities SS_b and SS_w without making use of the sum of squares identity. Basically, it will repeat the recursion approach developed in Section 3.1 of Chapter 4 to compute the sample mean and sample variance. Once the program computes the value of the F-statistic it will then, by using Program 3-8-3-A, compute the probability that an F random variable with $m - 1$ and $m(n - 1)$ degrees of freedom would exceed that value. This probability will then equal the p-value of the test data and the null hypothesis should be rejected at significance level α if α exceeds this p-value.

Example 8.2a An auto rental company is testing 3 brands of gasoline for mileage by using 18 more or less identical motors that are adjusted to run at a fixed speed. Of these 18, 6 are given gas G_i, $i = 1, 2, 3$. Each motor runs on 10 gallons of gasoline until it is out of fuel. The resulting total mileage obtained by the motors is as follows:

Gas G_1	Gas G_2	Gas G_3
220	244	254
252	236	272
238	258	232
246	242	238
260	221	256
224	230	250

Test the hypothesis that a car's mileage will not depend on which of the 3 types of gasoline is used.

Solution Running Program 8-2 yields the following output:

```
RUN
THIS PROGRAM COMPUTES THE VALUE OF THE F-STATISTIC AND ITS p-value IN A ONE WAY
ANOVA
ENTER THE NUMBER OF SAMPLES
? 3
ENTER THE SIZE OF THE SAMPLES
? 6
ENTER SAMPLE 1 ONE AT A TIME
? 220
? 252
? 238
? 246
? 260
? 224
ENTER SAMPLE 2 ONE AT A TIME
? 244
? 236
? 258
? 242
? 221
? 230
ENTER SAMPLE 3 ONE AT A TIME
? 254
? 272
? 232
? 238
? 256
? 250
SSw/(M*(N-1))= 203.3888
SSb/(M-1)= 249.0553
THE VALUE OF THE F-STATISTIC IS 1.224528
THE p-value IS .3177825
Ok
```

Thus the hypothesis that the brand of gasoline does not affect mileage is accepted at any significance level $\alpha \leq .317$. ∎

The preceding test of the null hypothesis that all sample means are equal was derived by going back to first principles—specifically by using the fact that the sample mean and sample variance from a normal population are independent with a constant multiple of the sample variance having a chi-square distribution with a degree of freedom equal to the number of terms in the sample minus 1. In fact, we could also have derived this test by utilizing the sum of squares identity. However, before presenting this we need first talk about degrees of freedom.

The number of degrees of freedom of a sum of squares is equal to the number of terms that are being squared minus the number of linear relations (constraints) between them. For example, the number of degrees of freedom associated with

$$SS_w = \sum_{i=1}^{m} \sum_{j=1}^{n} \left(X_{ij} - \overline{X}_{i \cdot} \right)^2$$

is nm, the number of terms $X_{ij} - \overline{X}_{i \cdot}$, minus the number of linear constraints between these terms. Now for each treatment (that is, sample) the sum of the deviations about the mean must equal 0—that is

$$\sum_{j=1}^{n} \left(X_{ij} - \overline{X}_{i\cdot} \right) = 0 \qquad \text{for each } i = 1, \ldots, m$$

Hence, we have imposed one linear constraint for each sample, and since there are m samples, we see that the number of degrees of freedom associated with SS_w is $nm - m = m(n-1)$.

We now state without proof the following useful result known alternatively as the Partition Theorem or as Cochran's Theorem.

The Partition Theorem

Let $Z_1, \ldots, Z_N$ be independent unit normal random variables and suppose that

$$\sum_{i=1}^{N} Z_i^2 = T_1 + \cdots + T_k$$

where T_i is a sum of squares having v_i degrees of freedom, $i = 1, \ldots, k$. Then the necessary and sufficient condition that $T_1, \ldots, T_k$ are independent chi-square random variables having respective degrees of freedom $v_1, \ldots, v_k$ is that

$$v_1 + \cdots + v_k = N \quad \blacksquare$$

Let us now show how the partition theorem can be applied to the present situation under discussion. Suppose that H_0 is true and let μ denote the common mean—that is, $\mu_i = \mu$, $i = 1, \ldots, m$. Then, under condition that H_0 is true, $Z_{ij} \equiv (X_{ij} - \mu)/\sigma$, $i = 1, \ldots, m$, $j = 1, \ldots, m$ will be independent unit normal random variables. Now the sum of squares identity, given by Equation 8.2.10, when applied to the Z_{ij} yields that

$$\sum_{i=1}^{m} \sum_{j=1}^{n} Z_{ij}^2 = nm\overline{Z}_{\cdot\cdot}^2 + n \sum_{i=1}^{m} \left(\overline{Z}_{i\cdot} - \overline{Z}_{\cdot\cdot} \right)^2 + \sum_{i=1}^{m} \sum_{j=1}^{n} \left(Z_{ij} - \overline{Z}_{i\cdot} \right)^2 \qquad (8.2.15)$$

However, since $Z_{ij} = (X_{ij} - \mu)/\sigma$ it follows that

$$\overline{Z}_{\cdot\cdot} = \frac{\overline{X}_{\cdot\cdot} - \mu}{\sigma}, \qquad \overline{Z}_{i\cdot} = \frac{\overline{X}_{i\cdot} - \mu}{\sigma}$$

and so

$$\overline{Z}_{i\cdot} - \overline{Z}_{\cdot\cdot} = \frac{\overline{X}_{i\cdot} - \overline{X}_{\cdot\cdot}}{\sigma}, \qquad Z_{ij} - \overline{Z}_{i\cdot} = \frac{X_{ij} - \overline{X}_{i\cdot}}{\sigma}$$

Thus we may rewrite Equation 8.2.15 as

$$\sum_{i=1}^{m} \sum_{j=1}^{m} Z_{ij}^2 = \frac{nm\left(\overline{X}_{\cdot\cdot} - \mu\right)^2}{\sigma^2} + \frac{n}{\sigma^2} \sum_{i=1}^{m} \left(\overline{X}_{i\cdot} - \overline{X}_{\cdot\cdot} \right)^2 + \frac{1}{\sigma^2} \sum_{i=1}^{m} \sum_{j=1}^{n} \left(X_{ij} - \overline{X}_{i\cdot} \right)^2$$

It now follows from the definition of degrees of freedom that

$$\frac{nm\left(\overline{X}.. - \mu\right)^2}{\sigma^2} \qquad \text{has 1 degree of freedom}$$

$$\frac{SS_b}{\sigma^2} = \frac{n}{\sigma^2} \sum_{i=1}^{m} \left(\overline{X}_i. - \overline{X}..\right)^2 \qquad \text{has } m-1 \text{ degrees of freedom}$$

$$\frac{SS_w}{\sigma^2} = \frac{1}{\sigma^2} \sum_{i=1}^{m} \sum_{j=1}^{n} \left(X_{ij} - \overline{X}_i.\right)^2 \qquad \text{has } nm-m \text{ degrees of freedom}$$

Since

$$nm = 1 + m - 1 + nm - m$$

we can thus conclude from the partition theorem that, when H_0 is true, SS_b/σ^2 and SS_w/σ^2 are independent chi-square random variables with $m-1$ and $m(n-1)$ degrees of freedom respectively; and thus $[SS_b/(m-1)]/SS_w/m(n-1)]$ has an F-distribution with parameters $m-1$ and $m(n-1)$.

3 ONE-WAY ANALYSIS OF VARIANCE WITH UNEQUAL SAMPLE SIZES

The model in the previous section supposed that there were an equal number of data points in each sample. Whereas this is certainly a desirable situation (see the Remark at the end of this section), it is not always possible to attain. So let us now suppose that we have m normal samples of respective sizes $n_1, n_2, \ldots, n_m$. That is, the data consist of the $\sum_{i=1}^{m} n_i$ independent random variables X_{ij}, $j = 1, \ldots, n_i$, $i = 1, \ldots, m$, where

$$X_{ij} \sim \mathcal{N}\left(\mu_i, \sigma^2\right)$$

Again we are interested in testing the hypothesis H_0 that all means are equal.

To derive a test for H_0 we start with the following sum of squares identity, which is easily verifiable:

$$\sum_{i=1}^{m} \sum_{j=1}^{n_i} X_{ij}^2 = \sum_{i=1}^{m} n_i \overline{X}..^2 + \underbrace{\sum_{i=1}^{m} n_i\left(\overline{X}_i. - \overline{X}..\right)^2}_{\substack{SS_b = \text{ between sample} \\ \text{sum of squares}}} + \underbrace{\sum_{i=1}^{m} \sum_{j=1}^{n_i} \left(X_{ij} - \overline{X}_i.\right)^2}_{\substack{SS_w = \text{ within sample} \\ \text{sum of squares}}}$$

Now if H_0 is true and μ is the common mean, then if we set

$$Z_{ij} = \frac{X_{ij} - \mu}{\sigma}$$

then the Z_{ij}, $j = 1, \ldots, n_i$, $i = 1, \ldots, m$ constitute a set of $\sum_{i=1}^{m} n_i$ independent unit normals. On applying this sum of squares identity to the Z_{ij} we obtain that

$$\sum_{i=1}^{m} \sum_{j=1}^{n_i} Z_{ij}^2 = \sum_{i=1}^{m} n_i \overline{Z}_{..}^2 + \sum_{i=1}^{m} n_i (\overline{Z}_{i.} - \overline{Z}_{..})^2 + \sum_{i=1}^{m} \sum_{j=1}^{n_i} (Z_{ij} - \overline{Z}_{i.})^2$$

$\underbrace{}_{\substack{\sum_{i=1}^{m} n_i \text{ degrees} \\ \text{of freedom}}}$ $\underbrace{}_{\substack{1 \text{ degree} \\ \text{of freedom}}}$ $\underbrace{}_{\substack{m - 1 \text{ degrees} \\ \text{of freedom}}}$ $\underbrace{}_{\substack{\sum_{i=1}^{m} n_i - m \text{ degrees} \\ \text{of freedom}}}$

Hence, from the partition theorem, we can conclude that when H_0 is true

$$\sum_{i=1}^{m} n_i (\overline{Z}_{i.} - \overline{Z}_{..})^2 = \frac{1}{\sigma^2} \sum_{i=1}^{m} n_i (\overline{X}_{i.} - \overline{X}_{..})^2 = \frac{SS_b}{\sigma^2}$$

and

$$\sum_{i=1}^{m} \sum_{j=1}^{n_i} (Z_{ij} - \overline{Z}_{i.})^2 = \frac{1}{\sigma^2} \sum_{i=1}^{m} \sum_{j=1}^{n_i} (X_{ij} - \overline{X}_{i.})^2 = \frac{SS_w}{\sigma^2}$$

are independent chi-square random variables with $m - 1$ and $\sum_{i=1}^{m} n_i - m$ degrees of freedom respectively; and thus, when H_0 is true

$$\frac{\dfrac{SS_b}{(m-1)}}{\dfrac{SS_w}{(\sum_i n_i - m)}} \sim F_{m-1, \sum_i^m n_i - m}$$

From this we see that the hypothesis that $\mu_1 = \mu_2 = \cdots = \mu_m$ should be rejected at significance level α whenever

$$\frac{\dfrac{SS_b}{(m-1)}}{\dfrac{SS_w}{(\sum_i n_i - m)}} > F_{\alpha, m-1, \sum_i^m n_i - m}$$

REMARK

When the samples are of different sizes we say that we are in the *unbalanced* case. Whenever possible it is advantageous to choose a balanced design over an unbalanced one. For one thing, the test statistic in a balanced design is relatively insensitive to slight departures from the assumption of equal population variances. (That is, the balanced design is more robust than the unbalanced one). A second advantage is that for fixed value of $\sum_{i=1}^{m} n_i$, the power of the test is maximized by the balanced design.

4   TWO-WAY ANALYSIS OF VARIANCE

Whereas the model of Sections 2 and 3 enabled us to study the effect of a single variable on an experiment, we can also study the effect of several variables on the same experiment. Let us consider an experiment whose resulting data can be expressed in matrix form

$$
\text{data} = \begin{bmatrix} X_{11} & X_{12} & \cdots & X_{1n} \\ \vdots & & & \\ X_{i1} & X_{i2} & \cdots & X_{in} \\ \vdots & & & \\ X_{m1} & X_{m2} & \cdots & X_{mn} \end{bmatrix}
$$

We shall suppose that the distribution of a given random variable X_{ij} depends on both its row and column. Specifically, we shall suppose that the X_{ij} are independent with

$$
X_{ij} \sim \mathcal{N}\left(\alpha_i + \beta_j, \sigma^2\right), \qquad i = 1, \ldots, m, \qquad j = 1, \ldots, n
$$

We will be interested in testing such null hypotheses as

$$
H_0 : \alpha_1 = \alpha_2 = \cdots = \alpha_m
$$

and

$$
H_0' = \beta_1 = \beta_2 = \cdots = \beta_n
$$

Thus, H_0 states that there is no row effect—that is, the distribution of a given random variable depends only on its column and not its row—and H_0' that there is no column effect.

Let

$$
\overline{X}_{i\cdot} = \sum_{j=1}^{n} \frac{X_{ij}}{n}
$$

$$
\overline{X}_{\cdot j} = \sum_{i=1}^{m} \frac{X_{ij}}{m}
$$

$$
\overline{X}_{\cdot\cdot} = \sum_{i=1}^{m} \frac{\overline{X}_{i\cdot}}{m} = \sum_{j=1}^{n} \frac{\overline{X}_{\cdot j}}{n} = \sum_{j=1}^{n} \sum_{i=1}^{m} \frac{X_{ij}}{nm}
$$

In words, $\overline{X}_{i\cdot}$ is the average of the data in the ith row, $\overline{X}_{\cdot j}$ is the average of the data in the jth column, and $\overline{X}_{\cdot\cdot}$ is the average of all data values.

We will find it useful to break up each data value X_{ij} as follows:

$$
X_{ij} = \underbrace{\overline{X}_{\cdot\cdot}}_{\substack{\text{grand} \\ \text{mean}}} + \underbrace{\overline{X}_{i\cdot} - \overline{X}_{\cdot\cdot}}_{\substack{\text{deviation due} \\ \text{to row effect}}} + \underbrace{\overline{X}_{\cdot j} - \overline{X}_{\cdot\cdot}}_{\substack{\text{deviation due} \\ \text{to column effect}}} + \underbrace{X_{ij} - \overline{X}_{i\cdot} - \overline{X}_{\cdot j} + \overline{X}_{\cdot\cdot}}_{\substack{\text{residual or} \\ \text{error term}}}
$$

Using the preceding representation, we can write

$$
\sum_{i=1}^{m} \sum_{j=1}^{n} X_{ij}^2 = \sum_{i=1}^{m} \sum_{j=1}^{n} \left[\overline{X}_{\cdot\cdot} + \left(\overline{X}_{i\cdot} - \overline{X}_{\cdot\cdot} \right) + \left(\overline{X}_{\cdot j} - \overline{X}_{\cdot\cdot} \right) \right.
$$

$$
\left. + \left(X_{ij} - \overline{X}_{i\cdot} - \overline{X}_{\cdot j} + \overline{X}_{\cdot\cdot} \right) \right]^2
$$

On expanding the right side of this equation, it can be shown that all the sums corresponding to cross-product terms will equal 0, and thus we obtain the following sum of squares identity.

The Sum of Squares Identity

$$\sum_{i=1}^{m}\sum_{j=1}^{n}X_{ij}^2 = \sum_{i=1}^{m}\sum_{j=1}^{n}\overline{X}_{..}^2 + \sum_{i=1}^{m}\sum_{j=1}^{n}\left(\overline{X}_{i\cdot} - \overline{X}_{..}\right)^2 \tag{8.4.1}$$

$$+ \sum_{i=1}^{m}\sum_{j=1}^{n}\left(\overline{X}_{\cdot j} - \overline{X}_{..}\right)^2 + \sum_{i=1}^{m}\sum_{j=1}^{n}\left(X_{ij} - \overline{X}_{i\cdot} - \overline{X}_{\cdot j} + \overline{X}_{..}\right)^2$$

To develop tests for H_0 or H_0', we will make use of the preceding sum of squares identity in conjunction with the partition theorem. To start, suppose that H_0 is true and so $\alpha_1 = \alpha_2 = \cdots = \alpha_m$ and let α equal this common value. Also, let

$$Z_{ij} = \frac{X_{ij} - \alpha - \beta_j}{\sigma} \tag{8.4.2}$$

and note that, under H_0, the Z_{ij} are independent unit normal random variables.

From Equation 8.4.2 we see that

$$\overline{Z}_{i\cdot} = \frac{\sum_{j=1}^{n}Z_{ij}}{n} = \frac{\overline{X}_{i\cdot} - \alpha - \beta_\cdot}{\sigma} \tag{8.4.3}$$

$$\overline{Z}_{\cdot j} = \frac{\sum_{i=1}^{m}Z_{ij}}{m} = \frac{\overline{X}_{\cdot j} - \alpha - \beta_j}{\sigma}$$

$$\overline{Z}_{..} = \frac{\sum_{j=1}^{n}\sum_{i=1}^{m}Z_{ij}}{nm} = \frac{\overline{X}_{..} - \alpha - \beta_\cdot}{\sigma}$$

where

$$\beta_\cdot = \sum_{j=1}^{n}\frac{\beta_j}{n}$$

The foregoing implies that

$$\overline{Z}_{i\cdot} - \overline{Z}_{..} = \frac{\overline{X}_{i\cdot} - \overline{X}_{..}}{\sigma}$$

$$\overline{Z}_{\cdot j} - \overline{Z}_{..} = \frac{\overline{X}_{\cdot j} - \beta_j - \overline{X}_{..} + \beta_\cdot}{\sigma}$$

$$Z_{ij} - \overline{Z}_{i\cdot} - \overline{Z}_{\cdot j} + \overline{Z}_{..} = \frac{X_{ij} - \overline{X}_{i\cdot} - \overline{X}_{\cdot j} + \overline{X}_{..}}{\sigma}$$

Hence, using the foregoing it follows, from the sum of squares identity Equation 8.4.1 when applied to the Z_{ij}, that

$$\sum_{i=1}^{m}\sum_{j=1}^{n}Z_{ij}^2 = \sum_{i=1}^{m}\sum_{j=1}^{n}\frac{\left(\overline{X}_{..} - \alpha - \beta_\cdot\right)^2}{\sigma^2} + \sum_{i=1}^{m}\sum_{j=1}^{n}\frac{\left(\overline{X}_{i\cdot} - \overline{X}_{..}\right)^2}{\sigma^2}$$

$$+ \sum_{i=1}^{m}\sum_{j=1}^{n}\frac{\left(\overline{X}_{\cdot j} - \beta_j - \overline{X}_{..} + \beta_\cdot\right)^2}{\sigma^2}$$

$$+ \sum_{i=1}^{m}\sum_{j=1}^{n}\frac{\left(X_{ij} - \overline{X}_{i\cdot} - \overline{X}_{\cdot j} + \overline{X}_{..}\right)^2}{\sigma^2}$$

Now it can be shown that

$$nm\frac{\left(\overline{X}.. - \alpha - \beta.\right)^2}{\sigma^2}$$ has 1 degree of freedom (8.4.4)

$$n\sum_{i=1}^{m}\frac{\left(\overline{X}_{i.} - \overline{X}..\right)^2}{\sigma^2}$$ has $m - 1$ degrees of freedom

$$m\sum_{j=1}^{n}\frac{\left(\overline{X}._{j} - \beta_{j} - \overline{X}.. + \beta.\right)^2}{\sigma^2}$$ has $n - 1$ degrees of freedom

$$\sum_{i=1}^{m}\sum_{j=1}^{n}\frac{\left(X_{ij} - \overline{X}_{i.} - \overline{X}._{j} + \overline{X}..\right)^2}{\sigma^2}$$ has $(n - 1)(m - 1)$ degrees of freedom

That the preceding degrees of freedom are as stated is fairly easily checked for the first three sums. The number of degrees of freedom of the final sum is determined by noting that linear constraints arise because all of the row and column sums must equal 0. That is,

$$\sum_{i}\left(X_{ij} - \overline{X}_{i.} - \overline{X}._{j} + \overline{X}..\right) = \sum_{i}\left(X_{ij} - \overline{X}._{j}\right) - \sum_{i}\left(\overline{X}_{i.} - \overline{X}..\right) = 0 - 0$$

$$\sum_{j}\left(X_{ij} - \overline{X}_{i.} - \overline{X}._{j} + \overline{X}..\right) = \sum_{j}\left(X_{ij} - \overline{X}_{i.}\right) - \sum_{j}\left(\overline{X}._{j} - \overline{X}..\right) = 0 - 0$$

Now, in a matrix with m rows and n columns, the requirement that all the row sums and column sums equal 0 is equivalent to requiring that the first $m - 1$ row sums equal 0, the first $n - 1$ column sums equal 0, and the sum of all the terms equals 0 (why?). This thus leads to $n - 1 + m - 1 + 1 = n + m - 1$ linear constraints, and so the number of degrees of freedom corresponding to the final sum of squares in Equation 8.4.4 is $nm - (n + m - 1) = (n - 1)(m - 1)$.

It now follows from the partition theorem that if we let

$$SS_r = \sum_{i=1}^{m}\sum_{j=1}^{n}\left(\overline{X}_{i.} - \overline{X}..\right)^2 = n\sum_{i=1}^{m}\left(\overline{X}_{i.} - \overline{X}..\right)^2$$

$$SS_e = \sum_{i=1}^{m}\sum_{j=1}^{n}\left(X_{ij} - \overline{X}_{i.} - \overline{X}._{j} + \overline{X}..\right)^2$$

then, when H_0 is true, SS_r/σ^2 and SS_e/σ^2 are independent chi-square random variables with $m - 1$ and $(n - 1)(m - 1)$ degrees of freedom respectively, and so

$$\frac{\dfrac{SS_r}{(m - 1)}}{\dfrac{SS_e}{(m - 1)(n - 1)}} \sim F_{m-1,(m-1)(n-1)}$$

Thus a significance level α test of

$$H_0: \alpha_1 = \alpha_2 = \cdots = \alpha_m$$

is to reject H_0 if

$$\frac{\dfrac{SS_r}{(m-1)}}{\dfrac{SS_e}{(m-1)(n-1)}} > F_{\alpha,\,m-1,\,(m-1)(n-1)}$$

and to accept otherwise.

We call SS_r, the between row sum of squares, and SS_e, the error sum of squares.

A similar analysis can be performed to test the hypothesis of no column effect

$$H_0' : \beta_1 = \beta_2 = \cdots = \beta_n$$

with the result that a significance level α test is to reject H_0 if

$$\frac{\dfrac{SS_c}{(n-1)}}{\dfrac{SS_e}{(m-1)(n-1)}} > F_{\alpha,\,n-1,\,(m-1)(n-1)}$$

where SS_c, the between column sum of squares, is given by

$$SS_c = \sum_{i=1}^{m} \sum_{j=1}^{n} \left(\overline{X}_{.j} - \overline{X}_{..} \right)^2 = m \sum_{j=1}^{n} \left(\overline{X}_{.j} - \overline{X}_{..} \right)^2$$

This is all summarized in the following table.

TABLE 8.4.1

Two-way ANOVA Table

Source of variation	Sum of squares	Degrees of freedom	Value of F-statistic
Row	$SS_r = n \sum_{i=1}^{m} (\overline{X}_{i.} - \overline{X}_{..})^2$	$m-1$	$\dfrac{SS_r/(m-1)}{SS_e/(m-1)(n-1)}$
Column	$SS_c = m \sum_{j=1}^{n} (\overline{X}_{.j} - \overline{X}_{..})^2$	$n-1$	$\dfrac{SS_c/(n-1)}{SS_e/(m-1)(n-1)}$
Error	$SS_e = \sum_{i=1}^{m} \sum_{j=1}^{n} (X_{ij} - \overline{X}_{i.} - \overline{X}_{.j} + \overline{X}_{..})^2$	$(m-1)(n-1)$	

Program 8-4 computes the values of the F-statistics and also their associated p-values.

Example 8.4a Five automobiles were used in an experiment designed to compare the mileage per gallon obtained by 3 competing brands of gasoline, and the following data resulted:

Auto	*Gasoline*		
	I	*II*	*III*
1	21.2	23.1	22.1
2	24.8	26.4	23.6
3	28.6	30.2	29
4	32	34.2	31.8
5	18	23.8	22

(a) Test the hypothesis that the gasoline used has no effect on mileage obtained?

(b) Does the auto make a difference?

Solution   Run Program 8-4:

```
RUN
THIS PROGRAM COMPUTES THE VALUES OF THE F-STATISTICS AND THEIR ASSOCIATED p-valu
es IN A TWO WAY ANOVA
ENTER THE NUMBER OF ROWS
? 5
ENTER THE NUMBER OF COLUMNS
? 3
ENTER ROW 1 ONE AT A TIME
? 21.12
? 23.1
? 22.1
ENTER ROW 2 ONE AT A TIME
? 24.8
? 26.4
? 23.6
ENTER ROW 3 ONE AT A TIME
? 28.6
? 30.2
? 29
ENTER ROW 4 ONE AT A TIME
? 32
? 34.2
? 31.8
ENTER ROW 5 ONE AT A TIME
? 18
? 23.8
? 22
THE VALUE OF THE F-STATISTIC FOR TESTING THAT THERE IS NO ROW EFFECT IS
54.96949
THE p-value FOR TESTING THAT THERE IS NO ROW EFFECT IS LESS THAN .0001
THE VALUE OF THE F-STATISTIC FOR TESTING THAT THERE IS NO COLUMN EFFECT IS
7.085873
THE p-value FOR TESTING THAT THERE IS NO COLUMN EFFECT IS 1.625574E-02
Ok
```

Hence, both the hypothesis that the gasoline does not affect mileage and that the automobile used is not important are rejected at any significance level above $\alpha = .016$. ■

Example 8.4b The following data[†] represent the number of different macroinvertebrate species collected at 6 stations, located in the vicinity of a thermal

---

[†] Taken from Wartz and Skinner, "A 12 year macroinvertebrate study in the vicinity of 2 thermal discharges to the Susquehanna River near York, Haven, PA," *Jour. of Testing and Evaluation*, Vol. 12, No. 3, May 1984, 157–163.

discharge, from 1970 to 1977.

Year	Station					
	1	2	3	4	5	6
1970	53	35	31	37	40	43
1971	36	34	17	21	30	18
1972	47	37	17	31	45	26
1973	55	31	17	23	43	37
1974	40	32	19	26	45	37
1975	52	42	20	27	26	32
1976	39	28	21	21	36	28
1977	40	32	21	21	36	35

To test the hypotheses that the data are unchanging (a) from year to year, and (b) from station to station, run Program 8-4:

```
RUN
THIS PROGRAM COMPUTES THE VALUES OF THE F-STATISTICS AND THEIR ASSOCIATED p-valu
es IN A TWO WAY ANOVA
ENTER THE NUMBER OF ROWS
? 8
ENTER THE NUMBER OF COLUMNS
? 6
ENTER ROW 1 ONE AT A TIME
? 53
? 35
? 31
? 37
? 40
? 43
ENTER ROW 2 ONE AT A TIME
? 36
? 34
? 17
? 21
? 30
? 18
ENTER ROW 3 ONE AT A TIME
? 47
? 37
? 17
? 31
? 45
? 26
ENTER ROW 4 ONE AT A TIME
? 55
? 31
? 17
? 23
? 43
? 37
ENTER ROW 5 ONE AT A TIME
? 40
? 32
? 19
? 26
? 45
? 37
ENTER ROW 6 ONE AT A TIME
? 52
? 42
? 20
? 27
? 26
? 32
ENTER ROW 7 ONE AT A TIME
```

324 ANALYSIS OF VARIANCE

```
? 39
? 28
? 21
? 21
? 36
? 28
ENTER ROW 8 ONE AT A TIME
? 40
? 32
? 21
? 21
? 36
? 35
THE VALUE OF THE F-STATISTIC FOR TESTING THAT THERE IS NO ROW EFFECT IS
  3.729852
THE p-value FOR TESTING THAT THERE IS NO ROW EFFECT IS 4.042626E-03
THE VALUE OF THE F-STATISTIC FOR TESTING THAT THERE IS NO COLUMN EFFECT IS
  22.47898
THE p-value FOR TESTING THAT THERE IS NO COLUMN EFFECT IS LESS THAN .0001
Ok
```

Thus both the hypotheses of no row effect and the one of no column effect would be rejected even at very small significance levels. ∎

5 TWO-WAY ANALYSIS OF VARIANCE WITH INTERACTION

In Section 4, we considered experiments in which the distribution of the observed data depended on two factors—which we called the "row" and "column" factor. Specifically, we supposed that the mean value of X_{ij}, the data value in row i and column j, can be expressed as the sum of two terms—one depending on the row of the element and one on the column. That is, we supposed that

$$X_{ij} \sim \mathcal{N}\left(\alpha_i + \beta_j, \sigma^2\right), \qquad i = 1, \ldots, m, \qquad j = 1, \ldots, n$$

However, one weakness of this model is that in supposing that the row and column effects are additive, it does not allow for the possibility of a row and column interaction.

For instance, in Example 8.4a we considered an experiment designed to compare the mileage per gallon obtained by three competing brands of gasoline in five different cars. In analyzing the results, we supposed that the incremental mileage obtained when using a given gasoline affected all cars equally. However, it is quite possible that a certain one of these gasolines could interact in a very strong manner with a particular car. Thus there could be a gasoline–car interaction that the model does not allow for.

To account for the possibility of a row and column interaction, set

$$\mu_{ij} = E\left[X_{ij}\right]$$

and break up μ_{ij} by letting

$$\bar{\mu}_{i\cdot} = \sum_{j=1}^{n} \frac{\mu_{ij}}{n}, \qquad \bar{\mu}_{\cdot j} = \sum_{i=1}^{m} \frac{\mu_{ij}}{m}$$

$$\bar{\mu} = \sum_{i=1}^{m} \sum_{j=1}^{n} \frac{\mu_{ij}}{nm} = \sum_{i=1}^{m} \frac{\mu_{i\cdot}}{m} = \sum_{j=1}^{n} \frac{\bar{\mu}_{\cdot j}}{n}$$

We call

$$\bar{\mu}_{i\cdot} - \bar{\mu} \qquad \text{the effect of row } i$$

$$\bar{\mu}_{\cdot j} - \bar{\mu} \qquad \text{the effect of column } j$$

Also, the quantity

$$\mu_{ij} - \left(\bar{\mu} + \bar{\mu}_{i\cdot} - \bar{\mu} + \bar{\mu}_{\cdot j} - \bar{\mu}\right) = \mu_{ij} - \bar{\mu}_{i\cdot} - \bar{\mu}_{\cdot j} + \bar{\mu}$$

is called the interaction effect between row i and row j. It is the amount by which the quantity μ_{ij} differs from the average mean value $(\bar{\mu})$ plus the increment due to row i $(\bar{\mu}_{i\cdot} - \bar{\mu})$ plus the increment due to column j $(\bar{\mu}_{\cdot j} - \bar{\mu})$; it is thus a measure of the departure from additivity of the row and column effects.

If we let

$$\alpha_i = \bar{\mu}_{i\cdot} - \bar{\mu}$$

$$\beta_j = \bar{\mu}_{\cdot j} - \bar{\mu}$$

$$\gamma_{ij} = \mu_{ij} - \bar{\mu}_{i\cdot} - \bar{\mu}_{\cdot j} + \bar{\mu}$$

we see that

$$\underbrace{\mu_{ij}}_{\substack{\text{mean of} \\ \text{value in} \\ \text{row } i, \\ \text{column } j}} = \underbrace{\bar{\mu}}_{\substack{\text{average} \\ \text{mean} \\ \text{value}}} + \underbrace{\alpha_i}_{\substack{\text{increment} \\ \text{due to} \\ \text{row}}} + \underbrace{\beta_j}_{\substack{\text{increment} \\ \text{due to} \\ \text{column}}} + \underbrace{\gamma_{ij}}_{\substack{\text{increment} \\ \text{due to} \\ \text{interaction} \\ \text{of row } i \\ \text{and column } j}}$$

It should be noted, and the verification is left as an exercise, that

$$\sum_{i=1}^{m} \alpha_i = \sum_{j=1}^{n} \beta_j = \sum_{i=1}^{m}\sum_{j=1}^{n} \gamma_{ij} = 0 \qquad (8.5.1)$$

As we shall see later, in order to be able to test the hypothesis that there is no row and column interaction—that is, that all $\gamma_{ij} = 0$—it will be necessary to have more than one observation for each pair of factors. So let us suppose that we have l observations for each row and column. That is, suppose that the data are $\{X_{ijk}, i = 1,\ldots, m, \ j = 1,\ldots, n, \ k = 1,\ldots, l\}$, where X_{ijk} is the kth observation in row i and column j. All observations will be assumed to be independent normal random variables with a common variance σ^2, and so the model is

$$X_{ijk} \sim \mathcal{N}\left(\bar{\mu} + \alpha_i + \beta_j + \gamma_{ij}, \sigma^2\right)$$

where

$$\sum_{i=1}^{m} \alpha_i = \sum_{j=1}^{n} \beta_j = \sum_{i=1}^{m}\sum_{j=1}^{n} \gamma_{ij} = 0$$

We will be interested in testing the hypotheses

$$H_0^r : \alpha_1 = \alpha_2 = \cdots = \alpha_m = 0$$

$$H_0^c : \beta_1 = \beta_2 = \cdots = \beta_n = 0$$

$$H_0^{int} : \gamma_{ij} = 0 \qquad \text{for all } i, j$$

That is, H_0^r and H_0^c are the hypothesis of no row and no column effect respectively, and H_0^{int} is the hypothesis that there is no row and column interaction.

To develop tests of the foregoing hypotheses, let

$$\overline{X}_{ij\cdot} = \frac{\sum_{k=1}^{l} X_{ijk}}{l} = \text{average of the elements in row } i, \text{ column } j$$

$$= \text{estimate of } \mu_{ij}$$

$$\overline{X}_{i\cdot\cdot} = \frac{\sum_{k=1}^{l}\sum_{j=1}^{n} X_{ijk}}{ln} = \text{average of all the elements in row } i$$

$$= \text{estimate of } \bar{\mu}_{i\cdot}$$

$$\overline{X}_{\cdot j\cdot} = \frac{\sum_{k=1}^{l}\sum_{i=1}^{m} X_{ijk}}{lm} = \text{average of all the elements in column } j$$

$$= \text{estimate of } \bar{\mu}_{\cdot j}$$

$$\overline{X}_{\cdots} = \frac{\sum_{k=1}^{l}\sum_{j=1}^{n}\sum_{i=1}^{m} X_{ijk}}{nml} = \text{average of all elements}$$

$$= \text{estimate of } \bar{\mu}$$

and note that

$$\overline{X}_{i\cdot\cdot} - \overline{X}_{\cdots} = \text{estimate of } \alpha_i$$

$$\overline{X}_{\cdot j\cdot} - \overline{X}_{\cdots} = \text{estimate of } \beta_j$$

$$\overline{X}_{ij\cdot} - \overline{X}_{i\cdot\cdot} - \overline{X}_{\cdot j\cdot} + \overline{X}_{\cdots} = \text{estimate of } \gamma_{ij}$$

Now write

$$X_{ijk} = \overline{X}_{\cdots} + (\overline{X}_{i\cdot\cdot} - \overline{X}_{\cdots}) + (\overline{X}_{\cdot j\cdot} - \overline{X}_{\cdots})$$

$$+ (\overline{X}_{ij\cdot} - \overline{X}_{i\cdot\cdot} - \overline{X}_{\cdot j\cdot} + \overline{X}_{\cdots}) + (X_{ijk} - \overline{X}_{ij\cdot})$$

Using the foregoing, we have that

$$\sum_k \sum_j \sum_i X_{ijk}^2 = \sum_k \sum_j \sum_i \left[\overline{X}_{\cdots} + (\overline{X}_{i\cdot\cdot} - \overline{X}_{\cdots}) + (\overline{X}_{\cdot j\cdot} - \overline{X}_{\cdots}) \right.$$

$$\left. + (\overline{X}_{ij\cdot} - \overline{X}_{i\cdot\cdot} - \overline{X}_{\cdot j\cdot} + \overline{X}_{\cdots}) + (X_{ijk} - \overline{X}_{ij\cdot}) \right]^2$$

With the help of some algebra, it is possible to show that, when expanding the right side of the foregoing and then summing, all the cross-product terms will equal 0. Thus, we obtain the following sum of squares identity:

$$\sum_k \sum_j \sum_i X_{ijk}^2 = \sum_k \sum_j \sum_i \overline{X}_{...}^2 + \sum_k \sum_j \sum_i (\overline{X}_{i..} - \overline{X}_{...})^2 \qquad (8.5.2)$$

$$+ \sum_k \sum_j \sum_i (\overline{X}_{.j.} - \overline{X}_{...})^2$$

$$+ \sum_k \sum_j \sum_i (\overline{X}_{ij.} - \overline{X}_{i..} - \overline{X}_{.j.} + \overline{X}_{...})^2$$

$$+ \sum_k \sum_j \sum_i (X_{ijk} - \overline{X}_{ij.})^2$$

If we let

$$SS_r = \sum_{k=1}^{l} \sum_{j=1}^{n} \sum_{i=1}^{m} (\overline{X}_{i..} - \overline{X}_{...})^2$$

$$= ln \sum_{i=1}^{m} (\overline{X}_{i..} - \overline{X}_{...})^2 = \text{row sum of squares}$$

$$SS_c = \sum_{k=1}^{l} \sum_{j=1}^{n} \sum_{i=1}^{m} (\overline{X}_{.j.} - \overline{X}_{...})^2$$

$$= ml \sum_{j=1}^{n} (\overline{X}_{.j.} - \overline{X}_{...})^2 = \text{column sum of squares}$$

$$SS_{int} = \sum_{k=1}^{l} \sum_{j=1}^{n} \sum_{i=1}^{m} (\overline{X}_{ij.} - \overline{X}_{i..} - \overline{X}_{.j.} + \overline{X}_{...})^2$$

$$= l \sum_{j=1}^{n} \sum_{i=1}^{m} (\overline{X}_{ij.} - \overline{X}_{i..} - \overline{X}_{.j.} + \overline{X}_{...})^2 = \text{interaction sum of squares}$$

$$SS_e = \sum_{k=1}^{l} \sum_{j=1}^{n} \sum_{i=1}^{m} (X_{ijk} - \overline{X}_{ij.})^2 = \text{error sum of squares}$$

$$SS_T \equiv \sum_{k=1}^{l} \sum_{j=1}^{n} \sum_{i=1}^{m} (X_{ijk} - \overline{X}_{...})^2 = \text{total sum of squares}$$

then the identity in Equation 8.5.2 can be written as

$$SS_T = SS_r + SS_c + SS_{int} + SS_e$$

The degrees of freedom associated with the terms on the right side of the sum of squares identity in Equation 8.5.2 are as follows:

$$lmn\overline{X}^2_{...} \qquad\qquad \text{has 1 degree of freedom} \qquad (8.5.3)$$

$$SS_r = ln\sum_{i=1}^{m}\left(\overline{X}_{i..} - \overline{X}_{...}\right)^2 \qquad \text{has } m-1 \text{ degrees of freedom}$$

$$SS_c = lm\sum_{j=1}^{n}\left(\overline{X}_{.j.} - \overline{X}_{...}\right)^2 \qquad \text{has } n-1 \text{ degrees of freedom}$$

$$SS_{int} = l\sum_{j=1}^{n}\sum_{i=1}^{m}\left(\overline{X}_{ij.} - \overline{X}_{i..}\right. \qquad \text{has } (n-1)(m-1)$$

$$\left. - \overline{X}_{.j.} + \overline{X}_{...}\right)^2 \qquad \text{degrees of freedom}$$

$$SS_e = \sum_{k=1}^{l}\sum_{j=1}^{n}\sum_{i=1}^{m}\left(X_{ijk} - \overline{X}_{ij.}\right)^2 \qquad \text{has } nm(l-1) \text{ degrees of freedom}$$

The argument that $\overline{X}^2_{...}$, SS_r, SS_c have the stated number of degrees of freedom is fairly straightforward and is left to the reader. The degrees of freedom of SS_{int} are determined by noting that for all i, j

$$\sum_{i}\left(\overline{X}_{ij.} - \overline{X}_{i..} - \overline{X}_{.j.} + \overline{X}_{...}\right) = \sum_{j}\left(\overline{X}_{ij.} - \overline{X}_{i..} - \overline{X}_{.j.} + \overline{X}_{...}\right)$$

$$= 0$$

and, as noted in Section 4, to require all m row sums and all n column sums to equal zero is equivalent to having $n+m-1$ linear constraints. Hence, the degrees of freedom of SS_{int} is $nm - (n+m-1) = (n-1)(m-1)$. The verification of the degrees of freedom given for SS_e is left as an exercise.

To test the hypothesis of no row and column interactions—that is

$$H_0^{int}: \gamma_{ij} = 0 \qquad \text{for all } i, j$$

we note that under H_0^{int} the random variable

$$Z_{ijk} \equiv \frac{\left(X_{ijk} - \bar{\mu} - \alpha_i - \beta_j\right)}{\sigma}$$

are, for all i, j, k, independent random variables having unit normal distributions. Also,

$$\overline{Z}_{ij.} = \frac{\overline{X}_{ij.} - \bar{\mu} - \alpha_i - \beta_j}{\sigma}$$

$$\overline{Z}_{i..} = \frac{\overline{X}_{i..} - \bar{\mu} - \alpha_i}{\sigma} \qquad \left(\text{since } \beta. = \sum_{j=1}^{n}\frac{\beta_j}{n} = 0\right)$$

$$\overline{Z}_{.j.} = \frac{\overline{X}_{.j.} - \bar{\mu} - \beta_j}{\sigma} \qquad \left(\text{since } \alpha. = \sum_{i=1}^{m}\frac{\alpha_i}{m} = 0\right)$$

$$\overline{Z}_{...} = \frac{\overline{X}_{...} - \bar{\mu}}{\sigma}$$

From this, it follows that

$$\overline{Z}_{ij\cdot} - \overline{Z}_{i\cdot\cdot} - \overline{Z}_{\cdot j\cdot} + \overline{Z}_{\cdots} = \frac{\overline{X}_{ij\cdot} - \overline{X}_{i\cdot\cdot} - \overline{X}_{\cdot j\cdot} + \overline{X}_{\cdots}}{\sigma}$$

$$Z_{ijk} - \overline{Z}_{ij\cdot} = \frac{X_{ijk} - \overline{X}_{ij\cdot}}{\sigma}$$

Hence, since $\sum_{k=1}^{l}\sum_{j=1}^{n}\sum_{i=1}^{m}Z_{ijk}^2$ has a chi-square distribution with lnm degrees of freedom, it will follow upon applying the sum of squares identity in Equation 8.5.2 to the Z_{ijk} and then using the partition theorem that, when H_0^{int} is true, SS_{int}/σ^2 and SS_e/σ^2 will be independent chi-square random variables with $(n-1)(m-1)$ and $nm(l-1)$ degrees of freedom respectively. Hence, when H_0^{int} is true $[SS_{int}/(n-1)(m-1)]/[SS_e/nm(l-1)]$ will have an F-distribution with $(n-1)(m-1)$ and $nm(l-1)$ degrees of freedom.

As analogous results can be shown to hold when testing $H_0^r : \alpha_i = 0$ for all i, and $H_0^c : \beta_j = 0$ for all j, we obtain the ANOVA table presented on page 330.

It should be noted that all of the preceding tests call for rejection only when their related F-statistic is large. The reason that only large (and not small) values call for rejection of the null hypothesis is that the numerator of the F-statistic will tend to be larger when H_0 is not true than when it is, whereas the distribution of the denominator will be the same whether or not H_0 is true. As an indication of this, it can be shown that

$$E\left[\frac{SS_r}{(m-1)}\right] = \sigma^2 + \frac{nl}{m-1}\sum_{i=1}^{m}\alpha_i^2$$

$$E\left[\frac{SS_c}{(n-1)}\right] = \sigma^2 + \frac{ml}{n-1}\sum_{j=1}^{n}\beta_j^2$$

$$E\left[\frac{SS_{int}}{(n-1)(m-1)}\right] = \sigma^2 + \frac{l}{(n-1)(m-1)}\sum_{j=1}^{n}\sum_{i=1}^{m}\gamma_{ij}^2$$

$$E\left[\frac{SS_e}{nm(l-1)}\right] = \sigma^2$$

Therefore, we can see that when the null hypothesis is not true (and so the parameters—either α's, β's, or γ's are not all zero), the mean value of the numerator of the F-statistic is increased whereas $E[SS_e/nm(l-1)]$ is always equal to σ^2 (and thus $SS_e/nm(l-1)$ can always be used to estimate σ^2).

Program 8-5 computes the values of the F-statistics and their associated p-values.

Example 8.5a The life of a particular type of generator is thought to be influenced by the material used in its construction and also by the temperature at the location where it is utilized. The following represents lifetime data on 24 generators made from 3 different types of materials and utilized at 2 different

TABLE 8.5.1

Two-way ANOVA with l Observations per Cell

Source of variation	Degrees of freedom	Sum of squares	F-statistic	Level α test
Row	$m-1$	$SS_r = ln \sum_{i=1}^{m} (\bar{X}_{i..} - \bar{X}_{...})^2$	$F_r = \dfrac{SS_r/(m-1)}{SS_e/nm(l-1)}$	Reject H_0^r if $F_r > F_{\alpha,\,m-1,\,nm(l-1)}$
Column	$n-1$	$SS_c = lm \sum_{j=1}^{n} (\bar{X}_{.j.} - \bar{X}_{...})^2$	$F_c = \dfrac{SS_c/(n-1)}{SS_e/nm(l-1)}$	Reject H_0^c if $F_c > F_{\alpha,\,n-1,\,nm(l-1)}$
Interaction	$(n-1)(m-1)$	$SS_{int} = l \sum_{j=1}^{n} \sum_{i=1}^{m} (\bar{X}_{ij.} - \bar{X}_{i..} - \bar{X}_{.j.} + \bar{X}_{...})^2$	$F_{int} = \dfrac{SS_{int}/(n-1)(m-1)}{SS_e/nm(l-1)}$	Reject H_0^{int} if $F_{int} > F_{\alpha,\,(n-1)(m-1),\,nm(l-1)}$
Error	$nm(l-1)$	$SS_e = \sum_{k=1}^{l} \sum_{j=1}^{n} \sum_{i=1}^{m} (X_{ijk} - \bar{X}_{ij.})^2$		

temperatures. Do the data indicate that the material and the temperature do indeed affect the lifetime of a generator? Is there evidence of an interaction effect?

	Temperature	
Material	10°C	18°C
1	135, 150	50, 55
	176, 85	64, 38
2	150, 162	76, 88
	171, 120	91, 57
3	138, 111	68, 60
	140, 106	74, 51

Solution We run Program 8-5:

```
RUN
THIS PROGRAM COMPUTES THE VALUES OF THE F-STATISTICS AND THEIR ASSOCIATED p-valu
es IN A TWO WAY ANOVA WITH L OBSERVATIONS IN EACH ROW-COLUMN CELL
ENTER THE NUMBER OF ROWS
? 3
ENTER THE NUMBER OF COLUMNS
? 2
ENTER THE NUMBER OF OBSERVATIONS IN EACH ROW-COLUMN CELL
? 4
ENTER THE 4 VALUES IN ROW 1 COLUMN 1 ONE AT A TIME
? 135? 150? 176? 85ENTER THE 4 VALUES IN ROW 1 COLUMN 2 ONE AT A TIME
? 50? 55? 64? 38ENTER THE 4 VALUES IN ROW 2 COLUMN 1 ONE AT A TIME
? 150? 162? 171? 120ENTER THE 4 VALUES IN ROW 2 COLUMN 2 ONE AT A TIME
? 76? 88? 91? 57ENTER THE 4 VALUES IN ROW 3 COLUMN 1 ONE AT A TIME
? 138? 111? 140? 106ENTER THE 4 VALUES IN ROW 3 COLUMN 2 ONE AT A TIME
? 68? 60? 74? 51THE VALUE OF THE F-STATISTIC FOR TESTING THAT THERE IS NO ROW
EFFECT IS 2.479762
THE p-value FOR TESTING THAT THERE IS NO ROW EFFECT IS .1092952
THE VALUE OF THE F-STATISTIC FOR TESTING THAT THERE IS NO COLUMN EFFECT IS
69.63223
THE p-value FOR TESTING THAT THERE IS NO COLUMN EFFECT IS LESS THAN .0001
THE VALUE OF THE F-STATISTIC FOR TESTING THAT THERE IS NO INTERACTION EFFECT IS
.6462455
THE p-value FOR TESTING THAT THERE IS NO INTERACTION EFFECT IS .5328998
Ok
```

Thus, we see that there is not sufficient evidence to rule out the possibility that the material used does not make any difference. However, it is clear that the temperature does make a difference. The hypothesis of no interaction is accepted. ■

PROBLEMS

1. A purification process for a chemical involves passing it, in solution, through a resin on which impurities are adsorbed. A chemical engineer wishing to test the efficiency of 3 different resins took a chemical solution and broke it into 15 batches. She tested each resin 5 times and then

measured the concentration of impurities after passing through the resins. Her data were as follows.

Concentration of Impurities		
Resin I	Resin II	Resin III
.046	.038	.031
.025	.035	.042
.014	.031	.020
.017	.022	.018
.043	.012	.039

Test the hypothesis that there is no difference in the efficiency of the resins.

2. It is desired to know what type of filter should be used over the screen of a cathode ray oscilloscope in order to have a radar operator easily pick out targets on the presentation. A test to accomplish this has been set up. A noise is first applied to the scope to make it difficult to pick out a target. A second signal, representing the target, is put into the scope, and its intensity is increased from zero until detected by the observer. The intensity setting at which the observer first notices the target signal is then recorded. This experiment is repeated 20 times with each filter. The numerical value of each reading listed in the table of data is proportional to the target intensity at the time the operator first detects the target.

Filter No. 1	Filter No. 2	Filter No. 3
90	88	95
87	90	95
93	97	89
96	87	98
94	90	96
88	96	81
90	90	92
84	90	79
101	100	105
96	93	98
90	95	92
82	86	85
93	89	97
90	92	90
96	98	87
87	95	90
99	102	101
101	105	100
79	85	84
98	97	102

Test at the 5 percent level of significance the hypothesis that the filters are the same.

3. Explain why we cannot test the hypothesis $H_0: \mu_1 = \mu_2 = \cdots = \mu_m$ by running t-tests on all the $\binom{m}{2}$ pairs of samples.

4. A machine shop contains 3 ovens that are used to heat metal specimens. They are all supposed to heat to the same temperature, and to test this hypothesis temperatures were noted on 5 separate heatings. The following data resulted:

Oven	Temperature
1	492.4, 493.6, 498.5, 488.6, 494
2	488.5, 485.3, 482, 479.4, 478
3	502.1, 492, 497.5, 495.3, 486.7

Do the ovens appear to operate at the same temperature? Test at the 5 percent level of significance. What is the p-value?

5. There are 4 standard chemical procedures used to determine the magnesium content in a certain chemical compound. Each procedure is used 4 times on a given compound with the following data resulting:

	Method		
1	2	3	4
76.42	80.41	74.20	86.20
78.62	82.26	72.68	86.04
80.40	81.15	78.84	84.36
78.20	79.20	80.32	80.68

Do the data indicate that the procedures yield equivalent results?

6. Explain how the partition theorem can be used to obtain the joint distribution of $\overline{X}$ and S^2 in normal sampling.

7. Five servings each of 3 different brands of processed meat were tested for fat content. The following data (in fat percentage per gram) resulted:

Brand	1	2	3
	32	41	36
Fat	34	32	37
content	31	33	30
	35	29	28
	33	35	33

Does the fat content differ depending on the brand?

8. Twenty overweight individuals, each more than 40 pound overweight, were randomly assigned to one of 2 diets. After 10 weeks, the total weight loss

(in pounds) of the individuals on each of the diets was as follows:

Weight Loss	
Diet 1	Diet 2
22.2	24.2
23.4	16.8
24.2	14.6
16.1	13.7
9.4	19.5
12.5	17.6
18.6	11.2
32.2	9.5
8.8	30.1
7.6	21.5

Test, at the 5 percent level of significance, the hypothesis that the two diets have equal effect.

9. A study has been made on pyrethrum flowers to determine the content of pyrethrin, a chemical used in insecticides. Four methods of extracting the chemical are used, and samples are obtained from flowers stored under three conditions, namely fresh flowers, flowers stored for one year, and flowers stored for one year but treated. It is assumed that there is no interaction present. The data are

Pyrethrin Content, Percent

Storage	Method			
Condition	A	B	C	D
1	1.35	1.13	1.06	0.98
2	1.40	1.23	1.26	1.22
3	1.49	1.46	1.40	1.35

(a) Do the methods of extraction appear to differ?
(b) Do the storage conditions affect the content? Test at the $\alpha = .01$ level of significance.

10. An experiment was devised to test the effects of running 3 different types of gasoline with 3 possible types of additive. The experiment called for 9 identical motors to be run with 5 gallons for each of the pairs of gasoline and additives. The following data resulted:

Mileage Obtained

Gasoline	Additive		
	1	2	3
1	124.1	131.5	127
2	126.4	130.6	128.4
3	127.2	132.7	125.6

(a) Test the hypothesis that the gasoline used does not affect the mileage.

(b) Test the hypothesis that the additives are equivalent.

(c) What assumptions are you making?

11. Suppose in Problem 8 that the 10 people placed on each diet consisted of 5 men and 5 women. The data being

	Diet 1	Diet 2
Women	7.6	19.5
	8.8	17.6
	12.5	16.8
	16.1	13.7
	18.6	21.5
Men	22.2	30.1
	23.4	24.2
	24.2	9.5
	32.2	14.6
	9.4	11.2

(a) Test the hypothesis that the diet has the same effect on men and women.

(b) Test the hypothesis that there is no interaction between sex and diet.

12. A researcher is interested in comparing the breaking strength of different laminated beams made from 3 different types of glue and 3 varieties of wood. To make the comparison, 5 beams of each of the 9 combinations were manufactured and then put under a stress test. The following table indicates the pressure readings at which each of the beams broke.

Glue Wood	G_1		G_2		G_3	
W_1	196	208	214	216	258	250
	247	216	235	240	264	248
	221		252		272	
W_2	216	228	215	217	246	247
	240	224	235	219	261	250
	236		241		255	
W_3	230	242	212	218	255	251
	232	244	216	224	261	258
	228		222		247	

(a) Test the hypothesis that the wood used does not affect the breaking strength.

(b) Test the hypothesis that the glue used does not affect the breaking strength.

(c) Test the hypothesis that the wood and glue effect is additive.

13. A study was made as to how the concentration of a certain drug in the blood, 24 hours after being injected, is influenced by age and sex. An

analysis of the blood samples of 40 people given the drug yielded the following concentrations (in milligrams per cubic centimeter).

	Age-group			
	11–25	*26–40*	*41–65*	*Over 65*
Male	52	52.5	53.2	82.4
	56.6	49.6	53.6	86.2
	68.2	48.7	49.8	101.3
	82.5	44.6	50.0	92.4
	85.6	43.4	51.2	78.6
Female	68.6	60.2	58.7	82.2
	80.4	58.4	55.9	79.6
	86.2	56.2	56.0	81.4
	81.3	54.2	57.2	80.6
	77.2	61.1	60.0	82.2

(a) Test the hypothesis that sexual gender does not affect the blood concentration.

(b) Test the hypothesis that age does not affect blood concentration.

(c) Test the hypothesis of no age and sex interaction.

14. Suppose, in Problem 10, that there has been some controversy about the assumption of no interaction between gasoline and additive used. To allow for the possibility of an interaction effect between gasoline and additive, it was decided to run 36 motors— 4 in each grouping. The following data resulted.

	Additive		
Gasoline	*1*	*2*	*3*
1	126.2	130.4	127
	124.8	131.6	126.6
	125.3	132.5	129.4
	127.0	128.6	130.1
2	127.2	142.1	129.5
	126.6	132.6	142.6
	125.8	128.5	140.5
	128.4	131.2	138.7
3	127.1	132.3	125.2
	128.3	134.1	123.3
	125.1	130.6	122.6
	124.9	133.0	120.9

(a) Do the data indicate an interaction effect?

(b) Do the gasolines appear to give equal results?

(c) Test whether or not there is an additive effect or whether all additives work equally well.

(d) What conclusions can you draw?

15. An experiment has been devised to test the hypothesis that an elderly person's memory retention can be improved by a set of "oxygen treatments." A group of scientists administered these treatments to men and women. The men and women were each randomly divided into 4 groups of 5 each, and the people in the ith group were given treatments over an $(i - 1)$ week interval, $i = 1, 2, 3, 4$. (The 2 groups not given any treatments served as "controls"). The treatments were set up in such a manner so that all individuals thought they were receiving the oxygen treatments for the total 3 weeks. After treatment ended, a memory retention test was administered. The results (with higher scores indicating higher memory retentions) were as follows.

Scores

	Number of Weeks of Oxygen Treatment			
	0	1	2	3
Men	42	39	38	42
	54	52	50	55
	46	51	47	39
	38	50	45	38
	51	47	43	51
Women	49	48	27	61
	44	51	42	55
	50	52	47	45
	45	54	53	40
	43	40	58	42

(a) Test the hypothesis that the length of treatment does not affect memory retention.
(b) Is there a sex difference?
(c) Test whether or not there is an interaction effect.
(d) A randomly chosen group of 5 elderly men, without receiving any oxygen treatment, were given the memory retention test. Their scores were 37, 35, 33, 39, 29. What conclusions can you draw?

16. In the model of Section 5 show that

$$\sum_{i=1}^{m} \alpha_i = \sum_{j=1}^{m} \beta_j = \sum_{j=1}^{n} \sum_{i=1}^{n} \gamma_{ij} = 0$$

17. For the two-way ANOVA model of Section 5, argue that the degrees of freedom are as given by Equation 8.5.3.

18. For the two way ANOVA model of Section 5 show that

$$E[SS_e] = nm(l - 1)\sigma^2$$

CHAPTER 9

Goodness of Fit and Nonparametric Testing

0 INTRODUCTION

In this chapter we consider two statistical topics whose treatments have been greatly affected by the advent of fast and inexpensive computer power—namely goodness of fit and nonparametric testing problems. In both situations a key element is the computation of the p-value of a given test statistic. The classical approach has been to approximate the distribution of the test statistic when the null hypothesis is true (usually by either a normal or chi-square distribution), and then compute the p-value by acting as if that were the actual distribution. However, it is now possible to determine the p-value, to a greater degree of accuracy, either by a simulation approach (useful in both goodness of fit and nonparametric problems) or by solving a set of recursive equations (as can be done in some nonparametric testing problems). We will delineate both the classical and the more modern approach in this chapter.

1 GOODNESS OF FIT TESTS

We are often interested in determining whether or not a particular probabilistic model is appropriate for a given random phenomenon. This determination often reduces to testing whether a given random sample comes from some specified, or partially specified, probability distribution. For example, we may a priori feel that the number of industrial accidents occurring daily at a particular plant should constitute a random sample from a Poisson distribution. This hypothesis can then be tested by observing the number of accidents over a sequence of days and then testing whether it is reasonable to suppose that the underlying distribution is Poisson. Statistical tests that determine whether a given probabilistic mechanism is appropriate are called *goodness of fit* tests.

338

The classical approach to obtaining a goodness of fit test is to partition the possible values of a random variable into a finite number of regions. One then observes a finite sequence of such random variables and compares the numbers of them that fall into each region with the theoretical expected numbers under the specified probability distribution.

2 GOODNESS OF FIT TESTS WHEN ALL PARAMETERS ARE SPECIFIED

Suppose that n independent random variables—$Y_1, \ldots, Y_n$—each taking on one of the values $1, 2, \ldots, k$—are to be observed and we are interested in testing the null hypothesis that $\{ p_i, \ i = 1, \ldots, k \}$ is the probability mass function of the Y_j. That is, if Y represents any of the Y_j, then the null hypothesis is

$$H_0 : P\{ Y = i \} = p_i \qquad i = 1, \ldots, k$$

whereas the alternative hypothesis is

$$H_1 : P\{ Y = i \} \neq p_i \qquad \text{for some } i = 1, \ldots, k$$

To test the foregoing hypothesis, let X_i, $i = 1, \ldots, k$, denote the number of the Y_j's that equal i. Then as each Y_j will independently equal i with probability $P\{ Y = i \}$, it follows that, under H_0, X_i is binomial with parameters n and p_i. Hence, when H_0 is true,

$$E[X_i] = np_i$$

and so $(X_i - np_i)^2$ will be an indication as to how likely it appears that p_i indeed equals the probability that $Y = i$. When this is large, say in relationship to np_i, then it is an indication that H_0 is not correct. Indeed such reasoning leads us to consider the following test statistic

$$T = \sum_{i=1}^{k} \frac{(X_i - np_i)^2}{np_i} \qquad (9.2.1)$$

and to reject the null hypothesis when T is large.

In order to determine the critical region, we need first specify a significance level α and then we must determine that critical value c such that

$$P_{H_0}\{ T \geq c \} = \alpha$$

That is, we need determine c so that the probability that the test statistic T is at least as large as c is, when H_0 is true, α. The test is then to reject the hypothesis, at the α level of significance, when $T \geq c$ and to accept when $T < c$.

It remains to determine c. The classical approach to doing so is to use the result that when n is large T will, when H_0 is true, approximately have (with the approximation becoming exact as n approaches infinity) a chi-square distribution with $k - 1$ degrees of freedom. Hence, for n large, c can be taken to equal $\chi^2_{\alpha,\,k-1}$; and so the approximate α level test is

$$\text{reject} \quad H_0 \quad \text{if} \quad T \geq \chi^2_{\alpha,\,k-1}$$
$$\text{accept} \quad H_0 \quad \text{otherwise}$$

If the observed value of T is $T = t$, then the preceding test is equivalent to rejecting H_0 if the significance level α is at least as large as the p-value given by

$$p\text{-value} = P_{H_0}\{T \geq t\}$$
$$\approx P\{\chi^2_{k-1} \geq t\}$$

where χ^2_{k-1} is a chi-square random variable with $k - 1$ degrees of freedom.

An accepted rule of thumb as to how large n need be for the foregoing to be a good approximation is that it should be large enough so that $np_i \geq 1$ for each i, $i = 1, \ldots, k$, and also at least 80 percent of the values np_i should exceed 5.

REMARKS

(a) A computationally simpler formula for T can be obtained by expanding the square in Equation 9.2.1 and using the results that $\sum_i p_i = 1$ and $\sum_i X_i = n$ (why is this true?):

$$T = \sum_{i=1}^{k} \frac{X_i^2 - 2np_i X_i + n^2 p_i^2}{np_i} \tag{9.2.2}$$

$$= \sum_i X_i^2 / np_i - 2\sum_i X_i + n\sum_i p_i$$

$$= \sum_i X_i^2 / np_i - n$$

(b) The intuitive reason why T, which depends on the k values $X_1, \ldots, X_k$, has only $k - 1$ degrees of freedom is that one degree of freedom is lost because of the linear relationship $\sum_i X_i = n$.

(c) Whereas the proof that, asymptotically, T has a chi-square distribution is advanced, it can be easily shown when $k = 2$. In this case, since $X_1 + X_2 = n$, and $p_1 + p_2 = 1$, we see that

$$T = \frac{(X_1 - np_1)^2}{np_1} + \frac{(X_2 - np_2)^2}{np_2}$$

$$= \frac{(X_1 - np_1)^2}{np_1} + \frac{(n - X_1 - n[1 - p_1])^2}{n(1 - p_1)}$$

$$= \frac{(X_1 - np_1)^2}{np_1} + \frac{(X_1 - np_1)^2}{n(1 - p_1)}$$

$$= \frac{(X_1 - np_1)^2}{np_1(1 - p_1)} \quad \text{since} \quad \frac{1}{p} + \frac{1}{1 - p} = \frac{1}{p(1 - p)}$$

However, X_1 is a binomial random variable with mean np_1 and variance $np_1(1 - p_1)$ and thus, by the normal approximation to the binomial, it follows that $(X_1 - np_1)/\sqrt{np_1(1 - p_1)}$ has, for large n, approximately a unit normal distribution; and so its square has approximately a chi-square distribution with 1 degree of freedom.

Example 9.2a In recent years, a correlation between mental and physical well-being has increasingly become accepted. An analysis of birthdays and death days of famous people could be used as further evidence in the study of this correlation. To use these data, we are supposing that being able to look forward to something betters a person's mental state; and that a famous person would probably look forward to his birthday because of the resulting attention, affection, and so on. If a famous person is in poor health and dying, then perhaps anticipating his birthday would "cheer him up and therefore improve his health and possibly decrease the chance that he will die shortly before his birthday." The data might therefore reveal that a famous person is less likely to die in the months before his birthday and more likely to die in the months afterward.

To test this, a sample of 1251 (deceased) Americans were randomly chosen from *Who Was Who in America*, and their birth and death days were noted. (The data are taken from D. Phillips, "Death Day and Birthday: An Unexpected Connection," in *Statistics: A Guide to the Unknown*, Holden-Day, 1972.) The data can be summarized in the table on page 342.

If the death day does not depend on the birthday, then it would seem that each of the 1251 individuals would be equally likely to fall in any of the 12 categories. Thus, let us test the null hypothesis

$$H_0 = p_i = \frac{1}{12} \qquad i = 1, \ldots, 12$$

Since $np_i = 1251/12 = 104.25$, the chi-square test statistic for this hypothesis is

$$T = \frac{(90)^2 + (100)^2 + (87)^2 + \cdots + (106)^2}{104.25} - 1251$$

$$= 17.192$$

The p-value is given by

$$p\text{-value} = P\{\chi_{11}^2 \geq 17.192\}$$

$$= 1 - .8977 = .1023 \qquad \text{by Program 3-8-1-A}$$

The results of this test leave us somewhat up in the air concerning the hypothesis that an approaching birthday, when regarded as something to be looked forward to, has a positive effect on an individual's health. This is so since whereas the data are not quite strong enough to reject (at the 10 percent level of significance) the null hypothesis that each $p_i = 1/12$, it is certainly suggestive of the possible falsity of this hypothesis. Indeed is it not possible that we are to blame by not having wisely chosen the null hypothesis to be

Number of Deaths Before, During, and After the Birth Month

	6 months before	5 months before	4 months before	3 months before	2 months before	1 month before	The month	1 month after	2 months after	3 months after	4 months after	5 months after
Number of deaths	90	100	87	96	101	86	119	118	121	114	113	106

$n = 1251$
$n/12 = 104.25$

tested? Perhaps having as many as 12 intervals makes it difficult to reject the hypothesis that all outcomes are equally likely—that is, perhaps the probability of rejection is not very large when many of the p_i are near $1/12$ and a few are above and a few below. What would have happened if we would have coded the data into 4 possible outcomes—say

$$
\begin{aligned}
\text{outcome } 1 &= -6, -5, -4 \\
\text{outcome } 2 &= -3, -2, -1 \\
\text{outcome } 3 &= 0, 1, 2 \\
\text{outcome } 4 &= 3, 4, 5
\end{aligned}
$$

That is, for instance, an individual whose death day occurred 3 months before his birthday would be placed in outcome 2. With this classification, the data would be as follows:

Outcome	Number of times occurring
1	277
2	283
3	358
4	333

$n = 1251$
$n/4 = 312.75$

The test statistic for testing $H_0 = p_i = 1/4$, $i = 1, 2, 3, 4$ is

$$
T = \frac{(277)^2 + (283)^2 + (358)^2 + (333)^2}{312.75} - 1251
$$

$$
= 14.775
$$

Hence, as $\chi^2_{.01,3} = 11.345$, the null hypothesis would be rejected even at the 1 percent level of significance. Indeed, using Program 3-8-1-A yields that

$$
p\text{-value} = P\{\chi_3^2 \geq 14.775\} = 1 - .998 = .002
$$

The foregoing analysis is, however, subject to the criticism that the null hypothesis was chosen after the data were observed. Indeed, while there is nothing incorrect about using a set of data to determine the "correct way" of phrasing a null hypothesis, the additional use of those data to test that very hypothesis is certainly questionable. Therefore, to be quite certain of the conclusion to be drawn from this example, it seems prudent to choose a second random sample—coding the values as before—and again test $H_0 : p_i = 1/4$, $i = 1, 2, 3, 4$, (see Problem 3). ■

Program 9-2-1 can be used to quickly calculate the value of T. To avoid round-off error, it computes T using Equation 9.2.1 rather than 9.2.2.

Example 9.2b A contractor who purchases a large number of fluorescent light bulbs has been told by the manufacturer that these bulbs are not of uniform quality but rather have been produced in such a way that each bulb produced will, independently, either be of quality level A, B, C, D, or E, with respective probabilities .15, .25, .35, .20, .05. However, the contractor feels that he is receiving too many type E (the lowest quality) bulbs, and so he decides to test the producer's claim by taking the time and expense to ascertain the quality of 30 such bulbs. Suppose that he discovers that of the 30 bulbs, 3 are of quality level A, 6 are of quality level B, 9 are of quality level C, 7 are of quality level D, and 5 are of quality level E. Do these data, at the 5 percent level of significance, enable the contractor to reject the producer's claim?

Solution We run Program 9-2-1:

```
RUN
THIS PROGRAM COMPUTES THE GOODNESS OF FIT TEST STATISTIC
ENTER THE NUMBER OF GROUPINGS
? 5
ENTER THE NULL HYPOTHESIS PROBABILITIES ONE AT A TIME
? .15
? .25
? .35
? .20
? .05
ENTER THE SAMPLE SIZE
? 30
ENTER THE NUMBERS THAT FALL IN EACH GROUPING ONE AT A TIME
? 3
? 6
? 9
? 7
? 5
TEST STATISTIC HAS VALUE      9.347619
Ok
```

The approximate *p*-value:

$$p\text{-value} = P_{H_0}\{T \geq 9.347619\}$$

$$\approx P\{\chi_4^2 \geq 9.347619\}$$

is obtained by running Program 3-8-1:

```
RUN
THIS PROGRAM COMPUTES THE PROBABILITY THAT A CHI-SQUARE RANDOM VARIABLE WITH N D
EGREES OF FREEDOM IS LESS THAN X
ENTER THE DEGREE OF FREEDOM PARAMETER
? 4
ENTER THE DESIRED VALUE OF X
? 9.347619
THE PROBABILITY IS .9472415
```

Hence, the approximate *p*-value is

$$p\text{-value} \approx 1 - .947 = .053$$

Thus the hypothesis would be accepted at the 5 percent level of significance (but, since it would have been rejected at all significance levels above .053, the contractor should certainly remain skeptical). ■

2.1   Determining the Critical Region by Simulation

From 1900 when Karl Pearson first showed that T has approximately (becoming exact as n approaches infinity) a chi-square distribution with $k - 1$ degrees of freedom, until quite recently, this approximation was the only means available for determining the p-value of the goodness of fit test. However, with the recent advent of inexpensive, fast, and easily available computational power a second, potentially more accurate, approach has become available. Namely, the use of simulation to obtain to a high level of accuracy the p-value of the test statistic.

The simulation approach is as follows. First, the value of T is determined —say $T = t$. Now to determine whether or not to accept H_0, at a given significance level α, we need to know the probability that T would be at least as large as t when H_0 is true. To determine this probability, we simulate n independent random variables $Y_1^{(1)}, \ldots, Y_n^{(1)}$ each having the probability mass function $\{ p_i, i = 1, \ldots, k \}$—that is

$$P\{ Y_j^{(1)} = i \} = p_i \qquad i = 1, \ldots, k, \qquad j = 1, \ldots, n$$

Now let

$$X_i^{(1)} = \text{number } j : Y_j^{(1)} = i$$

and set

$$T^{(1)} = \sum_{i=1}^{k} \frac{\left(X_i^{(1)} - np_i \right)^2}{np_i}$$

Now repeat this procedure by simulating a second set, independent of the first set, of n independent random variables $Y_1^{(2)}, \ldots, Y_n^{(2)}$ each having the probability mass function $\{ p_i, i = 1, \ldots, k \}$ and then, as for the first set, determining $T^{(2)}$. Repeating this a large number, say r, of times yields r independent random variables $T^{(1)}, T^{(2)}, \ldots, T^{(r)}$, each of which has the same distribution as does the test statistic T when H_0 is true. Hence, by the law of large numbers, the proportion of the T_i that are as large as t will be very nearly equal to the probability that T is as large as t when H_0 is true—that is,

$$\frac{\text{number } l : T^{(l)} \geq t}{r} \approx P_{H_0}\{ T \geq t \}$$

In fact, by letting r be large, the foregoing can be considered to be, with high probability, almost an equality. Hence, if that proportion is less than or equal to α, then the p-value, equal to the probability of observing a T as large as t when H_0 is true, is less than α and so H_0 should be rejected.

REMARKS

(a) In order to utilize the foregoing simulation approach to determine whether or not to accept H_0 when T is observed, we need specify how one can simulate, or generate, a random variable Y such that $P\{Y = i\} = p_i$, $i = 1, \ldots, k$. One way is as follows:

Step 1: Generate a random number U.

Step 2: If

$$p_1 + \cdots + p_{i-1} < U < p_1 + \cdots + p_i$$

set $Y = i$ (where $p_1 + \cdots + p_{i-1} \equiv 0$ when $i = 1$). That is,

$$U < p_1 \Rightarrow Y = 1$$

$$p_1 \le U < p_1 + p_2 \Rightarrow Y = 2$$

$$\vdots$$

$$p_1 + \cdots + p_{i-1} \le U < p_1 + p_i \Rightarrow Y = i$$

$$\vdots$$

$$p_1 + \cdots + p_{n-1} < U \Rightarrow Y = n$$

Since a random number is equivalent to a uniform $(0, 1)$ random variable, we have that

$$P\{a < U < b\} = b - a \qquad 0 < a < b < 1$$

and so

$$P\{Y = i\} = P\{p_1 + \cdots + p_{i-1} < U < p_1 + \cdots + p_i\} = p_i$$

(b) A significant question that remains is how many simulation runs are necessary. It has been shown that the value $r = 100$ is usually sufficient at the conventional 5 percent level of significance.[†]

Example 9.2c Let us reconsider the problem presented in Example 9.2b. A simulation study yielded the result

$$P_{H_0}\{T \le 9.52381\} = .95$$

and so the critical value should be 9.52381, which is remarkably close to $\chi^2_{.05, 4} = 9.488$ given as the critical value by the chi-square approximation. This is most interesting since the rule of thumb for when the chi-square approximation can be applied—namely, that each $np_i \ge 1$ and at least 80 percent of the np_i exceed 5—does not apply, thus raising the possibility that it is rather conservative. ∎

Program 9-2-2 can be utilized to determine the p-value.

To obtain more information as to how well the chi-square approximation performs, consider the following example.

Example 9.2d Consider an experiment having 6 possible outcomes whose probabilities are hypothesized to be .1, .1, .05, .4, .2, and .15. This is to be

[†]See Hope, A., "A Simplified Monte Carlo Significance Test Procedure," *J. of Royal Statist. Soc.*, B, 30, 582–598, 1968.

tested by performing 40 independent replications of the experiment. If the resultant number of times that each of the six outcomes occur is 3, 3, 5, 18, 4, 7, should the hypothesis be accepted?

Solution Running Program 9-2-1 yields the output:

```
THIS PROGRAM COMPUTES THE GOODNESS OF FIT TEST STATISTIC
ENTER THE NUMBER OF GROUPINGS
? 6
ENTER THE NULL HYPOTHESIS PROBABILITIES ONE AT A TIME
? .1
? .1
? .05
? .4
? .2
? .15
ENTER THE SAMPLE SIZE
? 40
ENTER THE NUMBERS THAT FALL IN EACH GROUPING ONE AT A TIME
? 3
? 3
? 5
? 18
? 4
? 7
TEST STATISTIC HAS VALUE     7.416667
```

To determine the *p*-value given by the chi-squared approximation, we now run Program 3-8-1-A. The result is that

$$P\{\chi_5^2 \leq 7.41667\} = .8088$$

and so

$$p\text{-value} \approx .1912$$

To check the foregoing we ran Program 9-2-2, using 10,000 simulation runs, with the following result:

```
RUN
THIS PROGRAM USES SIMULATION TO APPROXIMATE THE p-value IN THE GOODNESS OF FIT T
EST
Random number seed (-32768 to 32767)? 6324
ENTER THE NUMBER OF POSSIBLE VALUES
? 6
ENTER THE PROBABILITIES ONE AT A TIME
? .1
? .1
? .05
? .4
? .2
? .15
ENTER THE SAMPLE SIZE
? 40
ENTER THE DESIRED NUMBER OF SIMULATION RUNS
? 10000
ENTER THE VALUE OF THE TEST STATISTIC
? 7.416667
THE ESTIMATE OF THE p-value IS .179
```

Since the number of the 10^4 simulated test values that exceed 7.41667 is binomially distributed with parameters $n = 10^4$, $p = p$-value, it follows from the results of Section 3.3 of Chapter 5 that a 90 percent confidence interval for the p-value is

$$p\text{-value} \in .179 \pm 1.64\sqrt{(.179)(.821)/10^4}$$

or

$$p\text{-value} \in (.172713, .185287) \qquad \text{with 90 percent confidence}$$

Thus if we use .179 as our estimate of the actual p-value, then the chi-square estimate .1912 is too large by a factor of $100(.0122/.179) = 6.8$ percent. ∎

3 GOODNESS OF FIT TESTS WHEN SOME PARAMETERS ARE UNSPECIFIED

We can also perform goodness-of-fit tests of a null hypothesis that does not completely specify the probabilities $\{ p_i, \; i = 1, \ldots, k \}$. For instance, consider the situation previously mentioned in which one is interested in testing whether the number of accidents occurring daily in a certain industrial plant is Poisson distributed with some unknown mean λ. To test this hypothesis, suppose that the daily number of accidents is recorded for n days—let $Y_1, \ldots, Y_n$ be these data. To analyze these data we must first address the difficulty that the Y_i can assume an infinite number of possible values. However, this is easily dealt with by breaking up the possible values into a finite number k of regions and then considering the region in which each Y_i falls. For instance, we might say that the outcome of the number of accidents on a given day is in region 1 if there are 0 accidents, region 2 if there are 1 accident, region 3 if there are 2 or 3 accidents, region 4 if there are 4 or 5 accidents, and region 5 if there greater than 5 accidents. Hence, if the distribution is indeed Poisson with mean λ then

$$P_1 = P\{Y = 0\} = e^{-\lambda} \qquad\qquad\qquad (9.3.1)$$

$$P_2 = P\{Y = 1\} = \lambda e^{-\lambda}$$

$$P_3 = P\{Y = 2\} + P\{Y = 3\} = \frac{e^{-\lambda}\lambda^2}{2} + \frac{e^{-\lambda}\lambda^3}{6}$$

$$P_4 = P\{Y = 4\} + P\{Y = 5\} = \frac{e^{-\lambda}\lambda^4}{24} + \frac{e^{-\lambda}\lambda^5}{120}$$

$$P_5 = P\{Y > 5\} = 1 - e^{-\lambda} - \lambda e^{-\lambda} - \frac{e^{-\lambda}\lambda^2}{2} - \frac{e^{-\lambda}\lambda^3}{6} - \frac{e^{-\lambda}\lambda^4}{24} - \frac{e^{-\lambda}\lambda^5}{120}$$

The second difficulty we face in obtaining a goodness of fit test results from the fact that the mean value λ is not specified. Clearly, the intuitive thing to do is to assume that H_0 is true and then estimate it from the data—say $\hat{\lambda}$ is the estimate of λ—and then compute the test statistic

$$T = \sum_{i=1}^{k} \frac{(X_i - n\hat{p}_i)^2}{n\hat{p}_i}$$

where X_i is, as before, the number of Y_j that fall in region i, $i = 1, \ldots, k$, and $\hat{p}_i$ is the estimated probability of the event that Y_j falls in region i, which is determined by substituting $\hat{\lambda}$ for λ in the expression 9.3.1 for p_i.

In general, this approach can be utilized whenever there are unspecified parameters in the null hypothesis that are needed to compute the quantities p_i, $i = 1, \ldots, k$. Suppose now that there are m such unspecified parameters and that they are to be estimated by the method of maximum likelihood. It can then be proven that when n is large, the test statistic T will, when H_0 is true, approximately have a chi-square distribution with $k - 1 - m$ degrees of freedom. (In other words, one degree of freedom is lost for each parameter that needs to be estimated). The test is, therefore, to

$$\text{reject} \quad H_0 \quad \text{if} \quad T \geq \chi^2_{\alpha, \, k-1-m}$$

$$\text{accept} \quad H_0 \quad \text{otherwise}$$

An equivalent way of performing the foregoing is to first determine the value of the test statistic T, say $T = t$, and then compute

$$p\text{-value} = P\{\chi^2_{k-1-m} \geq t\}$$

The hypothesis would be rejected if $\alpha \geq p$-value.

Example 9.3a Suppose the weekly number of accidents over a 30-week period is as follows:

8,	0,	0,	1,	3,	4,	0,	2,	12,	5
1,	8,	0,	2,	0,	1,	9,	3,	4,	5
3,	3,	4,	7,	4,	0,	1,	2,	1,	2

Test the hypothesis that the number of accidents in a week has a Poisson distribution.

Solution Since the total number of accidents in the 30 weeks is 95, the maximum likelihood estimate of the mean of the Poisson distribution is

$$\hat{\lambda} = \frac{95}{30} = 3.16667$$

Since the estimate of $P\{Y = i\}$ is then

$$P\{Y = i\} \stackrel{\text{est}}{=} \frac{e^{-\hat{\lambda}}\hat{\lambda}^i}{i!}$$

we obtain, after some computation, that with the 5 regions as given in the beginning of this section

$$\hat{p}_1 = .04214$$
$$\hat{p}_2 = .13346$$
$$\hat{p}_3 = .43434$$
$$\hat{p}_4 = .28841$$
$$\hat{p}_5 = .10164$$

Using the data values $X_1 = 6$, $X_2 = 5$, $X_3 = 8$, $X_4 = 6$, $X_5 = 5$, an additional computation yields the test statistic value

$$T = \sum_{i=1}^{5} \frac{(X_i - 30\hat{p}_i)^2}{30\hat{p}_i} = 21.99156$$

To determine the p-value we run Program 3-8-1-A. This yields

$$p\text{-value} \approx P\{\chi_3^2 > 21.99\}$$
$$= 1 - .999936$$
$$= .000064$$

and so the hypothesis of an underlying Poisson distribution is rejected. (Clearly, there were too many weeks having 0 accidents for the hypothesis that the underlying distribution is Poisson with mean 3.167 to be tenable.) ∎

3.1   Tests of Independence in Contingency Tables

In this subsection we consider problems in which each member of a population can be classified according to two distinct characteristics—which we shall denote as the X-characteristic and the Y-characteristic. We suppose that there are r possible values for the X-characteristic and s for the Y-characteristic, and let for $i = 1, \ldots, r$, $j = 1, \ldots, s$

$$P_{ij} = P\{X = i, Y = j\}$$

That is, P_{ij} represents the probability that an arbitrary member of the population will have X-characteristic i and Y-characteristic j. The different members of the population will be assumed to be independent. Also, let

$$p_i = P\{X = i\} = \sum_{j=1}^{s} P_{ij} \qquad i = 1, \ldots, r$$

and

$$q_j = P\{Y = j\} = \sum_{i=1}^{r} P_{ij} \qquad j = 1, \ldots, s$$

That is, p_i is equal to the probability that an arbitrary member of the population will have X-characteristic i, and q_j that it will have Y-characteristic j.

We are interested in testing the hypothesis that a population member's X- and Y-characteristics are independent. That is, we are interested in testing

$$H_0 : P_{ij} = p_i q_j \qquad \text{for all} \qquad i = 1, \ldots, r$$
$$j = 1, \ldots, s$$

against the alternative

$$H_1 : P_{ij} \neq p_i q_j \qquad \text{for some } i, j \qquad i = 1, \ldots, r$$
$$j = 1, \ldots, s$$

To test this hypothesis suppose that n members of the population have been sampled with the result that N_{ij} of them have simultaneously had X-characteristic i and Y-characteristic j, $i = 1, \ldots, r$, $j = 1, \ldots, s$.

Since the quantities p_i, $i = 1, \ldots, r$, and q_j, $j = 1, \ldots, s$ are not specified by the null hypothesis, they must first be estimated. Now since

$$N_i = \sum_{j=1}^{s} N_{ij} \qquad i = 1, \ldots, r$$

represents the number of the sampled population members that have X-characteristic i, a natural (in fact, the maximum-likelihood) estimator of p_i is

$$\hat{p}_i = \frac{N_i}{n} \qquad i = 1, \ldots, r$$

Similarly, letting

$$M_j = \sum_{i=1}^{r} N_{ij} \qquad j = 1, \ldots, s$$

denote the number of sampled members having Y-characteristic j, the estimator for q_j is

$$\hat{q}_j = \frac{M_j}{n} \qquad j = 1, \ldots, s$$

At first glance, it may seem that we have had to use the data to estimate $r + s$ parameters. However, since the p_i's and q_j's have to sum to 1—that is, $\sum_{i=1}^{r} p_i = \sum_{j=1}^{s} q_j = 1$—we need estimate only $r - 1$ of the p's and $s - 1$ of the q's. (For instance, if r were equal to 2, then an estimate of p_1 would automatically provide an estimate of p_2 since $p_2 = 1 - p_1$.) Hence, we actually need estimate $r - 1 + s - 1 = r + s - 2$ parameters, and since each population member has $k = rs$ different possible values, it follows that the resulting test statistic will, for large n, have approximately a chi-square distribution with $rs - 1 - (r + s - 2) = (r - 1)(s - 1)$ degrees of freedom.

Finally, since

$$E[N_{ij}] = nP_{ij}$$
$$= np_i q_j \qquad \text{when } H_0 \text{ is true}$$

it follows that the test statistic is given by

$$T = \sum_{j=1}^{s} \sum_{i=1}^{r} \frac{(N_{ij} - n\hat{p}_i \hat{q}_j)^2}{n\hat{p}_i \hat{q}_j} = \sum_{j=1}^{s} \sum_{i=1}^{r} \frac{N_{ij}^2}{n\hat{p}_i \hat{q}_j} - n$$

and the approximate significance level α test is to

reject H_0 if $T \geq \chi^2_{\alpha,(r-1)(s-1)}$

accept H_0 otherwise

Equivalently, if the observed value of T is $T = t$, then the approximate test would reject if the p-value $\leq \alpha$ where

$$p\text{-value} = P_{H_0}\{T \geq t\}$$

$$\approx P\left\{\chi^2_{(r-1)(s-1)} \geq t\right\}$$

Program 9-3 will compute the value of T.

Example 9.3b A company operates 4 machines on 3 separate shifts daily. The following table, called a contingency table, presents the data, over a 6-month time period, concerning the machine breakdowns that resulted.

Number of Breakdowns

	Machine				
	A	B	C	D	Total per shift
Shift 1	10	12	6	7	35
Shift 2	10	24	9	10	53
Shift 3	13	20	7	10	50
Total per machine	33	56	22	27	138

Suppose we are interested in determining whether a machine's breakdown probability during a particular shift is influenced by that shift. In other words, we are interested in testing, for an arbitrary breakdown, whether the machine causing the breakdown and the shift on which the breakdown occurred are independent.

Using the table, we obtain upon running Program 9-3:

```
THIS PROGRAM COMPUTES THE TEST STATISTIC FOR TESTING FOR INDEPENDENCE IN A 2-WAY
CONTINGENCY TABLE
ENTER THE NUMBER OF ROWS
? 3
ENTER THE NUMBER OF COLUMNS
? 4
ENTER ROW 1 ONE AT A TIME
? 10
? 12
? 6
? 7
ENTER ROW 2 ONE AT A TIME
? 10
? 24
? 9
? 10
ENTER ROW 3 ONE AT A TIME
? 13
? 20
? 7
? 10
THE TEST STATISTIC VALUE IS T = 1.814778
```

We now run Program 3-8-1-A to obtain that the p-value is

$$p\text{-value} \approx P\left\{\chi^2_6 \geq 1.814788\right\}$$

$$= 1 - .0641$$

$$= .9359$$

and so the hypothesis that the machine causing a breakdown is independent of the shift when the breakdown occurs is accepted. ■

3.2 Using Simulation

Since the null hypothesis no longer completely specifies the probability model, the use of simulation to determine the p-value of the test statistic is somewhat trickier than before. The way it should be done is as follows.

(a) *The Model* Suppose that the null hypothesis is that the data values $Y_1, \ldots, Y_n$, constitute a random sample from a distribution that is specified up to a set of unknown parameters $\theta_1, \ldots, \theta_m$. Suppose also that when this hypothesis is true, the possible values of the Y_i are $1, \ldots, k$.

(b) *The Initial Step* Use the data to estimate the unknown parameters. Specifically, let $\hat{\theta}_j$ denote the value of the maximum likelihood estimator of θ_j, $j = 1, \ldots, m$. Now compute the value of the test statistic

$$T = \sum_{i=1}^{k} \frac{\left(X_i - \hat{E}[X_i]\right)^2}{\hat{E}[X_i]}$$

where X_i is the number of the data values that are equal to i, $i = 1, \ldots, k$, and $\hat{E}[X_i]$ is its expectation when the null hypothesis is true and $\theta_j = \hat{\theta}_j$, $j = 1, \ldots, m$. Let t denote the value of the test statistic T.

(c) *The Simulation Step* We will now do a series of simulations to estimate the p-value of the data. First note that all simulations are to be obtained by using the population distribution that results when the null hypothesis is true and θ_j is equal to its maximum likelihood estimate $\hat{\theta}_j$, $j = 1, \ldots, m$, determined in step (b).

Simulate a sample of size n from the aforementioned population distribution and let $\hat{\theta}_j(\text{sim})$ denote the maximum likelihood estimate of θ_j, $j = 1, \ldots, m$, based on the simulated data. Now determine the value of

$$T_{\text{sim}} = \sum_{i=1}^{k} \frac{\left(X_i - E_{\text{sim}}[X_i]\right)^2}{E_{\text{sim}}[X_i]}$$

where X_i, $i = 1, \ldots, k$, is the number of the simulated data values equal to i, and $E_{\text{sim}}[X_i]$ is the value of $E[X_i]$ when θ_j is equal to $\hat{\theta}_j(\text{sim})$, $j = 1, \ldots, m$.

The simulation step should then be repeated many times. The estimate of the p-value is then equal to the proportion of the values of T_{sim} that are at least as large as t.

Example 9.3c Let us reconsider Example 9.3a. The data given in this example resulted in the maximum likelihood estimate $\lambda = 3.16667$ and the test statistic value $T = 21.99156$. The simulation step now consists of simulating 30 independent Poisson random variables having mean 3.16667 and then computing the value of $\sum_{i=1}^{5}(X_i - 30p_i)^2/30p_i$ where X_i is the number of simulated values falling in region i, and p_i is the probability that a Poisson random variable with mean $\hat{\lambda}(\text{sim})$, equal to the average of the 30 simulated values, would fall into region i. The simulation step should be repeated many times and the estimated p-value is the proportion of times it results in a value at least as large as 21.99156. ∎

4 NONPARAMETRIC TESTS

In the following sections of this chapter we shall develop some hypotheses tests in situations where the data come from random samples whose underlying forms are not specified. That is, it will not be assumed that the underlying distribution is normal, or exponential, or any other given type. Because no particular parametric form for the underlying distribution is assumed, such tests are called *nonparametric*.

The strength of a nonparametric test resides in the fact that it can be applied without any assumption on the form of the underlying distribution. Of course if there is justification for assuming a particular parametric form, such as the normal, then the relevant parametric test should be employed.

5 THE SIGN TEST

Let $X_1, \ldots, X_n$ denote a sample from a continuous distribution F and suppose that we are interested in testing the hypothesis that the median of F, call it m, is equal to a specified value m_0. That is, consider a test of

$$H_0 : m = m_0 \qquad \text{versus} \qquad H_1 : m \neq m_0$$

where m is such that $F(m) = .5$.

This hypothesis can easily be tested by noting that each of the observations will, independently, be less than m_0 with probability $F(m_0)$. Hence, if we let

$$I_i = \begin{cases} 1 & \text{if } X_i \leq m_0 \\ 0 & \text{if } X_i > m_0 \end{cases}$$

then $I_1, \ldots, I_n$ are independent Bernoulli random variables with parameter $p = F(m_0)$; and so the null hypothesis is equivalent to stating that this Bernoulli parameter is equal to $\frac{1}{2}$. Now, if v is the observed value of $\sum_{i=1}^{n} I_i$—that is, if v is the number of data values less than m_0—then it follows from the results of Section 6 of Chapter 6 that the p-value of the test that this Bernoulli parameter is equal to $\frac{1}{2}$ is

$$p\text{-value} = 2 \min \left(P\{ \text{Bin} (n, 1/2) \leq v \}, P\{ \text{Bin} (n, 1/2) \geq v \} \right) \quad (9.5.1)$$

where $\text{Bin} (n, p)$ is a binomial random variable with parameters n and p.

However,

$$P\{\,\mathrm{Bin}\,(n,\,p)\geq v\,\} = P\{\,n - \mathrm{Bin}\,(n,\,p)\leq n - v\,\}$$
$$= P\{\,\mathrm{Bin}\,(n,1-p)\leq n - v\,\}\qquad(\text{why?})$$

and so we see from Equation 9.5.1 that the p-value is given by

$$p\text{-value} = 2\min\left(P\{\,\mathrm{Bin}\,(n,1/2)\leq v\,\},\ P\{\,\mathrm{Bin}\,(n,1/2)\leq n - v\,\}\right)\quad(9.5.2)$$

$$= \begin{cases} 2P\{\,\mathrm{Bin}\,(n,1/2)\leq v\,\} & \text{if } v \leq \dfrac{n}{2} \\[2mm] 2P\{\,\mathrm{Bin}\,(n,1/2)\leq n - v\,\} & \text{if } v \geq \dfrac{n}{2} \end{cases}$$

Since the value of $v = \sum_{i=1}^{n} I_i$ depends on the signs of the terms $X_i - m_0$, the foregoing is called the sign test.

Example 9.5a If a sample of size 200 contains 120 values that are less than m_0 and 80 values that are greater, what is the p-value of the test of the hypothesis that the median is equal to m_0?

Solution From Equation 9.5.2 the p-value is equal to twice the probability that binomial random variable with parameters 200, $\frac{1}{2}$ is less than or equal to 80. Running Program 3-1 yields the following:

```
THIS PROGRAM COMPUTES THE PROBABILITY THAT A BINOMIAL(n,p) VARIABLE IS LESS THAN
   OR EQUAL TO i
ENTER n
? 200
ENTER p
? .5
ENTER i
? 80
THE PROBABILITY IS 2.842415E-03
```

Thus the p-value is .00568 and so the hypothesis would be rejected at even the 1 percent level of significance. ∎

The sign test can also be used in situations analogous to ones in which the paired t-test was previously applied. For instance, let us reconsider Example 6.4c of Chapter 6, which is interested in testing whether or not a recently instituted industrial safety program has had an effect on the number of man-hours lost to accidents. For each of 10 plants, the data consisted of the pair X_i, Y_i which represented, respectively, the average weekly loss at plant i before and after the program. Letting $Z_i = X_i - Y_i$, $i = 1,\ldots,10$, it follows that if the program had not had any effect, then Z_i, $i = 1,\ldots,10$, would be a sample from a distribution whose median value is 0. Since the resulting values of Z_i—namely, 7.5, -2.3, 2.6, 3.7, 1.5, $-.5$, -1, 4.9, 4.8, 1.6—contain 3 whose sign is negative and 7 whose sign is positive, it follows that the hypothesis that the median of Z is 0 should be rejected at significance level α if

$$\sum_{i=0}^{3} \binom{10}{i} \left(\frac{1}{2}\right)^{10} \leq \frac{\alpha}{2}$$

Since

$$\sum_{i=0}^{3} \binom{10}{i} \left(\frac{1}{2}\right)^{10} = \frac{176}{1024} = .172$$

it follows that the hypothesis would be accepted at the 5 percent significance level (indeed, it would be accepted at all significance levels less than the p-value equal to .344).

Thus, the sign test does not enable us to conclude that the safety program has had any statistically significant effect, which is in contradiction to the result obtained in Example 6.4c, when it was assumed that the differences were normally distributed. The reason for this disparity is that the assumption of normality allows us to take into account not only the number of values greater than 0 (which is all the sign test considers) but also the magnitude of these values. The next test to be considered, while still being nonparametric, improves on the sign test by taking into account whether those values that most differ from the hypothesized median value m_0 tend to lie on one side of m_0—that is, whether they tend to be primarily bigger (or smaller) than m_0.

6   THE SIGNED RANK TEST

The sign test can be employed to test the hypothesis that the median of a continuous distribution F is equal to a specified value m_0. However, in many applications one is really interested in testing not only that the median is equal to m_0 but that the distribution is symmetric about m_0. That is, if X has distribution function F, then one is often interested in testing the hypothesis $H_0: P\{X < m_0 - a\} = P\{X > m_0 + a\}$ for all $a > 0$ (see Figure 9.6.1). Whereas the sign test could still be employed to test the foregoing hypothesis, it suffers in that it compares only the number of data values that are less than m_0 with the number that are greater than m_0 and does not take into account whether or not one of these sets tends to be further away from m_0 than the other. A nonparametric test that does take this into account is the so-called *signed rank* test. It is described as follows.

Let $Y_i = X_i - m_0$, $i = 1, \ldots, n$ and rank (that is, order) the absolute values $|Y_1|, |Y_2|, \ldots, |Y_n|$. Set, for $j = 1, \ldots, n$.

$$I_j = \begin{cases} 1 & \text{if the } j\text{th smallest value comes from a data value} \\ & \text{that is smaller than } m_0 \\ 0 & \text{otherwise} \end{cases}$$

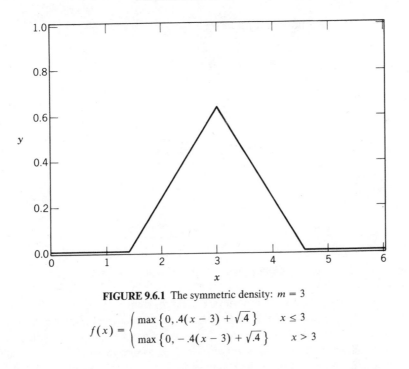

FIGURE 9.6.1 The symmetric density: $m = 3$

$$f(x) = \begin{cases} \max\{0, .4(x-3) + \sqrt{.4}\} & x \le 3 \\ \max\{0, -.4(x-3) + \sqrt{.4}\} & x > 3 \end{cases}$$

Now, whereas $\sum_{j=1}^{n} I_j$ represents the test statistic for the sign test, the signed rank test uses the statistic $T = \sum_{j=1}^{n} j I_j$. That is, like the sign test it considers those data values that are less than m_0, but rather than giving equal weight to each such value it gives larger weights to those data values that are furthest away from m_0.

Example 9.6a If $n = 4$, $m_0 = 2$ and the data values are $X_1 = 4.2$, $X_2 = 1.8$, $X_3 = 5.3$, $X_4 = 1.7$, then the rankings of $|X_i - 2|$ are .2, .3, 2.2, 3.3. Since the first of these values—namely .2—comes from the data point X_2, which is less than 2, it follows that $I_1 = 1$. Similarly, $I_2 = 1$, and I_3 and I_4 equal 0. Hence the value of the test statistic is $T = 1 + 2 = 3$. ∎

When H_0 is true, the mean and variance of the test statistic T are easily computed. This is accomplished by noting that, since the distribution of $Y_j = X_j - m_0$ is symmetric about 0, for any given value of $|Y_j|$, say $|Y_j| = y$, it is equally likely that either $Y_j = y$ or $Y_j = -y$. From this fact it can be seen that, under H_0, $I_1, \ldots, I_n$ will be independent random variables such that

$$P\{I_j = 1\} = \frac{1}{2} = P\{I_j = 0\} \qquad j = 1, \ldots, n$$

Hence, we can conclude that, under H_0

$$E[T] = E\left[\sum_{j=1}^{n} jI_j\right] \tag{9.6.1}$$

$$= \sum_{j=1}^{n} \frac{j}{2} = \frac{n(n+1)}{4}$$

$$\text{Var}\,(T) = \text{Var}\left(\sum_{j=1}^{n} jI_j\right) \tag{9.6.2}$$

$$= \sum_{j=1}^{n} j^2\,\text{Var}\,(I_j)$$

$$= \sum_{j=1}^{n} \frac{j^2}{4} = \frac{n(n+1)(2n+1)}{24}$$

where the fact that the variance of the Bernoulli random variable I_j is $\frac{1}{2}(1 - \frac{1}{2}) = \frac{1}{4}$ is used.

It can be shown that for moderately large values of n ($n > 25$ is often quoted as being sufficient) T will, when H_0 is true, have approximately a normal distribution with mean and variance as given by Equations 9.6.1 and 9.6.2. Although this approximation can be used to derive an approximate level α test of H_0 (which has been the usual approach until the recent advent of fast and cheap computational power), we shall not pursue this approach but rather will determine the p-value for given test data by an explicit computation of the relevant probabilities. This is accomplished as follows.

Suppose we desire a significance level α test of H_0. Since the alternative hypothesis is that the median is not equal to m_0, a two-sided test is called for. That is, if the observed value of T is equal to t, then H_0 should be rejected if either

$$P_{H_0}\{T \le t\} \le \frac{\alpha}{2} \quad \text{or} \quad P_{H_0}\{T \ge t\} \le \frac{\alpha}{2} \tag{9.6.3}$$

The p-value of the test data when $T = t$ is given by

$$p\text{-value} = 2\min\left(P_{H_0}\{T \le t\}, P_{H_0}\{T \ge t\}\right) \tag{9.6.4}$$

That is, if $T = t$, the signed rank test calls for rejection of the null hypothesis if the significance level α is at least as large as this p-value. The amount of computation necessary to compute the p-value can be reduced by utilizing the following equality (whose proof will be given at the end of the section):

$$P_{H_0}\{T \ge t\} = P_{H_0}\left\{T \le \frac{n(n+1)}{2} - t\right\}$$

Using Equation 9.6.4 the p-value is given by

$$p\text{-value} = 2\min\left(P_{H_0}\{T \le t\}, P_{H_0}\left\{T \le \frac{n(n+1)}{2} - t\right\}\right)$$

$$= 2P_{H_0}\{T \le t^*\}$$

where

$$t^* = \min\left(t, \frac{n(n+1)}{2} - t\right)$$

It remains to compute $P_{H_0}\{T \le t^*\}$. To do so, let $P_k(i)$ denote the probability, under H_0, that the signed rank statistic T will be less than or equal to i when the sample size is k. We will determine a recursive formula for $P_k(i)$ starting with $k = 1$. When $k = 1$ since there is only a single data value, which, when H_0 is true, is equally likely to be either less than or greater than m_0, it follows that T is equally likely to be either 0 or 1. Thus

$$P_1(i) = \begin{cases} 0 & i < 0 \\ \frac{1}{2} & i = 0 \\ 1 & i \ge 1 \end{cases} \tag{9.6.5}$$

Now suppose the sample size is k. To compute $P_k(i)$ we condition on the value of I_k as follows:

$$P_k(i) = P_{H_0}\left\{\sum_{j=1}^{k} jI_j \le i\right\}$$

$$= P_{H_0}\left\{\sum_{j=1}^{k} jI_j \le i | I_k = 1\right\} P_{H_0}\{I_k = 1\}$$

$$+ P_{H_0}\left\{\sum_{j=1}^{k} jI_j \le i | I_k = 0\right\} P_{H_0}\{I_k = 0\}$$

$$= P_{H_0}\left\{\sum_{j=1}^{k-1} jI_j \le i - k | I_k = 1\right\} P_{H_0}\{I_k = 1\}$$

$$+ P_{H_0}\left\{\sum_{j=1}^{k-1} jI_j \le i | I_k = 0\right\} P_{H_0}\{I_k = 0\}$$

$$= P_{H_0}\left\{\sum_{j=1}^{k-1} jI_j \le i - k\right\} P_{H_0}\{I_k = 1\} + P_{H_0}\left\{\sum_{j=1}^{k-1} jI_j \le i\right\} P_{H_0}\{I_k = 0\}$$

where the last equality utilized the independence of $I_1, \ldots, I_{k-1}$, and I_k (when H_0 is true). Now $\sum_{j=1}^{k-1} jI_j$ has the same distribution as the signed rank statistic of a sample of size $k - 1$, and since

$$P_{H_0}\{I_k = 1\} = P_{H_0}\{I_k = 0\} = \frac{1}{2}$$

we see that

$$P_k(i) = \frac{1}{2}P_{k-1}(i - k) + \frac{1}{2}P_{k-1}(i) \tag{9.6.6}$$

Starting with Equation 9.6.5, the recursion given by Equation 9.6.6 can be successfully employed to compute $P_2(\cdot)$ then $P_3(\cdot)$, and so on stopping when the desired value $P_n(t^*)$ has been obtained.

Example 9.6b For the data of Example 9.6a

$$t^* = \min\left(3, \frac{4 \cdot 5}{2} - 3\right) = 3$$

Hence the p-value is $2P_4(3)$, which is computed as follows

$$P_2(0) = \frac{1}{2}[P_1(-2) + P_1(0)] = \frac{1}{4}$$

$$P_2(1) = \frac{1}{2}[P_1(-1) + P_1(1)] = \frac{1}{2}$$

$$P_2(2) = \frac{1}{2}[P_1(0) + P_1(2)] = \frac{3}{4}$$

$$P_2(3) = \frac{1}{2}[P_1(1) + P_1(3)] = 1$$

$$P_3(0) = \frac{1}{2}[P_2(-3) + P_2(0)] = \frac{1}{8} \qquad [\text{since } P_2(-3) = 0]$$

$$P_3(1) = \frac{1}{2}[P_2(-2) + P_2(1)] = \frac{1}{4}$$

$$P_3(2) = \frac{1}{2}[P_2(-1) + P_2(2)] = \frac{3}{8}$$

$$P_3(3) = \frac{1}{2}[P_2(0) + P_2(3)] = \frac{5}{8}$$

$$P_4(0) = \frac{1}{2}[P_3(-4) + P_3(0)] = \frac{1}{16}$$

$$P_4(1) = \frac{1}{2}[P_3(-3) + P_3(1)] = \frac{1}{8}$$

$$P_4(2) = \frac{1}{2}[P_3(-2) + P_3(2)] = \frac{3}{16}$$

$$P_4(3) = \frac{1}{2}[P_3(-1) + P_3(3)] = \frac{5}{16} \quad\blacksquare$$

Program 9-6 will use in the recursion in Equations 9.6.5 and 9.6.6 to compute the p-value of the signed rank test data. The input needed is the sample size n and the value of test statistic T.

Example 9.6c Suppose we are interested in determining whether a certain population has an underlying probability distribution that is symmetric about 0. If a sample of size 20 from this population results in a signed rank test statistic of value 142, what conclusion can we draw at the 10 percent level of significance?

Solution Run Program 9-6:

```
RUN
THIS PROGRAM COMPUTES THE p-value FOR THE ONE SAMPLE SIGNED RANK TEST
ENTER THE SIZE OF THE SAMPLE
? 20
ENTER THE OBSERVED VALUE OF THE SUM OF THE SIGNED RANKS
? 142
THE p-value IS .1768532.
```

Thus the hypothesis that the population distribution is symmetric about 0 is accepted at the $\alpha = .10$ level of significance. ∎

We end this section with a proof of the equality

$$P_{H_0}\{T \geq t\} = P_{H_0}\left\{T \leq \frac{n(n+1)}{2} - t\right\}$$

To verify the foregoing, note first that $1 - I_j$ will equal 1 if the jth smallest value of $|Y_1|, \ldots, |Y_n|$ comes from a data value larger than m_0, and it will equal 0 otherwise. Hence, if we let

$$T^1 = \sum_{j=1}^{n} j(1 - I_j)$$

then T^1 will represent the sum of the ranks of the $|Y_j|$ that correspond to data values larger than m_0. By symmetry T^1 will, under H_0, have the same distribution as T. Now

$$T^1 = \sum_{j=1}^{n} j - \sum_{j=1}^{n} jI_j = \frac{n(n+1)}{2} - T$$

and so

$$P\{T \geq t\} = P\{T^1 \geq t\} \qquad \text{since } T \text{ and } T^1 \text{ have the same distribution}$$

$$= P\left\{\frac{n(n+1)}{2} - T \geq t\right\}$$

$$= P\left\{T \leq \frac{n(n+1)}{2} - t\right\}$$

REMARK ON TIES

Since we have assumed that the population distribution is continuous, there is no possibility of ties—that is, with probability 1, all observations will have different values. However, since in practice all measurements are quantized, ties are always a distinct possibility. If ties do occur, then the weights given to the values less than m_0 should be the average of the different weights they could have had if the values had differed slightly. For instance, if $m_0 = 0$ and the data values are $2, 4, 7, -5, -7$, then the ordered absolute values are $2, 4, 5, 7, 7$. Since 7 has rank both 4 and 5, the value of the test statistic T is $T = 3 + 4.5 = 7.5$. The p-value should be computed as when we assumed that all values were distinct. (Although technically this is not correct, the discrepancy is usually minor.)

7 THE TWO-SAMPLE PROBLEM

Suppose that one is considering two different methods for producing items having measurable characteristics with an interest in determining whether or not the two methods result in statistically identical items.

To attack this problem let $X_1, \ldots, X_n$ denote a sample of the measurable values of n items produced by method 1, and, similarly, let $Y_1, \ldots, Y_m$ be the corresponding value of m items produced by method 2. If we let F and G, both assumed to be continuous, denote the distribution functions of the two samples respectively, then the hypothesis we wish to test is $H_0 : F = G$.

One procedure for testing H_0—which is known by such names as the rank sum test, the Mann-Whitney test, or the Wilcoxon test—calls initially for ranking, or ordering, the $n + m$ data values $X_1, \ldots, X_n, Y_1, \ldots, Y_m$. Since we are assuming that F and G are continuous, this ranking will be unique—that is, there will be no ties. Give the smallest data value rank 1, the second smallest rank 2, ... and the $(n + m)$th smallest rank $n + m$. Now let, for $i = 1, \ldots, n$

$$R_i = \text{rank of the data value } X_i$$

The rank sum test utilizes the test statistic T equal to the sum of the ranks from the first sample—that is

$$T = \sum_{i=1}^{n} R_i$$

Example 9.7a An experiment designed to compare two treatments against corrosion yielded the following data in pieces of wire subjected to the two treatments.

Treatment 1	65.2, 67.1, 69.4, 78.2, 74, 80.3
Treatment 2	59.4, 72.1, 68, 66.2, 58.5

(The data represent the maximum depth of pits in units of one thousandth of an inch.) The ordered values are 58.5, 59.4, 65.2*, 66.2, 67.1*, 68, 69.4*, 72.1, 74*, 78.2*, 80.3* with an asterisk noting that the data value was from sample 1. Hence, the value of the test statistic is $T = 3 + 5 + 7 + 9 + 10 + 11 = 45$. ∎

Suppose that we desire a significance level α test of H_0. If the observed value of T is $T = t$, then H_0 should be rejected if either

$$P_{H_0}\{T \le t\} \le \frac{\alpha}{2} \quad \text{or} \quad P_{H_0}\{T \ge t\} \le \frac{\alpha}{2} \quad \quad (9.7.1)$$

That is, the hypothesis that the two samples are equivalent should be rejected if the sum of the ranks from the first sample is either too small or too large to be explained by chance.

Since, for integral t,

$$P\{T \ge t\} = 1 - P\{T < t\}$$
$$= 1 - P\{T \le t - 1\}$$

it follows from Equation 9.7.1 that H_0 should be rejected if either

$$P_{H_0}\{T \le t\} \le \frac{\alpha}{2} \quad \text{or} \quad P_{H_0}\{T \le t - 1\} \ge 1 - \frac{\alpha}{2} \quad \quad (9.7.2)$$

To compute the probabilities in Equation 9.7.2, let $P\{N, M, K\}$ denote the probability that the sum of the ranks of the first sample will be less than or equal to K when the sample sizes are N and M and H_0 is true. We will now determine a recursive formula for $P(N, M, K)$, which will then allow us to obtain the desired quantities $P(n, m, t) = P_{H_0}\{T \leq t\}$ and $P(n, m, t - 1)$.

To compute the probability that the sum of the ranks of the first sample is less than or equal to K when N and M are the sample sizes and H_0 is true, let us condition on whether the largest of the $N + M$ data values belongs to the first or second sample. If it belongs to the first sample, then the sum of the ranks of this sample is equal to $N + M$ plus the sum of the ranks of the other $N - 1$ values from the first sample. Hence this sum will be less than or equal to K if the sum of the ranks of the other $N - 1$ values is less than or equal to $K - (N + M)$. But since the remaining $N - 1 + M$—that is, all but the largest—values all come from the same distribution (when H_0 is true), it follows that the sum of the ranks of $N - 1$ of them will be less than or equal to $K - (N + M)$ with probability $P(N - 1, M, K - N - M)$. By a similar argument we can show that, given that the largest value is from the second sample, the sum of the ranks of the first sample will be less than or equal to K with probability $P(N, M - 1, K)$. Also, since the largest value is equally likely to be any of the $N + M$ values $X_1, \ldots, X_N, Y_1, \ldots, Y_M$, it follows that it will come from the first sample with probability $N/(N + M)$. Putting these together we thus obtain that

$$P(N, M, K) = \frac{N}{N + M} P(N - 1, M, K - N - M)$$

$$+ \frac{M}{N + M} P(N, M - 1, K) \qquad (9.7.3)$$

Starting with the boundary condition

$$P(1, 0, K) = \begin{cases} 0 & K \leq 0 \\ 1 & K > 0 \end{cases} \qquad P(0, 1, K) = \begin{cases} 0 & K < 0 \\ 1 & K \geq 0 \end{cases}$$

Equation 9.7.3 can be solved recursively to obtain $P(n, m, t - 1)$ and $P(n, m, t)$.

Example 9.7b Suppose we wanted to determine $P(2, 1, 3)$. We use Equation 9.7.3 as follows:

$$P(2, 1, 3) = \frac{2}{3} P(1, 1, 0) + \frac{1}{3} P(2, 0, 3)$$

and

$$P(1, 1, 0) = \frac{1}{2} P(0, 1, -2) + \frac{1}{2} P(1, 0, 0) = 0$$
$$P(2, 0, 3) = P(1, 0, 1)$$
$$= P(0, 0, 0) = 1$$

Hence,

$$P(2, 1, 3) = \frac{1}{3}$$

which checks since in order for the sum of the ranks of the two X-values to be less than or equal to 3, the largest of the values X_1, X_2, Y_1, must be Y_1, which, when H_0 is true, has probability $\frac{1}{3}$. ∎

Since the rank sum test calls for rejection when either

$$2P(n, m, t) \leq \alpha \quad \text{or} \quad \alpha \geq 2[1 - P(n, m, t - 1)]$$

it follows that the p-value of the test statistic when $T = t$ is

$$p\text{-value} = 2 \min \{ P(n, m, t), 1 - P(n, m, t - 1) \}$$

Program 9-7 uses the recursion in Equation 9.7.3 to compute the p-value for the rank sum test. The input needed are the sizes of the first and second samples and the sum of the ranks of the elements of the first sample. Whereas either sample can be designated as the first sample, the program will run fastest if the first sample is the one whose sum of ranks is smallest.

Example 9.7c In Example 9.7a, the sizes of the 2 samples are 5 and 6 respectively and the sum of the ranks of the first sample is 21. Running Program 9-7 yields the following output:

```
RUN
THIS PROGRAM COMPUTES THE p-value FOR THE TWO-SAMPLE RANK SUM TEST
THIS PROGRAM WILL RUN FASTEST IF YOU DESIGNATE AS THE FIRST SAMPLE THE SAMPLE HA
VING THE SMALLER SUM OF RANKS
ENTER THE SIZE OF THE FIRST SAMPLE
? 5
ENTER THE SIZE OF THE SECOND SAMPLE
? 6
ENTER THE SUM OF THE RANKS OF THE FIRST SAMPLE
? 21
THE p-value IS .1255411
Ok
```

Example 9.7d Suppose that in testing whether 2 production methods yield identical results, 9 items are produced using the first method and 13 using the second. If, among all 22 items, the sum of the ranks of the 9 items produced by method 1 is 72, what conclusions would you draw?

Solution Run Program 9-7:

```
RUN
THIS PROGRAM COMPUTES THE p-value FOR THE TWO-SAMPLE RANK SUM TEST
THIS PROGRAM WILL RUN FASTEST IF YOU DESIGNATE AS THE FIRST SAMPLE THE SAMPLE HA
VING THE SMALLER SUM OF RANKS
ENTER THE SIZE OF THE FIRST SAMPLE
? 9
ENTER THE SIZE OF THE SECOND SAMPLE
? 13
ENTER THE SUM OF THE RANKS OF THE FIRST SAMPLE
? 72
THE p-value IS 3.641591E-02
```

Thus the hypothesis of identical results would be rejected at the 5 percent level of significance. ∎

It remains to compute the value of the test statistic T. It is quite efficient to compute T directly by first using a standard computer science algorithm (such as quicksort) to sort, or order, the $n + m$ values. Another approach, easily programmed, although efficient for only small values of n and m, uses the following identity.

Proposition 9.7a

For $i = 1, \ldots, n$, $j = 1, \ldots, m$ let

$$W_{ij} = \begin{cases} 1 & \text{if } X_i > Y_j \\ 0 & \text{otherwise} \end{cases}$$

Then

$$T = \frac{n(n+1)}{2} + \sum_{i=1}^{n} \sum_{j=1}^{m} W_{ij}$$

Proof Consider the values $X_1, \ldots, X_n$ of the first sample and order them. Let $X_{(i)}$ denote the ith smallest, $i = 1, \ldots, n$. Now consider the rank of $X_{(i)}$ among all $n + m$ data values. This is given by

$$\text{rank of } X_{(i)} = i + \text{number } j: \ Y_j < X_{(i)}$$

Summing over i gives

$$\sum_{i=1}^{n} \text{rank of } X_{(i)} = \sum_{i=1}^{n} i + \sum_{i=1}^{n} (\text{number } j: \ Y_j < X_{(i)}) \qquad (9.7.4)$$

But since the order in which we add terms does not change the sum obtained, we see that

$$\sum_{i=1}^{n} \text{rank of } X_{(i)} = \sum_{i=1}^{n} \text{rank of } X_i = T \qquad (9.7.5)$$

$$\sum_{i=1}^{n} (\text{number } j: \ Y_j < X_{(i)}) = \sum_{i=1}^{n} (\text{number } j: \ Y_j < X_i)$$

Hence, from Equations 9.7.4 and 9.7.5 we obtain that

$$T = \sum_{i=1}^{n} i + \sum_{i=1}^{n} (\text{number } j: \ Y_j < X_i)$$

$$= \frac{n(n+1)}{2} + \sum_{i=1}^{n} \sum_{j=1}^{m} W_{ij}$$

7.1   The Classical Approximation and Simulation

The difficulty with employing the recursion in Equation 9.7.3 to compute the p-value of the two-sample sum of rank test statistic is that the amount of computation grows enormously as the sample sizes increase. For instance, if $n = m = 20$, then even if we choose the test statistic to be the smaller sum of ranks, since the sum of all the ranks is $1 + 2 + \cdots + 40 = 820$, it is possible that the test statistic could have a value as large as 410. Hence, there can be as many as $20 \times 20 \times 410 = 164{,}000$ values of $P(N, M, K)$ that would have to be computed to determine the p-value. Thus, for large sample sizes the approach based on the recursion in Equation 9.7.3 is not viable. Two approximate methods that can be utilized in such cases are (a) a classical method based on approximating the distribution of the test statistic, and (b) simulation.

(a) *The Classical Approximation*   When the null hypothesis is true and so $F = G$, it follows that all $n + m$ data values come from the same distribution and thus all $(n + m)!$ possible rankings of the values $X_1, \ldots, X_n, Y_1, \ldots, Y_m$ are equally likely. From this it follows that choosing the n rankings of the first sample is probabilistically equivalent to randomly choosing n of the (possible rank) values $1, 2, \ldots, n + m$. Therefore, using the results of Section 6 of Chapter 4, it follows that T has a mean and variance given by

$$E_{H_0}[T] = \frac{n(n + m + 1)}{2}$$

$$\mathrm{Var}_{H_0}(T) = \frac{nm(n + m + 1)}{12}$$

In addition, it can be shown that when both n and m are of moderate size (both being greater than 7 should suffice) T has, under H_0, approximately a normal distribution. Hence, when H_0 is true

$$\frac{T - \dfrac{n(n + m + 1)}{2}}{\sqrt{\dfrac{nm(n + m + 1)}{12}}} \doteq \mathcal{N}(0, 1) \qquad (9.7.6)$$

If we let d denote the absolute value of the difference between the observed value of T and its mean value given above, then based on Equation 9.7.6 the approximate p-value is

$$\text{p-value} = P_{H_0}\!\left\{ |T - E_{H_0}[T]| > d \right\}$$

$$\approx P\!\left\{ |Z| > d \Big/ \sqrt{\frac{nm(n + m + 1)}{12}} \right\} \qquad \text{where } Z \sim \mathcal{N}(0, 1)$$

$$= 2P\!\left\{ Z > d \Big/ \sqrt{\frac{nm(n + m + 1)}{12}} \right\}$$

Example 9.7e In Example 9.7a, $n = 5$, $m = 6$, and the test statistic's value is 21. Since

$$\frac{n(n + m + 1)}{2} = 30$$

$$\frac{nm(n + m + 1)}{12} = 30$$

we have that $d = 9$ and so

$$p\text{-value} \approx 2P\left\{ Z > \frac{9}{\sqrt{30}} \right\}$$

$$= 2P\{ Z > 1.643108 \}$$

$$= 2(1 - .9498) \quad \text{from Program 3-5-1-A}$$

$$= .1004$$

which can be compared with the exact value, as given in Example 9.7c, of .1225.

In Example 9.7d, $n = 9$, $m = 13$, and so

$$\frac{n(n + m + 1)}{2} = 103.5$$

$$\frac{nm(n + m + 1)}{12} = 224.25$$

Since $T = 72$ we have that

$$d = |72 - 103.5| = 31.5$$

Thus, the approximate p-value is

$$p\text{-value} \approx 2P\left\{ Z > \frac{31.5}{\sqrt{224.25}} \right\}$$

$$= 2P\{ Z > 2.103509 \}$$

$$= 2(1 - .9823) = .0354$$

which is quite close to the exact p-value (as given in Example 9.7d) of .0364.

Thus, in the two examples considered, the normal approximation worked quite well in the second example—where the guideline that both sample sizes should exceed 7 held—and not so well in the first example—where the guideline did not hold. ∎

    (b) *Simulation*   If the observed value of the test statistic is $T = t$, then the p-value is given by

$$p\text{-value} = 2 \min \left\{ P_{H_0}\{ T \geq t \}, P_{H_0}\{ T \leq t \} \right\}$$

We can approximate this value by continually simulating a random selection of n of the values $1, 2, \ldots, n + m$—noting on each occasion the sum of the n values. The value of $P_{H_0}\{ T \geq t \}$ can be approximated by the proportion of time that the sum obtained is greater than or equal to t, and similarly, $P_{H_0}\{T \le t\}$, by the proportion of time that it is less than or equal to t.

Program 9-7-1 approximates the p-value by performing the preceding simulation. It operates by continually summing the last n terms of a random permutation of the numbers $1, 2, \ldots, n + m$. (The method used to simulate a random permutation is explained in Example 12.1a of Chapter 12). The program will run most efficiently when the sample of smallest size is designated as the first sample.

Example 9.7f Running Program 9-7-1 for the data in Example 9.7a yields:

```
THIS PROGRAM APPROXIMATES THE p-value IN THE TWO-SAMPLE RANK SUM TEST BY A SIMUL
ATION STUDY
Random number seed (-32768 to 32767)? 1435
ENTER THE SIZE OF THE FIRST SAMPLE
? 5
ENTER THE SIZE OF THE SECOND SAMPLE
? 6
ENTER THE SUM OF THE RANKS OF THE FIRST SAMPLE
? 21
ENTER THE DESIRED NUMBER OF SIMULATION RUNS
? 1000
THE APPROXIMATE p-value IS .126
```

which is quite close to the exact value of .1225.

Running it for the data of Example 9.7d:

```
RUN
THIS PROGRAM APPROXIMATES THE p-value IN THE TWO-SAMPLE RANK SUM TEST BY A SIMUL
ATION STUDY
Random number seed (-32768 to 32767)? 9
ENTER THE SIZE OF THE FIRST SAMPLE
? 9
ENTER THE SIZE OF THE SECOND SAMPLE
? 13
ENTER THE SUM OF THE RANKS OF THE FIRST SAMPLE
? 72
ENTER THE DESIRED NUMBER OF SIMULATION RUNS
? 500
THE APPROXIMATE p-value IS .032
```

Once again the simulation result (though not in this case as accurate as the normal approximation) is quite close to the exact value of .0364. ∎

Both of the approximation methods work quite well. The normal approximation, when n and m both exceed 7, is usually quite accurate and requires almost no computational time. The simulation approach, on the other hand, can require a great deal of computational time. However, if an immediate answer is not required and great accuracy is desired, then simulation, by running a large number of cases, can be made accurate to an arbitrarily prescribed precision.

8 THE RUN TEST FOR RANDOMNESS

A basic assumption in much of statistics is that a set of data constitutes a random sample from some population. However, it is sometimes the case that the data are not generated by a truly random process but by one that may follow a trend or a type of cyclical pattern. In this section we will consider a test, called the run test, of the hypothesis H_0 that a given data set constitutes a random sample.

To begin let us suppose that each of the data values is either a 0 or a 1. That is, we shall assume that each data value can be dichotomized as being either a success or a failure. Let $X_1, \ldots, X_N$ denote the set of data. Any consecutive sequence of either 1's or 0's is called a *run*. For instance, the data set

$$1\ 0\ 0\ 1\ 1\ 1\ 0\ 0\ 1\ 0\ 1\ 1\ 1\ 1\ 0\ 1\ 0\ 0\ 0\ 0\ 1\ 1$$

contains 11 runs—6 runs of 1 and 5 runs of 0. Suppose that the data set $X_1, \ldots, X_N$ contains n 1's and m 0's, where $n + m = N$, and let R denote the number of runs. Now, if H_0 were true, then $X_1, \ldots, X_N$ would be equally likely to be any of the $N!/n!m!$ permutations of n 1's and m 0's; and therefore, given a total of n 1's and m 0's, it follows that, under H_0, the probability mass function of R, the number of runs is given by

$$P_{H_0}\{R = k\} = \frac{\text{number of permutations of } n \text{ 1's and } m \text{ 0's resulting in } k \text{ runs}}{\binom{n+m}{n}}$$

This number of permutations can be explicitly determined and it can be shown that

$$P_{H_0}\{R = 2k\} = 2\frac{\binom{m-1}{k-1}\binom{n-1}{k-1}}{\binom{m+n}{n}} \tag{9.8.1}$$

$$P_{H_0}\{R = 2k + 1\} = \frac{\left[\binom{m-1}{k-1}\binom{n-1}{k} + \binom{m-1}{k}\binom{n-1}{k-1}\right]}{\binom{n+m}{n}}$$

If the data contains n 1's and m 0's, then the run test calls for rejection of the hypothesis that the data constitutes a random sample if the observed number of runs is either too large or too small to be explained by chance. Specifically, if the observed number of runs is r, then the p-value of the run test is

$$p\text{-value} = 2 \min\left(P_{H_0}\{R \geq r\}, P_{H_0}\{R \leq r\}\right)$$

Program 9-8 uses Equation 9.8.1 to compute the p-value.

Example 9.8a The following is the result of the last 30 games played by an athletic team, with W signifying a win and L a loss.

$$W\ W\ W\ L\ W\ W\ L\ W\ W\ L\ W\ L\ W\ W\ L\ W\ W\ W\ W\ L\ W\ L\ W\ W\ W\ L\ W\ L\ W\ L$$

Are these data consistent with pure randomness?

Solution To test the hypothesis of randomness, note that the data, which consists of 20 W's and 10 L's, contains 20 runs. To see whether this justifies rejection, at say the 5 percent level of significance, we run Program 9-8.

```
RUN
THIS PROGRAM COMPUTES THE p-value FOR THE RUN TEST OF THE HYPOTHESIS THAT A DATA
 SET OF N ONES AND M ZEROES IS RANDOM
ENTER THE NUMBER OF ONES
? 20
ENTER THE NUMBER OF ZEROES
? 10
ENTER THE NUMBER OF RUNS
? 20
THE p-value IS 1.844835E-02
```

Therefore, the hypothesis of randomness would be rejected at the 5 percent level of significance. (The striking thing about these data is that the team always came back to win after losing a game, which would be quite unlikely if all outcomes containing 20 wins and 10 losses were equally likely.) ∎

The above can also be used to test for randomness when the data values are not just 0's and 1's. To test whether the data $X_1, \ldots, X_N$ constitutes a random sample, let s-med denote the sample median. Also let n denote the number of data values that are less than or equal to s-med and m the number that are greater. (Thus, if N is even and all data values are distinct then $n = m = N/2$.) Define $I_1, \ldots, I_N$ by

$$I_j = \begin{cases} 1 & \text{if } X_j \le \text{s-med} \\ 0 & \text{otherwise} \end{cases}$$

Now, if the original data constituted a random sample, then the number of runs in $I_1, \ldots, I_N$ would have a probability mass function given by Equation 9.8.1. Thus, it follows that we can use the preceding run test on the data values $I_1, \ldots, I_N$ to test that the original data are random.

Example 9.8b The lifetimes of 19 successively produced storage batteries is as follows:

$$145\ 152\ 148\ 155\ 176\ 134\ 184\ 132\ 145\ 162\ 165$$
$$185\ 174\ 198\ 179\ 194\ 201\ 169\ 182$$

The sample median is the 10th smallest value—namely, 169. The data indicating whether the successive values are less than or equal to or greater than 169 is as follows:

$$1\ 1\ 1\ 1\ 0\ 1\ 0\ 1\ 1\ 1\ 1\ 0\ 0\ 0\ 0\ 0\ 0\ 1\ 0$$

Hence, the number of runs is equal to 8. To determine if this value is statistically significant, we run Program 9-8 (with $n = 10$, $m = 9$).

```
RUN
THIS PROGRAM COMPUTES THE p-value FOR THE RUN TEST OF THE HYPOTHESIS THAT A DATA
 SET OF N ONES AND M ZEROES IS RANDOM
ENTER THE NUMBER OF ONES
? 10
ENTER THE NUMBER OF ZEROES
? 9
ENTER THE NUMBER OF RUNS
? 8
THE p-value IS .3571197
```

Thus the hypothesis of randomness is accepted. ∎

It can be shown that, when n and m are both large and H_0 is true, R will approximately have a normal distribution with mean and standard deviation given by

$$\mu = \frac{2nm}{n+m} + 1 \quad \text{and} \quad \sigma = \sqrt{\frac{2nm(2nm - n - m)}{(n+m)^2(n+m-1)}} \tag{9.8.2}$$

Therefore, when n and m are both large

$$P_{H_0}\{R \le r\} = P_{H_0}\left\{\frac{R-\mu}{\sigma} \le \frac{r-\mu}{\sigma}\right\}$$

$$\approx P\left\{Z \le \frac{r-\mu}{\sigma}\right\} \quad Z \sim \mathcal{N}(0,1)$$

$$= \Phi\left(\frac{r-\mu}{\sigma}\right)$$

and, similarly

$$P_{H_0}\{R \ge r\} \approx 1 - \Phi\left(\frac{r-\mu}{\sigma}\right)$$

Hence, for large n and m, the p-value of the run test for randomness is approximately given by

$$p\text{-value} \approx 2\min\left\{\Phi\left(\frac{r-\mu}{\sigma}\right), \ 1 - \Phi\left(\frac{r-\mu}{\sigma}\right)\right\}$$

where μ and σ are given by Equation 9.8.2, and r is the observed number of runs.

Example 9.8c Suppose that a sequence of 60 1's and 60 0's resulted in 75 runs. Since

$$\mu = 61 \quad \text{and} \quad \sigma = \sqrt{\frac{3540}{119}} = 5.454$$

we see that the approximate p-value is

$$p\text{-value} \approx 2\min\{\Phi(2.567), \quad 1 - \Phi(2.567)\}$$
$$= 2 \times (1 - .9949)$$
$$= .0102$$

On the other hand, by running Program 9-8, we obtain that the exact p-value is

$$p\text{-value} = .0130$$

If the number of runs was equal to 70 rather than 75, then the approximate p-value would be

$$p\text{-value} \approx 2[1 - \Phi(1.650)] = .0990$$

as opposed to the exact value of

$$p\text{-value} = .1189 \quad \text{(from Program 9-8)} \quad \blacksquare$$

PROBLEMS

1. According to the Mendelian theory of genetics a certain garden pea plant should produce either white, pink, or red flowers, with respective probabilities $\frac{1}{4}, \frac{1}{2}, \frac{1}{4}$. To test this theory a sample of 564 peas were studied with the result that 141 produced white, 291 produced pink, and 132 produced red flowers. Using the chi-square approximation, what conclusion would be drawn at the 5 percent level of significance?

2. To ascertain whether a certain die was fair, 1000 rolls of the die were recorded, with the following results:

Outcome	Number of occurrences
1	158
2	172
3	164
4	181
5	160
6	165

Test the hypothesis that the die is fair (that is, that $p_i = \frac{1}{6}, i = 1,\ldots,6$) at the 5 percent level of significance. Use the chi-square approximation.

3. Determine the birth and death dates of 100 famous individuals and, using the four-category approach of Example 9.2a, test the hypothesis that the

death month is not affected by the birth month. Use the chi-square approximation.

4. It is believed that the daily number of electrical power failures in a certain Midwestern city is a Poisson random variable with mean 4.2. Test this hypothesis if over 150 days the number of days having i power failures is as follows:

Failures	Number of days
0	0
1	5
2	22
3	23
4	32
5	22
6	19
7	13
8	6
9	4
10	4
11	0

5. Among 100 vacuum tubes tested, 41 had lifetimes less than 30 hours, 31 had lifetimes between 30 and 60 hours, 13 had lifetimes between 60 and 90 hours, and 15 had lifetimes greater than 90 hours. Are these data consistent with the hypothesis that a vacuum tube's lifetime is exponentially distributed with a mean of 50 hours?

6. The past output of a machine indicates that each unit it produces will be

top grade	with probability	.40
high grade	with probability	.30
medium grade	with probability	.20
low grade	with probability	.10

A new machine, designed to perform the same job, has produced 500 units with the following results.

top grade	234
high grade	117
medium grade	81
low grade	68

Can the difference in output be ascribed solely to chance?

7. The neutrino radiation from outer space was observed during several days. The frequencies of signals were recorded for each sidereal hour and are

as given below:

Frequency of Neutrino Radiation from Outer Space

Hour starting at	Frequency of signals	Hour starting at	Frequency of signals
0	24	12	29
1	24	13	26
2	36	14	38
3	32	15	26
4	33	16	37
5	36	17	28
6	41	18	43
7	24	19	30
8	37	20	40
9	37	21	22
10	49	22	30
11	51	23	42

Test whether the signals are uniformly distributed over the 24-hour period.

8. Neutrino radiation was observed over a certain period and the number of hours in which $0, 1, 2, \ldots$ signals were received was recorded.

Number of signals per hour	Number of hours with this frequency of signals
0	1924
1	541
2	103
3	17
4	1
5	1
6 or more	0

Test the hypothesis that the observations come from a population having a Poisson distribution with mean 0.3.

9. Use simulation to determine the p-value and compare it with the result you obtained using the chi-square approximation in Problem 1. Let the number of simulation runs be　　　.

(a) 100
(b) 500
(c) 1000

10. Use simulation to approximate the p-value and compare it with the result you obtained upon using the chi-square approximation in Problem 5. Use 100 simulation runs.

11. Use simulation, with 500 runs, to determine a 95 percent confidence interval for the true p-value of the data in Problem 2.

12. In Problem 4 test the hypothesis that the daily number of failures has a Poisson distribution.

13. A random sample of 500 families was classified by region and income (in units of $1000). The following data resulted:

Income	South	North
0–10	42	53
10–20	55	90
20–30	47	88
> 30	36	89

Determine the p-value of the test that a families' income and region are independent.

14. The following data relate the mother's age and the birthweight (in grams) of her child.

	Birthweight	
Maternal Age	Less than 2500 grams	More than 2500 grams
20 years or less	10	40
Greater than 20	15	135

Test the hypothesis that the baby's birthweight is independent of the mother's age.

15. Repeat Problem 14 if all the data values were doubled—that is, if the data were

20 80
30 270

16. The number of infant mortalities as a function of the baby's birthweight (in grams) for 72,730 live white births in New York in 1974 is as follows:

	Outcome at the end of 1 year	
Birthweight	Alive	Dead
Less than 2,500	4,597	618
Greater than 2,500	67,093	422

Test the hypothesis that the birthweight is independent of whether or not the baby survives its first year.

17. An experiment designed to study the relationship between hypertension and cigarette smoking yielded the following data.

	Nonsmoker	Moderate smoker	Heavy smoker
Hypertension	20	38	28
No hypertension	50	27	18

Test the hypothesis that whether or not an individual has hypertension is independent of how much that person smokes.

18. Use simulation to approximate the p-value in Problem 17.

19. Suppose that each item produced by method 1 is defective with probability p_1, whereas each produced by method 2 is defective with probability p_2. In order to test the hypothesis $H_0: p_1 = p_2$, n_i items have been produced by method i and k_i of them have been defective, $i = 1, 2$. Use the method of Section 3 to devise a test of H_0, and discuss the relationship between this test and the test presented in Section 6 of Chapter 6.

20. The following table shows the number of defective and acceptable items in samples taken both before and after the introduction of a modification in the manufacturing process.

	Defective	*Acceptable*
Before	25	218
After	9	103

Is this change significant at the .05 level?

21. A new medicine against hypertension was tested on 18 patients. After 40 days of treatment the following changes of the diastolic blood pressure were observed:

$$-5, \quad -1, \quad +2, \quad +8, \quad -25, \quad +1, \quad +5, \quad -12, \quad -16$$
$$-9, \quad -8, \quad -18, \quad -5, \quad -22, \quad +4, \quad -21, \quad -15, \quad -11$$

Use the sign test to determine if the medicine has an affect on blood pressure. What is the p-value?

22. An engineering firm is involved in selecting a computer system, and the choice has been narrowed to two manufacturers. The firm submits eight problems to the two computer manufacturers and has each manufacturer measure the number of seconds required to solve the design problem with the manufacturer's software. The times for the eight design problems are given below.

Design problem	1	2	3	4	5	6	7	8
Time with computer A	15	32	17	26	42	29	12	38
Time with computer B	22	29	1	23	46	25	19	47

Determine the p-value of the sign test when testing the hypothesis that there is no difference in the distribution of the time it takes the two types of software to solve problems.

23. Determine the p-value when using the sign rank statistic in Problems 21 and 22.

24. Twelve patients having high albumin content in their blood were treated with a medicine. Their blood content of albumin was measured before and after treatment. The measured values are

Blood Content of Albumin[a]

Patient N	Before treatment	After treatment
1	5.02	4.66
2	5.08	5.15
3	4.75	4.30
4	5.25	5.07
5	4.80	5.38
6	5.77	5.10
7	4.85	4.80
8	5.09	4.91
9	6.05	5.22
10	4.77	4.50
11	4.85	4.85
12	5.24	4.56

[a] Values given in grams per 100 ml.

Is the effect of the medicine significant at the 5 percent level?

(a) Use the sign test.

(b) Use the sign rank test.

25. An engineer claims that painting the exterior of a particular aircraft affects its cruising speed. To check this, the next 10 aircraft off the assembly line were flown to determine cruising speed prior to painting, and were then painted and reflown. The following data resulted.

| | Cruising speed (knots) | |
Aircraft	Not painted	Painted
1	426.1	416.7
2	418.4	403.2
3	424.4	420.1
4	438.5	431.0
5	440.6	432.6
6	421.8	404.2
7	412.2	398.3
8	409.8	405.4
9	427.5	422.8
10	441.2	444.8

Do the data uphold the engineer's claim?

26. Ten pairs of duplicate spectrochemical determinations for nickel are presented below. The readings in the first column were taken with one type of measuring instrument and those in the second column were taken with another type.

Sample	Duplicates	
1	1.94	2.00
2	1.99	2.09
3	1.98	1.95
4	2.07	2.03
5	2.03	2.08
6	1.96	1.98
7	1.95	2.03
8	1.96	2.03
9	1.92	2.01
10	2.00	2.12

Test the hypothesis, at the 5 percent level of significance, that the two measuring instruments give equivalent results.

27. Let $X_1, \ldots, X_n$ be a sample from the continuous distribution F having median m; and suppose we are interested in testing the hypothesis $H_0 : m = m_0$ against the one-sided alternative $H_1 : m > m_0$. Present the one-sided analog of the

    (a) Sign test
    (b) Sign rank test

    Explain, in both cases, how the p-value would be computed.

28. In a study of bilingual coding, 12 bilingual (French and English) college students are divided into two groups. Each group reads an article written in French, and each answers a series of 25 multiple-choice questions covering the content of the article. For one group the questions are written in French; the other takes the examination in English. The score (total correct) for the two groups is:

Examination in French	11	12	16	22	25	25
Examination in English	10	13	17	19	21	24

Is this evidence at the 5 percent significance level that there is difficulty in transferring information from one language to another?

29. Fifteen cities, of roughly equal size, are chosen for a traffic safety study. Eight of them are randomly chosen, and in these cities a series of newspaper articles dealing with traffic safety are run over a one-month period. The number of traffic accidents reported in the month following this campaign is as follows.

Treatment group	19	31	39	45	47	66	74	81
Control group	28	36	44	49	·52	52	60	

Determine the exact p-value when testing the hypothesis that the articles have not had any effect.

30. Determine the p-value in Problem 29 by
    (a) Using the normal approximation
    (b) Using a simulation study

31. The following are the burning times in seconds of floating smoke pots of two different types:

Type X		Type Y	
481	572	526	537
506	561	511	582
527	501	556	601
661	487	542	558
500	524	491	578

We are interested in testing the hypothesis that the burning time distributions are the same.

(a) Determine the exact p-value.
(b) Determine the p-value yielded by the normal approximation.
(c) Run a simulation study to estimate the p-value.

32. The m sample problem: Consider m independent random samples of respective sizes $n_1, \ldots, n_m$ from the respective population distributions $F_1, \ldots, F_m$; and consider the problem of testing $H_0: F_1 = F_2 = \cdots = F_m$. To devise a test let R_i denote the sum of the ranks of the n_i elements of sample i, $i = 1, \ldots, m$. Show that when H_0 is true
    (a) $E[R_i] = \dfrac{n_i(N + 1)}{2}$ where $N = \Sigma n_i$.
    (b) Using the foregoing, and drawing insight from the goodness of fit test statistic presented at the beginning of this chapter, determine an appropriate test statistic for H_0.
    (c) Explain how an algorithm that generates a random permutation of the integers $1, 2, \ldots, N$ can be employed in a simulation study to determine the p-value when using the statistic in (b) to test H_0.

33. A production run of 50 items resulted in 11 defectives, with the defectives occurring on the following items (where the items are numbered by their order of production): $8, 12, 13, 14, 31, 32, 37, 38, 40, 41, 42$. Can we conclude that the successive items did not constitute a random sample?

34. The following data represent the successive quality levels of 25 articles: $100, 110, 122, 132, 99, 96, 88, 75, 45, 211, 154, 143, 161, 142, 99, 111, 105, 133, 142, 150, 153, 121, 126, 117, 155$. Does it appear that these data are a random sample from some population?

35. Can we use the run test if we consider whether each data value is less than or greater than some predetermined value rather than the value s-med?

C H A P T E R 10

Life Testing

1 INTRODUCTION

In this chapter, we consider a population of items having lifetimes that are assumed to be independent random variables with a common distribution that is specified up to an unknown parameter. The problem of interest will be to use whatever data are available to estimate this parameter.

In Section 2 we introduce the concept of the hazard (or failure) rate function—a useful engineering concept that can be utilized to specify lifetime distributions. In Section 3, we suppose that the underlying life distribution is exponential and show how to obtain estimates (point, interval, and Bayesian) of its mean under a variety of sampling plans. In Section 4, we develop a test of the hypothesis that two exponentially distributed populations have a common mean. In Section 5, we consider two approaches to estimating the parameters of a Weibull distribution.

2 HAZARD RATE FUNCTIONS

Consider a positive continuous random variable X, that we interpret as being the lifetime of some item, having distribution function F and density f. The *hazard rate* (sometimes called the *failure rate*) function $\lambda(t)$ of F is defined by

$$\lambda(t) = \frac{f(t)}{1 - F(t)}$$

To interpret $\lambda(t)$, suppose that the item has survived for t hours and we desire the probability that it will not survive for an additional time dt. That is, consider $P\{X \in (t, t + dt)|X > t\}$. Now

$$P\{X \in (t, t + dt)|X > t\} = \frac{P\{X \in (t, t + dt), X > t\}}{P\{X > t\}}$$

$$= \frac{P\{X \in (t, t + dt)\}}{P\{X > t\}}$$

$$\approx \frac{f(t)}{1 - F(t)}dt$$

That is, $\lambda(t)$ represents the conditional probability intensity that an item of age t will fail in the next moment.

Suppose now that the lifetime distribution is exponential. Then, by the memoryless property of the exponential distribution it follows that the distribution of remaining life for a t-year old item is the same as for a new item. Hence $\lambda(t)$ should be constant, which is verified as follows:

$$\lambda(t) = \frac{f(t)}{1 - F(t)}$$

$$= \frac{\lambda e^{-\lambda t}}{e^{-\lambda t}}$$

$$= \lambda$$

Thus, the failure rate function for the exponential distribution is constant. The parameter λ is often referred to as the *rate* of the distribution.

We now show that the failure rate function $\lambda(t), t \geq 0$, uniquely determines the distribution F. To show this, note that by definition

$$\lambda(s) = \frac{f(s)}{1 - F(s)}$$

$$= \frac{\frac{d}{ds} F(s)}{1 - F(s)}$$

$$= \frac{d}{ds} \{ -\log[1 - F(s)] \}$$

Integrating both sides of this equation from 0 to t yields

$$\int_0^t \lambda(s) \, ds = -\log[1 - F(t)] + \log[1 - F(0)]$$

$$= -\log[1 - F(t)] \qquad (\text{since } F(0) = 0)$$

which implies that

$$1 - F(t) = \exp\left\{ -\int_0^t \lambda(s) \, ds \right\} \qquad (10.2.1)$$

Hence a distribution function of a positive continuous random variable can be specified by giving its hazard rate function. For instance, if a random variable has a linear hazard rate function—that is, if

$$\lambda(t) = a + bt$$

then its distribution function is given by

$$F(t) = 1 - e^{-at - bt^2/2}$$

and differentiation yields that its density is

$$f(t) = (a + bt)e^{-(at + bt^2/2)}, \qquad t \geq 0$$

When $a = 0$, the foregoing is known as the Rayleigh density function.

Example 10.2a One often hears that the death rate of a person that smokes is, at each age, twice that of a nonsmoker. What does this mean? Does it mean that a nonsmoker has twice the probability of surviving a given number of years as does a smoker of the same age?

Solution If $\lambda_s(t)$ denotes the hazard rate of a smoker of age t and $\lambda_n(t)$ that of a nonsmoker of age t, then the foregoing is equivalent to the statement that

$$\lambda_s(t) = 2\lambda_n(t)$$

The probability that an A-year-old nonsmoker will survive until age B, $A < B$ is

$$P\{A\text{-year-old nonsmoker reaches age } B\}$$

$$= P\{\text{nonsmoker's lifetime} > B | \text{nonsmoker's lifetime} > A\}$$

$$= \frac{1 - F_{\text{non}}(B)}{1 - F_{\text{non}}(A)}$$

$$= \frac{\exp\left\{-\int_0^B \lambda_n(t)\, dt\right\}}{\exp\left\{-\int_0^A \lambda_n(t)\, dt\right\}} \qquad \text{from Equation 10.2.1}$$

$$= \exp\left\{-\int_A^B \lambda_n(t)\, dt\right\}$$

whereas the corresponding probability for a smoker is, by the same reasoning,

$$P\{A\text{-year-old smoker reaches age } B\} = \exp\left\{-\int_A^B \lambda_s(t)\, dt\right\}$$

$$= \exp\left\{-2\int_A^B \lambda_n(t)\, dt\right\}$$

$$= \left[\exp\left\{-\int_A^B \lambda_n(t)\, dt\right\}\right]^2$$

In other words, of two individuals of the same age, one of whom is a smoker and other a nonsmoker, the probability that the smoker survives to any given age is the *square* (not one-half) of the corresponding probability for a nonsmoker. For instance, if $\lambda_n(t) = 1/20$, $t \leq 50 \leq 60$, then the probability that a 50-year-old nonsmoker reaches age 60 is $e^{-1/2} = .607$, whereas the corresponding probability for a smoker is $e^{-1} = .368$. ∎

REMARK ON TERMINOLOGY

We will say that X has failure rate function $\lambda(t)$ when more precisely we mean that the distribution function of X has failure rate function $\lambda(t)$.

3 THE EXPONENTIAL DISTRIBUTION IN LIFE TESTING

3.1 Simultaneous Testing-Stopping at the rth Failure

Suppose that we are testing items whose life distribution is exponential with unknown mean θ. We put n independent items simultaneously on test and stop the experiment when there have been a total of r, $r \leq n$, failures. The problem is to then use the observed data to estimate the mean θ.

The observed data will be the following:

$$Data: \quad x_1 \leq x_2 \leq \cdots \leq x_r, \qquad i_1, i_2, \ldots, i_r \qquad (10.3.1)$$

with the interpretation that the jth item to fail was item i_j and it failed at time x_j. Thus, if we let X_i, $i = 1, \ldots, n$ denote the lifetime of component i, then the data will be as given in Equation 10.3.1 if

$$X_{i_1} = x_1, X_{i_2} = x_2, \ldots, X_{i_r} = x_r$$

other $n - r$ of the X_j are all greater than x_r

Now the probability density of X_{i_j} is

$$f_{X_{i_j}}(x_j) = \frac{1}{\theta} e^{-x_j/\theta} \qquad j = 1, \ldots, r$$

and so, by independence, the joint probability density of X_{i_j}, $j = 1, \ldots, r$ is

$$f_{X_{i_1}, \ldots, X_{i_r}}(x_1, \ldots, x_r) = \prod_{j=1}^{r} \frac{1}{\theta} e^{-x_j/\theta}$$

Also the probability that the other $n - r$ of the X's are all greater than x_r is, again using independence,

$$P\{ X_j > x_r \text{ for } j \neq i_1 \text{ or } i_2 \cdots \text{ or } i_r \} = \left(e^{-x_r/\theta} \right)^{n-r}$$

Hence, we see that the *likelihood* of the observed data—call it $L(x_1, \ldots, x_r, i_1, \ldots, i_r)$—is, for $x_1 \leq x_2 \leq \cdots \leq x_r$

$$L(x_1, \ldots, x_r, i_1, \ldots, i_r) \qquad (10.3.2)$$

$$= f_{X_{i_1}, X_{i_2}, \ldots, X_{i_r}}(x_1, \ldots, x_r) P\{ X_j > x_r, j \neq i_1, \ldots, i_r \}$$

$$= \frac{1}{\theta} e^{-x_1/\theta} \cdots \frac{1}{\theta} e^{-x_r/\theta} \left(e^{-x_r/\theta} \right)^{n-r}$$

$$= \frac{1}{\theta^r} \exp \left\{ -\frac{\sum_{i=1}^{r} x_i}{\theta} - \frac{(n-r)x_r}{\theta} \right\}$$

REMARK

The likelihood in Equation 10.3.2 not only specifies that the first r failures occur at times $x_1 \leq x_2 \leq \cdots \leq x_r$ but also that the r items to fail were, in order, $i_1, i_2, \ldots, i_r$. If we only desired the density function of the first r failure times, then since there are $n(n - 1) \cdots (n - (r - 1)) = n!/(n - r)!$ possible (ordered) choices of the first r items to fail, it follows that the joint density is, for $x_1 \le x_2 \le \cdots \le x_r$

$$f(x_1, x_2, \ldots, x_r) = \frac{n!}{(n-r)!} \frac{1}{\theta^r} \exp\left\{-\frac{\sum_{i=1}^r x_i}{\theta} - \frac{(n-r)}{\theta} x_r\right\} \quad \blacksquare$$

To obtain the maximum likelihood estimator of θ, we take the logarithm of both sides of Equation 10.3.2. This yields

$$\log L(x_1, \ldots, x_r, i_1, \ldots, i_r) = -r \log \theta - \frac{\sum_{i=1}^r x_i}{\theta} - \frac{(n-r)x_r}{\theta}$$

and so

$$\frac{\partial}{\partial \theta} \log L(x_1, \ldots, x_r, i_1, \ldots, i_r) = -\frac{r}{\theta} + \frac{\sum_{i=1}^r x_i}{\theta^2} + \frac{(n-r)x_r}{\theta^2}$$

Equating to 0 and solving yields that $\hat{\theta}$, the maximum likelihood estimate, is given by

$$\hat{\theta} = \frac{\sum_{i=1}^r x_i + (n-r)x_r}{r}$$

Hence, if we let $X_{(i)}$ denote the time at which the ith failure occurs ($X_{(i)}$ is called the ith *order statistic*), then the maximum likelihood estimator of θ is

$$\hat{\theta} = \frac{\sum_{i=1}^r X_{(i)} + (n-r)X_{(r)}}{r} \tag{10.3.3}$$

$$= \frac{\tau}{r}$$

where τ, defined to equal the numerator in Equation 10.3.3, is called the total-time-on-test statistic. We call it this since the ith item to fail functions for a time $X_{(i)}$ (and then fails), $i = 1, \ldots, r$, whereas the other $n - r$ items function throughout the test (which lasts for a time $X_{(r)}$). Hence the sum of the times that all the items are on test is equal to τ.

To obtain a confidence interval for θ, we will determine the distribution of τ, the total time on test. Recalling that $X_{(i)}$ is the time of the ith failure, $i = 1, \ldots, r$, we will start by rewriting the expression for τ. To write an expression for τ, rather than summing the total time on test of each of the items, let us ask how much additional time on test was generated between each successive failure. That is, let us denote by Y_i, $i = 1, \ldots, r$, the additional time on test generated between the $(i-1)$st and ith failure. Now up to the first $X_{(1)}$ time units (as all n items are functioning throughout this interval) the total time on test is

$$Y_1 = nX_{(1)}$$

Between the first and second failures, there are a total of $n - 1$ functioning items and so

$$Y_2 = (n-1)(X_{(2)} - X_{(1)})$$

In general, we have

$$Y_1 = nX_{(1)}$$

$$Y_2 = (n - 1)(X_{(2)} - X_{(1)})$$

$$\vdots$$

$$Y_j = (n - j + 1)(X_{(j)} - X_{(j-1)})$$

$$\vdots$$

$$Y_r = (n - r + 1)(X_{(r)} - X_{(r-1)})$$

and

$$\tau = \sum_{j=1}^{r} Y_j$$

The importance of the foregoing representation for τ follows from the fact that the distributions of the Y_j's are easily obtained as follows. Since $X_{(1)}$, the time of the first failure, is the minimum of n independent exponential lifetimes, each having rate $1/\theta$, it follows from Proposition 3.6.1 of Chapter 3 that it is itself exponentially distributed with rate n/θ. That is, $X_{(1)}$ is exponential with mean θ/n; and so $nX_{(1)}$ is exponential with mean θ. Also, at the moment when the first failure occurs, the remaining $n - 1$ functioning items are, by the lack of memory property of the exponential, as good as new and so each will have an additional life that is exponential with mean θ; hence, the additional time until one of them fails is exponential with rate $(n - 1)/\theta$. That is, independent of $X_{(1)}$, $X_{(2)} - X_{(1)}$ is exponential with mean $\theta/(n - 1)$ and so $Y_2 = (n - 1)(X_{(2)} - X_{(1)})$ is exponential with mean θ. Indeed, continuing this argument leads us to the following conclusion:

$$Y_1, \ldots, Y_r \text{ are independent exponential} \tag{10.3.4}$$
$$\text{random variables each having mean } \theta$$

Hence, since the sum of independent and identically distributed exponential random variables has a gamma distribution (Corollary 3.7.2 of Chapter 3) we see that

$$\tau \sim \text{gamma}(r, 1/\theta)$$

That is, τ has a gamma distribution with parameters r and $1/\theta$. Equivalently, by recalling that a gamma random variable with parameters $(r, 1/\theta)$ is equivalent to $\theta/2$ times a chi-square random variable with $2r$ degrees of freedom (see Section 8.1 of Chapter 3), we obtain that

$$\frac{2\tau}{\theta} \sim \chi^2_{2r} \tag{10.3.5}$$

That is, $2\tau/\theta$ has a chi-square distribution with $2r$ degrees of freedom. Hence,

$$P\left\{ \chi^2_{1-\alpha/2, 2r} < 2\tau/\theta < \chi^2_{\alpha/2, 2r} \right\} = 1 - \alpha$$

and so a $100(1 - \alpha)$ percent confidence interval for θ is

$$\theta \in \left(\frac{2\tau}{\chi^2_{\alpha/2, 2r}}, \frac{2\tau}{\chi^2_{1-\alpha/2, 2r}} \right) \tag{10.3.6}$$

One-sided confidence intervals can be similarly obtained.

Example 10.3a A sample of 50 transistors are simultaneously put on a test that is to be ended when the 15th failure occurs. If the total time on test of all transistors is equal to 525 hours, determine a 95 percent confidence interval for the mean lifetime of a transistor. Assume that the underlying distribution is exponential.

Solution From Program or Table 3-8-1-B we obtain that

$$\chi^2_{.025, 30} = 46.98 \qquad \chi^2_{.975, 30} = 16.89$$

and so, using Equation 10.3.6 we can assert with 95 percent confidence that

$$\theta \in (22.35, 62.17) \quad \blacksquare$$

In testing a hypothesis about θ, we can use Equation 10.3.6 to determine the *p*-value of the test data. For instance, suppose we are interested in the one-sided test of

$$H_0 : \theta \geq \theta_0$$

against the alternative

$$H_1 : \theta < \theta_0$$

This can be tested by first computing the value of the test statistic $2\tau/\theta_0$—call this value v—and then computing the probability that a chi-square random variable with $2r$ degrees of freedom would be as small as v. This probability is the *p*-value in the sense that it represents the (maximal) probability that such a small value of $2\tau/\theta_0$ would have been observed if H_0 were true. The hypothesis should then be rejected at all significance levels at least as large as this *p*-value.

Example 10.3b A producer of batteries claims that the lifetimes of the items the manufactures are exponentially distributed with a mean life of at least 150 hours. To test this claim, 100 batteries are simultaneously put on a test that is slated to end when the 20th failure occurs. If, at the end of the experiment, the total test time of all the 100 batteries is equal to 1800, should the manufacturer's claim be accepted?

Since $2\tau/\theta_0 = 3600/150 = 24$ the *p*-value is

$$p\text{-value} = P\{\chi^2_{40} \leq 24\}$$

$$= .021 \qquad \text{from Program 3-8-1-A}$$

Hence, the manufacturer's claim should be rejected at the 5 percent level of significance (indeed at any significance level at least as large as .021). $\blacksquare$

It follows from Equation 10.3.5 that the accuracy of the estimator τ/r depends only on r and not on n, the number of items put on test. The importance of n resides in the fact that by choosing it large enough we can ensure that the test is, with high probability, of short duration. In fact, the moments of $X_{(r)}$, the time at which the test ends are easily obtained. Since, with $X_{(0)} \equiv 0$,

$$X_{(j)} - X_{(j-1)} = \frac{Y_j}{n - j + 1} \qquad j = 1, \ldots, r$$

it follows upon summing that

$$X_{(r)} = \sum_{j=1}^{r} \frac{Y_j}{n - j + 1}$$

Hence, from Equation 10.3.4 $X_{(r)}$ is the sum of r independent exponentials having respective means $\theta/n, \theta/(n-1), \ldots, \theta/(n-r+1)$. Using this, we see that

$$E[X_{(r)}] = \sum_{j=1}^{r} \frac{\theta}{n - j + 1} = \theta \sum_{j=n-r+1}^{n} \frac{1}{j} \qquad (10.3.7)$$

$$\mathrm{Var}(X_{(r)}) = \sum_{j=1}^{r} \left(\frac{\theta}{n - j + 1} \right)^2 = \theta^2 \sum_{j=n-r+1}^{n} \frac{1}{j^2}$$

where the second equality uses the fact that the variance of an exponential is equal to the square of its mean. For large n, we can approximate the preceding sums as follows:

$$\sum_{j=n-r+1}^{n} \frac{1}{j} \approx \int_{n-r+1}^{n} \frac{dx}{x} = \log\left(\frac{n}{n - r + 1} \right)$$

$$\sum_{j=n-r+1}^{n} \frac{1}{j^2} \approx \int_{n-r+1}^{n} \frac{dx}{x^2} = \frac{1}{n - r + 1} - \frac{1}{n} = \frac{r - 1}{n(n - r + 1)}$$

Thus, for instance, if in Example 10.3b the true mean life was 120 hours, then the expectation and variance of the length of the test are approximately given by

$$E[X_{(20)}] \approx 120 \log\left(\frac{100}{81} \right) = 25.29$$

$$\mathrm{Var}(X_{(20)}) \approx (120)^2 \frac{19}{100(81)} = 33.78$$

3.2 Sequential Testing

Suppose now that we have an infinite supply of items, each of whose lifetime is exponential with an unknown mean θ, which are to be tested sequentially, in that the first item is put on test and on its failure the second is put on test, and so on. That is, as soon as an item fails, it is immediately replaced on life test by the next item. We suppose that at some fixed time T the test ends.

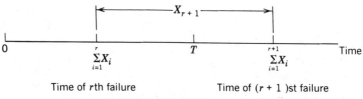

Time of rth failure   Time of $(r + 1)$st failure

FIGURE 10.3.1 r failures by time T

The observed data will consist of the following:

$$\text{Data:} \quad r, x_1, x_2, \ldots, x_r$$

with the interpretation that there have been a total of r failures with the ith item on test having functioned for a time x_i. Now the foregoing will be the observed data if

$$X_i = x_i, \quad i = 1, \ldots, r, \quad \sum_{i=1}^{r} x_i < T \qquad (10.3.8)$$

$$X_{r+1} > T - \sum_{i=1}^{r} x_i$$

where X_i is the functional lifetime of the ith item to be put in use. This follows since in order for there to be r failures, the rth failure must occur before time T—and so $\sum_{i=1}^{r} X_i < T$—and the functional life of the $(r + 1)$st item must exceed $T - \sum_{i=1}^{r} X_i$ (See Figure 10.3.1).

From Equation 10.3.8 we obtain that the likelihood of the data $r, x_1, \ldots, x_r$ is as follows:

$$f(r, x_1, \ldots, x_r | \theta)$$

$$= f_{X_1, \ldots, X_r}(x_1, \ldots, x_r) P\left\{ X_{r+1} > T - \sum_{i=1}^{r} x_i \right\}, \quad \sum_{i=1}^{r} x_i < T$$

$$= \frac{1}{\theta^r} e^{-\sum_{i=1}^{r} x_i / \theta} e^{-(T - \sum_{i=1}^{r} x_i)/\theta}$$

$$= \frac{1}{\theta^r} e^{-T/\theta}$$

Therefore,

$$\log f(r, x_1, \ldots, x_r | \theta) = -r \log \theta - \frac{T}{\theta}$$

and so

$$\frac{\partial}{\partial \theta} \log f(r, x_1, \ldots, x_r | \theta) = -\frac{r}{\theta} + \frac{T}{\theta^2}$$

On equating to 0 and solving, we obtain that the maximum likelihood estimate for θ is

$$\hat{\theta} = \frac{T}{r}$$

Since T is the total time on test of all items, it follows once again that the maximum likelihood estimate of the unknown exponential mean is equal to the total time on test divided by the number of observed failures in this time.

If we let $N(T)$ denote the number of failures by time T, then the maximum likelihood estimator of θ is $T/N(T)$. Suppose now that the observed value of $N(T)$ is $N(T) = r$. To determine a $100(1 - \alpha)$ percent confidence interval estimate for θ, we will first determine the values θ_L and θ_U, which are such that

$$P_{\theta_U}\{N(T) \geq r\} = \frac{\alpha}{2} \qquad P_{\theta_L}\{N(T) \leq r\} = \frac{\alpha}{2}$$

where by $P_\theta(A)$ we mean that we are computing the probability of the event A under the supposition that θ is the true mean. The $100(1 - \alpha)$ percent confidence interval estimate for θ is

$$\theta \in (\theta_L, \theta_U)$$

To understand why those values of θ for which either $\theta < \theta_L$ or $\theta > \theta_U$ are not included in the confidence interval, note that $P_\theta\{N(T) \geq r\}$ decreases and $P_\theta\{N(T) \leq r\}$ increases in θ (why?). Hence,

if $\theta < \theta_L$ then $P_\theta\{N(T) \leq r\} < P_{\theta_L}\{N(T) \leq r\} = \dfrac{\alpha}{2}$

if $\theta > \theta_U$ then $P_\theta\{N(T) \geq r\} < P_{\theta_U}\{N(T) \geq r\} = \dfrac{\alpha}{2}$

It remains to determine θ_L and θ_U. To do so note first that the event that $N(T) \geq r$ is equivalent to the statement that the rth failure occurs before or at time T. That is,

$$N(T) \geq r \Leftrightarrow X_1 + \cdots + X_r \leq T$$

and so

$$P_\theta\{N(T) \geq r\} = P_\theta\{X_1 + \cdots + X_r \leq T\}$$
$$= P\{\text{gamma}\,(r, 1/\theta) \leq T\}$$
$$= P\left\{\frac{\theta}{2}\chi^2_{2r} \leq T\right\}$$
$$= P\{\chi^2_{2r} \leq 2T/\theta\}$$

Hence, upon evaluating the foregoing at $\theta = \theta_U$, and using the fact that $P\{\chi^2_{2r} \leq \chi^2_{1-\alpha/2, 2r}\} = \alpha/2$, we obtain that

$$\frac{\alpha}{2} = P\left\{\chi^2_{2r} \leq \frac{2T}{\theta_U}\right\}$$

implying that

$$\frac{2T}{\theta_U} = \chi^2_{1-\alpha/2, 2r}$$

or

$$\theta_U = 2T/\chi^2_{1-\alpha/2, 2r}$$

Similarly, we can show that

$$\theta_L = 2T/\chi^2_{\alpha/2, 2r}$$

and thus the $100(1 - \alpha)$ percent confidence interval estimate for θ is

$$\theta \in \left(2T/\chi^2_{\alpha/2, 2r}, 2T/\chi^2_{1-\alpha/2, 2r}\right)$$

Example 10.3c If a one-at-a-time sequential test yields 10 failures in the fixed time of $T = 500$ hours, then the maximum likelihood estimate of θ is $500/10 = 50$ hours. A 95 percent confidence interval estimate of θ is

$$\theta \in \left(1000/\chi^2_{.025, 20}, 1000/\chi^2_{.975, 20}\right)$$

Running Program 3-8-1-B yields that

$$\chi^2_{.025, 20} = 34.17 \qquad \chi^2_{.975, 20} = 9.66$$

and so, with 95 percent confidence

$$\theta \in (29.27, 103.52). \quad \blacksquare$$

If we wanted to test the hypothesis

$$H_0 : \theta = \theta_0$$

against the alternative

$$H_1 : \theta \neq \theta_0$$

then we would first determine the value of $N(T)$. If $N(T) = r$ then the hypothesis would be rejected provided either

$$P_{\theta_0}\{N(T) \leq r\} \leq \frac{\alpha}{2} \qquad \text{or} \qquad P_{\theta_0}\{N(T) \geq r\} \leq \frac{\alpha}{2}$$

In other words, H_0 would be rejected at all significance levels greater than or equal to the p-value given by

$$p\text{-value} = 2\min\left(P_{\theta_0}\{N(T) \geq r\}, P_{\theta_0}\{N(T) \leq r\}\right)$$

$$p\text{-value} = 2\min\left(P_{\theta_0}\{N(T) \geq r\}, 1 - P_{\theta_0}\{N(T) \geq r + 1\}\right)$$

$$= 2\min\left(P\left\{\chi^2_{2r} \leq \frac{2T}{\theta_0}\right\}, 1 - P\left\{\chi^2_{2(r+1)} \leq \frac{2T}{\theta_0}\right\}\right)$$

The p-value for a one-sided test is similarly obtained.

The chi-square probabilities in the foregoing can be computed by making use of Program 3-8-1-A.

Example 10.3d A company claims that the mean lifetimes of the semiconductors it produces is at least 25 hours. To substantiate this claim, an independent testing service has decided to sequentially test, one at a time, the company's semiconductors for 600 hours. If 30 semiconductors failed during this period, what can we say about the validity of the company's claim? Test at the 10 percent level.

Solution This is a one-sided test of

$$H_0 : \theta \geq 25 \qquad \text{versus} \qquad H_1 : \theta < 25$$

The relevant probability for determining the *p*-value is the probability that there would have been as many as 30 failures if the mean life were 25. That is,

$$
\begin{aligned}
\text{p-value} &= P_{25}\{N(600) \geq 30\} \\
&= P\{\chi^2_{60} \leq 1200/25\} \\
&= .132 \qquad \text{from Program 3-8-1-A}
\end{aligned}
$$

Thus, H_0 would be accepted when the significance level is .10. ∎

3.3 Simultaneous Testing—Stopping by a Fixed Time

Suppose again that we are testing items whose life distributions are independent exponential random variables with a common unknown mean θ. As in Section 3.1, the n items are simultaneously put on test, but now we suppose that the test is to stop either at some fixed time T or whenever all n items have failed—whichever occurs first. The problem is to use the observed data to estimate θ.

The observed data will be as follows:

$$\text{Data:} \quad i_1, i_2, \ldots, i_r, \qquad x_1, x_2, \ldots, x_r$$

with the interpretation that the preceding results when the r items numbered $i_1, \ldots, i_r$ are observed to fail at respective times $x_1, \ldots, x_r$; and the other $n - r$ items have not failed by time T.

Since an item will not have failed by time T if and only if its lifetime is greater than T, we see that the likelihood of the foregoing data is

$$
\begin{aligned}
f(i_1, \ldots, i_r, x_1, \ldots, x_r) &= f_{X_{i_1}, \ldots, X_{i_r}}(x_1, \ldots, x_r) P\{X_j > T, \ j \neq i_1, \ldots, i_r\} \\
&= \frac{1}{\theta} e^{-x_1/\theta} \cdots \frac{1}{\theta} e^{-x_r/\theta} (e^{-T/\theta})^{n-r} \\
&= \frac{1}{\theta^r} \exp\left\{ -\frac{\sum_{i=1}^r x_i}{\theta} - \frac{(n-r)T}{\theta} \right\}
\end{aligned}
$$

To obtain the maximum likelihood estimates, take logs to obtain

$$\log f(i_1, \ldots, i_r, x_1, \ldots, x_r) = -r \log \theta - \frac{\sum_1^r x_i}{\theta} - \frac{(n-r)T}{\theta}$$

Hence,

$$\frac{\partial}{\partial \theta} \log f(i_1, \ldots, i_r, x_1, \ldots, x_r) = -\frac{r}{\theta} + \frac{\sum_1^r x_i + (n-r)T}{\theta^2}$$

Equating to 0 and solving yields that $\hat{\theta}$, the maximum likelihood estimate, is given by

$$\hat{\theta} = \frac{\sum_{i=1}^r x_i + (n-r)T}{r}$$

Hence, if we let R denote the number of items that fail by time T and let $X_{(i)}$ be the ith smallest of the failure times, $i = 1, \ldots, R$, then the maximum likelihood estimator of θ is

$$\hat{\theta} = \frac{\sum_{i=1}^{R} X_{(i)} + (n - R)T}{R}$$

Let τ denote the sum of the times that all items are on life test—that is, τ —is the total-time-on-test statistic. Then, as the R items that fail are on test for times $X_{(1)}, \ldots, X_{(R)}$ whereas the $n - R$ nonfailed items are all on test for time T, it follows that

$$\tau = \sum_{i=1}^{R} X_{(i)} + (n - R)T$$

and thus we can write the maximum likelihood estimator as

$$\hat{\theta} = \frac{\tau}{R}$$

In words, the maximum likelihood estimator of the mean life is (as in the life testing procedures of Sections 3.1 and 3.2) equal to the total time on test divided by the number of items observed to fail.

REMARK

As the reader may possibly have surmised, it turns out that for all possible life testing schemes for the exponential distribution, the maximum likelihood estimator of the unknown mean θ will always be equal to the total time on test divided by the number of observed failures. To see why this is true, consider *any* testing situation and suppose that the outcome of the data is that r items are observed to fail after having been on test for times $x_1, \ldots, x_r$ respectively and that s items have not yet failed when the test ends—at which time they had been on test for respective times $y_1, \ldots, y_s$. The likelihood of this outcome will be

$$\text{Likelihood} = K \frac{1}{\theta} e^{-x_1/\theta} \cdots \frac{1}{\theta} e^{-x_r/\theta} e^{-y_1/\theta} \cdots e^{-y_s/\theta} \qquad (10.3.9)$$

$$= \frac{K}{\theta^r} \exp \left\{ \frac{-\left(\sum_{i=1}^{r} x_i + \sum_{i=1}^{s} y_i \right)}{\theta} \right\}$$

where K, which is a function of the testing scheme and the data, does not depend on θ. (For instance, K may relate to a testing procedure in which the decision as to when to stop depends not only on the observed data but is allowed to be random). It follows from the foregoing that the maximum likelihood estimate of θ will be

$$\hat{\theta} = \frac{\sum_{i=1}^{r} x_i + \sum_{i=1}^{s} y_i}{r}. \qquad (10.3.10)$$

But $\sum_{i=1}^{r} x_i + \sum_{i=1}^{s} y_i$ is just the total-time-on-test statistic and so the maximum likelihood estimator of θ is indeed the total time on test divided by the number of observed failures in that time.

The distribution of τ/R is rather complicated for the life-testing scheme described in this section† and thus we will not be able to easily derive a confidence interval estimator for θ. Indeed, we will not further pursue this problem but rather will consider the Bayesian approach to estimating θ.

3.4 The Bayesian Approach

Suppose that items having independent and identically distributed exponential lifetimes with an unknown mean θ are put on life test. Then as noted in the remark given in Section 3.3, the likelihood of the data can be expressed as

$$f(\text{data}|\theta) = \frac{K}{\theta^r}e^{-t/\theta}$$

where t is the total time on test—that is, the sum of the time on test of all items used—and r is the number of observed failures for the given data.

Let $\lambda = 1/\theta$ denote the rate of the exponential distribution. In the Bayesian approach, it is more convenient to work with the rate λ rather than its reciprocal. From the foregoing we see that

$$f(\text{data}|\lambda) = K\lambda^r e^{-\lambda t}$$

If we suppose that prior to testing, we feel that λ is distributed according to the prior density $g(\lambda)$, then the posterior density of λ given the observed data is as follows:

$$f(\lambda|\text{data}) = \frac{f(\text{data}|\lambda)g(\lambda)}{\int f(\text{data}|\lambda)g(\lambda)\,d\lambda} \qquad (10.3.11)$$

$$= \frac{\lambda^r e^{-\lambda t}g(\lambda)}{\int \lambda^r e^{-\lambda t}g(\lambda)\,d\lambda}$$

The preceding posterior density becomes particularly convenient to work with when g is a gamma density function with parameters, say, (b, a)—that is, when

$$g(\lambda) = \frac{ae^{-a\lambda}(a\lambda)^{b-1}}{\Gamma(b)}, \qquad \lambda > 0$$

for some nonnegative constants a and b. Indeed for this choice of g we have from Equation 10.3.11 that

$$f(\lambda|\text{data}) = Ce^{-(a+t)\lambda}\lambda^{r+b-1}$$

$$= Ke^{-(a+t)\lambda}[(a+t)\lambda]^{b+r-1}$$

†For instance, for the scheme considered τ and R are not only both random but are also dependent.

where C and K do not depend on λ. As we recognize the preceding as the gamma density with parameters $(b + r, a + t)$ we can rewrite it as

$$f(\lambda|\text{data}) = \frac{(a + t)e^{-(a+t)\lambda}[(a + t)\lambda]^{b+r-1}}{\Gamma(b + r)} \qquad \lambda > 0$$

In other words, if the prior distribution of λ is gamma with parameters (b, a), then no matter what the testing scheme, the (posterior) conditional distribution of λ given the data is gamma with parameters $(b + R, a + \tau)$ where τ and R represent respectively the total-time-on-test statistic and the number of observed failures. Because the mean of a gamma random variable with parameters (b, a) is equal to b/a (see Section 7 of Chapter 3), we can conclude that $E[\lambda|\text{data}]$, the Bayes estimator of λ, is

$$E[\lambda|\text{data}] = \frac{b + R}{a + \tau}$$

Example 10.3e Suppose that 20 items having an exponential life distribution with an unknown rate λ are put on life test at various times. When the test is ended, there have been 10 observed failures—their lifetimes being (in hours) $5, 7, 6.2, 8.1, 7.9, 15, 18, 3.9, 4.6, 5.8$. The 10 items that did not fail had, at the time the test was terminated, been on test for times (in hours) $3, 3.2, 4.1, 1.8, 1.6, 2.7, 1.2, 5.4, 10.3, 1.5$. If prior to the testing it was felt that λ could be viewed as being a gamma random variable with parameters $(2, 20)$, what is the Bayes estimator of λ?

Solution Since

$$\tau = 116.1 \qquad R = 10$$

it follows that the Bayes estimate of λ is

$$E[\lambda|\text{data}] = \frac{12}{136.1} = .088 \quad \blacksquare$$

REMARK

As we have seen, the choice of a gamma prior distribution for the rate of an exponential distribution makes the resulting computations quite simple. Whereas, from an applied viewpoint, this is not a sufficient rationale, such a choice is often made with one justification being that the flexibility in fixing the two parameters of the gamma prior always enables one to reasonably approximate their true prior feelings.

4 A TWO-SAMPLE PROBLEM

A company has set up two separate plants to produce vacuum tubes. The company supposes that tubes produced at Plant I function for an exponentially distributed time with an unknown mean θ_1 whereas those produced at Plant II function for an exponentially distributed time with unknown mean θ_2.

To test the hypothesis that there is no difference between the two plants (at least in regard to the lifetimes of the tubes they produce), the company samples n tubes from Plant I and m from Plant II and then utilizes these tubes to determine their lifetimes. How can they thus determine whether the two plants are indeed identical?

If we let $X_1, \ldots, X_n$ denote the lifetimes of the n tubes produced at Plant I and $Y_1, \ldots, Y_m$ denote the lifetimes of the m tubes produced at Plant II, then the problem is to test the hypothesis that $\theta_1 = \theta_2$ when the X_i, $i = 1, \ldots, n$ are a random sample from an exponential distribution with mean θ_1 and the Y_i, $i = 1, \ldots, m$ are a random sample from an exponential distribution with mean θ_2. Moreover, the two samples are supposed to be independent.

To develop a test of the hypothesis that $\theta_1 = \theta_2$, let us begin by noting that $\sum_{i=1}^{n} X_i$ and $\sum_{i=1}^{m} Y_i$ (being the sum of independent and identically distributed exponentials) are independent gamma random variables with respective parameters $(n, 1/\theta_1)$ and $(m, 1/\theta_2)$. Hence, by the equivalence of the gamma and chi-square distribution it follows that

$$\frac{2}{\theta_1} \sum_{i=1}^{n} X_i \sim \chi^2_{2n}$$

$$\frac{2}{\theta_2} \sum_{i=1}^{m} Y_i \sim \chi^2_{2m}$$

Hence, it follows from the definition of the F distribution that

$$\frac{\left(\frac{2}{\theta_1} \sum_{i=1}^{n} X_i \right)}{2n} \Bigg/ \frac{\left(\frac{2}{\theta_2} \sum_{i=1}^{m} Y_i \right)}{2m} \sim F_{n,m}$$

That is, if $\overline{X}$ and $\overline{Y}$ are the two sample means respectively, then

$$\frac{\theta_2 \overline{X}}{\theta_1 \overline{Y}} \quad \text{has an } F \text{ distribution with } n \text{ and } m \text{ degrees of freedom}$$

Hence, when the hypothesis that $\theta_1 = \theta_2$ is true, we see that $\overline{X}/\overline{Y}$ has an F distribution with n and m degrees of freedom. This suggests the following test of the hypothesis that $\theta_1 = \theta_2$.

Test: $H_0: \theta_1 = \theta_2$ vs. alternative : $\theta_1 \neq \theta_2$

Step 1: Choose a significance level α.

Step 2: Determine the value of the test statistic $\overline{X}/\overline{Y}$—say its value is v.

Step 3: Compute $P\{F \leq v\}$ where $F \sim F_{n,m}$. If this probability is either less than $\alpha/2$ (which occurs when $\overline{X}$ is significantly less than $\overline{Y}$) or greater than $1 - \alpha/2$ (which occurs when $\overline{X}$ is significantly greater than $\overline{Y}$) then the hypothesis is rejected.

In other words, the p-value of the test data is given by

$$p\text{-value} = 2 \min \left(P\{ F \le v \}, 1 - P\{ F \le v \} \right)$$

Example 10.4a Test the hypothesis, at the 5 percent level of significance, that the lifetimes of items produced at two given plants have the same exponential life distribution if a sample of size 10 from the first plant has a total lifetime of 420 hours whereas a sample of 15 from the second plant has a total lifetime of 510 hours.

Solution The value of the test statistic $\overline{X}/\overline{Y}$ is $42/34 = 1.2353$. To compute the probability that an F random variable with parameters $10, 15$ is less than this value, we run Program 3-8-3-A.

```
RUN
THIS PROGRAM COMPUTES THE PROBABILITY THAT AN F RANDOM VARIABLE WITH  DEGREES
   FREEDOM N AND M IS LESS THAN X
ENTER THE FIRST DEGREE OF FREEDOM PARAMETER
? 10
ENTER THE SECOND DEGREE OF FREEDOM PARAMETER
? 15
ENTER THE DESIRED VALUE OF X
? 1.2353
THE PROBABILITY IS .6554172
```

Thus, we obtain that the p-value is equal to $2(1 - .6554) = .6892$ and so H_0 is accepted. ∎

5   THE WEIBULL DISTRIBUTION IN LIFE TESTING

Whereas the exponential distribution arises as the life distribution when the hazard rate function $\lambda(t)$ is assumed to be constant over time, there are many situations in which it is more realistic to suppose that $\lambda(t)$ either increases or decreases over time. One example of such a hazard rate function is given by

$$\lambda(t) = \alpha \beta t^{\beta-1} \qquad t > 0 \tag{10.5.1}$$

where α and β are positive constants. The distribution whose hazard rate function is given by Equation 10.5.1 is called the *Weibull* distribution with parameters (α, β). Note that $\lambda(t)$ increases when $\beta > 1$; decreases when $\beta < 1$; and is constant (reducing to the exponential) when $\beta = 1$.

The Weibull distribution function is obtained from Equation 10.5.1 as follows:

$$F(t) = 1 - \exp\left\{ - \int_0^t \lambda(s)\, ds \right\} \qquad t > 0$$

$$= 1 - \exp\left\{ -\alpha t^\beta \right\}$$

Differentiating yields its density function:

$$f(t) = \alpha \beta t^{\beta-1} \exp\left\{ -\alpha t^\beta \right\} \qquad t > 0 \tag{10.5.2}$$

This density is plotted for a variety of values of α and β in Figure 10.5.1.

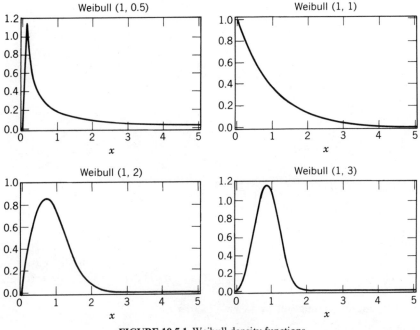

FIGURE 10.5.1 Weibull density functions

Suppose now that $X_1, \ldots, X_n$ are independent Weibull random variables each having parameters (α, β), which are assumed unknown. To estimate α and β, we can employ the maximum likelihood approach. Equation 10.5.2 yields the likelihood, given by

$$f(x_1, \ldots, x_n) = \alpha^n \beta^n x_1^{\beta-1} \cdots x_n^{\beta-1} \exp\left\{-\alpha \sum_{i=1}^{n} x_i^{\beta}\right\}$$

Hence,

$$\log f(x_1, \ldots, x_n) = n \log \alpha + n \log \beta + (\beta - 1) \sum_{i=1}^{n} \log x_i - \alpha \sum_{i=1}^{n} x_i^{\beta}$$

and

$$\frac{\partial}{\partial \alpha} \log f(x_1, \ldots, x_n) = \frac{n}{\alpha} - \sum_{i=1}^{n} x_i^{\beta}$$

$$\frac{\partial}{\partial \beta} \log f(x_1, \ldots, x_n) = \frac{n}{\beta} + \sum_{i=1}^{n} \log x_i - \alpha \sum_{i=1}^{n} x_i^{\beta} \log x_i$$

Equating to zero shows that the maximum likelihood estimates $\hat{\alpha}$ and $\hat{\beta}$ are the solutions of

$$\frac{n}{\hat{\alpha}} = \sum_{i=1}^{n} x_i^{\hat{\beta}}$$

$$\frac{n}{\hat{\beta}} + \sum_{i=1}^{n} \log x_i = \hat{\alpha} \sum_{i=1}^{n} x_i^{\hat{\beta}} \log x_i$$

or, equivalently

$$\hat{\alpha} = \frac{n}{\sum_{i=1}^{n} x_i^{\hat{\beta}}}$$

$$n + \hat{\beta} \log \left(\prod_{i=1}^{n} x_i \right) = \frac{n\hat{\beta} \sum_{i=1}^{n} x_i^{\hat{\beta}} \log x_i}{\sum_{i=1}^{n} x_i^{\hat{\beta}}}$$

This latter equation can then be solved numerically for $\hat{\beta}$, which will then also determine $\hat{\alpha}$. However, rather than pursuing this approach any further, let us consider a second approach, which is not only computationally easier but appears, as indicated by a simulation study, to yield more accurate estimates.

5.1   Parameter Estimation by Least Squares

Let $X_1, \ldots, X_n$ be a sample from the distribution

$$F(x) = 1 - e^{-\alpha x^{\beta}}, \qquad x \geq 0$$

Note that

$$\log(1 - F(x)) = -\alpha x^{\beta}$$

or

$$\log \left(\frac{1}{1 - F(x)} \right) = \alpha x^{\beta}$$

and so

$$\log \log \left(\frac{1}{1 - F(x)} \right) = \beta \log x + \log \alpha \qquad (10.5.3)$$

Now let $X_{(1)} < X_{(2)} < \cdots < X_{(n)}$ denote the ordered sample values—that is, for $i = 1, \ldots, n$

$$X_{(i)} = i\text{th smallest of } X_1, \ldots, X_n$$

and suppose that the data results in $X_{(i)}$ equaling $x_{(i)}$. If we were able to approximate the quantities $\log \log (1/[1 - F(x_{(i)})]$—say by the values $y_1, \ldots, y_n$—then from Equation 10.5.3, we could conclude that

$$y_i \approx \beta \log x_{(i)} + \log \alpha, \qquad i = 1, \ldots, n$$

We could then choose α and β to minimize the sum of the squared errors—that is, α and β are chosen to

$$\underset{\alpha, \beta}{\text{minimize}} \sum_{i=1}^{n} \left(y_i - \beta \log x_{(i)} - \log \alpha \right)^2$$

Indeed, using Proposition 7.2.1 of Chapter 7, we obtain that the preceding minimum is attained when $\alpha = \hat{\alpha}$, $\beta = \hat{\beta}$ where

$$\hat{\beta} = \frac{\sum_{i=1}^{n} y_i \log x_{(i)} - n \overline{\log x}\, \overline{y}}{\sum_{i=1}^{n} \left(\log x_{(i)} \right)^2 - n \left(\overline{\log x} \right)^2}$$

$$\log \hat{\alpha} = \overline{y} - \hat{\beta}\, \overline{\log x}$$

where

$$\overline{\log x} = \sum_{i=1}^{n} \left(\log x_{(i)} \right) \Big/ n \qquad \overline{y} = \sum_{i=1}^{n} y_i \Big/ n$$

To utilize the foregoing, we need to be able to determine values y_i that approximate $\log\log(1/[1 - F(x_{(i)})) = \log[-\log(1 - F(x_{(i)})],\ i = 1, \ldots, n.$ We now present two different methods for doing this.

Method 1: This method uses the fact that

$$E\left[F\left(X_{(i)} \right) \right] = \frac{i}{(n + 1)} \tag{10.5.5}$$

and then approximates $F(x_{(i)})$ by $E[F(X_{(i)})]$. Thus, this method calls for using

$$y_i = \log \left\{ -\log \left(1 - E\left[F\left(X_{(i)} \right) \right] \right) \right\} \tag{10.5.6}$$

$$= \log \left\{ -\log \left(1 - \frac{i}{(n + 1)} \right) \right\}$$

$$= \log \left\{ -\log \left(\frac{n + 1 - i}{n + 1} \right) \right\}$$

Method 2: This method uses the fact that

$$E\left[-\log \left(1 - F\left(X_{(i)} \right) \right) \right] = \frac{1}{n} + \frac{1}{n - 1} + \frac{1}{n - 2} \cdots + \frac{1}{n - i + 1} \tag{10.5.7}$$

and then approximates $-\log(1 - F(x_{(i)})$ by the foregoing. Thus, this second method calls for setting

$$y_i = \log \left[\frac{1}{n} + \frac{1}{(n - 1)} + \cdots + \frac{1}{(n - i + 1)} \right] \tag{10.5.8}$$

REMARKS

(a) It is not, at present, clear which method provides superior estimates of the parameters of the Weibull distribution, and extensive simulation studies will be necessary to determine this.

(b) Proofs of the equalities 10.5.5 and 10.5.7 (which hold whenever $X_{(i)}$ is the ith smallest of a sample of size n from any continuous distribution F) are outlined in Problems 28–30.

PROBLEMS

1. A random variable whose distribution function is given by
$$F(t) = 1 - \exp\{-\alpha t^\beta\}, \qquad t \geq 0$$
is said to have a Weibull distribution with parameters α, β. Compute its failure rate function.

2. If X and Y are independent random variables having failure rate functions $\lambda_x(t)$ and $\lambda_y(t)$, show that the failure rate function of $Z = \min(X, Y)$ is
$$\lambda_z(t) = \lambda_x(t) + \lambda_y(t)$$

3. The lung cancer rate of a t-year-old male smoker, $\lambda(t)$, is such that
$$\lambda(t) = .027 + .025\left(\frac{t - 40}{10}\right)^4 \qquad t \geq 40$$
Assuming that a 40-year-old male smoker survives all other hazards, what is the probability that he survives to (a) age 50, (b) age 60, without contracting lung cancer? In the foregoing we are assuming that he remains a smoker throughout his life.

4. Suppose the life distribution of an item has failure rate function $\lambda(t) = t^3$, $0 < t < \infty$.
   (a) What is the probability the item survives to age 2?
   (b) What is the probability that the item's life is between .4 and 1.4?
   (c) What is the mean life of the item?
   (d) What is the probability a 1-year-old item will survive to age 2?

5. A continuous life distribution is said to be an IFR (increasing failure rate) distribution if its failure rate function $\lambda(t)$ is nondecreasing in t.
   (a) Show that the gamma distribution with density
   $$f(t) = \lambda^2 t e^{-\lambda t} \qquad t > 0$$
   is IFR.
   (b) Show, more generally, that the gamma distribution with parameters α, λ is IFR whenever $\alpha \geq 1$.
   Hint: Write
   $$\lambda(t) = \left[\frac{\int_t^\infty \lambda e^{-\lambda s}(\lambda s)^{\alpha-1}\, ds}{\lambda e^{-\lambda t}(\lambda t)^{\alpha-1}}\right]^{-1}$$

6. Show that the uniform distribution on (a, b) is an IFR distribution.

7. For the model of Section 3.1, explain how the following figure can be used to show that
$$\tau = \sum_{j=1}^{r} Y_j$$

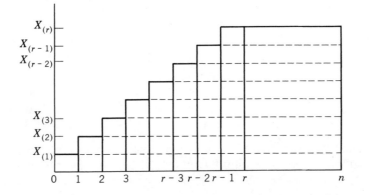

where

$$Y_j = (n - j + 1)(X_{(j)} - X_{(j-1)})$$

Hint: Argue that both τ and $\sum_{j=1}^{r} Y_j$ equal the total area of the foregoing figure.

8. When 30 transistors were simultaneously put on a life test that was to be terminated when the 10th failure occurred, the observed failure times were (in hours) 4.1, 7.3, 13.2, 18.8, 24.5, 30.8, 38.1, 45.5, 53, 62.2, Assume an exponential life distribution.

   (a) What is the maximum likelihood estimate of the mean life of a transistor?

   (b) Compute a 95 percent two-sided confidence interval for the mean life of a transistor.

   (c) Determine a value c that we can assert, with 95 percent confidence, is less than the mean transistor life.

   (d) Test at the $\alpha = .10$ level of significance the hypothesis that the mean lifetime is 7.5 hours against the alternative that it is not 7.5 hours.

9. Consider a test of $H_0: \theta = \theta_0$ versus $H_1: \theta \neq \theta_0$ for the model of Section 3.1. Suppose that the observed value of $2\tau/\theta_0$ is v. Show that the hypothesis should be rejected at significance level α whenever α is less than the p-value given by

$$p\text{-value} = 2 \min \left(P\{\chi_{2r}^2 < v\}, 1 - P\{\chi_{2r}^2 < v\} \right)$$

where χ_{2r}^2 is a chi-square random variable with r degrees of freedom.

10. Suppose 30 items are put on test that is scheduled to stop when the 8th failure occurs. If the failure times are, in hours, .35, .73, .99, 1.40, 1.45, 1.83, 2.20, 2.72, test at the 5 percent level of significance, the hypothesis that the mean life is equal to 10 hours. Assume that the underlying distribution is exponential.

11. Suppose that 20 items are to be put on test that is to be terminated when the 10th failure occurs. If the lifetime distribution is exponential with mean 10 hours, compute the following quantities.

(a) The mean length of the testing period.

(b) The variance of the testing period.

(c) Repeat the foregoing if not 20 but 40 items are put on test. Write a computer program to determine the exact values and compare these with their approximations as given at the end of Section 3.1.

12. Vacuum tubes produced at a certain plant are assumed to have an underlying exponential life distribution having an unknown mean θ. To estimate θ it has been decided to put a certain number n of tubes on test and to stop the test at the 10th failure. If the plant officials want the mean length of the testing period to be 3 hours when the value of θ is $\theta = 20$, approximately how large should n be?

13. A one-at-a-time sequential life testing scheme is scheduled to run for 300 hours. A total of 16 items fail within that time. Assuming an exponential life distribution with unknown mean θ (measured in hours):

(a) Determine the maximum likelihood estimate of θ.

(b) Test at the .05 level of significance the hypothesis that $\theta = 20$ against the alternative that $\theta \neq 20$.

(c) Determine a 95 percent confidence interval for θ.

14. Using the fact that a Poisson process results when the times between successive events are independent and identically distributed exponential random variables, show that

$$P\{X \geq n\} = F_{\chi^2_{2n}}(x)$$

when X is a Poisson random variable with mean $x/2$ and $F_{\chi^2_{2n}}$ is the chi-square distribution function with $2n$ degrees of freedom.

Hint: Use the results of Section 3.2.

15. From a sample of items having an exponential life distribution with unknown mean θ, items are tested in sequence. The testing continues until either the rth failure occurs or after a time T elapses.

(a) Determine the likelihood function.

(b) Verify that the maximum likelihood estimator of θ is equal to the total time on test of all items divided by the number of observed failures.

16. Verify that the maximum likelihood estimate corresponding to Equation 10.3.9 is given by Equation 10.3.10.

17. A testing laboratory has facilities to simultaneously life test 5 components. The lab tested a sample of 10 components from a common exponential distribution by initially putting 5 on test and then replacing any failed component by one still waiting to be tested. The test was designed to end either at 200 hours or when all 10 components had failed. If there were a total of 9 failures occurring at times 15, 28.2, 46, 62.2, 76, 86, 128, 153, 197, what is the maximum likelihood estimate of the mean life of a component?

18. Suppose that the remission time, in weeks, of leukemia patients that have undergone a certain type of chemotherapy treatment is an exponential random variable having an unknown mean θ. A group of 20 such patients are being monitored and, at present, their remission times are (in weeks) 1.2, 1.8*, 2.2, 4.1, 5.6, 8.4, 11.8*, 13.4*, 16.2, 21.7, 29*, 41, 42*, 42.4*, 49.3, 60.5, 61*, 94, 98, 99.2* where an * next to the data means that the patient's remission is still continuing, whereas an unstarred data point means that the remission ended at that time. What is the maximum likelihood estimate of θ?

19. In Problem 17 suppose that prior to the testing phase and based on past experience one felt that the value of $\lambda = 1/\theta$ could be thought of as the outcome of a gamma random variable with parameters 1, 100. What is the Bayes' estimate of λ?

20. What is the Bayes estimate of $\lambda = 1/\theta$ in Problem 18 if the prior distribution on λ is exponential with mean $1/30$?

21. The following data represent failure times, in minutes, for two types of electrical insulation subject to a certain voltage stress.

Type I	212, 88.5, 122.3, 116.4, 125, 132, 66
Type II	34.6, 54, 162, 49, 78, 121, 128

Test the hypothesis that the two sets of data come from the same exponential distribution.

22. Suppose that the life distributions of two types of transistors are both exponential. To test the equality of means of these two distributions, n_1 type 1 transistors are simultaneously put on a life test that is scheduled to end when there have been a total of r_1 failures. Similarly, n_2 type 2 transistors are simultaneously put on a life test that is to end when there have been r_2 failures.

(a) Using results from Section 3.1, show how the hypothesis that the means are equal can be tested by using a test statistic that, when the means are equal, has an F distribution with $2r_1$ and $2r_2$ degree of freedom.

(b) Suppose $n_1 = 20$, $r_1 = 10$ and $n_2 = 10$, $r_2 = 7$ with the following data resulting:
Type 1 Failures at Times:

$$10.4, 23.2, 31.4, 45, 61.1, 69.6, 81.3, 95.2, 112, 129.4$$

Type 2 Failures at Times:

$$6.1, 13.8, 21.2, 31.6, 46.4, 66.7, 92.4$$

What is the smallest significance level α for which the hypothesis of equal means would be rejected? (That is, what is the *p*-value of the test data?)

23. If X is a Weibull random variable with parameters (α, β) show that

$$E[X] = \alpha^{-1/\beta}\Gamma(1 + 1/\beta)$$

where $\Gamma(y)$ is the gamma function defined by

$$\Gamma(y) = \int_0^\infty e^{-x}x^{y-1}\,dx$$

Hint: Write

$$E[X] = \int_0^\infty t\alpha\beta t^{\beta-1}\exp\{-\alpha t^\beta\}\,dt$$

and make the change of variables

$$x = \alpha t^\beta, \qquad dx = \alpha\beta t^{\beta-1}\,dt$$

24. Show that if X is a Weibull random variable with parameters (α, β) then

$$\text{Var}(X) = \alpha^{-2/\beta}\left[\Gamma\left(1 + \frac{2}{\beta}\right) - \left(\Gamma\left(1 + \frac{1}{\beta}\right)\right)^2\right]$$

25. If the following is the sample data from a Weibull population having unknown parameters α and β, determine the least square estimates of these quantities, using either of the methods presented.

    Data: 15.4, 16.8, 6.2, 10.6, 21.4, 18.2, 1.6, 12.5, 19.4, 17

26. Show that if X is a Weibull random variable with parameters (α, β), then αX^β is an exponential random variable with mean 1.

27. If U is uniformly distributed on $(0, 1)$—that is, U is a random number—show that $(-(1/\alpha)\log U)^{1/\beta}$ is a Weibull random variable with parameters (α, β).

The next 3 problems will be concerned with verifying the Equations 10.5.5 and 10.5.7.

28. If X is a continuous random variable having distribution function F, show that

(a) $F(X)$ is uniformly distributed on $(0, 1)$
(b) $1 - F(X)$ is uniformly distributed on $(0, 1)$

29. Let $X_{(i)}$ denote the ith smallest of a sample of size n from a continuous distribution function F. Also, let $U_{(i)}$ denote the ith smallest from a sample of size n from a uniform $(0, 1)$ distribution.

(a) Argue that the density function of $U_{(i)}$ is given by

$$f_{U_{(i)}}(t) = \frac{n!}{(n-i)!(i-1)!}t^{i-1}(1-t)^{n-i} \qquad 0 < t < 1$$

Hint: In order for the ith smallest of n uniform $(0, 1)$ random variables to equal t, how many must be less than t and how many must be greater? Also, in how many ways can a set of n elements be broken into 3 subsets of respective sizes $i - 1$, 1, and $n - i$?

(b) Use (a) to show that $E[U_{(i)}] = i/(n + 1)$.

Hint: To evaluate the resulting integral, use the fact that the density in part (a) must integrate to 1.

(c) Use part (b) and Problem 28a to conclude that $E[F(X_{(i)})] = i/(n + 1)$.

30. If U is uniformly distributed on $(0, 1)$, show that $-\log U$ has an exponential distribution with mean 1. Now use Equation 10.3.7 and the results of the previous problems to establish Equation 10.5.7.

C H A P T E R 11

Quality Control

1 INTRODUCTION

Almost every manufacturing process results in some random variation in the items it produces. That is, no matter how stringently the process is being controlled, there is always going to be some variation between the items produced. This variation is called *chance variation* and is considered to be inherent to the process. However, there is another type of variation that sometimes appears. This variation, far from being inherent to the process, is due to some *assignable cause* and usually results in an adverse effect on the quality of the items produced. For instance, this latter variation may be caused by a faulty machine setting, or by poor quality of the raw materials presently being used, or by incorrect software, or human error, or any other of a large number of possibilities. When the only variation present is due to chance, and not to assignable cause, we say that the process is in control, and a key problem is to determine whether a process is in or is *out of control*.

The determination of whether a process is in or out of control is greatly facilitated by the use of *control charts*, which are determined by two numbers —the upper and lower control limits. To employ such a chart, the data generated by the manufacturing process are divided into subgroups and subgroup statistics—such as the subgroup average and subgroup standard deviation—are computed. When the subgroup statistic does not fall within the upper and lower control limit, we conclude that the process is out of control.

In Sections 2 and 3 we suppose that the successive items produced have measurable characteristics, with a mean and variance that are fixed when the process is in control. We show how to construct control charts based on subgroup averages (in Section 2) and on subgroup standard deviations (in Section 3). In Section 4 we suppose that rather than having a measurable characteristic, each item is judged by an *attribute*—that is, it is classified as either defective or nondefective. In Section 5 we suppose that each item produced has a random number of defects. In both of Sections 4 and 5 we show how to construct control charts that indicate changes in the production process. Finally, in Section 6 we consider more sophisticated control charts that consider not a single subgroup average but moving averages of such values.

2   CONTROL CHARTS FOR AVERAGE VALUES: THE $\bar{X}$-CONTROL CHART

Suppose that when the process is in control the successive items produced have measurable characteristics that are independent, normal random variables with mean μ and variance σ^2. However, due to special circumstances, suppose that the process may go out of control and start producing items having a different distribution. We would like to be able to recognize when this occurs so as to stop the process, find out what is wrong, and fix it.

Let $X_1, X_2, \ldots$ denote the measurable characteristics of the successive items produced. To determine when the process goes out of control, we start by breaking the data up into subgroups of some fixed size—call it n. The value of n is chosen so as to yield uniformity within subgroups. That is, n may be chosen so that all data items within a subgroup were produced on the same day, or on the same shift, or using the same settings, and so on. In other words, the value of n is chosen so that it is reasonable that a shift in distribution would occur between and not within subgroups. Typical values of n are 4, 5, or 6.

Let $\bar{X}_i$, $i = 1, 2, \ldots$ denote the average of the ith subgroup. That is,

$$\bar{X}_1 = \frac{(X_1 + \cdots + X_n)}{n}$$

$$\bar{X}_2 = \frac{(X_{n+1} + \cdots + X_{2n})}{n}$$

$$\bar{X}_3 = \frac{(X_{2n+1} + \cdots + X_{3n})}{n}$$

and so on. Since, when in control, each of the X_i have mean μ and variance σ^2, it follows that

$$E(\bar{X}_i) = \mu$$

$$\text{Var}(\bar{X}_i) = \frac{\sigma^2}{n}$$

and so

$$\frac{\bar{X}_i - \mu}{\sqrt{\dfrac{\sigma^2}{n}}} \sim \mathcal{N}(0, 1)$$

That is, if the process is in control throughout the production of subgroup i, then $\sqrt{n}\,(\bar{X}_i - \mu)/\sigma$ has a unit normal distribution. Now it follows that a unit normal random variable Z will almost always be between -3 and $+3$. (Indeed, $P\{-3 < Z < 3\} = .9973$). Hence, if the process is in control

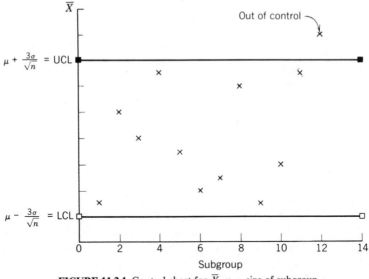

FIGURE 11.2.1 Control chart for $\overline{X}$, n = size of subgroup

throughout the production of the items in subgroup i, then we would certainly expect that

$$-3 < \sqrt{n} \, \frac{(\overline{X}_i - \mu)}{\sigma} < 3$$

or, equivalently, that

$$\mu - \frac{3\sigma}{\sqrt{n}} < \overline{X}_i < \mu + \frac{3\sigma}{\sqrt{n}}$$

The values

$$\text{UCL} \equiv \mu + \frac{3\sigma}{\sqrt{n}}$$

and

$$\text{LCL} \equiv \mu - \frac{3\sigma}{\sqrt{n}}$$

are called respectively the *upper* and *lower control limits*.

The $\overline{X}$-control chart, which is designed to detect a change in the average value of an item produced, is obtained by plotting the successive subgroup averages $\overline{X}_i$ and declaring that the process is out of control the first time $\overline{X}_i$ does not fall between LCL and UCL (see Figure 11.2.1).

Example 11.2a A manufacturer produces steel shafts having diameters that should be normally distributed with mean 3 mm and standard deviation

.1 mm. Successive samples of 4 shafts have yielded the following sample averages in millimeters.

Sample	$\bar{X}$	Sample	$\bar{X}$
1	3.01	6	3.02
2	2.97	7	3.10
3	3.12	8	3.14
4	2.99	9	3.09
5	3.03	10	3.20

What conclusion should be drawn?

When in control the successive diameters have mean $\mu = 3$ and standard deviation $\sigma = .1$ and so with $n = 4$ the control limits are

$$\text{LCL} = 3 - \frac{3(.1)}{\sqrt{4}} = 2.85 \qquad \text{UCL} = 3 + \frac{3(.1)}{\sqrt{4}} = 3.15$$

Because sample number 10 falls above the upper control limit, it appears that there is reason to suspect that the mean diameter of shafts now differs from 3. (Clearly, judging from the results of Samples 6 through 10 it appears to have increased beyond 3!) ∎

REMARKS

(a) The foregoing supposes that when the process is in control the underlying distribution is normal. However, even if this is not the case, by the central limit theorem it follows that the subgroup averages should have a distribution that is roughly normal and so would be unlikely to differ from its mean by more than 3 standard deviations.

(b) It is frequently the case that we do not determine the measurable qualities of all the items produced but only those of a randomly chosen subset of items. If this is so then it is natural to select, as a subgroup, items that are produced at roughly the same time.

It is important to note that even when the process is in control there is a chance—namely, .0027—that a subgroup average will fall outside the control limit and so one would incorrectly stop the process and hunt for the nonexistent source of trouble.

Let us now suppose that the process has just gone out of control by a change in the mean value of an item from μ to $\mu + a$ where $a > 0$. How long will it take (assuming things do not change again) until the chart will indicate that the process is now out of control? To answer this note that a subgroup average will be within the control limits if

$$-3 < \sqrt{n}\,\frac{(\bar{X} - \mu)}{\sigma} < 3$$

or, equivalently, if

$$-3 - \frac{a\sqrt{n}}{\sigma} < \sqrt{n}\,\frac{(\bar{X} - \mu)}{\sigma} - \frac{a\sqrt{n}}{\sigma} < 3 - \frac{a\sqrt{n}}{\sigma}$$

or

$$-3 - \frac{a\sqrt{n}}{\sigma} < \sqrt{n}\,\frac{(\overline{X} - \mu - a)}{\sigma} < 3 - \frac{a\sqrt{n}}{\sigma}$$

Hence, since $\overline{X}$ is normal with mean $\mu + a$ and variance σ^2/n—and so $\sqrt{n}\,(\overline{X} - \mu - a)/\sigma$ has a unit normal distribution—the probability that it will fall within the control limits is

$$P\left\{-3 - \frac{a\sqrt{n}}{\sigma} < Z < 3 - \frac{a\sqrt{n}}{\sigma}\right\} = \Phi\left(3 - \frac{a\sqrt{n}}{\sigma}\right) - \Phi\left(-3 - \frac{a\sqrt{n}}{\sigma}\right)$$

$$\approx \Phi\left(3 - \frac{a\sqrt{n}}{\sigma}\right)$$

and so the probability that it falls outside is approximately $1 - \Phi(3 - a\sqrt{n}/\sigma)$. For instance if the subgroup size is $n = 4$, then an increase in the mean value of 1 standard deviation—that is, $a = \sigma$—will result in the subgroup average falling outside of the control limits with probability $1 - \Phi(1) = .1587$. As each subgroup average will independently fall outside the control limits with probability $1 - \Phi(3 - a\sqrt{n}/\sigma)$, it follows that the number of subgroups that will be needed to detect this shift has a geometric distribution with mean $\{1 - \Phi(3 - a\sqrt{n}/\sigma)\}^{-1}$. (In the case mentioned before with $n = 4$, the number of subgroups one would have to chart to detect a change in the mean of 1 standard deviation has a geometric distribution with mean 6.3.)

2.1   Case of Unknown μ and σ

If one is just starting up a control chart and does not have reliable historical data, then μ and σ would not be known and would have to be estimated. To do so we employ k of the subgroups where k should be chosen so that $k \geq 20$ and $nk \geq 100$. If $\overline{X}_i,\ i = 1, \ldots, k$ is the average of the ith subgroup, then it is natural to estimate μ by $\overline{\overline{X}}$, the average of these subgroup averages. That is,

$$\overline{\overline{X}} = \frac{(\overline{X}_1 + \cdots + \overline{X}_k)}{k}$$

To estimate σ let S_i denote the sample standard deviation of the ith subgroup, $i = 1, \ldots, k$. That is,

$$S_1 = \sqrt{\sum_{i=1}^{n} \frac{(X_i - \overline{X}_1)^2}{(n-1)}}$$

$$S_2 = \sqrt{\sum_{i=1}^{n} \frac{(X_{n+i} - \overline{X}_2)^2}{(n-1)}}$$

$$\vdots$$

$$S_k = \sqrt{\sum_{i=1}^{n} \frac{(X_{(k-1)n+i} - \overline{X}_k)^2}{(n-1)}}$$

Let

$$\bar{S} = (S_1 + \cdots + S_k)/k$$

The statistic $\bar{S}$ will not be an unbiased estimator of σ—that is, $E[\bar{S}] \neq \sigma$. To transform it into an unbiased estimator we must first compute $E[\bar{S}]$, which is accomplished as follows:

$$E[\bar{S}] = \frac{(E[S_1] + \cdots + E[S_k])}{k} \qquad (11.2.1)$$

$$= E[S_1]$$

where the last equality follows since $S_1, \ldots, S_k$ are independent and identically distributed (and thus have the same mean). To compute $E[S_1]$ we make use of the following fundamental result about normal samples—namely, that

$$\frac{(n-1)S_1^2}{\sigma^2} = \sum_{i=1}^{n} \frac{(X_i - \bar{X})^2}{\sigma^2} \sim \chi_{n-1}^2 \qquad (11.2.2)$$

Now it is not difficult to show (see Problem 3) that

$$E[\sqrt{Y}] = \frac{\sqrt{2}\,\Gamma\left(\frac{n}{2}\right)}{\Gamma\left[\frac{(n-1)}{2}\right]} \qquad \text{when } Y \sim \chi_{n-1}^2 \qquad (11.2.3)$$

Since

$$E\left[\sqrt{\frac{(n-1)S_1^2}{\sigma^2}}\right] = \sqrt{n-1}\,\frac{E[S_1]}{\sigma}$$

we see from Equations 11.2.2 and 11.2.3 that

$$E[S_1] = \frac{\sqrt{2}\,\Gamma\left(\frac{n}{2}\right)\sigma}{\sqrt{n-1}\,\Gamma\left[\frac{(n-1)}{2}\right]}$$

Hence, if we set

$$c(n) = \frac{\sqrt{2}\,\Gamma\left(\frac{n}{2}\right)}{\sqrt{n-1}\,\Gamma\left[\frac{(n-1)}{2}\right]}$$

then it follows from Equation 11.2.1 that $\bar{S}/c(n)$ is an unbiased estimator of σ.

The following table presents the values of $c(n)$ for $n = 2$ through $n = 10$.

TABLE 11.2.1

Values of $c(n)$

$c(\ 2) = .7978849$
$c(\ 3) = .8862266$
$c(\ 4) = .9213181$
$c(\ 5) = .9399851$
$c(\ 6) = .9515332$
$c(\ 7) = .9593684$
$c(\ 8) = .9650309$
$c(\ 9) = .9693103$
$c(10) = .9726596$

TECHNICAL REMARK

In determining the values in Table 11.2.1, the computation of $\Gamma(n/2)$ and $\Gamma[(n-1)/2]$ was based on the recursive formula

$$\Gamma(a) = (a - 1)\Gamma(a - 1)$$

which was established in Section 7 of Chapter 3. This recursion yields that, for integer n,

$$\Gamma(n) = (n - 1)(n - 2) \cdots 3 \cdot 2 \cdot 1 \cdot \Gamma(1)$$

$$= (n - 1)! \qquad \text{since } \Gamma(1) = \int_0^\infty e^{-x}\, dx = 1$$

The recursion also yields that

$$\Gamma\left(\frac{n+1}{2}\right) = \left(\frac{n-1}{2}\right)\left(n - \frac{3}{2}\right) \cdots \frac{3}{2} \cdot \frac{1}{2} \cdot \Gamma\left(\frac{1}{2}\right)$$

with

$$\Gamma\left(\frac{1}{2}\right) = \int_0^\infty e^{-x} x^{-1/2}\, dx$$

$$= \int_0^\infty e^{-y^2/2} \frac{\sqrt{2}}{y}\, y\, dy \qquad \text{by } x = \frac{y^2}{2} \qquad dx = y\, dy$$

$$= \sqrt{2} \int_0^\infty e^{-y^2/2}\, dy$$

$$= 2\sqrt{\pi}\, \frac{1}{\sqrt{2\pi}} \int_0^\infty e^{-y^2/2}\, dy$$

$$= 2\sqrt{\pi}\, P[N(0, 1) > 0]$$

$$= \sqrt{\pi} \qquad \blacksquare$$

The preceding estimates for μ and σ make use of all k subgroups and thus are reasonable only if the process has remained in control throughout. To check this, we compute the control limits based on these estimates of μ and σ, namely

$$LCL = \bar{\bar{X}} - \frac{3\bar{S}}{\sqrt{n}\,c(n)}$$ (11.2.4)

$$UCL = \bar{\bar{X}} + \frac{3\bar{S}}{\sqrt{n}\,c(n)}$$

We now check that each of the subgroup averages $\bar{X}_i$ falls within these lower and upper limits. Any subgroup whose average value does not fall within is removed (we suppose that the process was temporarily out of control) and the estimates are recomputed. We then again check that all the remaining subgroup averages fall within the control limits. If not then they are removed and so on. Of course if too many of the subgroup averages fall outside the control limits, then it is clear that no control has yet been established.

Example 11.2b Let us reconsider Example 11.2a under the new supposition that the process is just beginning and so μ and σ are unknown. Also suppose that the sample standard deviations were as follows:

	$\bar{X}$	S		$\bar{X}$	S
1	3.01	.12	6	3.02	.08
2	2.97	.14	7	3.10	.15
3	3.12	.08	8	3.14	.16
4	2.99	.11	9	3.09	.13
5	3.03	.09	10	3.20	.16

Since $\bar{\bar{X}} = 3.067$, $\bar{S} = .122$, $c(4) = .9213$, the control limits are

$$LCL = 3.067 - \frac{3(.122)}{(2 \times .9213)} = 2.868$$

$$UCL = 3.067 + \frac{3(.122)}{(2 \times .9213)} = 3.266$$

Since all the $\bar{X}_i$ fall within these limits, we suppose that the process is in control with $\mu = 3.067$ and $\sigma = \bar{S}/c(4) = .1324$.

Suppose now that the values of the items produced are supposed to fall within the specifications 3 ± 0.1. Assuming that the process remains in control, and that the foregoing are accurate estimates of the true mean and standard deviation, what proportion of the items will meet the desired specifications?

To answer the foregoing we note that when $\mu = 3.067$ and $\sigma = .1324$

$$P\{2.9 \leq X \leq 3.1\} = P\left\{\frac{2.9 - 3.067}{.1324} \leq \frac{X - 3.067}{.1324} \leq \frac{3.1 - 3.067}{.1324}\right\}$$

$$= \Phi(.2492) - \Phi(-1.2613)$$

$$= .5984 - (1 - .8964)$$

$$= .4948$$

Hence, 49 percent of the items produced will meet the specifications.   ■

REMARKS

(a) The estimator $\overline{\overline{X}}$ is equal to the average of all nk measurements and is thus the obvious estimator of μ. However, it may not immediately be clear why the sample standard deviation of all the nk measurements, namely

$$S \equiv \sqrt{\sum_{i=1}^{nk} \frac{\left(X_i - \overline{\overline{X}}\right)^2}{nk - 1}}$$

is not used as the initial estimator of σ. The reason it is not is that the process may not have been in control throughout the first k subgroups, and thus this latter estimator could be far away from the true value. Also, it often happens that a process goes out of control by an occurrence that results in a change of its mean value μ while leaving its standard deviation unchanged. In such a case the subgroup sample deviations would still be estimators of σ, whereas the entire sample standard deviation would not. Indeed, even in the case where the process appears to be in control throughout, the estimator of σ presented is preferred over the sample standard deviation S. The reason for this is that we cannot be certain that the mean has not changed throughout this time. That is, even though all the subgroup averages fall within the control limits, and so we have concluded that the process is in control, there is no assurance that there are no assignable causes of variation present (which might have resulted in a change in the mean that has not yet been picked up by the chart). It merely means that for practical purposes it pays to act as if the process was in control and let it continue to produce items. However, since we realize that some assignable cause of variation might be present, it has been argued that $\overline{S}/c(n)$ is a "safer" estimator than the sample standard deviation. That is, though it is not quite as good when the process has really been in control throughout, it could be a lot better if there had been some small shifts in the mean.

(b) In the past an estimator of σ based on subgroup ranges—defined as the difference between the largest and smallest value in the subgroup—has been employed. This was done so as to keep the necessary computations simple (it is clearly much easier to compute the range than it is to compute the subgroup's sample standard deviation). However, with modern-day computational power this should no longer be a consideration, and since the standard deviation estimator both has smaller variance than the range estimator and is more robust (in the sense that it would still yield a reasonable estimate of the population standard deviation even when the underlying distribution is not normal), we will not consider the latter estimator in this text.

3 S-CONTROL CHARTS

The $\overline{X}$-control charts presented in the previous section are designed to pick up changes in the population mean. In cases where one is also concerned about possible changes in the population variance, we can utilize an S-control chart.

As before suppose that, when in control, the items produced have a measurable characteristic that is normally distributed with mean μ and variance σ^2. If S_i is the sample standard deviation for the ith subgroup, that is

$$S_i = \sqrt{\sum_{j=1}^{n} \frac{\left(X_{(i-1)n+j} - \overline{X}_i\right)^2}{(n-1)}}$$

then, as was shown in Section 2.1,

$$E[S_i] = c(n)\sigma \qquad (11.3.1)$$

In addition,

$$\begin{aligned} \text{Var}(S_i) &= E[S_i^2] - (E[S_i])^2 \qquad (11.3.2) \\ &= \sigma^2 - c^2(n)\sigma^2 \\ &= \sigma^2[1 - c^2(n)] \end{aligned}$$

where the next to last equality follows from Equation 11.2.2 and the fact that the expected value of a chi-square random variable is equal to its degrees of freedom parameter.

On using the fact that, when in control, S_i has the distribution of a constant (equal to $\sigma/\sqrt{n-1}$) times the square root of a chi-square random variable with $n-1$ degrees of freedom, it can be shown that S_i will, with probability near to 1, be within 3 standard deviations of its mean. That is,

$$P\left\{E[S_i] - 3\sqrt{\text{Var}(S_i)} < S_i < E[S_i] + 3\sqrt{\text{Var}(S_i)}\right\} \approx .99$$

Thus, using the formulas 11.3.1 and 11.3.2 for $E[S_i]$ and $\text{Var}(S_i)$, it is natural to set the upper and lower control limits for the S chart by

$$\text{UCL} = \sigma\left[c(n) + 3\sqrt{1 - c^2(n)}\right] \qquad (11.3.3)$$

$$\text{LCL} = \sigma\left[c(n) - 3\sqrt{1 - c^2(n)}\right]$$

The successive values of S_i should be plotted to make certain they fall within the upper and lower control limits. When a value falls outside, the process should be stopped and declared to be out of control.

When one is just starting up a control chart and σ is unknown, it can be estimated from $\bar{S}/c(n)$. Using the foregoing, the estimated control limits would then be

$$\text{UCL} = \bar{S}\left[1 + 3\sqrt{1/c^2(n) - 1}\right] \qquad (11.3.4)$$

$$\text{LCL} = \bar{S}\left[1 - 3\sqrt{1/c^2(n) - 1}\right]$$

As in the case of starting up an $\bar{X}$ control chart, it should then be checked that the k subgroup standard deviations $S_1, S_2, \ldots, S_k$ all fall within these control limits. If any of them falls outside, then those subgroups should be discarded and $\bar{S}$ recomputed.

Example 11.3a The following are the $\bar{X}$ and S values for 20 subgroups of size 5 for a recently started process.

Subgroup	$\bar{X}$	S	Subgroup	$\bar{X}$	S	Subgroup	$\bar{X}$	S
1	35.1	4.2	8	38.4	5.1	15	43.2	3.5
2	33.2	4.4	9	35.7	3.8	16	41.3	8.2
3	31.7	2.5	10	27.2	6.2	17	35.7	8.1
4	35.4	3.2	11	38.1	4.2	18	36.3	4.2
5	34.5	2.6	12	37.6	3.9	19	35.4	4.1
6	36.4	4.5	13	38.8	3.2	20	34.6	3.7
7	35.9	3.4	14	34.3	4.0			

Since $\bar{\bar{X}} = 35.94$, $\bar{S} = 4.35$, $c(5) = .9400$, we see from Equations 11.2.4 and 11.3.4 that the preliminary upper and lower control limits for $\bar{X}$ and S are as

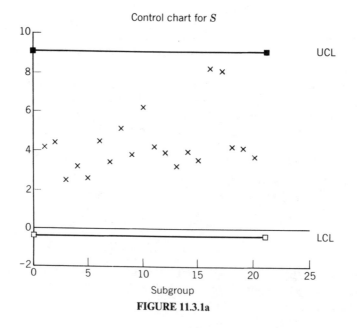

Control chart for S

FIGURE 11.3.1a

Control chart for $\overline{X}$

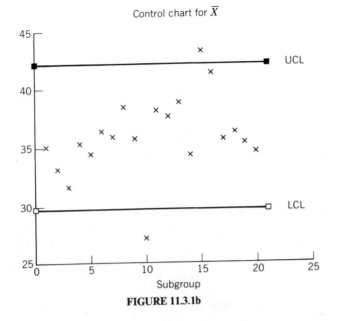

FIGURE 11.3.1b

follows:

$$UCL(\overline{X}) = 42.149$$

$$LCL(\overline{X}) = 29.731$$

$$UCL(S) = 9.087$$

$$LCL(S) = -.386$$

The control charts for $\overline{X}$ and S with the preceding control limits are shown in Figures 11.3.1a and 11.3.1b. Since $\overline{X}_{10}$ and $\overline{X}_{15}$ fall outside the $\overline{X}$ control limits, these subgroups must be eliminated and the control limits recomputed. We leave the necessary computations as an exercise. ∎

4 CONTROL CHART FOR FRACTION DEFECTIVE

The $\overline{X}$ and S control charts can be used when the data are measurements whose values can vary continuously over a region. There are also situations in which the items produced have quality characteristics that are classified as either being defective or nondefective. Control charts can also be constructed in this latter situation.

Let us suppose that when the process is in control each item produced will independently be defective with probability p. If we let X denote the number of defective items in a subgroup of n items, then assuming control, X will be a binomial random variable with parameters (n, p). If $F = X/n$ is the fraction of the subgroup that is defective, then assuming the process is in control, its mean and standard deviation are given by

$$E[F] = \frac{E[X]}{n} = \frac{np}{n} = p$$

$$\sqrt{\text{Var}(F)} = \sqrt{\frac{\text{Var}(X)}{n^2}} = \sqrt{\frac{np(1-p)}{n^2}} = \sqrt{\frac{p(1-p)}{n}}$$

Hence, when the process is in control the fraction defective in a subgroup of size n should, with high probability, be between the limits

$$\text{LCL} = p - 3\sqrt{\frac{p(1-p)}{n}} \qquad \text{UCL} = p + 3\sqrt{\frac{p(1-p)}{n}}$$

The subgroup size n is usually much larger than the typical values of between 4 and 10 used in $\overline{X}$ and S charts. The main reason for this is that if p is small and n is not of reasonable size, then most of the subgroups will have zero defects even when the process goes out of control. Thus it would take longer than it would if n were chosen so that np were not close to zero to detect a shift in quality.

To start such a control chart it is, of course, necessary first to estimate p. To do so, choose k of the subgroups, where again one should try to take $k \geq 20$, and let F_i denote the fraction of the ith subgroup that are defective. The estimate of p is given by $\overline{F}$ defined by

$$\overline{F} = \frac{F_1 + \cdots + F_k}{k}$$

Since nF_i is equal to the number of defectives in subgroup i, we see that $\overline{F}$ can also be expressed as

$$\overline{F} = \frac{nF_1 + \cdots + nF_k}{nk}$$

$$= \frac{\text{total number of defectives in all the subgroups}}{\text{number of items in the subgroups}}$$

In other words, the estimate of p is just the proportion of items inspected that are defective.

The upper and lower control limits are now given by

$$\text{LCL} = \overline{F} - 3\sqrt{\frac{\overline{F}(1-\overline{F})}{n}} \qquad \text{UCL} = \overline{F} + 3\sqrt{\frac{\overline{F}(1-\overline{F})}{n}}$$

It should now be checked whether the subgroup fractions $F_1, F_2, \ldots, F_k$ fall within these control limits. If some of them fall outside, then the corresponding subgroups should be eliminated and $\overline{F}$ recomputed.

Example 11.4a Successive samples of 50 screws are drawn from the hourly production of an automatic screw machine, with each screw being rated as being either acceptable or defective. This is done for 20 such samples with the following data resulting:

Subgroup	Defectives	F	Subgroup	Defectives	F
1	6	.12	11	1	.02
2	5	.10	12	3	.06
3	3	.06	13	2	.04
4	0	.00	14	0	.00
5	1	.02	15	1	.02
6	2	.04	16	1	.02
7	1	.02	17	0	.00
8	0	.00	18	2	.04
9	2	.04	19	1	.02
10	1	.02	20	2	.04

We can compute the trial control limits as follows:

$$\bar{F} = \frac{\text{total number defectives}}{\text{total number items}} = \frac{34}{1000} = .034$$

and so

$$\text{UCL} = .034 + 3\sqrt{\frac{(.034)(.968)}{50}} = .1109$$

$$\text{LCL} = .034 - 3\sqrt{\frac{(.034)(.966)}{50}} = -.0429$$

Since the proportion of defectives in the first subgroup falls outside the upper control limit, we eliminate that subgroup and recompute $\bar{F}$ as

$$\bar{F} = \frac{34 - 6}{950} = .0295$$

The new upper and lower control limits are $.0295 \pm \sqrt{(.0295)(1 - .0295)/50}$, or

$$\text{LCL} = -.0423 \qquad \text{UCL} = .1013$$

Since the remaining subgroups all have fraction defectives that fall within the control limits, we can accept that, when in control, the fraction of defective items in a subgroup should be below .1013. ∎

REMARK

It should be noted that we are attempting to detect any change in quality even when this change results in improved quality. That is, we regard the process as being "out of control" even when the probability of a defective item decreases. The reason for this is that it is important to notice any change in quality, for either better or worse, so as to be able to evaluate the reason for the change. In other words, if an improvement in product quality occurs, then it is important to analyze the production process to determine the reason for the improvement (what are we doing right?).

5 CONTROL CHART FOR NUMBER OF DEFECTS

In this section we consider situations in which the data consist of the number of defects in a unit that consists of an item or group of items. For instance, it could be the number of defective rivets in an airplane wing, or the number of defective computer chips that are produced daily by a given company. As it is often the case that there are a large number of possible things that can be defective, with each of these having a small probability of actually being defective, it is probably reasonable to assume that the resulting number of defects has a Poisson distribution[†]. So let us suppose that, when the process is in control, the number of defects per unit has a Poisson distribution with mean λ.

If we let X_i denote the number of defects in the ith unit, then, since the variance of a Poisson random variable is equal to its mean, we have that, when the process is in control

$$E[X_i] = \lambda \qquad \text{Var}(X_i) = \lambda$$

Hence, when in control each X_i should with high probability be within $\lambda \pm 3\sqrt{\lambda}$, and so the upper and lower control limits are given by

$$\text{UCL} = \lambda + 3\sqrt{\lambda} \qquad \text{LCL} = \lambda - 3\sqrt{\lambda}$$

As before when the control chart is started and λ is unknown, a sample of k units should be used to estimate λ by

$$\overline{X} = (X_1 + \cdots + X_k)/k$$

This results in trial control limits

$$\overline{X} + 3\sqrt{\overline{X}} \qquad \text{and} \qquad \overline{X} - 3\sqrt{\overline{X}}$$

If all the X_i, $i = 1, \ldots, k$ fall within these limits, then we suppose that the process is in control with $\lambda = \overline{X}$. If some fall outside then these points are eliminated and we recompute $\overline{X}$, and so on.

In situations where the mean number of defects per item (or per day) is small, one should combine items (days) and use as data the number of defects in a given number—say n—of items (or days). Since the sum of independent Poisson random variables remains a Poisson random variable, the data values will be Poisson distributed with a larger mean value λ. Such combining of items is useful when the mean number of defects per item is less than 25.

To obtain a feel for the advantage in combining items, suppose that the mean number of defects per item is 4 when the process is under control; and suppose that something occurs that results in this value changing from 4 to 6, that is, an increase of 1 standard deviation occurs. Let us see how many items will be produced, on average, until the process is declared out of control when the successive data consist of the number of defects in n items.

[†]See Section 2 of Chapter 3 for a theoretical explanation.

Since the number of defects in a sample of n items is, when under control, Poisson distributed with mean and variance equal to $4n$, the control limits are $4n \pm 3\sqrt{4n}$ or $4n \pm 6\sqrt{n}$. Now if the mean number of defects per item changes to 6, then a data value will be Poisson with mean $6n$ and so the probability that it will fall outside the control limits—call it $p(n)$—is given by

$$p(n) = P\{Y > 4n + 6\sqrt{n}\} + P\{Y < 4n - 6\sqrt{n}\}$$

when Y is Poisson with mean $6n$. Now

$$p(n) \approx P\{Y > 4n + 6\sqrt{n}\}$$

$$= P\left\{\frac{Y - 6n}{\sqrt{6n}} > \frac{6\sqrt{n} - 2n}{\sqrt{6n}}\right\}$$

$$\approx P\left\{Z > \frac{6\sqrt{n} - 2n}{\sqrt{6n}}\right\} \qquad \text{where } Z \sim N(0,1)$$

$$= 1 - \Phi\left(\sqrt{6} - 2\sqrt{\frac{n}{6}}\right)$$

As each data value will be outside the control limits with probability $p(n)$, it follows that the number of data values needed to obtain one outside the limits is a geometric random variable with parameter $p(n)$, and thus has mean $1/p(n)$. Finally, since there are n items for each data value, it follows that the number of items produced before the process is seen to be out of control has mean value $n/p(n)$, or using Equation 11.5.1, we obtain that

average number of items produced while out of control

$$= \frac{n}{1 - \Phi\left(\sqrt{6} - \sqrt{\frac{2n}{3}}\right)}$$

We plot this for various n:

TABLE 11.5.1

n	Average number of items
1	19.6
2	20.66
3	19.80
4	19.32
5	18.80
6	18.18
7	18.13
8	18.02
9	18
10	18.18
11	18.33
12	18.51

Since larger values of n are better when the process is in control (because the average number of items produced before the process is incorrectly said to be out of control is approximately $n/.0027$), it is clear from Table 115.1 that one should combine at least 9 of the items. This would mean that each data value (equal to the number of defects in the combined set) would have mean at least $9 \times 4 = 36$.

Example 11.5a The following represent the number of defects discovered at the factory on successive units of 10 cars each:

Car	Defects	Car	Defects	Car	Defects
1	141	8	95	15	94
2	162	9	76	16	68
3	150	10	68	17	95
4	111	11	63	18	81
5	92	12	74	19	102
6	74	13	103	20	73
7	85	14	81		

Does it appear that the production process was in control throughout?

Since $\overline{X} = 94.4$ it follows that the trial control limits are

$$LCL = 94.4 - 3\sqrt{94.4} = 65.25$$

$$UCL = 94.4 + 3\sqrt{94.4} = 123.55$$

Since the first three data values are larger than UCL, they are removed and the sample mean recomputed. This yields

$$\overline{X} = \frac{(94.4)20 - (141 + 162 + 150)}{17} = 84.41$$

and so the new trial control limits are

$$LCL = 84.41 - 3\sqrt{84.41} = 56.85$$

$$UCL = 84.41 + 3\sqrt{84.41} = 111.97$$

At this point since all remaining 17 data values fall within the limits, we could declare that the process is now in control with a mean value of 84.41. However, because it seems that the mean number of defects was initially high before settling into control, it seems quite plausible that the data value X_4 also originated before the process was in control. Thus, it would seem prudent in this situation to also eliminate X_4 and recompute. Based on the remaining 16 data values we obtain that

$$\overline{X} = 82.56$$

$$LCL = 82.56 - 3\sqrt{82.56} = 55.30$$

$$UCL = 82.56 + 3\sqrt{82.56} = 109.82$$

and so it appears that the process is now in control with a mean value of 82.56. ∎

6 OTHER CONTROL CHARTS FOR DETECTING CHANGES IN THE POPULATION MEAN

The major weakness of the $\overline{X}$ control chart presented in Section 2 is that it is relatively insensitive to small changes in the population mean. That is, when such a change occurs, since each plotted value is based on only a single subgroup and so tends to have a relatively large variance, it takes, on average, a large number of plotted values to detect the change. One way to remedy this weakness is to allow each plotted value to depend not only on the most recent subgroup average but on some of the other subgroup averages as well. Two approaches for doing this that have been found to be quite effective are based on (1) moving averages and (2) exponentially weighted moving averages.

6.1 Moving-Average Control Charts

The moving-average control chart of span size k is obtained by continually plotting the average of the k most recent subgroups. That is, the moving average at time t, call it M_t, is defined by

$$ M_t = \frac{\left[\overline{X}_t + \overline{X}_{t-1} + \cdots + \overline{X}_{t-k+1} \right]}{k} $$

where $\overline{X}_i$ is the average of the values of subgroup i. The successive computations can be easily performed by noting that

$$ kM_t = \overline{X}_t + \overline{X}_{t-1} + \cdots + \overline{X}_{t-k+1} $$

and, substituting $t + 1$ for t,

$$ kM_{t+1} = \overline{X}_{t+1} + \overline{X}_t + \cdots + \overline{X}_{t-k+2} $$

Subtraction now yields that

$$ kM_{t+1} - kM_t = \overline{X}_{t+1} - \overline{X}_{t-k+1} $$

or

$$ M_{t+1} = M_t + \frac{\left[\overline{X}_{t+1} - \overline{X}_{t-k+1} \right]}{k} $$

In words, the moving average at time $t + 1$ is equal to the moving average at time t plus $1/k$ times the difference between the newly added and the deleted value in the moving average. For values of t less than k, M_t is defined as the average of the first t subgroups. That is,

$$M_t = \frac{[\overline{X}_1 + \cdots + \overline{X}_t]}{t} \qquad \text{if } t < k$$

Suppose now that when the process is in control the successive values come from a normal population with mean μ and variance σ^2. Therefore, if n is the subgroup size, it follows that $\overline{X}_i$ is normal with mean μ and variance σ^2/n. From this we see that the average of m of the $\overline{X}_i$ will be normal with mean μ and variance given by $\text{Var}(\overline{X}_i)/m = \sigma^2/nm$ and, therefore, when the process is in control

$$E[M_t] = \mu$$

$$\text{Var}(M_t) = \begin{cases} \sigma^2/nt & \text{if } t < k \\ \sigma^2/nk & \text{otherwise} \end{cases}$$

Because a normal random variable is almost always within 3 standard deviations of its mean, we have the following upper and lower control limits for M_t.

$$\text{UCL} = \begin{cases} \mu + 3\sigma/\sqrt{nt} & \text{if } t < k \\ \mu + 3\sigma/\sqrt{nk} & \text{otherwise} \end{cases}$$

$$\text{LCL} = \begin{cases} \mu - 3\sigma/\sqrt{nt} & \text{if } t < k \\ \mu - 3\sigma/\sqrt{nk} & \text{otherwise} \end{cases}$$

In other words, aside from the first $k - 1$ moving averages, the process will be declared out of control whenever a moving average differs from μ by more than $3\sigma/\sqrt{nk}$.

Example 11.6a To see how well a moving-average control chart can detect a small change in the population mean, the author considered a population that, when in control, is known to be normal with mean 10 and variance $\sigma^2 = 4$. He supposed that there was a change in the mean value from 10 to 11 (an increase of $.5\sigma$) and then simulated 25 subgroup averages (of size $n = 5$) from this changed distribution. That is, he simulated the values of $\overline{X}_i$, $i = 1, 2, \ldots, 25$ that are normally distributed with mean 11 and variance 4/5. Table 11.6.1 presents the 25 values along with the moving averages based on span size $k = 8$ as well as the upper and lower control limits. In this table the LCL for $t > 8$ is 9.051318, whereas the UCL is 10.94868.

As the reader can see, the first moving average to fall outside its control limits occurred at time 11, with other such occurrences at times 12, 13, 14, 16, and 25. (It is interesting to note that the usual control chart—that is, the moving-average with $k = 1$—would have declared the process out of control at time 7 since $\overline{X}_7$ was so large. However, this is the only point where this chart would have indicated a lack of control; (see Figure 11.6.1).

TABLE 11.6.1

t	$\overline{X}_t$	M_t	LCL	UCL
1	9.617728	9.617728	7.316719	12.68328
2	10.25437	9.936049	8.102634	11.89737
3	9.876195	9.913098	8.450807	11.54919
4	10.79338	10.13317	8.658359	11.34164
5	10.60699	10.22793	8.8	11.2
6	10.48396	10.2706	8.904554	11.09545
7	13.33961	10.70903	8.95815	11.01419
8	9.462969	10.55328	9.051318	10.94868
9	10.14556	10.61926	⋮	⋮
10	11.66342	10.79539		
*11	11.55484	11.00634		
*12	11.26203	11.06492		
*13	12.31473	11.27839		
*14	9.220009	11.1204		
15	11.25206	10.85945		
*16	10.48662	10.98741		
17	9.025091	10.84735		
18	9.693386	10.6011		
19	11.45989	10.58923		
20	12.44213	10.73674		
21	11.18981	10.59613		
22	11.56674	10.88947		
23	9.869849	10.71669		
24	12.11311	10.92		
*25	11.48656	11.22768		

* = Out of control.

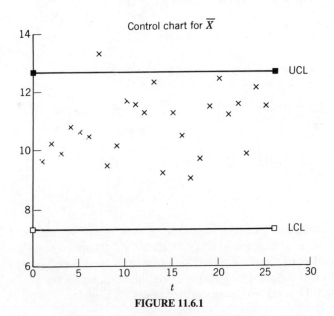

FIGURE 11.6.1

There is an inverse relationship between the size of the change in the mean value that one wants to guard against and the appropriate moving average span size k. That is, the smaller this change is the larger k ought to be.

6.2   Exponentially Weighted Moving-Average Control Charts

The moving-average control chart of Section 6.1 considered at each time t a weighted average of all subgroup averages up that time, with the k most recent values being given weight $1/k$ and the others given weight 0. Since this appears to be a most effective procedure for detecting small changes in the population mean, it raises the possibility that other sets of weights might also be successfully employed. One set of weights that is often utilized is obtained by decreasing the weight of each earlier subgroup average by a constant factor.
Let

$$W_t = \alpha \overline{X}_t + (1 - \alpha) W_{t-1} \qquad (11.6.1)$$

where α is a constant between 0 and 1, and where

$$W_0 = \mu$$

The sequence of values W_t, $t = 0, 1, 2, \ldots$ is called an exponentially weighted moving average. To understand why it has been given that name, note that if we continually substitute for the W term on the right side of Equation 11.6.1, we obtain that

$$W_t = \alpha \overline{X}_t + (1 - \alpha)\left[\alpha \overline{X}_{t-1} + (1 - \alpha) W_{t-2}\right] \qquad (11.6.2)$$

$$= \alpha \overline{X}_t + \alpha(1 - \alpha)\overline{X}_{t-1} + (1 - \alpha)^2 W_{t-2}$$

$$= \alpha \overline{X}_t + \alpha(1 - \alpha)\overline{X}_{t-1} + (1 - \alpha)^2\left[\alpha \overline{X}_{t-2} + (1 - \alpha) W_{t-3}\right]$$

$$= \alpha \overline{X}_t + \alpha(1 - \alpha)\overline{X}_{t-1} + \alpha(1 - \alpha)^2 \overline{X}_{t-2} + (1 - \alpha)^3 W_{t-3}$$

$$\vdots$$

$$= \alpha \overline{X}_t + \alpha(1 - \alpha)\overline{X}_{t-1} + \alpha(1 - \alpha)^2 \overline{X}_{t-2}$$

$$+ \cdots + \alpha(1 - \alpha)^{t-1}\overline{X}_1 + (1 - \alpha)^t \mu$$

where the foregoing used the fact that $W_0 = \mu$. Thus we see from Equation 11.6.2 that W_t is a weighted average of all the subgroup averages up to time t, giving weight α to the most recent subgroup and then successively decreasing the weight of earlier subgroup averages by the constant factor $1 - \alpha$, and then giving weight $(1 - \alpha)^t$ to the in-control population mean.

The smaller the value of α, the more even are the successive weights. For instance, if $\alpha = .1$ then the initial weight is .1 and the successive weights decrease by the factor .9; that is, the weights are $.1, .09, .081, .073, .066, .059$, and so on. On the other hand, if one chooses, say, $\alpha = .4$, then the successive weights are $.4, .24, .144, .087, .052, \ldots$ Since the successive weights $\alpha(1 - \alpha)^{i-1}$, $i = 1, 2, \ldots$, can be written as

$$\alpha(1 - \alpha)^{i-1} = \bar{\alpha}e^{-\beta i}$$

where

$$\bar{\alpha} = \frac{\alpha}{(1 - \alpha)} \qquad \beta = -\log(1 - \alpha)$$

we say that the successively older data values are "exponentially weighted" (see Figure 11.6.2).

To compute the mean and variance of the W_t, recall that, when in control, the subgroup averages $\bar{X}_i$ are independent normal random variables each having mean μ and variance σ^2/n. Therefore, using Equation 11.6.2 we see that

$$E[W_t] = \mu\left[\alpha + \alpha(1 - \alpha) + \alpha(1 - \alpha)^2 + \cdots + \alpha(1 - \alpha)^{t-1} + (1 - \alpha)^t\right]$$

$$= \frac{\mu\alpha\left[1 - (1 - \alpha)^t\right]}{1 - (1 - \alpha)} + \mu(1 - \alpha)^t$$

$$= \mu$$

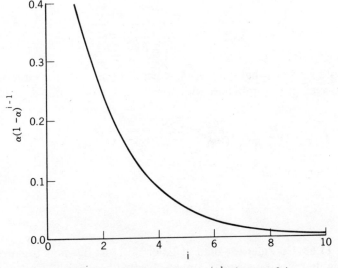

FIGURE 11.6.2 Plot of $\alpha(1 - \alpha)^{i-1}$ when $\alpha = 0.4$

To determine the variance we again use Equation 11.6.2:

$$\text{Var}(W_t) = \frac{\sigma^2}{n}\left\{\alpha^2 + [\alpha(1-\alpha)]^2 + [\alpha(1-\alpha)^2]^2 + \cdots + [\alpha(1-\alpha)^{t-1}]^2\right\}$$

$$= \frac{\sigma^2}{n}\alpha^2[1 + \beta + \beta^2 + \cdots + \beta^{t-1}] \quad \text{where } \beta = (1-\alpha)^2$$

$$= \frac{\sigma^2\alpha^2[1-(1-\alpha)^{2t}]}{n[1-(1-\alpha)^2]}$$

$$= \frac{\sigma^2\alpha[1-(1-\alpha)^{2t}]}{n(2-\alpha)}$$

Hence, when t is large we see that, provided that the process has remained in control throughout,

$$E[W_t] = \mu$$

$$\text{Var}(W_t) \approx \frac{\sigma^2\alpha}{n(2-\alpha)} \quad \text{since } (1-\alpha)^{2t} \approx 0$$

Thus the upper and lower control limits for W_t are given by

$$\text{UCL} = \mu + 3\sigma\sqrt{\frac{\alpha}{n(2-\alpha)}}$$

$$\text{LCL} = \mu - 3\sigma\sqrt{\frac{\alpha}{n(2-\alpha)}}$$

It should be noted that the preceding control limits are the same as those in a moving-average control chart with span k (after the initial k values) when

$$\frac{3\sigma}{\sqrt{nk}} = 3\sigma\sqrt{\frac{\alpha}{n(2-\alpha)}}$$

or, equivalently, when

$$k = \frac{2-\alpha}{\alpha} \quad \text{or} \quad \alpha = \frac{2}{k+1}$$

Example 11.6b Consider the data of Example 11.6a but now use an exponentially weighted moving-average control chart with $\alpha = 2/9$. This gives rise to the following data set:

t	$\overline{X}_t$	W_t	t	$\overline{X}_t$	W_t
1	9.617728	9.915051	14	9.220009	10.84522
2	10.25437	9.990456	15	11.25206	10.93563
3	9.867195	9.963064	16	10.48662	10.83585
4	10.79338	10.14758	17	9.025091	10.43346
5	10.60699	10.24967	18	9.693386	10.269
6	10.48396	10.30174	19	11.45989	10.53364
*7	13.33961	10.97682	*20	12.44213	10.95775
8	9.462969	10.64041	*21	11.18981	11.00932
9	10.14556	10.53044	*22	11.56674	11.13319
10	11.66342	10.78221	23	9.869849	10.85245
*11	11.55484	10.95391	*24	12.11311	11.13259
*12	11.26203	11.02238	*25	11.48656	11.21125
*13	12.31473	11.30957			

* = Out of control

Since

$$UCL = 10.94868$$
$$LCL = 9.051318$$

we see that the process could be declared out of control as early as $t = 7$ (Figure 11.6.3). ∎

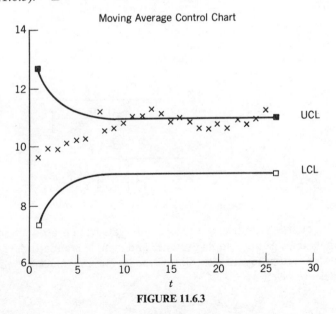

FIGURE 11.6.3

PROBLEMS

1. Assume that items produced are supposed to be normally distributed with mean 35 and standard deviation 3. To monitor this process subgroups of

size 5 are sampled. If the following represents the averages of the first 20 subgroups, does it appear that the process was in control?

Subgroup No.	$\overline{X}$	Subgroup No.	$\overline{X}$
1	34.0	11	35.8
2	31.6	12	35.8
3	30.8	13	34.0
4	33.0	14	35.0
5	35.0	15	33.8
6	32.2	16	31.6
7	33.0	17	33.0
8	32.6	18	33.2
9	33.8	19	31.8
10	35.8	20	35.6

2. Suppose that a process is in control with $\mu = 14$ and $\sigma = 2$. An $\overline{X}$ control chart based on subgroups of size 5 is employed. If a shift in the mean of 2.2 units occurs, what is the probability that the next subgroup average will fall outside the control limits? On average, how many subgroups will have to be looked at in order to detect this shift?

3. If Y has a chi-square distribution with $n - 1$ degrees of freedom, show that

$$E[\sqrt{Y}] = \sqrt{2} \, \frac{\Gamma(n/2)}{\Gamma[(n - 1)/2]}$$

Hint: Write

$$E[\sqrt{Y}] = \int_0^\infty \sqrt{y} \, f_{\chi^2_{n-1}}(y) \, dy$$

$$= \int_0^\infty \sqrt{y} \, \frac{e^{-y/2} y^{(n-1)/2-1} \, dy}{2^{(n-1)/2} \Gamma\left[\dfrac{(n - 1)}{2}\right]}$$

$$= \int_0^\infty \frac{e^{-y/2} y^{n/2-1} \, dy}{2^{(n-1)/2} \Gamma\left[\dfrac{(n - 1)}{2}\right]}$$

Now make the transformation $x = y/2$.

4. Samples of size 5 are taken at regular intervals from a production process, and the values of the sample averages and sample standard deviations are calculated. Suppose that the sum of the $\overline{X}$ and S values for the first 25 samples are given by

$$\sum \overline{X}_i = 357.2 \qquad \sum S_i = 4.88$$

(a) Determine the control limits for $\overline{X}$ control chart.
(b) Suppose that the measurable values of the items produced are supposed to be within the limits 14.3 ± 0.45. Assuming that the process remains in control with a mean and variance that is approximately

equal to the estimates derived, what percentage of the items produced will fall within the specification limits?

5. Determine the revised $\overline{X}$ and S control limits for the data in Example 11.3a.

6. In Problem 4 determine the control limits for an S control chart.

7. The following are $\overline{X}$ and S values for 20 subgroups of size 5.

Subgroup	$\overline{X}$	S	Subgroup	$\overline{X}$	S
1	33.8	5.1	11	29.7	5.1
2	37.2	5.4	12	31.6	5.3
3	40.4	6.1	13	38.4	5.8
4	39.3	5.5	14	40.2	6.4
5	41.1	5.2	15	35.6	4.8
6	40.4	4.8	16	36.4	4.6
7	35.0	5.0	17	37.2	6.1
8	36.1	4.1	18	31.3	5.7
9	38.2	7.3	19	33.6	5.5
10	32.4	6.6	20	36.7	4.2

(a) Determine trial control limits for an $\overline{X}$ control chart.
(b) Determine trial control limits for an S control chart.
(c) Does it appear that the process was in control throughout?
(d) If your answer in part (c) is no, suggest values for upper and lower control limits to be used with succeeding subgroups.
(e) If each item is supposed to have a value within 35 ± 10, what is your estimate of the percentage of items that will fall within this specification?

8. Control charts for $\overline{X}$ and S are maintained on the shear strength of spot welds. After 30 subgroups of size 4, $\Sigma \overline{X}_i = 12{,}660$ and $\Sigma S_i = 500$. Assume that the process is in control.

(a) What are the $\overline{X}$ control limits?
(b) What are the S control limits?
(c) Estimate the standard deviation for the process.
(d) If the minimum specification for this weld is 400 pounds, what percentage of the welds will not meet the minimum specification?

9. Control charts for $\overline{X}$ and S are maintained on resistors (in ohms). The subgroup size is 4. The values of $\overline{X}$ and S are computed for each subgroup. After 20 subgroups, $\Sigma \overline{X}_i = 8620$, and $\Sigma S_i = 450$.

(a) Compute the values of the limits for the $\overline{X}$ and S charts.
(b) Estimate the value of σ on the assumption that the process is in statistical control.
(c) If the specification limits are 430 ± 30, what conclusions can you draw regarding the ability of the process to produce items within these specifications?

(d) If μ is increased by 60, what is the probability of a subgroup average falling outside the control limits?

10. The following data refer to the amounts by which the diameters of $\frac{1}{4}$-inch ball bearings differ from $\frac{1}{4}$-inch in units of .001 inches. The subgroup size is $n = 5$.

Subgroup			Data values		
1	2.5	0.5	2.0	−1.2	1.4
2	0.2	0.3	0.5	1.1	1.5
3	1.5	1.3	1.2	−1.0	0.7
4	0.2	0.5	−2.0	0.0	−1.3
5	−0.2	0.1	0.3	−0.6	0.5
6	1.1	−0.5	0.6	0.5	0.2
7	1.1	−1.0	−1.2	1.3	0.1
8	0.2	−1.5	−0.5	1.5	0.3
9	−2.0	−1.5	1.6	1.4	0.1
10	−0.5	3.2	−0.1	−1.0	−1.5
11	0.1	1.5	−0.2	0.3	2.1
12	0.0	−2.0	−0.5	0.6	−0.5
13	−1.0	−0.5	−0.5	−1.0	0.2
14	0.5	1.3	−1.2	−0.5	−2.7
15	1.1	0.8	1.5	−1.5	1.2

(a) Set up trial control limits for $\overline{X}$ and S control charts.
(b) Does the process appear to have been in control throughout the sampling?
(c) If the answer to (b) is no, construct revised control limits.

11. Samples of $n = 6$ items are taken from a manufacturing process at regular intervals. A normally distributed quality characteristic is measured and $\overline{X}$ and S values are calculated for each sample. After 50 subgroups have been analyzed, we have

$$\sum_{i=1}^{50} \overline{X}_i = 970 \quad \text{and} \quad \sum_{i=1}^{50} S_i = 85$$

(a) Compute the control limit for the $\overline{X}$ and S control charts. Assume that all points on both charts plot within the control limits.
(b) If the specification limits are 19 ± 4.0, what are your conclusions regarding the ability of the process to produce items conforming to specifications?

12. The following data present the number of defective bearing and seal assemblies in samples of size 100.

Sample number	Number of defectives	Sample number	Number of defectives
1	5	11	4
2	2	12	10
3	1	13	0
4	5	14	8
5	9	15	3
6	4	16	6
7	3	17	2
8	3	18	1
9	2	19	6
10	5	20	10

Does it appear that the process was in control throughout? If not, determine revised control limits if possible.

13. The following data represent the results of inspecting all personal computers produced at a given plant during the last 12 days.

Day	Number of units	Number defective
1	80	5
2	110	7
3	90	4
4	80	9
5	100	12
6	90	10
7	80	4
8	70	3
9	80	5
10	90	6
11	90	5
12	110	7

Does the process appear to have been in control? Determine control limits for future production.

14. Suppose that when a process is in control each item will be defective with probability .04. Suppose that your control chart calls for taking daily samples of size 500. What is the probability that, if the probability of a defective item should suddenly shift to .08, your control chart would detect this shift on the next sample?

15. The following data represent the number of defective chips produced on the last 15 days: 121, 133, 98, 85, 101, 78, 66, 82, 90, 78, 85, 81, 100, 75, 89. Would you conclude that the process has been in control throughout these 15 days? What control limits would you advise using for future production?

16. Surface defects have been counted on 25 rectangular steel plates, and the data are shown below. Set up a control chart. Does the process producing the plates appear to be in statistical control?

Plate numbers	Number of defects	Plate number	Number of defects
1	2	14	10
2	3	15	2
3	4	16	2
4	3	17	6
5	1	18	5
6	2	19	4
7	5	20	6
8	0	21	3
9	2	22	7
10	5	23	0
11	1	24	2
12	7	25	4
13	8		

17. The following data represent 25 successive subgroup averages and moving averages of span size 5 of these subgroup averages. The data are generated by a process that, when in control, produces normally distributed items having mean 30 and variance 40. The subgroups are of size 4. Would you judge that the process has been in control throughout?

$\overline{X}_t$	M_t
35.62938	35.62938
39.13018	37.37978
29.45974	34.73976
32.5872	34.20162
30.06041	33.37338
26.54353	31.55621
37.75199	31.28057
26.88128	30.76488
32.4807	30.74358
26.7449	30.08048
34.03377	31.57853
32.93174	30.61448
32.18547	31.67531
35.80945	32.34106
30.9136	33.1748
30.54829	32.47771
36.39414	33.17019
27.62703	32.2585
34.02624	31.90186
27.81629	31.2824
26.99926	30.57259
32.44703	29.78317
38.53433	31.96463
28.53698	30.86678
28.65725	31.03497

18. The following data give subgroup averages and moving averages of these values. The span of the moving averages is $k = 8$. When in control the subgroup averages are normally distributed with mean 50 and variance 5. What can you conclude?

$\overline{X}_t$	M_t
50.79806	50.79806
46.21413	48.50609
51.85793	49.62337
50.27771	49.78696
53.81512	50.59259
50.67635	50.60655
51.39083	50.71859
51.65246	50.83533
52.15607	51.00508
54.57523	52.05022
53.08497	52.2036
55.02968	52.79759
54.25338	52.85237
50.48405	52.82834
50.34928	52.69814
50.86896	52.6002
52.03695	52.58531
53.23255	52.41748
48.12588	51.79759
52.23154	51.44783

19. Redo Problem 17 by employing an exponential weighted moving-average control chart with $\alpha = 1/3$.

20. Analyze the data of Problem 18 with an exponential weighted moving-average control chart having $\alpha = 2/9$.

21. Explain why a moving-average control chart with span size k must use different control limits for the first $k = 1$ moving averages, whereas an exponentially weighted moving-average control chart can use the same control limits throughout.

Hint: Argue that $\text{Var}(M_t)$ decreases in t, whereas $\text{Var}(W_t)$ increases, and explain why this is relevant.

CHAPTER 12

Simulation

1 INTRODUCTION

Suppose that $X_1, \ldots, X_n$ is a sample from a normal distribution having mean μ_x and variance σ_x^2 and $Y_1, \ldots, Y_m$ is an independent sample from a normal population having mean μ_y and variance σ_y^2. Suppose that these parameters are unknown and we are interested in testing the hypothesis that $\mu_x = \mu_y$.

Since $\overline{X} = \sum_{i=1}^{n} X_i/n$ is normal with mean μ_x and variance σ_x^2/n and $\overline{Y} = \sum_{i=1}^{m} Y_i/m$ is normal with mean μ_y and variance σ_y^2/m, it follows, by the independence of the two samples, that

$$\frac{\overline{X} - \overline{Y} - (\mu_x - \mu_y)}{\sqrt{\dfrac{\sigma_x^2}{n} + \dfrac{\sigma_y^2}{m}}} \sim \mathcal{N}(0, 1)$$

and so when $\mu_x = \mu_y$

$$\frac{\overline{X} - \overline{Y}}{\sqrt{\dfrac{\sigma_x^2}{n} + \dfrac{\sigma_y^2}{m}}} \sim \mathcal{N}(0, 1)$$

Since σ_x^2 and σ_y^2 are unknown, it thus seems reasonable to approximate them by the sample variances

$$S_x^2 = \frac{\sum_{i=1}^{n} (X_i - \overline{X})^2}{n - 1}, \qquad S_y^2 = \frac{\sum_{i=1}^{m} (Y_i - \overline{Y})^2}{m - 1}$$

and then accept the hypothesis that $\mu_x = \mu_y$ if the statistic

$$W \equiv \frac{\overline{X} - \overline{Y}}{\sqrt{\dfrac{S_x^2}{n} + \dfrac{S_y^2}{m}}}$$

is in absolute value not too large. To determine how large W need be to reject the hypothesis, we could choose a significance level α and then act as if W were actually a unit normal random variable. That is, we would accept the

436

hypothesis that $\mu_x = \mu_y$ if

$$-z_{\alpha/2} < \frac{\overline{X} - \overline{Y}}{\sqrt{\dfrac{S_x^2}{n} + \dfrac{S_y^2}{m}}} < z_{\alpha/2}$$

However, we must be aware of the fact that the probability of rejection when the hypothesis $\mu_1 = \mu_2$ is true will not exactly equal α (though it should be close when n and m are large) but will also depend on the values of σ_x^2 and σ_y^2. Now suppose we wanted to know the actual probability of rejection when the hypothesis is true for particular values of σ_x^2 and σ_y^2—say for $\sigma_x^2 = a$ and $\sigma_y^2 = b$. Clearly, to determine this, we would need to compute the distribution of $(\overline{X} - \overline{Y})/\sqrt{(S_x^2/n) + (S_y^2/m)}$ when $\mu_x = \mu_y$, $\sigma_x^2 = a$, $\sigma_y^2 = b$. However, since this is analytically very complicated, how can we proceed?

One possibility is to determine (to a high level of approximation) this distribution via a simulation study. To start, simulate n independent normal random variables having mean 0 and variance a—call them $X_1^{(1)}, \ldots, X_n^{(1)}$—and then simulate a second sequence, independent of the first, of m normal random variables having mean 0 and variance b—call them $Y_1^{(1)}, \ldots, Y_m^{(1)}$. Now compute

$$W^{(1)} = \frac{\overline{X}^{(1)} - \overline{Y}^{(1)}}{\sqrt{\dfrac{S_x^{(1)^2}}{n} + \dfrac{S_y^{(1)^2}}{m}}}$$

where

$$\overline{X}^{(1)} = \frac{\sum_{i=1}^n X_i^{(1)}}{n}, \qquad \overline{Y}^{(1)} = \frac{\sum_{i=1}^m Y_i^{(1)}}{m}$$

$$S_x^{(1)^2} = \frac{\sum_{i=1}^n \left(X_i^{(1)} - \overline{X}^{(1)} \right)^2}{n - 1}, \qquad S_y^{(1)^2} = \frac{\sum_{i=1}^m \left(Y_i^{(1)} - \overline{Y}^{(1)} \right)^2}{m - 1}$$

Now generate a second set of $n + m$ independent normals—$X_1^{(2)}, \ldots, X_n^{(2)}$, $Y_1^{(2)}, \ldots, Y_m^{(2)}$—where $X_i^{(2)}$, $i = 1, \ldots, n$ has mean 0 and variance a and $Y_i^{(2)}$, $i = 1, \ldots, m$ has mean 0 and variance b. Then set

$$W^{(2)} = \frac{\overline{X}^{(2)} - \overline{Y}^{(2)}}{\sqrt{\dfrac{S_x^{(2)^2}}{n} + \dfrac{S_y^{(2)^2}}{m}}}$$

where $\overline{X}^{(2)}, \overline{Y}^{(2)}, S_x^{(2)^2}, S_y^{(2)^2}$ are defined analogously as in the first set. Keep on doing this until r, a fixed number, of independent sets have been generated and the quantities $W^{(1)}, W^{(2)}, \ldots, W^{(r)}$ determined. Now $W^{(1)}, \ldots, W^{(r)}$ will constitute a sequence of independent random variables each having the same distribution as W when $\sigma_x^2 = a$, $\sigma_y^2 = b$. Hence, we can estimate F, the distribution function of W when $\sigma_x^2 = a$, $\sigma_y^2 = b$ by

$$F(x) = P_{a,b}\{W \leq x\}$$

$$\approx \frac{\text{Number } j : W^{(j)} \leq x}{r}$$

By the law of large numbers, the foregoing approximation will become exact as r approaches ∞. This approach to estimating the distribution function of a random variable is called the *Monte-Carlo Simulation* approach.

Clearly there remains the problem of how to generate, or *simulate*, random variables having a specified distribution. The first step in doing this is to be able to generate random variables from a uniform distribution on $(0, 1)$; such variates are called random numbers. To generate such numbers most computers have a built-in subroutine, called a random number generator, whose output is a sequence of pseudo random numbers. This is a sequence of numbers that is, for all practical purposes, indistinguishable from a sample from the uniform $(0, 1)$ distribution. Most random number generators start with an initial value X_0, called the seed, and then recursively compute values by specifying positive integers a, c, and m, and then letting

$$X_{n+1} = (aX_n + c) \text{ modulo } m \qquad n \geq 0$$

where the foregoing means that $aX_n + c$ is divided by m and the remainder is taken as the value of X_{n+1}. Thus each X_n is either $0, 1, \ldots, m - 1$ and the quantity X_n/m is taken as an approximation to a uniform $(0, 1)$ random variable. It can be shown that, subject to suitable choices for a, c, and m, the foregoing gives rise to a sequence of numbers that look as if they were generated from independent uniform $(0, 1)$ random variables.

As our starting point in the simulation of random variables from an arbitrary distribution, we shall suppose that we can simulate from the uniform $(0, 1)$ distribution and we shall use the term "random numbers" to mean independent random variables from this distribution.

Example 12.1a Generating a Random Permutation Suppose we are interested in generating a permutation of the integers $1, 2, \ldots, n$ that is such that all $n!$ possible orderings are equally likely. Starting with any initial permutation we will accomplish this after $n - 1$ steps where at each step we will interchange the positions of two of the numbers of the permutation. Throughout, we will keep track of the permutation by letting $X(i)$, $i = 1, \ldots, n$ denote the number currently in position i. The algorithm operates as follows:

1. Consider any arbitrary permutation and let $X(i)$ denote the element in position i, $i = 1, \ldots, n$. (For instance, we could take $X(i) = i$, $i = 1, \ldots, n$).
2. Generate a random variable N_n that is equally likely to equal any of the values $1, 2, \ldots, n$.
3. Interchange the values of $X(N_n)$ and $X(n)$. The value of $X(n)$ will now remain fixed. [For instance, suppose $n = 4$ and initially $X(i) = i$, $i = $

1, 2, 3, 4. If $N_4 = 3$, then the new permutation is $X(1) = 1$, $X(2) = 2$, $X(3) = 4$, $X(4) = 3$, and element 3 will remain in position 4 throughout.]

4. Generate a random variable N_{n-1} that is equally likely to be either 1, 2, ..., $n - 1$.

5. Interchange the values of $X(N_{n-1})$ and $X(n - 1)$. [If $N_3 = 1$, then the new permutation is $X(1) = 4$, $X(2) = 2$, $X(3) = 1$, $X(4) = 3$.]

6. Generate N_{n-2}, which is equally likely to be either 1, 2, ..., $n - 2$.

7. Interchange the values of $X(N_{n-2})$ and $X(2)$. [If $N_2 = 1$, then the new permutation is $X(1) = 2$, $X(2) = 4$, $X(3) = 1$, $X(4) = 3$ and this is the final permutation.]

8. Generate N_{n-3}, and so on. The algorithm continues until N_2 is generated and after the next interchange the resulting permutation is the final one.

∎

To implement this algorithm, it is necessary to be able to generate a random variable that is equally likely to be any of the values 1, 2, ..., k. To accomplish this, let U denote a random number—that is, U is uniformly distributed on $(0, 1)$, and note that kU is uniform on $(0, k)$. Hence,

$$P\{i - 1 < kU < i\} = \frac{1}{k} \qquad i = 1, \ldots, k$$

and so if we take $N_k = [kU] + 1$, where $[x]$ is the integer part of x (that is, it is the largest integer less than or equal to x), then N_k will have the desired distribution.

The algorithm can now be succinctly written as follows:

Step 1: Let $X(1), \ldots, X(n)$ be any permutation of 1, 2, ..., n. (For instance, we can set $X(i) = i$, $i = 1, \ldots, n$.)
Step 2: Let $I = n$.
Step 3: Generate a random number U and set $N = [IU] + 1$.
Step 4: Interchange the values of $X(N)$ and $X(I)$.
Step 5: Reduce the value of I by 1 and if $I > 1$ go to Step 3.
Step 6: $X(1), \ldots, X(n)$ is the desired randomly generated permutation.

The foregoing algorithm for generating a random permutation is extremely useful to a statistician. For suppose that a statistician is developing an experiment to compare the effects of m different treatments on a set of n subjects. He decides to split the subjects into m different groups of respective sizes $n_1, n_2, \ldots, n_m$ where $\sum_{i=1}^{m} n_i = n$ with the members of the ith group to receive treatment i. To eliminate any bias in the assignment of subjects to treatments (for instance, it would cloud the meaning of the experimental results if it turned out that all the "best" subjects had been put in the same group), it is imperative that the assignment of a subject to a given group be done "at random." How is this to be accomplished?[†]

A simple and efficient procedure is to arbitrarily number the subjects 1 through n and then generate a random permutation $X(1), \ldots, X(n)$ of $1, 2, \ldots, n$. Now assign subjects $X(1), X(2), \ldots, X(n_1)$ to be in group 1, $X(n_1 + 1), \ldots, X(n_1 + n_2)$ to be in group 2, and in general group j is to consist of subjects numbered $X(n_1 + n_2 + \cdots + n_{j-1} + k)$, $k = 1, \ldots, n_j$.

Program 12-1 uses the foregoing algorithm to generate a random permutation. When the program asks the user to input a random number seed, he should enter any number within the prescribed limits. For instance, suppose we wanted a random permutation of the numbers 1 through 15. We run Program 12-1-1:

```
RUN
THIS PROGRAM GENERATES A RANDOM PERMUTATION OF THE NUMBERS 1,2,...,N
ENTER THE VALUE OF N
? 15
Random number seed (-32768 to 32767)? 3542
THE RANDOM PERMUTATION IS AS FOLLOWS
 3
13
 1
 8
 2
 9
 4
12
 7
10
15
14
 6
 5
11
Ok
```

Example 12.1b Draft Lottery On December 1, 1969, officials of the U.S. Selective Service System placed 366 allegedly identical capsules in an urn. Within each capsule was a slip of paper on which was written one of the days of the year (January 1 through December 31 with February 29 included). The capsules were then to be randomly withdrawn with the order of selection determining the vulnerability to the draft of all males of draft age. Moreover, it was announced that all those whose birthdays were among the first 120 days drawn were almost certain to be drafted.

It turned out that of the first 120 days withdrawn the number of them that fell into the successive months of the year, starting with January, were

$$8, 7, 4, 8, 9, 11, 12, 13, 10, 9, 12, 17$$

When these results were announced there was an immediate controversy over

---

[†]When $m = 2$, another technique for randomly dividing the subjects was presented in Example 3.4d of Chapter 3. The preceding procedure is faster but requires more space than the one of Example 3.4d.

whether the method used actually led to a random selection. Indeed, many people claimed that it could not have been truly random since of the first 120 days there were certain months that had too many days selected and others too few for the selection to have been random. What do you think?

Solution To determine whether the result is consistent with the hypothesis of a random selection, consider a random sample of 120 of the 366 days of the year and let $X(i)$ denote the number of days from month i in the sample, $i = 1, \ldots, 12$. Also, let $N(i)$ denote the number of days in month i—that is, $\{ N(i),\ i = 1, \ldots, 12 \} = \{ 31, 29, 31, 30, 31, 30, 31, 31, 30, 31, 30, 31 \}$.

Now, consider a given month i and let

$$I_j = \begin{cases} 1 & \text{if day } j \text{ of month } i \text{ is in the sample} \\ 0 & - \end{cases}$$

Therefore,

$$X(i) = \sum_{j=1}^{N(i)} I_j$$

and as each of the 366 days will be in the random sample with probability $120/366$ we see that

$$E[X(i)] = \sum_{j=1}^{N(i)} E[I_j] = N(i)\frac{120}{366}$$

Hence, in analogy with the chi-square goodness of fit test presented in Section 2 of Chapter 9, a natural statistic to consider in testing whether the sample is random is

$$S = \sum_{i=1}^{12} \frac{\left(X(i) - N(i)\frac{120}{366} \right)^2}{\frac{120}{366}N(i)}$$

When this statistic is computed for the resulting values of the $X(i)$—namely $8, 7, 4, 8, 9, 11, 12, 13, 10, 9, 12, 17$—it turns out that $S = 11.80716$. Hence, to see how unlikely such a result would have been if the method of selection were truly random, we need determine $P\{ S > 11.80716 \}$. However, whereas one might at first suppose that S would have approximately a chi-square distribution (again in analogy with the chi-square goodness of fit tests presented in Chapter 9), this is not the case. The reason is that the selection is being done without replacement and thus the relevant distribution of $X(1), \ldots, X(12)$ is not multinomial (as it would be if the sampling were with replacement). In fact, it is just about impossible to analytically determine the distribution of S. However, what we can do is determine it via a simulation. In fact, what we did do was to simulate a random choice of 120 of the 366 days of the year (by considering the final 120 elements in a random permutation of $1, 2, \ldots, 366$) and then compute the number of days falling into each of the 12 months. The statistic S was then derived. This was repeated 1000 times and it turned out that in 92 of these simulations S exceeded 11.80716. Hence, we can assert with great confidence that

$$P\{S > 11.80716\} \approx .092$$

Hence, if the draft selection were truly random, then 9.2 percent of the time one could expect an S-value at least as large as that obtained. From this it follows that we cannot conclude, at any significance level $\alpha < .092$, that the selection was not random.

It should be noted that if we would have incorrectly assumed that S has approximately a chi-square distribution with 11 degrees of freedom, then we would have come to the conclusion that an S-value as large as the one obtained would occur with probability approximately equal to .39 (since $P\{\chi_{11}^2 > 11.807\} \approx .38$) rather than the correct value (as indicated by the simulation study) of .092. ∎

2 GENERAL TECHNIQUES FOR SIMULATING CONTINUOUS RANDOM VARIABLES

In this section we present two general methods for simulating continuous random variables.

2.1 The Inverse Transformation Method

A general method for simulating a random variable having a continuous distribution—called the *Inverse Transformation Method*—is based on the following proposition.

Proposition 12.2.1

Let U be a uniform $(0, 1)$ random variable. For any continuous distribution function F if we define the random variable X by

$$X = F^{-1}(U)$$

then the random variable X has distribution function F. [$F^{-1}(u)$ is defined to equal that value x for which $F(x) = u$.]

Proof

$$F_X(a) = P\{X \le a\} \qquad (12.2.1)$$
$$= P\{F^{-1}(U) \le a\}$$

Now, since $F(x)$ is a monotone function, it follows that $F^{-1}(U) \le a$ is equivalent to $F(F^{-1}(U)) \le F(a)$. Since $F[F^{-1}(U)] = U$, we obtain from Equation 12.2.1 that

$$F_X(a) = P\{U \le F(a)\}$$
$$= F(a) \quad ∎$$

It follows from Proposition 12.2.1 that we can simulate a random variable X having a continuous distribution function F by generating a random number U and then setting $X = F^{-1}(U)$.

Example 12.2a Simulating an Exponential Random Variable If $F(x) = 1 - e^{-x}$, then $F^{-1}(u)$ is that value of x such that

$$1 - e^{-x} = u$$

or

$$x = -\log(1 - u)$$

Hence, if U is a uniform $(0, 1)$ variable, then

$$F^{-1}(U) = -\log(1 - U)$$

is exponentially distributed with mean 1. Since $1 - U$ is also uniformly distributed on $(0, 1)$, it follows that $-\log U$ is exponential with mean 1. Since cX is exponential with mean c when X is exponential with mean 1, it follows that $-c \log U$ is exponential with mean c. ∎

The results of Example 12.2a can also be utilized to simulate a gamma random variable.

Example 12.2b Simulating a Gamma (n, λ) Random Variable To simulate from a gamma distribution with parameters (n, λ), when n is an integer, we use the fact that the sum of n independent exponential random variables each having rate λ has this distribution. Hence, if $U_1, \ldots, U_n$ are independent uniform $(0, 1)$ random variables,

$$X = -\sum_{i=1}^{n} \frac{1}{\lambda} \log U_i = -\frac{1}{\lambda} \log\left(\prod_{i=1}^{n} U_i\right)$$

has the desired distribution. ∎

2.2   The Rejection Method

Suppose that we have a method for simulating a random variable having density function $g(x)$. We can use this as the basis for simulating from the continuous distribution having density $f(x)$ by simulating Y from g and then accepting this simulated value with a probability proportional to $f(Y)/g(Y)$.

Specifically, let c be a constant such that

$$\frac{f(y)}{g(y)} \leq c \qquad \text{for all } y$$

We then have the following technique for simulating a random variable having density f.

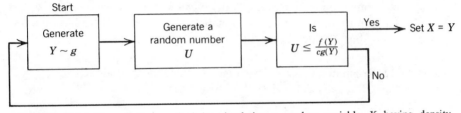

FIGURE 12.2.1 The rejection method for simulating a random variable X having density function f

Rejection Method

Step 1: Simulate Y having density g and simulate a random number U.
Step 2: If $U \le f(Y)/cg(Y)$, set $X = Y$. Otherwise return to Step 1.
　　The rejection method is pictorially expressed in Figure 12.2.1
　　We now prove that the rejection method works.

Proposition 12.2.2

The random variable X generated by the Rejection Method has density function f.

Proof Let X be the value obtained and let N denote the number of necessary iterations. Then

$$P\{X \le x\} = P\{Y_N \le x\}$$

$$= P\left\{Y \le x \middle| U \le \frac{f(Y)}{cg(Y)}\right\}$$

$$= \frac{P\left\{Y \le x, U \le \frac{f(Y)}{cg(Y)}\right\}}{K}$$

where $K = P\{U \le f(Y)/cg(Y)\}$. Now the joint density function of Y and U is, by independence,

$$f(y, u) = g(y) \qquad 0 < u < 1$$

and so using the foregoing we have

$$P\{X \le x\} = \frac{1}{K} \iint_{\substack{y \le x \\ 0 \le u \le \frac{f(y)}{cg(y)}}} g(y)\, du\, dy \qquad (12.2.2)$$

$$= \frac{1}{K} \int_0^x \int_0^{f(y)/cg(y)} du\, g(y)\, dy$$

$$= \frac{1}{cK} \int_0^x f(y)\, dy$$

Letting x approach ∞ and using the fact that f is a density gives

$$1 = \frac{1}{cK} \int_0^\infty f(y)\, dy = \frac{1}{cK}$$

Hence, from Equation 12.2.2 we obtain that

$$P\{X \leq x\} = \int_0^x f(y)\, dy$$

which completes the proof.

REMARKS

(a) It should be noted that the way in which we "accept the value Y with probability $f(Y)/cg(Y)$" is by generating a uniform $(0,1)$ random variable and then accepting Y if $U \leq f(Y)/cg(Y)$.

(b) Since each iteration will, independently, result in an accepted value with probability $P\{U \leq f(Y)/cg(Y)\} = K = 1/c$, it follows that the number of iterations has a geometric distribution with mean c.

Example 12.2c Simulating a Normal Random Variable To simulate a unit normal random variable Z (that is, one with mean 0 and variance 1), note first that the absolute value of Z has probability density function

$$f(x) = \frac{2}{\sqrt{2\pi}} e^{-x^2/2} \qquad 0 < x < \infty \tag{12.2.3}$$

We will start by simulating from the preceding density function by using the rejection method with g being the exponential density function with mean 1—that is

$$g(x) = e^{-x} \qquad 0 < x < \infty$$

Now, note that

$$\frac{f(x)}{g(x)} = \sqrt{2/\pi}\, \exp\left\{ \frac{-(x^2 - 2x)}{2} \right\} \tag{12.2.4}$$

$$= \sqrt{2/\pi}\, \exp\left\{ \frac{-(x^2 - 2x + 1)}{2} + \frac{1}{2} \right\}$$

$$= \sqrt{2e/\pi}\, \exp\left\{ \frac{-(x - 1)^2}{2} \right\}$$

$$\leq \sqrt{2e/\pi}$$

Hence, we can take $c = \sqrt{2e/\pi}$ and so, from Equation 12.2.4,

$$\frac{f(x)}{cg(x)} = \exp\left\{ \frac{-(x - 1)^2}{2} \right\}$$

Therefore, using the rejection method we can simulate the absolute value of a unit normal random variable as follows:

(a) Generate independent random variables Y and U, Y being exponential with rate 1 and U being uniform on $(0,1)$.

(b) If $U \le \exp\{-(Y-1)^2/2\}$ set $X = Y$. Otherwise return to (a).

Once we have simulated a random variable X having density function as in Equation 12.2.3, we can then generate a unit normal random variable Z by letting Z be equally likely to be either X or $-X$.

In Step (b), the value Y is accepted if $U \le \exp\{-(Y-1)^2/2\}$, which is equivalent to $-\log U \ge (Y-1)^2/2$. However, in Example 12.2a it was shown that $-\log U$ is exponential with rate 1, and so Steps (a) and (b) are equivalent to

(a′) Generate independent exponentials with rate 1, Y_1 and Y_2.

(b′) If $Y_2 \ge (Y_1 - 1)^2/2$, set $X = Y_1$. Otherwise return to (a).

Suppose now that the foregoing results in Y_1 being accepted—and so we know that Y_2 is larger than $(Y_1 - 1)^2/2$. By how much does the one exceed the other? To answer this recall that Y_2 is exponential with rate 1, and so, given that it exceeds some value, the amount by which Y_2 exceeds $(Y_1 - 1)^2/2$ [that is, its "additional life" beyond the time $(Y_1 - 1)^2/2$] is (by the memoryless property) also exponentially distributed with rate 1. That is, when we accept Step (b′), we obtain not only X (the absolute value of a unit normal) but by computing $Y_2 - (Y_1 - 1)^2/2$ we can also generate an exponential random variable (independent of X) having rate 1.

Hence, summing up, we have the following algorithm that generates an exponential with rate 1 and an independent unit normal random variable.

Step 1: Generate Y_1, an exponential random variable with rate 1.

Step 2: Generate Y_2, an exponential random variable with rate 1.

Step 3: If $Y_2 - (Y_1 - 1)^2/2 > 0$ set $Y = Y_2 - (Y_1 - 1)^2/2$ and go to Step 4. Otherwise go to Step 1.

Step 4: Generate a random number U and set

$$Z = \begin{cases} Y_1 & \text{if } U \le 1/2 \\ -Y_1 & \text{if } U > 1/2 \end{cases}$$

The random variables Z and Y generated by the foregoing are independent with Z being normal with mean 0 and variance 1 and Y being exponential with rate 1. (If we want the normal random variable to have mean μ and variance σ^2, just take $\mu + \sigma Z$). ∎

REMARKS

(a) Since $c = \sqrt{2e/\pi} \approx 1.32$, the foregoing requires a geometric distributed number of iterations of Step 2 with mean 1.32.

(b) If we want to generate a sequence of unit normal random variables, then we can use the exponential random variable Y obtained in Step 3 as the initial exponential needed in Step 1 for the next normal to be generated. Hence, on the average, we can simulate a unit normal by generating 1.64 $(= 2 \times 1.32 - 1)$ exponentials and computing 1.32 squares.

Program 12-2 uses the foregoing to simulate a sample of normal random variables. The user needs to input the mean, variance, and size of the sample, as well as specifying a random number seed. As an illustration suppose we wanted 5 normal random variables having mean 3 and variance 4. To obtain them we run Program 12-2.

```
RUN
THIS PROGRAM SIMULATES NORMAL RANDOM VARIABLES
Random number seed (-32768 to 32767)? 4361
ENTER THE MEAN
? 3
ENTER THE VARIANCE
? 4
HOW MANY NORMALS DO YOU WANT?
? 5
 5.608636
 2.226122
 3.858154
 2.168643
 3.011912
Ok
```

Example 12.2d Simulating a Chi-Square Random Variable The chi-square distribution with n degrees of freedom is the distribution of $\chi_n^2 = Z_1^2 + \cdots + Z_n^2$ where Z_i, $i = 1, \ldots, n$ are independent unit normals. Now it was shown in Section 8.1 of Chapter 3 that $Z_1^2 + Z_2^2$ has an exponential distribution with rate $\frac{1}{2}$. Hence, when n is even—say $n = 2k$—χ_{2k}^2 has a gamma distribution with parameters $(k, \frac{1}{2})$. Hence, $-2 \log (\prod_{i=1}^{k} U_i)$ has a chi-square distribution with $2k$ degrees of freedom. We can simulate a chi-square random variable with $2k + 1$ degrees of freedom by first simulating a unit normal random variable Z and then adding Z^2 to the foregoing. That is,

$$\chi_{2k+1}^2 = Z^2 - 2 \log \left(\prod_{i=1}^{k} U_i \right)$$

where $Z, U_1, \ldots, U_n$ are independent with Z being a unit normal and the others being uniform $(0, 1)$ random variables. ∎

3 SIMULATING FROM DISCRETE DISTRIBUTIONS

All of the general methods for simulating random variables from continuous distributions have analogs in the discrete case. For instance, if we want to simulate a random variable X having probability mass function

$$P\{X = x_j\} = P_j, \qquad j = 0, 1, \ldots, \qquad \sum_j P_j = 1$$

We can use the following discrete time analog of the inverse transform technique.

To simulate X for which $P\{X = x_j\} = P_j$ let U be uniformly distributed over $(0, 1)$, and set

$$X = \begin{cases} x_1 & \text{if} \quad U < P_1 \\ x_2 & \text{if} \quad P_1 < U < P_1 + P_2 \\ \vdots \\ x_j & \text{if} \quad \displaystyle\sum_{1}^{j-1} P_i < U < \sum_{i}^{j} P_i \\ \vdots \end{cases}$$

Since

$$P\{X = x_j\} = P\left(\sum_{1}^{j-1} P_i < U < \sum_{1}^{j} P_i \right) = P_j$$

we see that X has the desired distribution.

Example 12.3a The Geometric Distribution Suppose that independent trials each of which results in a "success" with probability p, $0 < p < 1$, are continually performed until a success occurs. Letting X denote the number of necessary trials, then

$$P\{X = i\} = (1 - p)^{i-1} p \qquad i \geq 1$$

which is seen by noting that $X = i$ if the first $i - 1$ trials are all failures and the ith is a success. The random variable X is said to be a geometric random variable with parameter p. Since

$$\sum_{i=1}^{j-1} P\{X = i\} = 1 - P\{X > j - 1\}$$

$$= 1 - P\{\text{first } j - 1 \text{ are all failures}\}$$

$$= 1 - (1 - p)^{j-1} \qquad j \geq 1$$

we can simulate such a random variable by generating a random number U and then setting X equal to that value j for which

$$1 - (1 - p)^{j-1} < U < 1 - (1 - p)^j$$

or, equivalently, for which

$$(1 - p)^j < 1 - U < (1 - p)^{j-1}$$

Since $1 - U$ has the same distribution as U, we can thus define X by

$$X = \min \left\{ j : (1 - p)^j < U \right\}$$
$$= \min \left\{ j : j \log (1 - p) < \log U \right\}$$
$$= \min \left\{ j : j > \frac{\log U}{\log (1 - p)} \right\}$$

where the inequality changed sign since $\log (1 - p)$ is negative [since $\log (1 - p) < \log 1 = 0$]. Using the notation $[x]$ for the integer part of x (that is, $[x]$ is the largest integer less than or equal to x), we can write

$$X = 1 + \left[\frac{\log U}{\log (1 - p)} \right] \quad \blacksquare$$

As in the continuous case, special simulation techniques have been developed for the more common discrete distributions. We now present certain of these.

Example 12.3b Simulating a Binomial Random Variable A binomial (n, p) random variable can be most easily simulated by recalling that it can be expressed as the sum of n independent Bernoulli random variables. That is, if $U_1, \ldots, U_n$ are independent uniform $(0, 1)$ variables, then letting

$$X_i = \begin{cases} 1 & \text{if } U_i < p \\ 0 & \text{otherwise} \end{cases}$$

it follows that $X \equiv \sum_{i=1}^{n} X_i$ is a binomial random variable with parameters n and p. $\quad \blacksquare$

Example 12.3c Simulating a Poisson Random Variable To simulate a Poisson random variable with mean λ, generate independent uniform $(0, 1)$ random variables $U_1, U_2, \ldots$ stopping at

$$N = \min \left\{ n : \prod_{i=1}^{n} U_i < e^{-\lambda} \right\}$$

The random variable $X \equiv N - 1$ has the desired distribution. That is, if we continue generating random numbers until their product falls below $e^{-\lambda}$, then the number required, minus 1, is Poisson with mean λ.

That $X \equiv N - 1$ is indeed a Poisson random variable having mean λ can perhaps be most easily seen by noting that

$$X + 1 = \min \left\{ n : \prod_{i=1}^{n} U_i < e^{-\lambda} \right\}$$

is equivalent to

$$X = \max \left\{ n : \prod_{i=1}^{n} U_i \geq e^{-\lambda} \right\} \qquad \text{where } \prod_{i=1}^{0} U_i \equiv 1$$

or, taking logarithms, to

$$X = \max \left\{ n : \sum_{i=1}^{n} \log U_i \geq -\lambda \right\}$$

or

$$X = \max \left\{ n : \sum_{i=1}^{n} - \log U_i \leq \lambda \right\}$$

However, $-\log U_i$ is exponential with rate 1 and so X can be thought of as being the maximum number of exponentials having rate 1 that can be summed and still be less than λ. But by recalling that the times between successive events of a Poisson process having rate 1 are independent exponentials with rate 1, it follows that X is equal to the number of events by time λ of a Poisson process having rate 1; and thus X has a Poisson distribution with mean λ (Section 6.1 of Chapter 3). ■

Program 12-3 utilizes the above to simulate Poisson random variables.

Example 12.3d Simulate 5 independent Poisson random variables having mean 50.

Solution Run Program 12-3:

```
RUN
THIS PROGRAM SIMULATES N INDEPENDENT POISSON RANDOM VARIABLES EACH HAVING MEAN L
AMBDA
ENTER THE MEAN VALUE LAMBDA
? 50
HOW MANY VARIATES DO YOU WANT?
? 5
Random number seed (-32768 to 32767)? 7542
THE 5 VALUES ARE AS FOLLOWS
 56
 55
 42
 43
 44
Ok     ■
```

4 VARIANCE REDUCTION TECHNIQUES

Let $X_1, \ldots, X_n$ have a given joint distribution and suppose we are interested in computing

$$\theta \equiv E\big[g(X_1, \ldots, X_n) \big]$$

where g is some specified function. It sometimes turns out that it is extremely difficult to analytically compute the foregoing, and when such is the case we can attempt to use simulation to estimate θ. This is done as follows: generate $X_1^{(1)}, \ldots, X_n^{(1)}$ having the same joint distribution as $X_1, \ldots, X_n$ and set

$$Y_1 = g\left(X_1^{(1)}, \ldots, X_n^{(1)} \right)$$

Now simulate a second set of random variables (independent of the first set) $X_1^{(2)}, \ldots, X_n^{(2)}$ having the distribution of $X_1, \ldots, X_n$ and set

$$Y_2 = g\left(X_1^{(2)}, \ldots, X_n^{(2)} \right)$$

Continue this until you have generated k (some predetermined number) sets and so have also computed $Y_1, Y_2, \ldots, Y_k$. Now, $Y_1, \ldots, Y_k$ are independent and identically distributed random variables each having the same distribution of $g(X_1, \ldots, X_n)$. Thus, if we let $\overline{Y}$ denote the average of these k random variables—that is

$$\overline{Y} = \sum_{i=1}^{k} \frac{Y_i}{k}$$

then

$$E[\overline{Y}] = \theta$$

$$E\left[(\overline{Y} - \theta)^2 \right] = \text{Var}(\overline{Y})$$

Hence, we can use $\overline{Y}$ as an estimate of θ. Since the expected square of the difference between $\overline{Y}$ and θ is equal to the variance of $\overline{Y}$ we would like this quantity to be as small as possible. [In the preceding situation, $\text{Var}(\overline{Y}) = \text{Var}(Y_i)/k$, which is usually not known in advance but must be estimated from the generated values $Y_1, \ldots, Y_n$]. We now present two general techniques for reducing the variance of our estimator.

4.1 Use of Antithetic Variables

In the foregoing situation, suppose that we have generated Y_1 and Y_2, which are identically distributed random variables having mean θ. Now

$$\text{Var}\left(\frac{Y_1 + Y_2}{2} \right) = \frac{1}{4}\left[\text{Var}(Y_1) + \text{Var}(Y_2) + 2\,\text{Cov}(Y_1, Y_2) \right]$$

$$= \frac{\text{Var}(Y_1)}{2} + \frac{\text{Cov}(Y_1, Y_2)}{2}$$

Hence it would be advantageous (in the sense that the variance would be reduced) if Y_1 and Y_2 rather than being independent were negatively correlated. To see how we could arrange this, let us suppose that the random variables $X_1, \ldots, X_n$ are independent and, in addition, that each is simulated via the inverse transform technique. That is, X_i is simulated from $F_i^{-1}(U_i)$ where U_i is a random number and F_i is the distribution of X_i. Hence, Y_1 can be expressed as

$$Y_1 = g\left(F_1^{-1}(U_1), \ldots, F_n^{-1}(U_n) \right)$$

Now, since $1 - U$ is also uniform over $(0, 1)$ whenever U is a random number (and is negatively correlated with U), it follows that Y_2 defined by

$$Y_2 = g\big(F_1^{-1}(1 - U_1), \ldots, F_n^{-1}(1 - U_n)\big)$$

will have the same distribution as Y_1. Hence, if Y_1 and Y_2 were negatively correlated, then generating Y_2 by this means would lead to a smaller variance than if it were generated by a new set of random numbers. (In addition, there is a computational savings since rather than having to generate n additional random numbers, we need only subtract each of the previous n from 1). Although we cannot, in general, be certain that Y_1 and Y_2 will be negatively correlated, this often turns out to be the case and indeed it can be proven that it will be so whenever g is a monotonic function.

Example 12.4a Simulating the Reliability Function Consider a system of n components, each of which is either functioning or failed. Letting

$$x_i = \begin{cases} 1 & \text{if component } i \text{ works} \\ 0 & \text{otherwise} \end{cases}$$

we call $\mathbf{x} = (x_1, \ldots, x_n)$ the state vector. Suppose also that there is a nondecreasing function $\phi(x_1, \ldots, x_m)$ such that

$$\phi(x_1, \ldots, x_n) = \begin{cases} 1 & \text{if the system works under state vector } x_1, \ldots, x_n \\ 0 & \text{otherwise} \end{cases}$$

The function $\phi(x_1, \ldots, x_n)$ is called the structure function.

Some common structure functions are the following:

(a) *The series structure:* For the series structure

$$\phi(x_1, \ldots, x_n) = \min_i x_i$$

The series system will work only if all of its components function.

(b) *The parallel structure:* For the parallel structure

$$\phi(x_1, \ldots, x_n) = \max_i x_i$$

Hence, the parallel system will work if at least one of its components works.

(c) *The k-of-n system:* The structure function

$$\phi(x_1, \ldots, x_n) = \begin{cases} 1 & \text{if } \sum_{i=1}^{n} x_i \geq k \\ 0 & \text{otherwise} \end{cases}$$

is called a *k-of-n* structure function. Since $\sum_{i=1}^{n} x_i$ represents the number of functioning components, a k-of-n system will work if at least k of the n components are working.

It should be noted that a series system is an n-of-n system, whereas a parallel system is a l-of-n system.

(d) *The bridge structure:* A five-component system for which

$$\phi(x_1, x_2, x_3, x_4, x_5) = \max(x_1 x_3 x_5, x_2 x_3 x_4, x_1 x_4, x_2 x_5)$$

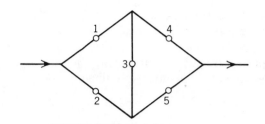

FIGURE 12.4.1 The bridge structure

is said to have a bridge structure. Such a system can be represented schematically by Figure 12.4.1. The idea of the diagram is that the system will function if a signal can go, from left to right, through the system. The signal can go through any given node i provided that component i is functioning. We leave it as an exercise for the reader to verify the formula given for the bridge structure function.

Let us suppose now that the state of component i—call it X_i—$i = 1, \ldots, n$ are independent random variables such that

$$P\{ X_i = 1\} = p_i = 1 - P\{ X_i = 0\} \qquad i = 1, \ldots, n$$

Let

$$r(p_1, \ldots, p_n) = P\{ \phi(X_1, \ldots, X_n) = 1\}$$
$$= E[\phi(X_1, \ldots, X_n)]$$

The function $r(p_1, \ldots, p_n)$ is called the *reliability* function. It represents the probability that the system will work when the components are independent with component i functioning with probability p_i, $i = 1, \ldots, n$.

For a series system

$$r(p_1, \ldots, p_n) = P\{ X_i = 1 \quad \text{for all } i = 1, \ldots, n\}$$
$$= \prod_{i=1}^{n} P\{ X_i = 1\}$$
$$= \prod_{i=1}^{n} p_i$$

and for a parallel system

$$r(p_1, \ldots, p_n) = P\{ X_i = 1 \quad \text{for at least one } i, i = 1, \ldots, n\}$$
$$= 1 - P\{ X_i = 0 \quad \text{for all } i = 1, \ldots, n\}$$
$$= 1 - \prod_{i=1}^{n} P(X_i = 0)$$
$$= 1 - \prod_{i=1}^{n} (1 - p_i)$$

However, for most systems it remains a formidable problem to compute the reliability function (even for such small systems as a 5-of-10 system or the bridge system it can be quite tedious to compute). So let us suppose that for a given nondecreasing structure function ϕ and given probabilities $p_1, \ldots, p_n$, we are interested in using simulation to estimate

$$r(p_1, \ldots, p_n) = E[\phi(X_1, \ldots, X_n)]$$

Now, we can simulate the X_i by generating uniform random numbers $U_1, \ldots, U_n$ and then setting

$$X_i = \begin{cases} 1 & \text{if } U_i < p_i \\ 0 & \text{otherwise} \end{cases}$$

Hence, we see that

$$\phi(X_1, \ldots, X_n) = k(U_1, \ldots, U_n)$$

where k is a decreasing function of $U_1, \ldots, U_n$. Hence

$$\text{Cov}(k(\mathbf{U}), k(\mathbf{1} - \mathbf{U})) \leq 0$$

and so the antithetic variable approach of using $U_1, \ldots, U_n$ to generate both $k(U_1, \ldots, U_n)$ and $k(1 - U_1, \ldots, 1 - U_n)$ results in a smaller variance than if an independent set of random numbers was used to generate the second value of k. ∎

4.2 Control Variates

Again suppose we want to use simulation to estimate $E[g(\mathbf{X})]$ where $\mathbf{X} = (X_1, \ldots, X_n)$. But now suppose that for some function f the expected value of $f(\mathbf{X})$ is known—say $E[f(\mathbf{X})] = \mu$. Then for any constant a we can also use

$$W = g(\mathbf{X}) + a[f(\mathbf{X}) - \mu]$$

as an estimator of $E[g(\mathbf{X})]$. Now

$$\text{Var}(W) = \text{Var}[g(\mathbf{X})] + a^2 \text{Var}[f(\mathbf{X})] + 2a\,\text{Cov}[g(\mathbf{X}), f(\mathbf{X})] \quad (12.4.1)$$

Simple calculus shows that the foregoing is minimized when

$$a = \frac{-\text{Cov}[f(\mathbf{X}), g(\mathbf{X})]}{\text{Var}[f(\mathbf{X})]} \quad (12.4.2)$$

and for this value of a

$$\text{Var}(W) = \text{Var}[g(\mathbf{X})] - \frac{[\text{Cov}[f(\mathbf{X}), g(\mathbf{X})]]^2}{\text{Var}[f(\mathbf{X})]} \quad (12.4.3)$$

Unfortunately, neither $\text{Var}[f(\mathbf{X})]$ nor $\text{Cov}[f(\mathbf{X}), g(\mathbf{X})]$ is usually known so we cannot usually obtain the foregoing reduction in variance. One approach in practice is to guess at these values and hope the resulting W does indeed have smaller variance than does $g(\mathbf{X})$, whereas a second possibility is to use the simulated data to estimate these quantities.

Example 12.4b Suppose, as in Example 12.4a, that we want to use simulation to estimate

$$r(p_1,\ldots,p_n) = E\left[\phi(X_1,\ldots,X_n)\right]$$

where

$$X_i = \begin{cases} 1 & \text{if } U_i < p_i \\ 0 & \text{otherwise} \end{cases}$$

Since

$$E[X_i] = p_i$$

we have that

$$E\left[\sum_{i=1}^{n} X_i\right] = \sum_{i=1}^{n} p_i$$

Because $\sum_{i=1}^{n} X_i$ and $\phi(X_1,\ldots,X_n)$ are both increasing functions of the X_i, it seems intuitive (and can be proven) that they are positively correlated. Thus, rather than using the average of the values $\phi(X_1,\ldots,X_n)$ to estimate the reliability function, we could try estimators of the form

$$\phi(X_1,\ldots,X_n) - c\sum_{i=1}^{n}(X_i - p_i)$$

where $c > 0$. ∎

PROBLEMS

1. The following algorithm will generate a random permutation of the elements $1, 2, \ldots, n$. It is somewhat faster than the one presented in Example 12.1a but is such that no position is fixed until the algorithm ends.

   In this algorithm, $P(i)$ can be interpreted as the element in position i.
   Step 1: Set $k = 1$
   Step 2: Set $P(1) = 1$
   Step 3: If $k = n$, stop. Otherwise, let $k = k + 1$.
   Step 4: Generate a random number U and let

   $$P(k) = P([kU] + 1)$$
   $$P([kU] + 1) = k$$
   Go to Step 3.

   (a) Explain in words what the algorithm is doing.
   (b) Show that at iteration k—that is, when the value of $P(k)$ is initially set—that $P(1), P(2), \ldots, P(k)$ is a random permutation of $1, 2, \ldots, k$.
   Hint: Use induction and argue that

   $$P_k\{i_1, i_2, \ldots, i_{j-1}, k, i_j, \ldots, i_{k-2}, i\}$$
   $$= P_{k-1}\{i_1, i_2, \ldots, i_{j-1}, i, i_j, \ldots, i_{k-2}\}\frac{1}{k}$$
   $$= \frac{1}{k!} \text{ by the induction hypothesis.}$$

Program 12-5, which uses the foregoing algorithm, is presented in the Table of Programs.

2. Develop a technique for simulating a random variable having density function

$$f(x) = \begin{cases} e^{2x} & -\infty < x < 0 \\ e^{-2x} & 0 < x < \infty \end{cases}$$

3. Give a technique for simulating a random variable having the probability density function

$$f(x) = \begin{cases} \dfrac{1}{2}(x-2) & 2 \le x \le 3 \\ \dfrac{1}{2}\left(2 - \dfrac{x}{3}\right) & 3 < x \le 6 \\ 0 & \text{otherwise} \end{cases}$$

4. Simulate n random variables having the distribution of Problem 3 and compare the sample mean with the true mean of the distribution. Do it for $n = 10, 20, 50$.

5. Present a method to simulate a random variable having distribution function

$$F(x) = \begin{cases} 0 & x \le -3 \\ \dfrac{1}{2} + \dfrac{x}{6} & -3 < x < 0 \\ \dfrac{1}{2} + \dfrac{x^2}{32} & 0 < x \le 4 \\ 1 & x > 4 \end{cases}$$

6. Use the inverse transformation method to present an approach for generating a random variable from the Weibull distribution

$$F(t) = 1 - e^{-\alpha t^{\beta}} \qquad t \ge 0$$

7. Give a method for simulating a random variable having failure rate function

(a) $\lambda(t) = c$
(b) $\lambda(t) = ct$
(c) $\lambda(t) = ct^2$
(d) $\lambda(t) = ct^3$

8. In the following, F is the distribution function

$$F(x) = x^n \qquad 0 < x < \infty$$

(a) Give a method for simulating a random variable having distribution F that uses only a single random number.

(b) Let $U_1, \ldots, U_n$ be independent random numbers. Show that

$$P\{\max{(U_1, \ldots, U_n)} \leq x\} = x^n$$

(c) Use part (b) to give a second method of simulating a random variable having distribution F.

(d) Write two computer programs to simulate 100 random variables from F—the first one based on the results of part (a) and the second using the method of part (c). Run the programs to see which is quicker when $n = 2$, $n = 3$, $n = 10$, $n = 20$.

9. Suppose it is relatively easy to simulate from F_i for each $i = 1, \ldots, n$. How can we simulate from

(a)
$$F(x) = \prod_{i=1}^{n} F_i(x)$$

(b)
$$F(x) = 1 - \prod_{i=1}^{n} [1 - F_i(x)]$$

10. Give a method for simulating a random variable whose failure rate function is

$$\lambda(t) = c_1 + c_2 t + c_3 t^2 + c_4 t^3$$

Hint: Make use of the solutions of Problems 7 and 9. Problem 2 of Chapter 10 is also relevant.

11. Suppose we have a method to simulate random variables from the distributions F_1 and F_2. Explain how to simulate from the distribution

$$F(x) = pF_1(x) + (1 - p)F_2(x) \qquad 0 < p < 1$$

Give a method for simulating from

$$F(x) = \begin{cases} \frac{1}{3}(1 - e^{-3x}) + \frac{2}{3}x & 0 < x \leq 1 \\ \frac{1}{3}(1 - e^{-3x}) + \frac{2}{3} & x > 1 \end{cases}$$

12. In Section 2.2, we simulated the absolute value of a unit normal by using the rejection procedure on exponential random variables with rate 1. This raises the question of whether we could obtain a more efficient algorithm by using a different exponential density—that is, we could use the density $g(x) = \lambda e^{-\lambda x}$. Show that the mean number of iterations needed in the rejection scheme is minimized when $\lambda = 1$.

13. Use the rejection method with $g(x) = 1, 0 < x < 1$, to determine an algorithm for simulating a random variable having density function

$$f(x) = \begin{cases} 60x^3(1 - x)^2 & 0 < x < 1 \\ 0 & \text{otherwise} \end{cases}$$

14. Explain how you could use random numbers to approximate $\int_0^1 k(x)\, dx$ where $k(x)$ is an arbitrary function.
Hint: If U is uniform on $(0, 1)$, what is $E[k(U)]$?

15. Write a program to use simulation to approximate the following integrals.

(a)
$$\int_0^1 e^x \, dx$$

(b)
$$\int_0^\infty e^{-x^2/2} \, dx$$

Compare the approximation, using 100 random numbers, with the actual answer.

Hint: In part (b), if we make the change of variables $y = 1/(x+1)$, $dx = -(1/y^2) \, dy$, we see that

$$\int_0^\infty e^{-x^2/2} \, dx = \int_0^1 \frac{e^{-\frac{1}{2}[(1-y)/y]^2}}{y^2} \, dy$$

16. Use simulation to approximate

$$\int_0^1 \int_0^1 e^{(x+y)^2} \, dy \, dx$$

17. Let X be a random variable on $[0, 1]$ whose density is $f(x)$. Show that we can estimate $\int_0^1 g(x) \, dx$ by simulating X and then taking as our estimate $g(X)/f(X)$. This method, called *importance sampling*, tries to choose f similar in shape to g so that $g(X)/f(X)$ has a small variance.

18. Consider a 6-out-of-10 system in which component i functions with probability $.05 + (i - 1)/10$.

(a) Simulate such a system 100 times and use the results of this simulation to estimate the probability that such a system will function.

(b) Repeat part (a) but now use the antithetic variable approach (so only 500 rather than 1000 random numbers are needed).

(c) Compare the preceding results with the actual answer. (You may want to write a computer program to determine the true probability.)

19. Verify that the minimum of Equation 12.4.1 occurs when a is given by Equation 12.4.2.

20. Verify that the minimal value of Equation 12.4.1 is as given by Equation 12.4.3.

Appendix of Programs

Program	*What it Computes*
3-1	The Binomial Distribution Function

```
LIST
10 PRINT"THIS PROGRAM COMPUTES THE PROBABILITY THAT A BINOMIAL(n,p) RANDOM VARIA
BLE IS LESS THAN OR EQUAL TO i"
20 PRINT "ENTER n"
30 INPUT N
40 PRINT "ENTER p"
50 INPUT P
60 PRINT "ENTER i"
70 INPUT I
80 S=(1-P)^N
90 IF S=0 GOTO 180
100 A=P/(1-P)
110 T=S
120 IF I=0 GOTO 390
130 FOR K=0 TO I-1
140 S=S*A*(N-K)/(K+1)
150 T=T+S
160 NEXT K
170 GOTO 390
180 J=I
190 IF J>N*P THEN J=INT(N*P)
200 FOR K=1 TO J
210 L=L+LOG(N+1-K)-LOG(J+1-K)
220 NEXT K
230 L=L+J*LOG(P)+(N-J)*LOG(1-P)
240 L=EXP(L)
250 B=(1-P)/P
260 F=1
270 FOR K=1 TO J
280 F=F*B*(J+1-K)/(N-J+K)
290 T=T+F
300 NEXT K
310 IF J=I GOTO 380
320 C=1/B
330 F=1
340 FOR K=1 TO I-J
350 F=F*C*(N+1-J-K)/(J+K)
360 T=T+F
370 NEXT K
380 T=(T+1)*L
390 PRINT "THE PROBABILITY IS";T
400 END
```

3-2	The Poisson Distribution Function

```
10 PRINT "THIS PROGRAM COMPUTES THE PROBABILITY THAT A POISSON RANDOM VARIABLE
        IS LESS THAN OR EQUAL TO i"
20 PRINT "ENTER THE MEAN OF THE RANDOM VARIABLE"
30 INPUT C
40 PRINT "ENTER THE DESIRED VALUE OF i"
50 INPUT I
60 S=EXP(-C)
70 IF S=0 GOTO 150
```

```
80 T=S
90 IF I=0 GOTO 340
100 FOR K=0 TO I-1
110 S=S*C/(K+1)
120 T=T+S
130 NEXT K
140 GOTO 340
150 J=I
160 IF J>C THEN J=INT(C)
170 FOR K=1 TO J
180 FAC=FAC+LOG(K)
190 NEXT K
200 L=-C-FAC+J*LOG(C)
210 L=EXP(L)
220 F=1
230 FOR K=1 TO J
240 F=F*(J+1-K)/C
250 T=T+F
260 NEXT K
270 IF J=I GOTO 330
280 F=1
290 FOR K=1 TO I-J
300 F=F*C/(K+J)
310 T=T+F
320 NEXT K
330 T=(T+1)*L
340 PRINT "THE PROBABILITY THAT A POISSON RANDOM VARIABLE WITH MEAN " C "IS LESS
 THAN OR EQUAL TO" I "IS";T
350 END
Ok
```

3-4   A Random Subset

```
10 PRINT "THIS PROGRAM GENERATES A RANDOM SUBSET OF SIZE K FROM THE SET 1,2,...N
"
20 PRINT "ENTER THE VALUE OF N"
30 INPUT N
40 PRINT "ENTER THE VALUE OF K"
50 INPUT K
60 RANDOMIZE
70 PRINT "THE RANDOM SUBSET CONSISTS OF THE FOLLOWING" K "VALUES"
80 I=1
90 FOR J=1 TO N
100 IF RND < (K-S)/(N-J+1) THEN PRINT , I :S=S+1
110 I=I+1
120 NEXT
130 END
Ok
```

3-5-1-A   The Unit Normal Distribution Function

```
10 PRINT "THIS PROGRAM COMPUTES THE PROBABILITY THAT A UNIT NORMAL RANDOM VARIAB
LE IS LESS THAN X"
20 PRINT "ENTER THE DESIRED VALUE OF X"
30 INPUT X
40 U=ABS(X)
50 IF U>4 GOTO 180
60 Y=U^2
70 I=U
80 FOR J=1 TO 40
90 U=-U*Y*(2*J-1)/(2*J*(2*J+1))
100 I=I+U
110 NEXT
120 I=I/SQR(2*3.14159)
130 IF X<0 GOTO 160
140 PRINT "THE PROBABILITY IS".5+I
150 GOTO 220
160 PRINT "THE PROBABILITY IS".5-I
170 GOTO 220
180 IF X<0 GOTO 210
190 PRINT "THE PROBABILITY IS GREATER THAN" 1-10^-4
200 GOTO 220
210 PRINT "THE PROBABILITY IS LESS THAN" 10^-4
220 END
```

3-5-1-B   The Inverse of the Unit Normal Distribution Function

```
10 PRINT "FOR A GIVEN INPUT a, 0<a<.5, THIS PROGRAM COMPUTES THE VALUE z   SUCH
THAT THE   PROBABILITY THAT A UNIT NORMAL EXCEEDS z IS EQUAL TO a"
20 PRINT "ENTER THE DESIRED VALUE OF a"
30 INPUT A
40 T=SQR(-2*LOG(A))
50 C=2.515517
60 D=.802853
70 E=.010328
80 F=1.432788
90 G=.189269
100 H=.001308
110 Z=T-(C+D*T+E*T^2)/(1+F*T+G*T^2+H*T^3)
120 PRINT "THE VALUE IS" Z
130 END
```

3-8-1-A   The Chi-Square Distribution Function

```
10 PRINT "THIS PROGRAM COMPUTES THE PROBABILITY THAT A CHI-SQUARE RANDOM VARIABL
E WITH N DEGREES OF FREEDOM IS LESS THAN X"
20 PRINT "ENTER THE DEGREE OF FREEDOM PARAMETER"
30 INPUT N
40 S=(N-1)/2
50 PRINT "ENTER THE DESIRED VALUE OF X"
60 INPUT X
70 M=X/2
80 D=X/2-N/2+1/3
90 D=D-.04/N
100 IF N=1 GOTO 160
110 IF S=M GOTO 180
120 H=S/M
130 X=(1-H*H+2*H*LOG(H))/(1-H)^2
140 X=D*SQR((1+X)/M)
150 GOTO 190
160 X=D*SQR(2/M)
170 GOTO 190
180 X=D/SQR(M)
190 U=ABS(X)
200 IF U>4 GOTO 330
210 Y=U^2
220 I=U
230 FOR J=1 TO 40
240 U=-U*Y*(2*J-1)/(2*J*(2*J+1))
250 I=I+U
260 NEXT
270 I=I/SQR(2*3.14159)
280 IF X<0 GOTO 310
290 PRINT "THE PROBABILITY IS".5+I
300 GOTO 370
310 PRINT "THE PROBABILITY IS".5-I
320 GOTO 370
330 IF X<0 GOTO 360
340 PRINT "THE PROBABILITY IS GREATER THAN" 1-10^-4
350 GOTO 370
360 PRINT "THE PROBABILITY IS LESS THAN" 10^-4
370 END
Ok
```

3-8-1-B   The Inverse of the Chi-Square Distribution Function

```
10 PRINT "FOR A GIVEN INPUT a, 0<a<.5, THIS PROGRAM COMPUTES THE VALUE
chisq(a,n) SUCH THAT THE PROBABILITY THAT A CHI SQUARE RANDOM VARIABLE WITH n
DEGREES OF   FREEDOM EXCEEDS chisq(a,n) IS EQUAL TO a"
20 PRINT "ENTER THE DEGREE OF FREEDOM PARAMETER n"
30 INPUT N
40 PRINT "ENTER THE DESIRED VALUE OF a"
50 INPUT A
60 T=SQR(-2*LOG(A))
70 C=2.515517
80 D=.802853
90 E=.010328
100 F=1.432788
```

```
110 G=.189269
120 H=.001308
130 X=T-(C+D*T+E*T^2)/(1+F*T+G*T^2+H*T^3)
140 W=N*(X*SQR(2/(9*N))+1-2/(9*N))^3
150 B=1
160 K=N/2
170 IF K=INT(K) GOTO 230
180 FOR I=1 TO K-1/2
190 B=B*(K-I)
200 NEXT
210 B=B*SQR(3.14159)
220 GOTO 260
230 FOR I=1 TO K-1
240 B=B*I
250 NEXT
260 M=W/2
270 S=(N-1)/2
280 D=M-N/2+1/3
290 D=D-.04/N
300 IF N=1 GOTO 360
310 IF S=M GOTO 380
320 H=S/M
330 X=(1-H*H+2*H*LOG(H))/(1-H)^2
340 X=D*SQR((1+X)/M)
350 GOTO 390
360 X=D*SQR(2/M)
370 GOTO 390
380 X=D/SQR(M)
390 U=ABS(X)
400 IF U>4 GOTO 530
410 Y=U^2
420 I=U
430 FOR J=1 TO 40
440 U=-U*Y*(2*J-1)/(2*J*(2*J+1))
450 I=I+U
460 NEXT
470 I=I/SQR(2*3.14159)
480 IF X<0 GOTO 510
490 E=.5+I
500 GOTO 570
510 E=.5-I
520 GOTO 570
530 IF X<0 GOTO 560
540 E=.9999
550 GOTO 570
560 E=.0001
570 E=1-A-E
580 B=2^(-N/2)*B^-1*EXP(-W/2)*W^(N/2-1)
590 W=W+E/B
600 PRINT "THE VALUE IS" W
610 END
```

3-8-2-A   The *t*-Distribution Function

```
10 PRINT "THIS PROGRAM COMPUTES THE PROBABILITY THAT A t-RANDOM VARIABLE WITH N
DEGREES OF FREEDOM  IS LESS THAN X"
20 PRINT "ENTER THE DEGREES OF FREEDOM"
30 INPUT N
40 PRINT "ENTER THE VALUE OF X"
50 INPUT X
60 A=N-2/3+1/(10*N)
70 B=LOG(1+X^2/N)/(N-5/6)
80 IF X>0 THEN X=A*SQR(B) ELSE X=-A*SQR(B)
90 U=ABS(X)
100 IF U>4 GOTO 230
110 Y=U^2
120 I=U
130 FOR J=1 TO 40
140 U=-U*Y*(2*J-1)/(2*J*(2*J+1))
150 I=I+U
160 NEXT
170 I=I/SQR(2*3.14159)
180 IF U>X GOTO 210
190 PRINT "THE PROBABILITY IS".5+I
```

```
200 GOTO 270
210 PRINT "THE PROBABILITY IS".5-I
220 GOTO 270
230 IF X<0 GOTO 260
240 PRINT "THE PROBABILITY IS GREATER THAN" 1-10^-4
250 GOTO 270
260 PRINT "THE PROBABILITY IS LESS THAN" 10^-4
270 END
```

3-8-2-B   The Inverse of the *t*-Distribution Function

```
10 PRINT "FOR A GIVEN INPUT a, 0<a<.5, THIS PROGRAM COMPUTES THE VALUE t(a,n)
 SUCH THAT THE PROBABILITY THAT A t-RANDOM VARIABLE WITH n DEGREES OF FREEDOM
 EXCEEDS t(a,n) IS EQUAL TO a"
20 PRINT "ENTER THE DEGREES OF FREEDOM PARAMETER n"
30 INPUT N
40 PRINT "ENTER THE DESIRED VALUE OF a"
50 INPUT A
60 T=SQR(-2*LOG(A))
70 C=2.515517
80 D=.802853
90 E=.010328
100 F=1.432788
110 G=.189269
120 H=.001308
130 X=T-(C+D*T+E*T^2)/(1+F*T+G*T^2+H*T^3)
140 W=X+(X+X^3)/(4*N) + (5*X^5+16*X^3+3*X)/(96*N^2) + (3*X^7+19*X^5+17*X^3-15*X)
/(384*N^3)
150 PRINT "THE VALUE IS" W
160 END
```

3-8-3-A   The *F*-Distribution Function

```
10 PRINT "THIS PROGRAM COMPUTES THE PROBABILITY THAT AN F RANDOM VARIABLE WITH
DEGREES OF FREEDOM N AND M IS LESS THAN X"
20 PRINT "ENTER THE FIRST DEGREE OF FREEDOM PARAMETER"
30 INPUT N
40 PRINT "ENTER THE SECOND DEGREE OF FREEDOM PARAMETER"
50 INPUT M
60 PRINT "ENTER THE DESIRED VALUE OF X"
70 INPUT X
80 S=(M-1)/2
90 T=(N-1)/2
100 K=(N+M)/2-1
110 P=M/(N*X+M)
120 Q=1-P
130 D=S+1/6-(K+1/3)*P+.02*(Q/(S+.5)-P/(T+.5)+(Q-.5)/(K+1))
140 A=S/(K*P)
150 B=T/(K*Q)
160 IF A=0 THEN C=1 ELSE C=(1-A*A+2*A*LOG(A))/(1-A)^2
170 IF B=0 THEN E=1 ELSE E=(1-B*B+2*B*LOG(B))/(1-B)^2
180 X=D*SQR((1+Q*C+P*E)/((K+1/6)*P*Q))
190 U=ABS(X)
200 IF U>4 GOTO 310
210 Y=U^2
220 I=U
230 FOR J=1 TO 40
240 U=-U*Y*(2*J-1)/(2*J*(2*J+1))
250 I=I+U
260 NEXT
270 I=I/SQR(2*3.14159)
280 IF X<0 THEN Z=.5-I ELSE Z=.5+I
290 PRINT "THE PROBABILITY IS";Z
300 GOTO 350
310 IF X<0 GOTO 340
320 PRINT "THE PROBABILITY IS GREATER THEN .9999"
330 GOTO 350
340 PRINT "THE PROBABILITY IS LESS THAN .0001"
350 END
```

4-3   The Sample Mean, Sample Variance, and Sample Standard
                       Deviation

```
LIST
10 PRINT "THIS PROGRAM COMPUTES THE SAMPLE MEAN, SAMPLE VARIANCE, AND SAMPLE STA
NDARD DEVIATION OF A DATA SET"
20 PRINT "ENTER THE SAMPLE SIZE"
30 INPUT N
40 PRINT "ENTER THE DATA VALUES ONE AT A TIME"
50 INPUT M
60 FOR J=1 TO N-1
70 INPUT X
80 A=M
90 M = M + (X-M)/(J+1)
100 S = (1-1/J)*S + (J+1)*(M-A)^2
110 NEXT J
120 PRINT "SAMPLE MEAN IS";M
130 PRINT "SAMPLE VARIANCE IS";S
140 PRINT "SAMPLE STANDARD DEVIATION IS";SQR(S)
150 END
Ok
```

5-3-1   Confidence Interval for a Normal Mean Whose Variance Is
                       Unknown

```
10 PRINT "THIS PROGRAM COMPUTES A 100(1-a)% CONFIDENCE INTERVAL FOR THE MEAN OF
           A NORMAL POPULATION WHEN THE VARIANCE IS UNKNOWN"
20 PRINT "ENTER THE SAMPLE SIZE"
30 INPUT N
40 PRINT "ENTER THE DATA VALUES ONE AT A TIME"
50 INPUT M
60 FOR J=1 TO N-1
70 INPUT X
80 A=M
90 M=M+(X-M)/(J+1)
100 S=(1-1/J)*S+(J+1)*(M-A)^2
110 NEXT J
120 PRINT "ENTER THE VALUE OF a"
130 INPUT A
140 PRINT "IS A TWO-SIDED INTERVAL DESIRED? ENTER 1 IF THE ANSWER IS YES AND 0
           IF NO"
150 INPUT I
160 IF I=0 GOTO 220
170 B=A/2
180 L=N-1
190 GOSUB 350
200 PRINT "THE"100*(1-A)"% CONFIDENCE INTERVAL FOR THE MEAN IS
        ("M-T*SQR(S)/SQR(N)","M+T*SQR(S)/SQR(N)")"
210 GOTO 310
220 B=A
230 L=N-1
240 GOSUB 350
250 PRINT "IS THE ONE-SIDED CONFIDENCE INTERVAL TO BE UPPER OR LOWER? ENTER 1
           FOR UPPER AND 0 FOR LOWER"
260 INPUT K
270 IF K=0 GOTO 300
280 PRINT "THE" 100*(1-A)"% UPPER CONFIDENCE INTERVAL FOR THE MEAN IS
           ("M-T*SQR(S)/SQR(N)",INFINITY)"
290 GOTO 310
300 PRINT "THE" 100*(1-A)"% LOWER CONFIDENCE INTERVAL FOR THE MEAN IS
           (-INFINITY,"M+T*SQR(S)/SQR(N)")"
310 PRINT "IS ANOTHER CONFIDENCE INTERVAL DESIRED? IF YES ENTER 1.
           IF NO ENTER 0."
320 INPUT Y
330 IF Y=1 GOTO 120
340 GOTO 450
350 W=SQR(-2*LOG(B))
360 C=2.515517
370 D=.802853
380 E=.010328
390 F=1.432788
400 G=.189269
410 H=.001308
```

```
420 Z=W-(C+D*W+E*W^2)/(1+F*W+G*W^2+H*W^3)
430 T=Z+(Z+Z^3)/(4*L)+(5*Z^5+16*Z^3+3*Z)/(96*L^2)+(3*Z^7+19*Z^5+17*Z^3-15*Z)/(38
4*L^3)
440 RETURN
450 END
Ok
```

5-3-2-A   Confidence Interval for the Difference of Two Normal Means
                    When Variances Are Known

```
10 PRINT "THIS PROGRAM COMPUTES A 100(1-a)% CONFIDENCE INTERVAL FOR THE
DIFFERENCE OF MEANS IN TWO NORMAL POPULATIONS HAVING KNOWN VARIANCES"
20 FOR I=1 TO 2
30 PRINT "ENTER THE SIZE OF SAMPLE" I
40 INPUT N(I)
50 PRINT "ENTER THE SAMPLE" I "DATA VALUES ONE AT A TIME"
60 FOR J=1 TO N(I)
70 INPUT ;X
80 V(I)=V(I)+X
90 NEXT
100 PRINT "ENTER THE POPULATION VARIANCE OF SAMPLE" I
110 INPUT C(I)
120 NEXT
130 W=SQR(C(1)/N(1)+C(2)/N(2))
140 U=V(1)/N(1)-V(2)/N(2)
150 PRINT "ENTER THE VALUE OF a"
160 INPUT A
170 PRINT "IS A TWO-SIDED INTERVAL DESIRED? ENTER 1 IF THE ANSWER IS YES AND 0
IF NO"
180 INPUT I
190 IF I=0 GOTO 240
200 B=A/2
210 GOSUB 360
220 PRINT "THE"100*(1-A)"% CONFIDENCE INTERVAL  IS ("U-Z*W","U+Z*W")"
230 GOTO 320
240 B=A
250 GOSUB 360
260 PRINT "IS THE ONE-SIDED CONFIDENCE INTERVAL TO BE UPPER OR LOWER? ENTER 1
FOR UPPER  AND 0 FOR LOWER"
270 INPUT K
280 IF K=0 GOTO 310
290 PRINT "THE" 100*(1-A)"% UPPER CONFIDENCE INTERVAL  IS ("U-Z*W",INFINITY)"
300 GOTO 320
310 PRINT "THE" 100*(1-A)"% LOWER CONFIDENCE INTERVAL  IS (-INFINITY,"U+Z*W")"
320 PRINT "IS ANOTHER CONFIDENCE INTERVAL DESIRED? IF YES ENTER 1. IF NO
ENTER 0."
330 INPUT Y
340 IF Y=1 GOTO 150
350 GOTO 450
360 T=SQR(-2*LOG(B))
370 C=2.515517
380 D=.802853
390 E=.010328
400 F=1.432788
410 G=.189269
420 H=.001308
430 Z=T-(C+D*T+E*T^2)/(1+F*T+G*T^2+H*T^3)
440 RETURN
450 END
```

5-3-2-B   Confidence Interval for the Difference of Two Normal Means
                    Having Unknown but Equal Variances

```
10 PRINT "THIS PROGRAM COMPUTES A 100(1-a)% CONFIDENCE INTERVAL FOR THE DIFFEREN
CE OF MEANS IN TWO NORMAL POPULATIONS HAVING UNKNOWN BUT EQUAL VARIANCES"
20 K=K+1
30 PRINT "ENTER THE SIZE OF SAMPLE NUMBER" K
40 INPUT N(K)
50 PRINT "ENTER THE SAMPLE"K"DATA VALUES ONE AT A TIME"
60 INPUT M
```

```
70 FOR J=1 TO N(K)-1
80 INPUT X
90 A=M
100 M=M+(X-M)/(J+1)
110 S=(1-1/J)*S+(J+1)*(M-A)^2
120 NEXT J
130 M(K)=M
140 S(K)=S
150 IF K=1 GOTO 20
160 U=SQR((1/N(1)+1/N(2))*((N(1)-1)*S(1)+(N(2)-1)*S(2))/(N(1)+N(2)-2))
170 L=N(1)+N(2)-2
180 PRINT "ENTER THE VALUE OF a"
190 INPUT A
200 PRINT "IS A TWO-SIDED INTERVAL DESIRED? ENTER 1 IF THE ANSWER IS YES AND 0 I
F NO"
210 INPUT I
220 IF I=0 GOTO 270
230 B=A/2
240 GOSUB 390
250 PRINT "THE"100*(1-A)"% CONFIDENCE INTERVAL  IS ("M(1)-M(2)-T*U","M(1)-M(2)+T
*U")"
260 GOTO 350
270 B=A
280 GOSUB 390
290 PRINT "IS THE ONE-SIDED CONFIDENCE INTERVAL TO BE UPPER OR LOWER? ENTER 1 FO
R UPPER AND 0 FOR LOWER"
300 INPUT K
310 IF K=0 GOTO 340
320 PRINT "THE" 100*(1-A)"% UPPER CONFIDENCE INTERVAL  IS ("M(1)-M(2)-T*U",INFIN
ITY)"
330 GOTO 350
340 PRINT "THE" 100*(1-A)"% LOWER CONFIDENCE INTERVAL  IS (-INFINITY,"M(1)-M(2)+
T*U")"
350 PRINT "IS ANOTHER CONFIDENCE INTERVAL DESIRED? IF YES ENTER 1. IF NO ENTER 0
."
360 INPUT Y
370 IF Y=1 GOTO 180
380 GOTO 490
390 W=SQR(-2*LOG(B))
400 C=2.515517
410 D=.802853
420 E=.010328
430 F=1.432788
440 G=.189269
450 H=.001308
460 Z=W-(C+D*W+E*W^2)/(1+F*W+G*W^2+H*W^3)
470 T=Z+(Z+Z^3)/(4*L)+(5*Z^5+16*Z^3+3*Z)/(96*L^2)+(3*Z^7+19*Z^5+17*Z^3-15*Z)/(38
4*L^3)
480 RETURN
490 END
```

6-3-2   The p-Value of the One-sample t-Test

```
10 PRINT "THIS PROGRAM COMPUTES THE p-value WHEN TESTING THAT A NORMAL POPULATIO
N WHOSE VARIANCE IS UNKNOWN HAS MEAN EQUAL TO MU-ZERO."
20 PRINT "ENTER THE VALUE OF MU-ZERO"
30 INPUT MU
40 PRINT "ENTER THE SAMPLE SIZE"
50 INPUT N
60 PRINT "ENTER THE DATA VALUES ONE AT A TIME"
70 INPUT; M
80 FOR J=1 TO N-1
90 INPUT; D
100 A=M
110 M=M+(D-M)/(J+1)
120 S=(1-1/J)*S+(J+1)*(M-A)^2
130 NEXT J
140 X=SQR(N)*(M-MU)/SQR(S)
150 PRINT "THE VALUE OF THE t-STATISTIC IS"X
160 N=N-1
170 A=N-2/3+1/(10*N)
180 B=LOG(1+X^2/N)/(N-5/6)
190 IF X>0 THEN X=A*SQR(B) ELSE X=-A*SQR(B)
200 U=ABS(X)
210 IF U>4 GOTO 430
220 Y=U^2
230 I=U
```

```
240 FOR J=1 TO 40
250 U=-U*Y*(2*J-1)/(2*J*(2*J+1))
260 I=I+U
270 NEXT
280 I=I/SQR(2*3.14159)
290 PRINT "IS THE ALTERNATIVE HYPOTHESIS TWO-SIDED? ENTER 1 IF YES AND 0 IF NO"
300 INPUT TWO
310 IF TWO =0 GOTO 340
320 PRINT "THE p-value IS" 1-2*I
330 GOTO 570
340 PRINT "IS THE ALTERNATIVE THAT THE MEAN EXCEEDS MU-ZERO OR THAT IT IS LESS?
 ENTER 1 IN THE FORMER CASE AND 0 IN THE LATTER"
350 INPUT AL
360 IF AL=0 GOTO 400
370 IF X<0 GOTO 410
380 PRINT "THE p-value IS" .5-I
390 GOTO 570
400 IF X<0 GOTO 380
410 PRINT "THE p-value IS" .5+I
420 GOTO 570
430 PRINT "IS THE ALTERNATIVE HYPOTHESIS TWO-SIDED? ENTER 1 IF YES AND 0 IF NO"
440 INPUT A
450 IF A=0 GOTO 480
460 PRINT "THE p-value IS LESS THAN" .0001
470 GOTO 570
480 PRINT "IS THE ALTERNATIVE THAT THE MEAN EXCEEDS MU-ZERO OR THAT IT IS LESS?
ENTER 1 IN THE FORMER CASE AND 0 IN THE LATTER"
490 INPUT B
500 IF B=0 GOTO 550
510 IF X<0 GOTO 530
520 GOTO 460
530 PRINT "THE p-value IS GREATER THAN " 1-10^-4
540 GOTO 570
550 IF X<0 GOTO 460
560 GOTO 530
570 END
```

6-4-1   The Test Statistic for Testing Equality of Two Normal Means When the Variances Are Known

```
10 PRINT "THIS PROGRAM COMPUTES THE VALUE OF THE TEST STATISTIC IN TESTING THAT
TWO NORMAL MEANS ARE EQUAL WHEN THE VARIANCES ARE KNOWN"
20 PRINT "ENTER THE SAMPLE SIZES"
30 INPUT N,M
40 PRINT "ENTER THE SAMPLE VARIANCES"
50 INPUT C1,C2
60 PRINT "ENTER THE FIRST SAMPLE ONE AT A TIME"
70 FOR I=1 TO N
80 INPUT ;X
90 SX=SX+X
100 NEXT
110 PRINT "ENTER THE SECOND SAMPLE ONE AT A TIME"
120 FOR I=1 TO M
130 INPUT ;Y
140 SY=SY+Y
150 NEXT
160 PRINT "THE VALUE OF THE TEST STATISTIC IS", (SX/N-SY/M)/SQR(C1/N+C2/M)
170 END
Ok
```

6-4-2   The p-Value of the Two-sample t-Test

```
10 PRINT "THIS PROGRAM COMPUTES THE p-value WHEN TESTING THAT TWO NORMAL POPULAT
IONS HAVING EQUAL BUT UNKNOWN VARIANCES HAVE A COMMON MEAN"
20 FOR J=1 TO 2
30 PRINT "ENTER THE SIZE OF SAMPLE" J
40 INPUT N(J)
50 PRINT "ENTER SAMPLE" J "ONE AT A TIME"
60 INPUT M(J)
70 FOR I=1 TO N(J)-1
80 INPUT X
90 A=M(J)
100 M(J)=M(J)+(X-M(J))/(I+1)
```

```
110 S(J)=(1-1/I)*S(J)+(I+1)*(M(J)-A)^2
120 NEXT I
130 NEXT J
140 R=(N(1)-1)*S(1)+(N(2)-1)*S(2)
150 R=R*(1/N(1)+1/N(2))/(N(1)+N(2)-2)
160 X=(M(1)-M(2))/SQR(R)
170 PRINT "THE VALUE OF THE t-STATISTIC IS"X
180 N=N(1)+N(2)-2
190 A=N-2/3+1/(10*N)
200 B=LOG(1+X^2/N)/(N-5/6)
210 IF X>0 THEN X=A*SQR(B) ELSE X=-A*SQR(B)
220 U=ABS(X)
230 IF U>4 GOTO 450
240 Y=U^2
250 I=U
260 FOR J=1 TO 40
270 U=-U*Y*(2*J-1)/(2*J*(2*J+1))
280 I=I+U
290 NEXT
300 I=I/SQR(2*3.14159)
310 PRINT "IS THE ALTERNATIVE HYPOTHESIS TWO-SIDED? ENTER 1 IF YES AND 0 IF NO"
320 INPUT TWO
330 IF TWO =0 GOTO 360
340 PRINT "THE p-value IS" 1-2*I
350 GOTO 590
360 PRINT "IS THE ALTERNATIVE THAT THE MEAN OF SAMPLE ONE EXCEEDS THE MEAN OF SA
MPLE TWO  OR THAT IT IS LESS?  ENTER 1 IN THE FORMER CASE AND 0 IN THE LATTER"
370 INPUT AL
380 IF AL=0 GOTO 420
390 IF X<0 GOTO 430
400 PRINT "THE p-value IS" .5-I
410 GOTO 590
420 IF X<0 GOTO 400
430 PRINT "THE p-value IS" .5+I
440 GOTO 590
450 PRINT "IS THE ALTERNATIVE HYPOTHESIS TWO-SIDED? ENTER 1 IF YES AND 0 IF NO"
460 INPUT A
470 IF A=0 GOTO 500
480 PRINT "THE p-value IS LESS THAN" .0001
490 GOTO 590
500 PRINT "IS THE ALTERNATIVE THAT THE MEAN OF SAMPLE ONE EXCEEDS THE MEAN OF SA
MPLE TWO  OR THAT IT IS LESS? ENTER 1 IN THE FORMER CASE AND 0 IN THE LATTER"
510 INPUT B
520 IF B=0 GOTO 570
530 IF X<0 GOTO 550
540 GOTO 480
550 PRINT "THE p-value IS GREATER THAN " 1-10^-4
560 GOTO 590
570 IF X<0 GOTO 480
580 GOTO 550
590 END
```

6-6-1   The *p*-Value in the Fisher-Irwin Test

```
10 PRINT "THIS PROGRAM COMPUTES THE p-value FOR THE TEST DATA IN THE FISHER-IRWI
N TEST"
20 PRINT "ENTER THE SIZE OF THE FIRST SAMPLE"
30 INPUT N1
40 PRINT "ENTER THE SIZE OF THE SECOND SAMPLE"
50 INPUT N2
60 PRINT "ENTER THE TOTAL NUMBER OF FAILURES"
70 INPUT K
80 PRINT "ENTER THE NUMBER OF FAILURES IN THE FIRST SAMPLE"
90 INPUT X1
100 P=1
110 FOR J=0 TO X1-1
120 P=P*(N1-J)/(N1+N2-J)
130 NEXT J
140 FOR J=0 TO K-X1-1
150 P=P*(N2-J)/(N1+N2-X1-J)
160 NEXT J
170 U=N1-X1
180 IF K<N1 THEN U=K-X1
190 D=X1
200 IF K>N2 THEN D=X1-K+N2
210 IF U<D GOTO 300
220 F=1
```

```
230 FOR J=1 TO D
240 F=F*(X1+1-J)*(N2-K+X1+1-J)/((N1-X1+J)*(K-X1+J))
250 T=T+F
255 NEXT J
260 GOTO 350
300 F=1
310 FOR J=1 TO U
320 F=F*(N1+1-X1-J)*(K+1-X1-J)/((X1+J)*(N2-K+X1+J))
330 T=T+F
340 NEXT J
350 C=1
353 FOR J=1 TO X1
356 C=C*(K+1-J)/(X1+1-J)
359 NEXT J
370 V1=(T+1)*P*C
380 V2=1-T*P*C
390 V=V1
400 IF V>V2 THEN V=V2
410 PRINT "THE p-value IS"2*V
420 END
```

7-2   Simple Linear Regression Solutions

```
LIST
10 PRINT "THIS PROGRAM COMPUTES THE LEAST SQUARES ESTIMATORS AND RELATED STATIS
ICS IN SIMPLE LINEAR REGRESSION MODELS"
20 PRINT "ENTER THE NUMBER OF DATA PAIRS n"
30 INPUT N
40 PRINT "ENTER THE n  SUCCESSIVE PAIRS x,Y ONE PAIR AT A TIME"
50 FOR I=1 TO N
60 INPUT X,Y
70 TXY=TXY+X*Y
80 TXX=TXX+X*X
90 TYY=TYY+Y*Y
100 MX=MX+X
110 MY=MY+Y
120 NEXT
130 SXY=TXY-MX*MY/N
140 SXX=TXX-MX*MX/N
150 SYY=TYY-MY*MY/N
160 B=SXY/SXX
170 A=(MY-B*MX)/N
180 SSR=(SXX*SYY-SXY*SXY)/SXX
190 PRINT "THE LEAST SQUARES ESTIMATORS ARE  AS FOLLOWS"
200 PRINT "A = " A
210 PRINT "B = " B
220 PRINT "THE ESTIMATED REGRESSION LINE IS  Y = "A" + "B" x"
230 PRINT "DO YOU WANT OTHER COMPUTED VALUES? ENTER 1 IF YES AND 0 IF NO."
240 INPUT Z
250 IF Z=0 GOTO 320
260 PRINT "S(x,Y) =" SXY
270 PRINT "S(x,x) =" SXX
280 PRINT "S(Y,Y) =" SYY
290 PRINT "SSR =" SSR
300 PRINT "THE AVERAGE x VALUE IS"  MX/N
310 PRINT "THE SUM OF THE SQUARES OF THE x VALUES IS" TXX
320 END
```

7-10   Multiple Linear Regression Solutions

```
LIST
10 PRINT "THIS PROGRAM COMPUTES THE LEAST SQUARES ESTIMATES OF THE COEFFICIENTS
AND THE   SUM OF SQUARES OF THE RESIDUALS IN MULTIPLE LINEAR REGRESSION"
20 PRINT "IT BEGINS BY COMPUTING THE INVERSE OF THE X-TRANSPOSE*X MATRIX"
30 PRINT "ENTER THE NUMBER OF ROWS OF THE X-MATRIX"
40 INPUT N
50 PRINT "ENTER THE NUMBER OF COLUMNS OF THE X-MATRIX"
60 INPUT P
70 DIM A(N,P)
80 FOR I=1 TO N
90 PRINT "ENTER ROW" I "ONE AT A TIME"
100 FOR J=1 TO P-1
110 INPUT ;A(I,J)
120 NEXT J
```

```
130 INPUT A(I,P)
140 NEXT I
150 DIM X(P,2*P)
160 FOR I=1 TO P
170 FOR J=1 TO I
180 FOR K=1 TO N
190 X(I,J)=X(I,J)+A(K,I)* A(K,J)
200 NEXT K
210 NEXT
220 NEXT
230 FOR I=1 TO P
240 FOR J=I TO P
250 X(I,J)=X(J,I)
260 NEXT
270 NEXT
280 FOR K=1 TO P
290 X(K,P+K)=1
300 NEXT K
310 FOR K=1 TO P
320 IF X(K,K)=0 GOTO 510
330 C=1/X(K,K)
340 FOR J=1 TO 2*P
350 X(K,J)=X(K,J)*C
360 NEXT J
370 FOR I=1 TO K-1
380 D=X(I,K)
390 FOR J=1 TO 2*P
400 X(I,J)=X(I,J)-D*X(K,J)
410 NEXT J
420 NEXT I
430 FOR I=K+1 TO P
440 D=X(I,K)
450 FOR J=1 TO 2*P
460 X(I,J)=X(I,J)-D*X(K,J)
470 NEXT J
480 NEXT I
490 NEXT K
500 GOTO 600
510 FOR I=K+1 TO P
520 IF X(I,K)=0 GOTO 570
530 FOR J=1 TO 2*P
540 X(K,J)=X(K,J)+X(I,J)
550 NEXT J
560 GOTO 340
570 NEXT I
580 PRINT "THE INVERSE DOES NOT EXIST"
590 GOTO 900
600 PRINT "THE INVERSE MATRIX IS AS FOLLOWS"
610 FOR I=1 TO P
620 FOR J=1 TO P-1
630 PRINT X(I,P+J);
640 NEXT J
650 PRINT X(I,2*P)
660 PRINT
670 NEXT I
680 PRINT "ENTER THE RESPONSE VALUES ONE AT A TIME"
690 DIM Y(N)
700 FOR I=1 TO N
710 INPUT ;Y(I)
720 NEXT
730 DIM XTY(P)
740 FOR I=1 TO P
750 FOR K=1 TO N
760 XTY(I)=XTY(I)+A(K,I)*Y(K)
770 NEXT
780 NEXT
790 DIM B(P)
800 FOR I=0 TO P-1
810 FOR J=1 TO P
820 B(I)=B(I)+X(I+1,P+J)*XTY(J)
830 NEXT
840 NEXT
850 PRINT "THE ESTIMATES OF THE REGRESSION COEFFICIENTS ARE AS FOLLOWS"
860 PRINT
870 FOR I=0 TO P-1
880 PRINT, "B("I")="B(I)
890 NEXT
900 FOR I=1 TO N
```

```
910 YTY=YTY+(Y(I))^2
920 NEXT
930 FOR I=0 TO P-1
940 BTXTY=BTXTY+B(I)*XTY(I+1)
950 NEXT
960 PRINT "THE SUM OF SQUARES OF THE RESIDUALS IS   SS(R) =" YTY-BTXTY
970 END
Ok
```

8-2   The *p*-Values in a One-way ANOVA

```
10 PRINT "THIS PROGRAM COMPUTES THE VALUE OF THE F-STATISTIC AND ITS p-value IN
A ONE WAY ANOVA"
20 PRINT "ENTER THE NUMBER OF SAMPLES"
30 INPUT M
40 PRINT "ENTER THE SIZE OF THE SAMPLES"
50 INPUT N
60 DIM M(M)
70 FOR I=1 TO M
80 PRINT "ENTER SAMPLE" I "ONE AT A TIME"
90 INPUT A
100 FOR J=1 TO N-1
110 INPUT X
120 B=A
130 A=A+(X-A)/(J+1)
140 S=(1-1/J)*S+(J+1)*(A-B)^2
150 NEXT J
160 SS=SS+S
170 M(I)=A
180 NEXT I
190 DEN=SS/M
200 PRINT "SSw/(M*(N-1))=";DEN
210 T=M(1)
220 FOR K=1 TO M-1
230 A=T
240 T=T+(M(K+1)-T)/(K+1)
250 V=(1-1/K)*V+(K+1)*(T-A)^2
260 NEXT K
270 NUM=V*N
280 PRINT "SSb/(M-1)=";NUM
290 PRINT "THE VALUE OF THE F-STATISTIC IS";NUM/DEN
300 Y=N
310 N=M-1
320 M=M*(Y-1)
330 X=NUM/DEN
340 S=(M-1)/2
350 T=(N-1)/2
360 K=(N+M)/2-1
370 P=M/(N*X+M)
380 Q=1-P
390 D=S+1/6-(K+1/3)*P+.02*(Q/(S+.5)-P/(T+.5)+(Q-.5)/(K+1))
400 A=S/(K*P)
410 B=T/(K*Q)
420 IF A=0 THEN C=1 ELSE C=(1-A*A+2*A*LOG(A))/(1-A)^2
430 IF B=0 THEN E=1 ELSE E=(1-B*B+2*B*LOG(B))/(1-B)^2
440 X=D*SQR((1+Q*C+P*E)/(((K+1/6)*P*Q))
450 U=ABS(X)
460 IF U>4 GOTO 570
470 Y =U^2
480 I=U
490 FOR J=1 TO 40
500 U=-U*Y*(2*J-1)/(2*J*(2*J+1))
510 I=I+U
520 NEXT
530 I=I/SQR(2*3.14159)
540 IF X<0 THEN Z=.5+I ELSE Z=.5-I
550 PRINT "THE p-value IS";Z
560 GOTO 610
570 IF X<0 GOTO 600
580 PRINT "THE p-value IS LESS THAN .0001"
590 GOTO 610
600 PRINT "THE p-value IS GREATER THAN .9999"
610 END
```

8-4   The *p*-Values in a Two-way ANOVA

```
LIST
10 PRINT "THIS PROGRAM COMPUTES THE VALUES OF THE F-STATISTICS AND THEIR ASSOCIA
TED p-values IN A TWO WAY ANOVA"
20 PRINT "ENTER THE NUMBER OF ROWS"
30 INPUT M
40 PRINT "ENTER THE NUMBER OF COLUMNS"
50 INPUT N
60 DIM X(M,N)
70 DIM R(M)
80 DIM C(N)
90 FOR I=1 TO M
100 PRINT "ENTER ROW" I "ONE AT A TIME"
110 FOR J=1 TO N
120 INPUT X(I,J)
130 R(I)=R(I)+X(I,J)
140 C(J)=C(J)+X(I,J)
150 NEXT J
160 G=G+R(I)
170 NEXT I
180 G=G/(N*M)
190 FOR I=1 TO M
200 R(I)=R(I)/N
210 SSR=SSR+(R(I)-G)^2
220 NEXT I
230 FOR J=1 TO N
240 C(J)=C(J)/M
250 SSC=SSC+(C(J)-G)^2
260 NEXT J
270 FOR I=1 TO M
280 FOR J=1 TO N
290 SSE=SSE+(X(I,J)-R(I)-C(J)+G)^2
300 NEXT J
310 NEXT I
320 F(1)=N*(N-1)*SSR/SSE
330 F(2)=M*(M-1)*SSC/SSE
360 N(1)=M-1
370 N(2)=N-1
380 FOR H=1 TO 2
382 IF H=1 THEN A$="ROW" ELSE A$="COLUMN"
390 PRINT "THE VALUE OF THE F-STATISTIC FOR TESTING THAT THERE IS NO " A$ " EFFE
CT IS";F(H)
400 N=N(H)
410 M=N(1)* N(2)
420 X=F(H)
430 S=(M-1)/2
440 T=(N-1)/2
450 K=(N+M)/2-1
460 P=M/(N*X+M)
470 Q=1-P
480 D=S+1/6-(K+1/3)*P+.02*(Q/(S+.5)-P/(T+.5)+(Q-.5)/(K+1))
490 A=S/(K*P)
500 B=T/(K*Q)
510 IF A=0 THEN C=1 ELSE C=(1-A*A+2*A*LOG(A))/(1-A)^2
520 IF B=0 THEN E=1 ELSE E=(1-B*B+2*B*LOG(B))/(1-B)^2
530 X=D*SQR((1+Q*C+P*E)/((K+1/6)*P*Q))
540 U=ABS(X)
550 IF U>4 GOTO 660
560 Y=U^2
570 I=U
580 FOR J=1 TO 40
590 U=-U*Y*(2*J-1)/(2*J*(2*J+1))
600 I=I+U
610 NEXT
620 I=I/SQR(2*3.14159)
630 IF X<0 THEN Z=.5+I ELSE Z=.5-I
640 PRINT "THE p-value FOR TESTING THAT THERE IS NO " A$ " EFFECT IS";Z
650 GOTO 700
660 IF X<0 GOTO 690
670 PRINT "THE p-value FOR TESTING THAT THERE IS NO " A$ " EFFECT IS LESS THAN .
0001"
680 GOTO 700
690 PRINT "THE p-value FOR TESTING THAT THERE IS NO " A$ " EFFECT IS GREATER THA
N .9999"
700 NEXT H
710 END
Ok
```

8-5   The *p*-Values in a Two-way ANOVA with Possible Interaction

```
10 PRINT "THIS PROGRAM COMPUTES THE VALUES OF THE F-STATISTICS AND THEIR ASSOCIA
TED p-values IN A TWO WAY ANOVA WITH L OBSERVATIONS IN EACH ROW-COLUMN CELL"
20 PRINT "ENTER THE NUMBER OF ROWS"
30 INPUT M
40 PRINT "ENTER THE NUMBER OF COLUMNS"
50 INPUT N
60 PRINT "ENTER THE NUMBER OF OBSERVATIONS IN EACH ROW-COLUMN CELL"
70 INPUT L
80 DIM X(M,N,L)
90 DIM Y(M,N)
100 DIM R(M)
110 DIM C(N)
120 FOR I=1 TO M
130 FOR J=1 TO N
140 PRINT "ENTER THE" L "VALUES IN ROW" I "COLUMN" J "ONE AT A TIME"
150 FOR K=1 TO L
160 INPUT; X(I,J,K)
170 Y(I,J)=Y(I,J)+X(I,J,K)
180 NEXT K
190 R(I)=R(I)+Y(I,J)
200 C(J)=C(J)+Y(I,J)
210 NEXT J
220 G=G+R(I)
230 NEXT I
240 G=G/(M*N*L)
250 A=1/L
260 B=1/(N*L)
270 C= 1/(M*L)
280 FOR I=1 TO M
290 R(I)=B*R(I)
300 SSR=SSR+(R(I)-G)^2
310 NEXT
320 SSR=SSR*L*N
330 FOR J=1 TO N
340 C(J)=C*C(J)
350 SSC=SSC+(C(J)-G)^2
360 NEXT
370 SSC=SSC*L*M
380 FOR I=1 TO M
390 FOR J=1 TO N
400 Y(I,J)=A*Y(I,J)
410 SSINT=SSINT+(Y(I,J)-R(I)-C(J)+G)^2
420 NEXT
430 NEXT
440 SSINT=SSINT*L
450 FOR I=1 TO M
460 FOR J=1 TO N
470 FOR K=1 TO L
480 SSE=SSE+(X(I,J,K)-Y(I,J))^2
490 NEXT
500 NEXT
510 NEXT
520 F(1)=N*M*(L-1)*SSR/((M-1)*SSE)
530 F(2)=N*M*(L-1)*SSC/((N-1)*SSE)
540 F(3)=N*M*(L-1)*SSINT/((N-1)*(M-1)*SSE)
550 IF H=1 THEN A$="ROW" ELSE A$="COLUMN"
560 N(1)=M-1
570 N(2)=N-1
580 N(3)=(N-1)*(M-1)
590 M=(N(1)+1)*(N(2)+1)*(L-1)
600 FOR H=1 TO 3
610 IF H=1 THEN A$="ROW"
620 IF H=2 THEN A$="COLUMN"
630 IF H=3 THEN A$="INTERACTION"
640 PRINT "THE VALUE OF THE F-STATISTIC FOR TESTING THAT THERE IS NO "A$ " EFFEC
T IS";F(H)
650 N=N(H)
660 X=F(H)
670 S=(M-1)/2
680 T=(N-1)/2
690 K=(N+M)/2-1
700 P=M/(N*X+M)
710 Q=1-P
720 D=S+1/6-(K+1/3)*P+.02*(Q/(S+.5)-P/(T+.5)+(Q-.5)/(K+1))
```

```
730 A=S/(K*P)
740 B=T/(K*Q)
750 IF A=0 THEN C=1 ELSE C=(1-A*A+2*A*LOG(A))/(1-A)^2
760 IF B=0 THEN E=1 ELSE E=(1-B*B+2*B*LOG(B))/(1-B)^2
770 X=D*SQR((1+Q*C+P*E)/((K+1/6)*P*Q))
780 U=ABS(X)
790 IF U>4 GOTO 900
800 Y=U^2
810 I=U
820 FOR J=1 TO 40
830 U=-U*Y*(2*J-1)/(2*J*(2*J+1))
840 I=I+U
850 NEXT
860 I=I/SQR(2*3.14159)
870 IF X<0 THEN Z=.5+I ELSE Z=.5-I
880 PRINT "THE p-value FOR TESTING THAT THERE IS NO " A$ " EFFECT IS";Z
890 GOTO 940
900 IF X<0 GOTO 930
910 PRINT "THE p-value FOR TESTING THAT THERE IS NO " A$ " EFFECT IS LESS THAN .
0001"
920 GOTO 940
930 PRINT "THE p-value FOR TESTING THAT THERE IS NO " A$ " EFFECT IS GREATER THA
N .9999"
940 NEXT H
950 END
Ok
```

9-2-1   The Goodness of Fit Test Statistic

```
10 PRINT "THIS PROGRAM COMPUTES THE GOODNESS OF FIT TEST STATISTIC"
20 PRINT "ENTER THE NUMBER OF GROUPINGS"
30 INPUT K
40 DIM P(K)
50 PRINT "ENTER THE NULL HYPOTHESIS PROBABILITIES ONE AT A TIME"
60 FOR I=1 TO K
70 INPUT P(I)
80 NEXT I
90 PRINT "ENTER THE SAMPLE SIZE"
100 INPUT N
110 DIM X(K)
120 PRINT "ENTER THE NUMBERS THAT FALL IN EACH GROUPING ONE AT A TIME"
130 FOR I=1 TO K
140 INPUT X(I)
150 NEXT I
160 FOR I=1 TO K
170 S=S+(X(I)-N*P(I))^2/(N*P(I))
180 NEXT I
190 PRINT "TEST STATISTIC HAS VALUE",S
200 END
Ok
```

9-2-2   A Simulation Approximation to the *p*-Value in Goodness of Fit

```
LIST
10 PRINT "THIS PROGRAM USES SIMULATION TO APPROXIMATE THE p-value IN THE GOODNES
S OF FIT TEST"
20 RANDOMIZE
30 PRINT "ENTER THE NUMBER OF POSSIBLE VALUES"
40 INPUT N
50 DIM Q(N)
60 PRINT "ENTER THE PROBABILITIES ONE AT A TIME"
70 FOR I = 1 TO N
80 INPUT P
90 Q(I)=Q(I-1)+P
100 NEXT I
110 PRINT "ENTER THE SAMPLE SIZE"
120 INPUT D
130 DIM X(N)
140 PRINT "ENTER THE DESIRED NUMBER OF SIMULATION RUNS"
150 INPUT R
160 PRINT "ENTER THE VALUE OF THE TEST STATISTIC"
170 INPUT W
```

```
180 FOR K=1 TO R
190 FOR L=1 TO N
200 X(L)=0
210 NEXT L
220 FOR J=1 TO D
230 U=RND
240 FOR I=1 TO N
250 IF Q(I-1)<U THEN 270
260 NEXT I
270 IF U>Q(I) THEN  260
280 X(I)=X(I)+1
290 NEXT J
300 S=0
310 FOR I=1 TO N
320 S=S+(X(I)-D*(Q(I)-Q(I-1)))^2/(D*(Q(I)-Q(I-1)))
330 NEXT I
340 IF S>=W THEN C=C+1
350 NEXT K
360 PRINT "THE ESTIMATE OF THE p-value IS" C/R
370 END
Ok
```

9-3   The Test Statistic for Independence in a Contingency Table

```
10 PRINT "THIS PROGRAM COMPUTES THE TEST STATISTIC FOR TESTING FOR INDEPENDENCE
IN A 2-WAY CONTINGENCY TABLE"
20 PRINT "ENTER THE NUMBER OF ROWS"
30 INPUT N
40 PRINT "ENTER THE NUMBER OF COLUMNS"
50 INPUT M
60 DIM X(N,M)
70 DIM R(N)
80 DIM C(M)
90 FOR I=1 TO N
100 PRINT "ENTER ROW" I "ONE AT A TIME"
110 FOR J=1 TO M
120 INPUT X(I,J)
130 R(I)=R(I)+X(I,J)
140 C(J)=C(J)+X(I,J)
150 NEXT
160 S=S+R(I)
170 NEXT
180 FOR I=1 TO N
190 FOR J=1 TO M
200 U=U+(X(I,J)-R(I)*C(J)/S)^2/(R(I)*C(J))
210 NEXT
220 NEXT
230 PRINT "THE TEST STATISTIC HAS VALUE T=";U*S
240 END
Ok
```

9-6   The p-Value in the One-sample Signed Rank Test

```
10 PRINT "THIS PROGRAM COMPUTES THE p-value FOR THE ONE SAMPLE SIGNED RANK TEST"
20 PRINT "ENTER THE SIZE OF THE SAMPLE"
30 INPUT N
40 PRINT "ENTER THE OBSERVED VALUE OF THE SUM OF THE SIGNED RANKS"
50 INPUT V
60 IF N*(N+1)/2-V < V THEN V=N*(N+1)/2-V
70 DIM P(V)
80 P(0)=.5
90 FOR I=1 TO V
100 P(I)=1
110 NEXT
120 FOR K=2 TO N
130 FOR I=V TO 0 STEP -1
140 IF K>I THEN P(I)=.5*P(I) ELSE P(I)=.5*(P(I-K)+P(I))
150 NEXT I
160 NEXT K
170 PRINT "THE p-value IS"; 2*P(V)
180 END
Ok
```

9-7   The *p*-Value in the Two-sample Rank Sum Test

```
10 PRINT "THIS PROGRAM COMPUTES THE p-value FOR THE TWO-SAMPLE RANK SUM TEST"
20 PRINT "THIS PROGRAM WILL RUN FASTEST IF YOU DESIGNATE AS THE FIRST SAMPLE THE
 SAMPLE HAVING THE SMALLER SUM OF RANKS"
30 PRINT "ENTER THE SIZE OF THE FIRST SAMPLE"
40 INPUT N
50 PRINT "ENTER THE SIZE OF THE SECOND SAMPLE"
60 INPUT M
70 PRINT "ENTER THE SUM OF THE RANKS OF THE FIRST SAMPLE"
80 INPUT T
90 DIM P(N,M,T+1)
100 FOR I=1 TO N
110 FOR  K=I*(I+1)/2 TO T
120 P(I,0,K)=1
130 NEXT
140 NEXT
150 FOR K=1 TO T+1
160 FOR J=1 TO M
170 P(0,J,K-1)=1
180 NEXT
190 NEXT
200 FOR I=1 TO N
210 FOR J=1 TO M
220 FOR K=1 TO T
230 IF K<(I+J) THEN P(I,J,K)= (J/(I+J))*P(I,J-1,K) ELSE P(I,J,K)=(I/(I+J))*P(I-1
,J,K-I-J) +(J/(I+J))*P(I,J-1,K)
240 NEXT
250 NEXT
260 NEXT
270 IF P(N,M,T) < 1-P(N,M,T-1) THEN V=P(N,M,T) ELSE V=1-P(N,M,T-1)
280 PRINT "THE p-value IS" 2*V
290 END
Ok
```

9-7-1   Simulation Approximation to the *p*-Value in Rank Sum Test

```
10 PRINT "THIS PROGRAM APPROXIMATES THE p-value IN THE TWO-SAMPLE RANK SUM TEST
BY A SIMULATION STUDY"
20 RANDOMIZE
30 PRINT "ENTER THE SIZE OF THE FIRST SAMPLE"
40 INPUT N(1)
50 PRINT "ENTER THE SIZE OF THE SECOND SAMPLE"
60 INPUT N(2)
70 PRINT "ENTER THE SUM OF THE RANKS OF THE FIRST SAMPLE"
80 INPUT T
90 PRINT "ENTER THE DESIRED NUMBER OF SIMULATION RUNS"
100 INPUT M
110 N=N(1)+N(2)
120 DIM X(N)
130 NUM=NUM+1
140 S=0
150 FOR I=1 TO N
160 X(I)=I
170 NEXT I
180 FOR I=1 TO N(1)
190 R=INT((N+1-I)*RND)+1
200 S=S+X(R)
210 X(R)=X(N+1-I)
220 NEXT I
230 IF S<=T THEN C(1)=C(1)+1
240 IF S>=T THEN C(2)=C(2)+1
250 IF NUM < M GOTO 130
260 IF C(1)>C(2) THEN C(1)=C(2)
270 PRINT "THE APPROXIMATE p-value IS" 2*C(1)/M
280 END
```

9-8   The *p*-Value for the Run Test for Randomness

```
LIST
10 PRINT "THIS PROGRAM COMPUTES THE p-value FOR THE RUN TEST OF THE HYPOTHESIS
THAT A DATA SET OF N ONES AND M ZEROES IS RANDOM"
20 PRINT "ENTER THE NUMBER OF ONES"
30 INPUT N
```

```
40 PRINT "ENTER THE NUMBER OF ZEROES"
50 INPUT M
60 PRINT "ENTER THE NUMBER OF RUNS"
70 INPUT R
80 IF M<N GOTO 120
90 L=N
100 N=M
110 M=L
120 C=1
130 FOR J=1 TO M
140 C=C*J/(N+J)
150 NEXT
160 S=N+M
170 A=1
180 IF INT(R/2) = R/2 GOTO 260
190 FOR I=1 TO (R-1)/2
200 B=A*(S/I -2)
210 X=X+2*A+B
220 A=A*(N-I)*(M-I)/I^2
230 NEXT
240 P=B
250 GOTO 330
260 X=2
270 FOR I=1 TO (R-2)/2
280 B=A*(S/I -2)
290 A=A*(N-I)*(M-I)/I^2
300 X=X+B+2*A
310 NEXT
320 P=2*A
330 X=X*C
340 P=P*C
350 IF (P+1-X)<X THEN X=P+1-X
360 PRINT "THE p-value IS" 2*X
370 END
Ok
```

12-1　　　　A Random Permutation

```
10 PRINT "THIS PROGRAM GENERATES A RANDOM PERMUTATION OF THE NUMBERS 1,2,...,N"
20 PRINT "ENTER THE VALUE OF N"
30 INPUT N
40 DIM X(N)
50 FOR I=1 TO N
60 X(I)=I
70 NEXT I
80 RANDOMIZE
90 FOR I=1 TO N-1
100 R=INT((N+1-I)*RND)+1
110 W=X(R)
120 X(R)=X(N+1-I)
130 X(N+1-I)=W
140 NEXT I
150 PRINT "THE RANDOM PERMUTATION IS AS FOLLOWS"
160 FOR J=1 TO N
170 PRINT X(J)
180 NEXT J
190 END
```

12-2　　　　Simulated Values for Normal Random Variables

```
LIST
10 PRINT "THIS PROGRAM SIMULATES NORMAL RANDOM VARIABLES"
20 RANDOMIZE
30 PRINT "ENTER THE MEAN"
40 INPUT MU
50 PRINT "ENTER THE VARIANCE"
60 INPUT C
70 D=SQR(C)
80 PRINT "HOW MANY NORMALS DO YOU WANT?"
90 INPUT N
100 U=-LOG(RND)
110 FOR I=1 TO N
120 W=-LOG(RND)
130 U=U-(W-1)^2/2
140 IF U>0 THEN GOTO 170
```

```
150 U=-LOG(RND)
160 GOTO 120
170 IF RND>.5 THEN PRINT MU+W*D ELSE PRINT MU-W*D
180 NEXT I
190 END
Ok
```

12-3   Simulated Values for Poisson Random Variables

```
LIST
10 PRINT "THIS PROGRAM SIMULATES N INDEPENDENT POISSON RANDOM VARIABLES EACH HAV
ING MEAN LAMBDA"
20 PRINT "ENTER THE MEAN VALUE LAMBDA"
30 INPUT L
40 PRINT "HOW MANY VARIATES DO YOU WANT?"
50 INPUT N
60 RANDOMIZE
70 A=EXP(-L)
80  PRINT "THE" N "VALUES ARE AS FOLLOWS"
90 FOR I=1 TO N
100 J=0
110 U=RND
120 IF U<A GOTO 160
130 U=U*RND
140 J=J+1
150 GOTO 120
160 PRINT J
170 NEXT
180 END
Ok
```

12-5   A Random Permutation

```
LIST
10 PRINT "THIS PROGRAM GENERATES A RANDOM PERMUTATION OF THE NUMBERS 1,2,...,N"
20 PRINT "ENTER THE VALUE OF N"
30 INPUT N
40 DIM P(N)
50 RANDOMIZE
60 P(1)=1
70 FOR I=2 TO N
80 U=RND
90 P(I)=P(INT(I*U+1))
100 P(INT(I*U+1))=I
110 NEXT I
120 PRINT "THE RANDOM PERMUTATION IS AS FOLLOWS"
130 FOR J=1 TO N
140 PRINT P(J)
150 NEXT J
160 END
Ok
```

Inv   The Inverse of a Square Matrix

```
LIST
10 PRINT "THIS PROGRAM COMPUTES THE INVERSE OF A SQUARE MATRIX"
20 PRINT "ENTER THE NUMBER OF ROWS OF THE MATRIX"
30 INPUT N
40 DIM X(N,2*N)
50 FOR I=1 TO N
60 PRINT "ENTER ROW" I "ONE ELEMENT AT A TIME"
70 FOR J=1 TO N-1
80 INPUT; X(I,J)
90 NEXT J
100 INPUT X(I,N)
110 NEXT I
120 FOR K=1 TO N
130 X(K,N+K)=1
140 NEXT K
150 FOR K=1 TO N
160 IF X(K,K)=0 GOTO 350
```

```
170 C=1/X(K,K)
180 FOR J=1 TO 2*N
190 X(K,J)=X(K,J)*C
200 NEXT J
210 FOR I=1 TO K-1
220 D=X(I,K)
230 FOR J=1 TO 2*N
240 X(I,J)=X(I,J)-D*X(K,J)
250 NEXT J
260 NEXT I
270 FOR I=K+1 TO N
280 D=X(I,K)
290 FOR J=1 TO 2*N
300 X(I,J)=X(I,J)-D*X(K,J)
310 NEXT J
320 NEXT I
330 NEXT K
340 GOTO 440
350 FOR I=K+1 TO N
360 IF X(I,K)=0 GOTO 410
370 FOR J=1 TO 2*N
380 X(K,J)=X(K,J)+X(I,J)
390 NEXT J
400 GOTO 180
410 NEXT I
420 PRINT "THE INVERSE DOES NOT EXIST"
430 GOTO 520
440 PRINT "THE INVERSE MATRIX IS AS FOLLOWS"
450 FOR I=1 TO N
460 FOR J=1 TO N-1
470 PRINT X(I,N+J);
480 NEXT J
490 PRINT X(I,2*N)
500 PRINT
510 NEXT I
520 END
Ok
```

Appendix of Tables

TABLE A3.5.1

Unit Normal Distribution Function:

$$\Phi(x) = \frac{1}{\sqrt{2\pi}} \int_{-\infty}^{x} e^{-y^2/2} \, dy$$

x	0.00	0.01	0.02	0.03	0.04	0.05	0.06	0.07	0.08	0.09
0.0	0.5000	0.5040	0.5080	0.5120	0.5160	0.5199	0.5239	0.5279	0.5319	0.5359
0.1	0.5398	0.5438	0.5478	0.5517	0.5557	0.5596	0.5636	0.5675	0.5714	0.5753
0.2	0.5793	0.5832	0.5871	0.5910	0.5948	0.5987	0.6026	0.6064	0.6103	0.6141
0.3	0.6179	0.6217	0.6255	0.6293	0.6331	0.6368	0.6406	0.6443	0.6480	0.6517
0.4	0.6554	0.6591	0.6628	0.6664	0.6700	0.6736	0.6772	0.6808	0.6844	0.6879
0.5	0.6915	0.6950	0.6985	0.7019	0.7054	0.7088	0.7123	0.7157	0.7190	0.7224
0.6	0.7257	0.7291	0.7324	0.7357	0.7389	0.7422	0.7454	0.7486	0.7517	0.7549
0.7	0.7580	0.7611	0.7642	0.7673	0.7704	0.7734	0.7764	0.7794	0.7823	0.7852
0.8	0.7881	0.7910	0.7939	0.7967	0.7995	0.8023	0.8051	0.8078	0.8106	0.8133
0.9	0.8159	0.8186	0.8212	0.8238	0.8264	0.8289	0.8315	0.8340	0.8365	0.8389
1.0	0.8413	0.8438	0.8461	0.8485	0.8508	0.8531	0.8554	0.8577	0.8599	0.8621
1.1	0.8643	0.8665	0.8686	0.8708	0.8729	0.8749	0.8770	0.8790	0.8810	0.8830
1.2	0.8849	0.8869	0.8888	0.8907	0.8925	0.8944	0.8962	0.8980	0.8997	0.9015
1.3	0.9032	0.9049	0.9066	0.9082	0.9099	0.9115	0.9131	0.9147	0.9162	0.9177
1.4	0.9192	0.9207	0.9222	0.9236	0.9251	0.9265	0.9279	0.9292	0.9306	0.9319
1.5	0.9332	0.9345	0.9357	0.9370	0.9382	0.9394	0.9406	0.9418	0.9429	0.9441
1.6	0.9452	0.9463	0.9474	0.9484	0.9495	0.9505	0.9515	0.9525	0.9535	0.9545
1.7	0.9554	0.9564	0.9573	0.9582	0.9591	0.9599	0.9608	0.9616	0.9625	0.9633
1.8	0.9641	0.9649	0.9656	0.9664	0.9671	0.9678	0.9686	0.9693	0.9699	0.9706
1.9	0.9713	0.9719	0.9726	0.9732	0.9738	0.9744	0.9750	0.9756	0.9761	0.9767
2.0	0.9772	0.9778	0.9783	0.9788	0.9793	0.9798	0.9803	0.9808	0.9812	0.9817
2.1	0.9821	0.9826	0.9830	0.9834	0.9838	0.9842	0.9846	0.9850	0.9854	0.9857
2.2	0.9861	0.9864	0.9868	0.9871	0.9875	0.9878	0.9881	0.9884	0.9887	0.9890
2.3	0.9893	0.9896	0.9898	0.9901	0.9904	0.9906	0.9909	0.9911	0.9913	0.9916
2.4	0.9918	0.9920	0.9922	0.9925	0.9927	0.9929	0.9931	0.9932	0.9934	0.9936
2.5	0.9938	0.9940	0.9941	0.9943	0.9945	0.9946	0.9948	0.9949	0.9951	0.9952
2.6	0.9953	0.9955	9.9956	0.9957	0.9959	0.9960	0.9961	0.9962	0.9963	0.9964
2.7	0.9965	0.9966	0.9967	0.9968	0.9969	0.9970	0.9971	0.9972	0.9973	0.9974
2.8	0.9974	0.9975	0.9976	0.9977	0.9977	0.9978	0.9979	0.9979	0.9980	0.9981
2.9	0.9981	0.9982	0.9982	0.9983	0.9984	0.9984	0.9985	0.9985	0.9986	0.9986
3.0	0.9987	0.9987	0.9987	0.9988	0.9988	0.9989	0.9989	0.9989	0.9990	0.9990
3.1	0.9990	0.9991	0.9991	0.9991	0.9992	0.9992	0.9992	0.9992	0.9993	0.9993
3.2	0.9993	0.9993	0.9994	0.9994	0.9994	0.9994	0.9994	0.9995	0.9995	0.9995
3.3	0.9995	0.9995	0.9995	0.9996	0.9996	0.9996	0.9996	0.9996	0.9996	0.9997
3.4	0.9997	0.9997	0.9997	0.9997	0.9997	0.9997	0.9997	0.9997	0.9997	0.9998

TABLE A3.8.1

Values of $x^2_{\alpha,n}$

n	$\alpha = 0.995$	$\alpha = 0.99$	$\alpha = 0.975$	$\alpha = 0.95$	$\alpha = 0.05$	$\alpha = 0.025$	$\alpha = 0.01$	$\alpha = 0.005$
1	0.0000393	0.000157	0.000982	0.00393	3.841	5.024	6.635	7.879
2	0.0100	0.0201	0.0506	0.103	5.991	7.378	9.210	10.597
3	0.0717	0.115	0.216	0.352	7.815	9.348	11.345	12.838
4	0.207	0.297	0.484	0.711	9.488	11.143	13.277	14.860
5	0.412	0.554	0.831	1.145	11.070	12.832	13.086	16.750
6	0.676	0.872	1.237	1.635	12.592	14.449	16.812	18.548
7	0.989	1.239	1.690	2.167	14.067	16.013	18.475	20.278
8	1.344	1.646	2.180	2.733	15.507	17.535	20.090	21.955
9	1.735	2.088	2.700	3.325	16.919	19.023	21.666	23.589
10	2.156	2.558	3.247	3.940	18.307	20.483	23.209	25.188
11	2.603	3.053	3.816	4.575	19.675	21.920	24.725	26.757
12	3.074	3.571	4.404	5.226	21.026	23.337	26.217	28.300
13	3.565	4.107	5.009	5.892	22.362	24.736	27.688	29.819
14	4.075	4.660	5.629	6.571	23.685	26.119	29.141	31.319
15	4.601	5.229	6.262	7.261	24.996	27.488	30.578	32.801
16	5.142	5.812	6.908	7.962	26.296	28.845	32.000	34.267
17	5.697	6.408	7.564	8.672	27.587	30.191	33.409	35.718
18	6.265	7.015	8.231	9.390	28.869	31.526	34.805	37.156
19	6.844	7.633	8.907	10.117	30.144	32.852	36.191	38.582
20	7.434	8.260	9.591	10.851	31.410	34.170	37.566	39.997
21	8.034	8.897	10.283	11.591	32.671	35.479	38.932	41.401
22	8.643	9.542	10.982	12.338	33.924	36.781	40.289	42.796
23	9.260	10.196	11.689	13.091	35.172	38.076	41.638	44.181
24	9.886	10.856	12.401	13.484	36.415	39.364	42.980	45.558
25	10.520	11.524	13.120	14.611	37.652	40.646	44.314	46.928
26	11.160	12.198	13.844	15.379	38.885	41.923	45.642	48.290
27	11.808	12.879	14.573	16.151	40.113	43.194	46.963	49.645
28	12.461	13.565	15.308	16.928	41.337	44.461	48.278	50.993
29	13.121	14.256	16.047	17.708	42.557	45.772	49.588	52.336
30	13.787	14.953	16.791	18.493	43.773	46.979	50.892	53.672

Other Chi-Square Probabilities:

$x^2_{0.9,9} = 4.2$ $P\{x^2_{16} < 14.3\} = 0.425$ $P\{x^2_{11} < 17.1875\} = 0.8976$.

TABLE A3.8.2

Values of $t_{\alpha, n}$

n	$\alpha = 0.10$	$\alpha = 0.05$	$\alpha = 0.025$	$\alpha = 0.01$	$\alpha = 0.005$
1	3.078	6.314	12.706	31.821	63.657
2	1.886	2.920	4.303	6.965	9.925
3	1.638	2.353	3.182	4.541	5.841
4	1.533	2.132	2.776	3.474	4.604
5	1.476	2.015	2.571	3.365	4.032
6	1.440	1.943	2.447	3.143	3.707
7	1.415	1.895	2.365	2.998	3.499
8	1.397	1.860	2.306	2.896	3.355
9	1.383	1.833	2.262	2.821	3.250
10	1.372	1.812	2.228	2.764	3.169
11	1.363	1.796	2.201	2.718	3.106
12	1.356	1.782	2.179	2.681	3.055
13	1.350	1.771	2.160	2.650	3.012
14	1.345	1.761	2.145	2.624	2.977
15	1.341	1.753	2.131	2.602	2.947
16	1.337	1.746	2.120	2.583	2.921
17	1.333	1.740	2.110	2.567	2.898
18	1.330	1.734	2.101	2.552	2.878
19	1.328	1.729	2.093	2.539	2.861
20	1.325	1.725	2.086	2.528	2.845
21	1.323	1.721	2.080	2.518	2.831
22	1.321	1.717	2.074	2.508	2.819
23	1.319	1.714	2.069	2.500	2.807
24	1.318	1.711	2.064	2.492	2.797
25	1.316	1.708	2.060	2.485	2.787
26	1.315	1.706	2.056	2.479	2.779
27	1.314	1.703	2.052	2.473	2.771
28	1.313	1.701	2.048	2.467	2.763
29	1.311	1.699	2.045	2.462	2.756
∞	1.282	1.645	1.960	2.326	2.576

Other t Probabilities:

$P\{T_8 < 2.541\} = 0.9825$ $P\{T_8 < 2.7\} = 0.9864$ $P\{T_{11} < 0.7635\} = 0.77$ $P\{T_{11} < 0.934\} = 0.81$ $P\{T_{11} < 1.66\} = 0.94$ $P\{T_{12} < 2.8\} = 0.984$.

TABLE A3.8.3

Values of $F_{0.05, n, m}$

$m = $ Degrees of freedom for denominator	$n = $ Degrees of freedom for numerator				
	1	2	3	4	5
1	161	200	216	225	230
2	18.50	19.00	19.20	19.20	19.30
3	10.10	9.55	9.28	9.12	9.01
4	7.71	6.94	6.59	6.39	6.26
5	6.61	5.79	5.41	5.19	5.05
6	5.99	5.14	4.76	4.53	4.39
7	5.59	4.74	4.35	4.12	3.97
8	5.32	4.46	4.07	3.84	3.69
9	5.12	4.26	3.86	3.63	3.48
10	4.96	4.10	3.71	3.48	3.33
11	4.84	3.98	3.59	3.36	3.20
12	4.75	3.89	3.49	3.26	3.11
13	4.67	3.81	3.41	3.18	3.03
14	4.60	3.74	3.34	3.11	2.96
15	4.54	3.68	3.29	3.06	2.90
16	4.49	3.63	3.24	3.01	2.85
17	3.45	3.59	3.20	2.96	2.81
18	4.41	3.55	3.16	2.93	2.77
19	4.38	3.52	3.13	2.90	2.74
20	4.35	3.49	3.10	2.87	2.71
21	4.32	3.47	3.07	2.84	2.68
22	4.30	3.44	3.05	2.82	2.66
23	4.28	3.42	3.03	2.80	2.64
24	4.26	3.40	3.01	2.78	2.62
25	4.24	3.39	2.99	2.76	2.60
30	4.17	3.32	2.92	2.69	2.53
40	4.08	3.23	2.84	2.61	2.45
60	4.00	3.15	2.76	2.53	2.37
120	3.92	3.07	2.68	2.45	2.29
∞	3.84	3.00	2.60	2.37	2.21

Other F Probabilities:
$F_{0.1, 7, 5} = 0.337$ $P\{F_{7,7} < 1.376\} = 0.316$ $P\{F_{20, 14} < 2.461\} = 0.911$ $P\{F_{9, 4} < 0.5\} = 0.1782$.

Answers to Selected Problems

CHAPTER 1

19. 0.3281   21. 0.6154, 0.3692   26. 0.0709   27. 0.825, 0.2303
31. 0.0703, 0.0703, 0.1416, 0.8594, 0.0313, 0.0313, 0.0625, 0.9375

CHAPTER 2

1. 0.5, 0.2778, 0.1389, 0.0595, 0.0198   6. 0.3834, 0.6321   7. 0.3292
8. 0.0183   21. 1.833   24. $a = 0.6$, $b = 1.2$ 28. 68.284
29. 148, 21   31. 56.156   33. $\text{Var}(X) = 1.25$ 35. 10.5, 8.75
38. $\text{Var}(X) = 0.0764$ 39. $\text{Var}(X_1) = 1.234$, $\text{Var}(X_2) = 0.25$
40. $\text{Var}(X) = 0.067$

CHAPTER 3

1. 0.8208   2. 0.9421   3. 0.2668   4. 0.4219   5. $p > 2/3$
12. 0.3935, 0.3033, 0.0902   13. 0.8908   14. 0.0511, 0.1229
19. 0.3630   24. 0.7977, 0.6827, 0.3695, 0.9522, 0.1587   25. 0.00023
26. 0.0956, < 0.00194   27. 2139.4   28. 0.0456   31. 0.9772
32. 0.6076   34. 0.0001   37. 0.3679, 0.6065   38. 0.2865
41. 0.7127, 0.0235   42. 0.2531

CHAPTER 4

4. 0.5364, 0.8326, 1.1774, $1/g^2$ 10. median = 0.4126   20. 0.0152, 0.1024,
0.00121   21. 0.00885   24. 0.1782

CHAPTER 5

9. (1.085, 1.315)   10. (1.072, 1.328)   11. 1.3136   13. (331.06,
336.93), (330.01, 337.98)   14. (39.55, 44.45)   16. (2013.9, 2111.6), (1996.0,
2129.5), 2022.4   17. (93.94, 103.40)   19. 3.453   21. (−11.18, −8.82)
22. (−11.12, −8.88)   23. (−74.97, 41.97)   24. 32.23, 112.3, 153.81, 69.6
25. (0.00206, 0.00529)   26. (53.2, 269.8)   31. it will be less than 0.108
32. (0.096, 0.244), (0.073, 0.267)   33. 0.67, 2024   34. (21.1, 74.2)
40. 7.64, 8.3   43. (182.57, 185.07)

CHAPTER 6

Unless otherwise mentioned, the answer given is the p-value of the relevant test.

1. 0.001   2. 0.31   3. 0.002   4. $n = 6$ 6. 0.005   7. 0.14
8. 0.38   9. 0.04   11. 0.28   12. 0.103   13. 0.67
14. 0.100013   15. 0.122   16. 0.044   17. 0.25   20. 0.023
21. < 0.0001   22. 0.4   23. 0.006   25. 0.0008   26. 0.18
27. 0.14   28. 0.16   32. 0.13   33. 0.47   34. 0.06

CHAPTER 7

2. 147.34   3. 0.045   4. 2439.8   5. 85.22   7. 105,659, (54,518,
308,521)   9. p-value = 0.016, (42.93, 49.95)   11. p-value = 0.035, 12.6,
(11.99, 13.22)   12. 4.125   18. p-value = 0.026, 2459.7, (2186, 2237)
19. 10340, (0.997, 1.005), (0.983, 1.018), 0.9998   20. 2.4998, 0.004
23. p-value < 0.0001, 144.37, 4.17, 0.76   24. $m = 0.07$, $A = 50.86$
25. $t = 22.5$, $s = 1.1$ 26. 1.22   27. 0.66   28. 216448.1
32. 85.5   34. 31.29, 30.35   38. 2.45   39. $SS_R = 202$, p-values =
0.46, 0.37, 0.125   40. 238.0, 3.9   41. 2309.6, 28.8   43. 225.70, 20.07
44. 4.645, 0.615   45. p-value = 0.81, 135.41, 17.24 and 51.27

CHAPTER 8

The answer given is the relevant p-value

1. 0.954   2. 0.727   4. 0.00245   5. 0.0043   7. 0.849
8. 0.9998   9. row: 0.0014 column: 0.0170   10. row: 0.706 column: 0.0214
11. row: 0.1001 column: 0.9995 interaction: 0.1065   12. row: 0.9867 column:
< 0.0001 interaction: 0.0445   13. row: 0.028 column: < 0.0001 interaction:
0.0278   14. row: 0.0003 column: 0.0003 interaction: 0.00005   15. row:
0.3815 column: 0.7611 interaction: 0.9497 additional: 0.0065

CHAPTER 9

The answer given is the p-value.

1. 0.648   2. 0.824   4. 0.143   5. 0.55   6. 0.00004
7. 0.0066   8. < 0.00001   12. 0.55   13. 0.095   14. 0.052
15. 0.007   16. < 10^{-6} 17. 0.0003   20. 0.48   21. 0.096
22. 1   23. 0.0047, 0.742   24. 0.0654, 0.0424   25. 0.02, 0.0039
26. 0.0195   28. 0.699   29. 0.866   30. 0.93   31. 0.4358, 0.2266
33. 0.009   34. 0.14

CHAPTER 10

3. 0.726, 0.118, 0.004, 2 10^{-6} 4. 0.0183, 0.6109, 1.277, 0.0235   8. reject
10. accept   12. 65   17. 84.556   18. 58.567   19. 0.0116
20. 0.0177   21. p-value = 0.684   22. p-value = 0.089

CHAPTER 11

1. No   2. On average, 3.39   4. 97 percent   7. 91 percent
8. 11 percent   9. (c) 78 percent will meet specifications (d) 0.97   11. 97.5
percent will meet specifications   12. yes   13. yes   14. 0.86
15. LCL = 57.5, UCL = 112.9   16. no

Index

(continued from front)